Management of Track and Field Injuries

Gian Luigi Canata • Pieter D'Hooghe
Kenneth J. Hunt
Gino M. M. J. Kerkhoffs
Umile Giuseppe Longo
Editors

Management of Track and Field Injuries

ISAKOS
International Society of Arthroscopy, Knee Surgery and Orthopaedic Sports Medicine

Springer

Editors
Gian Luigi Canata
Centre of Sports Traumatology
Koelliker Hospital
Torino
Italy

Kenneth J. Hunt
Orthopaedic Surgery
University of Colorado Health
Aurora, CO
USA

Umile Giuseppe Longo
Orthopaedics and Traumatology Unit
University Campus Biomedico
Rome
Italy

Pieter D'Hooghe
Orthopedic Surgery, Aspire Zone
ASPETAR Qatar Orthopaedic and
Sports Medicine Hospital
Doha
Qatar

Gino M. M. J. Kerkhoffs
Department of Orthopaedic Surgery
Amsterdam UMC
University of Amsterdam
Amsterdam Movement Sciences
Amsterdam
The Netherlands

Academic Center for Evidence-based
Sports medicine (ACES)
Amsterdam
The Netherlands

Amsterdam Collaboration for
Health and Safety in Sports
(ACHSS), AMC/VUmc IOC
Research Center
Amsterdam
The Netherlands

ISBN 978-3-030-60215-4 ISBN 978-3-030-60216-1 (eBook)
https://doi.org/10.1007/978-3-030-60216-1

This Springer imprint is published by the registered company Springer Nature Switzerland AG
The registered company address is: Gewerbestrasse 11, 6330 Cham, Switzerland

Foreword

WORLD
ATHLETICS™

Running, walking, jumping, and throwing form the basis and key components of several sports and physical activity in general. Indeed, almost everyone has practised Athletics once in their life at school, college, or later, and enjoyed watching or taking part in competition in Athletics. As a result, Athletics is the number one sport at the Summer Olympic Games.

Whether at recreational or elite level, regular practice of Athletics is sometimes associated with musculoskeletal injuries; some of them being event-specific. Because of its unique universal nature, medicine of Athletics also faces a paradoxical situation where top performers in this sport often live and train in countries where sports medicine is either underdeveloped or simply does not exist.

Therefore, it is important for coaches, sports physicians, orthopaedic surgeons, and physiotherapists to know about the basics and the latest developments in the Management of Track and Field Injuries.

This is what the present book is about and, as the Director of the Health and Science Department at World Athletics and a former member of the ISAKOS, I strongly support such a publication which will for sure help to raise awareness and disseminate knowledge on Athletics injuries among health professionals.

Stéphane Bermon
Monaco, France
Director – World Athletics Health and Science Department

The original version of this book was revised. Book title has been updated. The correction to this book is available at https://doi.org/10.1007/978-3-030-60216-1_35

Presidential Foreword

Track and field can definitely be considered the purest form of athletic competition. Individuals that compete to see who is the fastest, can jump the highest, or throw the furthest. I competed at Javelin, more in my Junior days, but was privileged to be coached by Klaus Wolfermann, the West German Olympic winner in 1972 by the smallest margin: ever 2 cm. He was so passionate about his sport that it rubbed off on me.

It is astonishing that the records just kept getting broken and at some stage they had to change the Javelin because the stadiums were getting too small. The athletes keep pushing their bodies to the absolute limits but sometimes these limits are overstepped and it leads to injury.

It is our job to make sure we keep our athletes healthy not just for the time that they are competing but also in the long term. I find it incredible that so many surgeons work on this tirelessly, giving up time to make sure we learn from each other but also train the next generation. They love our athletes, are passionate about what they are doing and, by challenging one another, they keep setting the bar higher.

Part of the book is dedicated to basic science of the musculoskeletal system, which is essential if we want to treat these injuries scientifically. Anything we do must be based on sound science and research. This book has managed to bring together a diverse group of world experts, which is what ISAKOS is all about: sharing knowledge from all corners of the globe.

I thank all the authors and congratulate them on a fantastic book that will ultimately lead to better treatment for our athletes - from professionals to weekend warriors - so that they can live a full and active life for many years.

Like Klaus Wolfermann, world record holder that had enough passion to coach some young kid in the art of throwing the Javelin, these surgeons devote their time to help even the most junior surgeon to constantly improve themselves in what they do.

Willem van der Merwe
ISAKOS President
San Ramon, CA, USA

Acknowledgements

This book on Track and Field injuries is a result of a great teamwork. International experts have cooperated dedicating time and energy to share their knowledge in the respective fields of interest.

We thank all the authors that have accepted with passion and enthusiasm to be part of this project.

Like in every other undertaking, even the publication of a scientific book requires multiple supporting energies.

We thank Prof. Jon Karlsson: he has been present every time we needed help.

Jari Dahmen must be commended for his extraordinary help checking every chapter and cooperating daily with Catena Cottone and Valentina Casale.

We also thank the ISAKOS team and our President Willem Van Der Merwe for their great support: ISAKOS has made possible the production of our book, and we are proud of being part of this great international scientific community.

We also thank the Springer team that has backed us in the production of this book that is dedicated to the entire world of Athletics.

Gian Luigi Canata
Pieter D'Hooghe
Kenneth J. Hunt
Gino M. M. J. Kerkhoffs
Umile Giuseppe Longo

Contents

Introduction

Track and Field has a great historical background and is a fascinating individual sport which couples competition against adversaries with a continuous research of self-improvement.

Competitive athletic spirit was first portrayed in the Mycenaean period, as a representation of two runners on a vase from Cyprus shows.

Running, jumping, discus and javelin throwing were important events in the ancient games.

All the historic Panhellenic games, including the Olympic ones, were related to myth and the Greek concept of perfection: *kalos kai agathos*. Competing athletes could become heroes, emulating Odysseus as Homer reported in the Iliad. The runner Leonidas of Rhodes was deified, having won twelve Olympic crowns in four consecutive Olympic games.

Competitions were later gradually associated with preliminary selection, training, and professional coaches with experience in training, diet, and medicine: the beginning of sports medicine.

Over the last decades biological, physiological, and biomechanical knowledge has greatly evolved, improving both the prevention and the management of acute and chronic injuries.

As an individual sport, track and field requires top level performances in any competition, as well as a constant training. As a consequence, there are still nowadays high risks to develop overuse pathologies, even if training should never be more strenuous than the athlete can endure without injury.

I practiced triple jump in Torino under the presidency of Primo Nebiolo, who became IAAF President in 1981 and greatly promoted Athletics development worldwide, keeping at the same time his role of CUS Torino president and transmitting his enthusiasm and love for track and field to his athletes. In the meantime great athletes like Livio Berruti, gold medallist in 200 mt and twice world recordman that same day at the Rome 1960 Olympic games, Peppe Gentile, Sara Simeoni, Maurizio Damilano, Marcello Fiasconaro, and Pietro Mennea were there, extraordinary living examples to follow. Worldwide renowned trainers like Elio Locatelli, Renato Canova, Claudio Gaudino, Sandro Damilano, and Steve Banner were on the field every day.

This book derives from the everlasting love for Athletics and the enthusiastic support of the coeditors and the authors of the chapters, all renowned experts in the field.

The result is an updated presentation of the current knowledge about the injuries in track and field, covering specific aspects of running, jumping, and throwing pathologies and stressing the importance of preventive measures.

We hope that this book will help all those involved in Athletics to improve the safety of a wonderful sport connected to our historical values.

Gian Luigi Canata
Centre of Sports Traumatology
Koelliker Hospital
Torino
Italy

Part I

Anatomy, Physiology and Biomechanics

1 The Burden and Epidemiology of Injury in Track and Field

Pascal Edouard

1.1 Introduction

Track and field (athletics) is an Olympic sport composed of several different disciplines (www.worldathletics.org/our-sport): sprints, hurdles, jumps, throws, combined events, middle and long distances, marathon, and race walking. It is internationally governed by the World Athletics (www.worldathletics.org), founded in 1912, and previously called International Association of Athletics Federations (IAAF). There are currently 214 members federations (countries or territories) affiliated to World Athletics, which places World Athletics among the world's largest sporting organizations. Based on the number of athletes, this is the first sport at the Olympic Games; for example, at the 2016 Olympics Games athletes registered for track and field represented 21% of all registered athletes (second sport was aquatics with 13%, and then, other sports represented less than 5% of athletes) [1].

As for many sports, the practice of track and field leads to a risk of injuries [2]. Indeed, all these track and field disciplines involved the musculoskeletal system (i.e., muscle, tendon, bone, cartilage, ligament, and soft tissue). When the load resulting from the practice exceeds the capabilities of the musculoskeletal system, there is a risk of failure of the musculoskeletal structure resulting in an injury. Injury has a negative impact on practice, because it can decrease training participation, decrease performance, and lead to pain [3]. Even if the injury is a minor anatomical lesion or leads to minor resounding on practice, there will be an impact, on the musculoskeletal (e.g., imbalance between injured and uninjured sides) and psychological (e.g., lack of confidence or fear of recurrence) aspects. All the consequences can not only affect the sports practice, but can also have a negative impact on other domains of life (e.g., social, professional, family, school, financial) in the short or long term [2].

Taken into account the number of athletes practicing track and field whatever their levels in addition to the risk of injuries, the prevention of injuries in track and field represents an important area for athletes and all stakeholders, such as coaches, health professionals, family, sports scientists, managers, sponsors, and international and national governing bodies [2, 4–6]. In order to reach this injury prevention challenge, Van

P. Edouard (✉)
Inter-university Laboratory of Human Movement Science (LIBM EA 7424), University of Lyon, University Jean Monnet, Saint Etienne, France

Department of Clinical and Exercise Physiology, Sports Medicine Unit, University Hospital of Saint-Etienne, Faculty of Medicine, Saint-Etienne, France

European Athletics Medical & Anti Doping Commission, European Athletics Association (EAA), Lausanne, Switzerland
e-mail: Pascal.Edouard@univ-st-etienne.fr

G. L. Canata et al. (eds.), *Management of Track and Field Injuries*,
https://doi.org/10.1007/978-3-030-60216-1_1

Mechelen et al. [7] described a four-step methodological sequence of evidence-based in injury prevention. The first step of this sequence consists in understanding the extent of the problem and describes the incidence and severity of injuries. This fundamental first step is of interest since it allows having a clear basis of the magnitude of the problem. It is also useful for long-term monitoring and for comparison if prevention measures are implemented. In addition, for clinical practice, it can help health professionals by anticipating the most frequent injuries and thus the need for medical provision. Thus, having a clear knowledge of the epidemiology of injuries is of great interest for injury prevention in track and field.

Given the impact of the data collection methodology on the quality of the data and thus the resulting information [8, 9], a great attention should be done to methodology of epidemiological studies in order to interpret results. The study design, the definition of injury and its characteristics, the exposure, the data collection procedures, and data analyses are key points of the methodology of epidemiological studies [5, 6, 10, 11]. To date, there is a consensual method for injury data collection during championships that has been developed by the International Olympic Committee (IOC) [12] and used in track and field at the IAAF World Championships in Athletics [13–17], the European Athletics Championships [18–21], and the French national championships [22]. This methodology has provided reliable and comparable data for this particular context of international championships [8, 23]. However, if we broaden the focus to the whole track and field season, we find that only a few studies exist and that they use different methods [4, 24–33], which does not allow a true comparison of the data, and could explain why injury data should now be presented separately between championships and whole season. A method was developed in 2014 at a consensus meeting of international and national athletics federations [11], and the IOC recently updated a consensus statement on methods for recording and reporting of epidemiological data on injury and illness in sport 2020 [10] that are expected to implement long-term cohort follow-ups over one or more seasons with a comparison between studies.

1.2 Injuries during Championships

1.2.1 Injuries during International Track and Field Championships

Injury data have been collected at a number of major championships following the IOC consensus methods for multi-event championships [12]. At each event, physicians and/or physiotherapists from the national medical teams and the local organizing committee prospectively collected new injuries occurring among athletes registered in the championships based on the same injury definitions (i.e., medical attention injury) and classifications and using a paper-based report form. This allowed description of the number, incidence, and characteristics of injuries in this context. These injury surveillance studies have allowed the collection of a large amount of data by combining all together these data. Indeed, a total of 2191 injuries were collected from 20 international championships from 2007 to 2019 among 19,066 registered athletes (unpublished data). This resulting in a clear vision of injuries that athletes can suffer during international championships [34–38].

The injury rates varied with sex and disciplines [35, 37]. From 14 international championships between 2007 and 2014, the number of injuries per 1000 registered athletes was significantly higher for male than female athletes (110.3 ± 6.8 vs. 88.5 ± 6.7 injuries per 1000 registered athletes, respectively; relative risk = 1.25 (confidence interval 95%: 1.13 to 1.32)) [35]. The injury location varied with sex: Male athletes suffered more injuries of the thigh, the lower leg, and the hip/groin than female athletes [35]. The injury type also varied according to sex: Male athletes suffered more muscle injuries than female athletes, while female athletes suffered more stress fractures than male athletes [35]. The injury rate also varied between disciplines, with a higher injury rate in combined events, marathon, and long-distance running [37]. Injury characteristics significantly varied between disciplines for location, type, cause, and severity, in both male and female athletes: Thigh muscle injuries were the main injury diagnoses in sprints, hurdles, jumps, combined events and race walking, lower leg muscle injuries in marathon, lower leg skin injury in middle and long distances, and trunk muscle and lower leg muscle injuries in throws [37]. The first injury was hamstring muscle injury (about 17% of all injuries), with higher proportion in sprints and other disciplines requiring sprint capabilities [36]. A summary of the key findings regarding injuries occurring during international track and field championships is presented in Table 1.1.

For three of the international championships studied, data collection on athletes' health was extended to the 4 weeks before the championships [16, 17, 20]. It was found that about 30% of the athletes participating in these studies reported an injury complaint in this preparation period, including a third who had to decrease their training load and about 4% who could not practice at all [16, 17, 20]. These injury complaints appeared to be overuse injuries mainly because there was a gradual onset and they existed for more than 4 weeks. These results support that an important proportion of high-level athletes are living and training with an injury complaint, suggesting that injury unfortunately is part of the athletes' life, and even more supporting the need for injury prevention.

1.2.2 Injuries during National Track and Field Championships

The methods used during international track and field championships [12, 13] have also been used for national championships. This allows providing information for athletes with a level just below the international level.

During the French national track and field outdoor championships, such injury surveillance studies have been carried out since 2014. From 2014 to 2019, the incidence was about 50 injuries per 1000 registered athletes, the thigh was the first injury location (about 30% of all injuries), and muscle was the first injury type (about 30% of all injuries), and explosive disciplines (i.e., combined events, sprints, hurdles, and jumps) were those accounting for the most important number of injuries (unpublished data).

During the 2010 French combined event championships, an incidence of 477 injuries per 1000 registered athletes was reported and the most common diagnosis was muscle injury to the thigh (18%) [22].

During 3 years of Penn Relay Carnival, Opar et al. [39] reported an incidence of 10 injuries per 1000 registered athletes. Hamstring muscle strain was the most prevalent injury accounting for 24% of injuries, with higher rates in male than female athletes [39].

During the 2016 track and field Olympic trials, Bigouette et al. [40] reported an incidence of 60 injuries per 1000 registered athletes. Hamstring strains were the most prevalent injuries with about 17% of all injuries, and jumps and long distances were the disciplines with the most number of injuries per registered athletes.

1.2.3 Conclusion Injuries During Championships

Although such context of championships represents few days in the season (3 to 9 days compared to the other 357 to 363 days), this represents the goal of the season for athletes and their stakeholders, and injuries have a negative impact on the performance [38]. Therefore, it is of interest to have a clear view of the "risks" in this very important period. All these studies provide an interesting and relevant overview of the injuries during track and field championships, especially for high-level athletes (Table 1.2). One of the learnings is that injury number, incidence, and characteristics varied with sex and disciplines; it is therefore important to analyze and provide such information separately by sex and disciplines. All these data allow athletes and all stakeholders

Table 1.1 Key points regarding injuries occurring during international track and field championships

	Sprints	Hurdles	Jumps	Throws	Combined events	Long distances	Middle distances	Marathon	Race walking
Male athletes									
Percentage of all injuries	24	9	16	6	8	11	11	9	7
Number of injuries per 1000 registered athletes	95	106	98	47	235	106	124	156	115
Podium of the injury diagnosis (number of injuries per 1000 registered athletes)									
1	Thigh muscle (44.4)	Thigh muscle (34.6)	Thigh muscle (22.6)	Trunk muscle (6.0), lower leg muscle (6.0)	Thigh muscle (42.7)	Lower leg skin (32.8)	Lower leg skin (24.4)	Lower leg muscle (29.1)	Thigh muscle (35.4)
2	Lower leg muscle (9.3)	Hip and groin muscle (9.3)	Ankle ligament (8.6)	Hip and groin muscle (5.2)	Achilles tendon (18.3), ankle ligament (18.3)	Lower leg muscle (15.1)	Upper extremity skin (9.5)	Thigh muscle (25.5)	Trunk muscle (13.4)
3	Hip and groin muscle (4.6)	Lower leg skin (5.3), lower leg muscle (5.3), knee skin (5.3)	Lower leg muscle (6.6)			Knee skin (12.6)	Foot skin (8.5)	Foot skin (14.5)	Lower leg muscle (11.5), foot skin (11.5)
Female athletes									
Percentage of all injuries	26	10	12	5	11	11	14	9	2
Number of injuries per 1000 registered athletes	75	83	52	32	212	85	128	119	42
Podium of the injury diagnosis (number of injuries per 1000 registered athletes)									

1	Thigh muscle (24.0)	Thigh muscle (15.5)	Thigh muscle (8.7)	Knee tendon (3.0), lower leg muscle (3.0), trunk muscle (3.0)	Thigh muscle (45.6)	Lower leg skin (25.9)	Foot skin (18.9)	Lower leg muscle (19.4)	Foot others (10.6)
2	Upper extremity skin (4.3)	Knee skin (9.9)	Lower leg muscle (4.4), Achilles tendon (4.4)		Ankle ligament (22.8)	Thigh muscle (13.7), lower leg muscle (13.7)	Knee skin (8.1)	Foot skin (17.2)	Thigh muscle (7.0)
3	Trunk muscle (3.8)	Upper extremity skin (8.5)			Lower leg muscle (16.3), trunk ligament (16.3)		Thigh muscle (6.7), Achilles tendon (6.7)	Knee ligament (8.6), trunk muscle (8.6)	

The data presented in this table are from the article by Edouard et al. [37] and have been collected during 14 international championships between 2007 and 2018

Table 1.2 Key points regarding injury characteristics occurring during the whole season

	Sprints	Hurdles	Jumps	Throws	Combined events	Middle and long distances and marathon
Main injuries	Thigh and hamstring muscle injuries	Thigh and hamstring muscle injuries	Thigh and hamstring muscle injuries	Shoulder and elbow injuries	Thigh muscle injuries	Lower leg injuries
	Achilles tendinopathy	Lower leg injuries	Achilles and patellar tendinopathy	Low back pain	Back injuries	Achilles tendinopathy
	Back injuries		Knee injuries		Upper extremity injuries	Overuse knee injuries
			Ankle sprain		Achilles and patellar tendinopathy	Stress fracture
			Low back pain			

around them having a clear basis and information to orient injury prevention approach toward these championships. However, there is a need to continue these data collections in other populations of athletes for reaching an understanding of injury epidemiology in all athletes practicing track and field whatever their age and level.

1.3 Injuries During the Whole Season

The whole season represents a significantly larger period in the athletes' life and practice than championships. And this also represents a significantly higher period of exposure to the risk of injuries. However, information on injuries in track and field during the whole season is not as important as for the championships. Methodological issues are probably one explanation of the fact that there are few studies during the whole season [4].

1.3.1 Injuries During the Whole Season in National-Level Athletes

Below are summarized the main results of three studies collecting injury data over one season in national-level athletes. This is not an exhaustive report of the scientific literature, but these results present an overview of the current knowledge on this population.

In a questionnaire-based retrospective study of 147 national-level athletes over about 12 months of training, D'Souza [26] reported that 61% of athletes had at least one injury during the season. The locations and types of injuries varied by event, with a high prevalence of shin splints in middle- and long-distance runners, ankle injuries in throwers, and thigh injuries in jumpers.

In another questionnaire-based retrospective study of 95 national-level athletes over about 12 months of training, Bennell and Crossley [24] reported that 76% of athletes had at least one injury during the season, with an incidence of 3.9 injuries per 1000 h of track and field practice. The main injuries were stress fractures (20.5%), hamstring muscle injuries (14.2%), and knee overuse injuries (12.6%). Overuse was the most frequent cause (72%). The mode of onset varied by event: more sudden injuries in the explosive events (sprints, hurdles, jumps, and combined events) and more gradual injuries in the endurance events (middle distance, marathon) and background training.

In a prospective study of 292 national-level athletes over 12 months, Jacobsson et al. [30] reported that 68% of those studied had at least one injury during the season and the injury incidence was 3.6 per 1000 h of track and field practice. Of the injuries, 96% were caused by overuse, and 51% evolved for more than 3 weeks. The main locations were the Achilles tendon, the foot and ankle, the thigh and hip, and the lower leg. The main complaints were hamstring injury among sprinters and jumpers, Achilles tendinopathy and shin splints among middle-distance runners, and lower back pain among throwers.

Although the methods (i.e., study design, injury definition, and data collection) were not similar between these studies, it seems that there are similar and consistent results on injury prevalence, incidence, and characteristics. Between 61 and 76% of the national-level athletes had at least one injury during the entire track and field season [24, 26, 30]. The incidence was reported as 3.6–3.9 injuries per 1000 h of track and field practice [24, 26, 30]. The location and type of injuries varied according to the disciplines, with a high prevalence of Achilles tendinopathy and "shin splints" in middle and long distances, ankle injuries and low back pain in throwers, and thigh and hamstring muscle injuries in sprinters and jumpers [24, 26, 30]. The injury mode of onset was more sudden in explosive disciplines and more gradual in endurance disciplines [24]. Overuse was the most frequent cause of track and field injury (72–96%) [24, 30].

1.3.2 Injuries During the Whole Season in Specific Population

Other studies provided an overview of the magnitude of the problem in specific population.

In combined events, in a prospective study over four athletic seasons (1994–1998) of 69 selected French combined event athletes, Edouard et al. [29] reported 39 injuries in 14 heptathletes and 47 injuries in 18 decathletes. The injury rate per 100 athletes per season for the heptathletes and the decathletes was 33 and 30, respectively. Of the injuries suffered, 41% affected the tendons and 23% affected the muscles. The most common diagnoses were knee tendinopathy (14%), followed by lower leg muscle injuries (13%), thigh muscle injuries (11%), and Achilles tendinopathy (11%). The causes of injuries were mainly overuse (49%) or acute trauma (43%).

In pole vault, in a prospective study of 140 pole vaulters over two seasons, Rebella et al. [41] reported an incidence of 26.4 injuries per 100 athletes, with ankle sprains representing a third of the cases. In a second prospective study of 150 pole vaulters over one season, Rebella [42] reported an incidence of 7.9 injuries per 1000 athlete exposure, with most injuries being in the low back pain, hamstring, and lower leg.

In youth and junior elite athletes, in a prospective cohort study of 70 athletes over 30 weeks, Carragher et al. [43] reported that 77% of athletes had at least one injury during the period, 44% at least one acute injury, and 53% at least one overuse injury. The prevalence of injury was similar between male and female athletes, but varied between explosive and endurance disciplines: higher prevalence of injuries in explosive than endurance disciplines. The prevalence of acute injuries was higher in explosive than endurance disciplines, while prevalence of overuse injuries was similar between both discipline categories. The main injury diagnoses of acute injuries were lower leg strain/tear in male endurance athletes (25%), trunk muscle cramps/spasms in male explosive athletes (31.6%), and hamstring strain/tear in female explosive athletes (21.1%). The main injury diagnoses of overuse injuries were knee tendinopathy in male endurance athletes (29.4%), lower leg muscle cramps in female endurance athletes (28.6%), and hamstring muscle cramps/spasms in both male explosive athletes (40.0%) and female explosive athletes (21.1%).

These are maybe not the only studies reporting information on injuries in specific track and field populations, but these studies provide some relevant insights that could help to orient injury prevention strategies by taking into account all the spectrum of specificities of track and field.

1.3.3 Characteristics of Injuries According to Disciplines During the Whole Season

Although studies used different definitions of injuries and injury characteristics, and the results are often only descriptive (no comparison), it seems that the injury characteristics (location and/or diagnosis) are quite constant over studies and clearly varied according to disciplines [24, 26, 30, 31, 44, 45]. In summary, these studies reported that athletes participating in sprints suffered more of thigh/hamstring [24, 26, 30, 31, 44, 45], Achilles tendon [30, 45], and/or back [26]; in hurdles: thigh [24] and/or lower leg [26]; in middle and long distances: lower leg [24, 26, 30, 31], foot/ankle/Achilles tendon [30, 31, 44, 45], back/hip [44], hamstring [45], and/or knee [24, 31, 45]; in jumps: thigh/hamstring [24, 26, 30, 31], knee [26], back [24], and/or Achilles [30, 31, 45]; in throws: back [26, 30, 31, 45], upper extremity [45], ankle [26], and/or knee [30, 31]; and in combined events: thigh [24, 30, 31], back [24], upper extremity [45], knee [31], and/or foot/ankle/Achilles [30]. This could be interpreted (probably with some caution) as specific disciplines lead to specific constraints and injuries whatever the circumstances and population [37].

1.3.4 Conclusions on Injuries During the Whole Season

There are currently and to our knowledge only few studies reporting injury data during the whole track and field season. This justifies increasing efforts on performing prospective injury surveillance studies on different populations of track and field athletes. However, the currently available results provide some relevant inputs to orient athletes and their stakeholders toward injury prevention strategies.

1.4 Conclusion

In light of all these results, it can be first said that we are beginning to identify and detail extent of the problem, especially among elite high-level populations taking part in major international championships. Although the data on injuries over the whole season come from only a few studies using different methodologies, it provides a first basis to move forward to prevention, and it supports the need for further studies. Thus, further epidemiological injury data collections would still seem to be relevant and necessary. Second, we can say that track and field is composed of several disciplines with different physical, mechanical, technical, and psychological demands, which lead to different constraints on the musculoskeletal system, and consequently different injuries according to these disciplines. The overall picture that has been shown in the present chapter is that the most common injury problems experienced are hamstring muscle injuries (especially in sprints, hurdles, and jumps), Achilles tendinopathies (in sprints, middle and long distances, and jumps), knee overuse injuries (in sprints, middle and long distances), shin splints and/or stress fractures (in sprints, middle and long distances), ankle sprains (in jumps), and low back pain (in jumps and throws).

References

1. Soligard T, Steffen K, Palmer D, Alonso JM, Bahr R, Lopes AD, et al. Sports injury and illness incidence in the Rio de Janeiro 2016 Olympic Summer Games: A prospective study of 11274 athletes from 207 countries. Br J Sports Med. 2017;51:1265–71.
2. Edouard P, Morel N, Serra J-M, Pruvost J, Oullion R, Depiesse F. Prévention des lésions de l'appareil locomoteur liées à la pratique de l'athlétisme sur piste. Revue des données épidémiologiques. Sci Sports. 2011;26:307–15.
3. Bolling C, Delfino Barboza S, van Mechelen W, Pasman HR. How elite athletes, coaches, and physiotherapists perceive a sports injury. Transl Sport Med. 2019;2:17–23.
4. Edouard P, Branco P, Alonso J-M. Challenges in Athletics injury and illness prevention: implementing prospective studies by standardised surveillance. Br J Sports Med. 2014;48:481–2.
5. Edouard P, Alonso JM, Jacobsson J, Depiesse F, Branco P, Timpka T. Injury prevention in athletics: the race has started and we are on track! New Stud Athl. 2015;30:69–78.
6. Edouard P, Alonso JM, Jacobsson J, Depiesse F, Branco P, Timpka T. On your marks, get set, go! A flying start to prevent injuries. Aspetar Sport Med J. 2019;8:210–3.
7. van Mechelen W, Hlobil H, Kemper HCG. Incidence, severity, aetiology and prevention of sports injuries. A review of concepts. Sports Med. 1992;14:82–99.
8. Edouard P, Branco P, Alonso JM, Junge A. Methodological quality of the injury surveillance system used in international athletics championships. J Sci Med Sport. 2016;19:984–9.

9. Ekegren CL, Gabbe BJ, Finch CF. Sports injury surveillance systems: a review of methods and data quality. Sport Med. 2016;46:49–65.
10. Bahr R, Clarsen B, Derman W, Dvorak J, Emery CA, Finch CF, et al. International Olympic Committee consensus statement: methods for recording and reporting of epidemiological data on injury and illness in sport 2020 (including STROBE Extension for Sport Injury and Illness Surveillance (STROBE-SIIS)). Br J Sports Med. 2020;54:1–18.
11. Timpka T, Alonso J-M, Jacobsson J, Junge A, Branco P, Clarsen B, et al. Injury and illness definitions and data collection procedures for use in epidemiological studies in Athletics (track and field): consensus statement. Br J Sports Med. 2014;48:483–90.
12. Junge A, Engebretsen L, Alonso JM, Renström P, Mountjoy M, Aubry M, et al. Injury surveillance in multi-sport events: the International Olympic Committee approach. Br J Sports Med. 2008;42:413–21.
13. Alonso J-M, Junge A, Renstrom P, Engebretsen L, Mountjoy M, Dvorak J. Sports injuries surveillance during the 2007 IAAF World Athletics Championships. Clin J Sport Med. 2009;19:26–32.
14. Alonso JM, Tscholl PM, Engebretsen L, Mountjoy M, Dvorak J, Junge A. Occurrence of injuries and illnesses during the 2009 IAAF World Athletics Championships. Br J Sports Med. 2010;44:1100–5.
15. Alonso J-M, Edouard P, Fischetto G, Adams B, Depiesse F, Mountjoy M. Determination of future prevention strategies in elite track and field: analysis of Daegu 2011 IAAF Championships injuries and illnesses surveillance. Br J Sports Med. 2012;46:505–14.
16. Alonso J-MJ-M, Jacobsson J, Timpka T, Ronsen O, Kajenienne A, Dahlström Ö, et al. Preparticipation injury complaint is a risk factor for injury: a prospective study of the Moscow 2013 IAAF Championships. Br J Sports Med. 2015;49:1118–24.
17. Timpka T, Jacobsson J, Bargoria V, Périard JD, Racinais S, Ronsen O, et al. Preparticipation predictors for championship injury and illness: cohort study at the Beijing 2015 International Association of Athletics Federations World Championships. Br J Sports Med. 2017;51:272–7.
18. Edouard P, Depiesse F, Hertert P, Branco P, Alonso J-MM. Injuries and illnesses during the 2011 Paris European Athletics Indoor Championships. Scand J Med Sci Sport. 2013;23:e213–8.
19. Edouard P, Depiesse F, Branco P, Alonso JM. Analyses of Helsinki 2012 European athletics championships injury and illness surveillance to discuss elite athletes risk factors. Clin J Sport Med. 2014;24:409–15.
20. Edouard P, Jacobsson J, Timpka T, Alonso JM, Kowalski J, Nilsson S, et al. Extending in-competition Athletics injury and illness surveillance with preparticipation risk factor screening: a pilot study. Phys Ther Sport. 2015;16:98–106.
21. Edouard P, Alonso J, Depiesse F, Branco P. Understanding injuries during the European athletics championships: an epidemiological injury surveillance study. New Stud Athl. 2014:7–9.
22. Edouard P, Samozino P, Escudier G, Baldini A, Morin JB. Injuries in youth and national combined events championships. Int J Sport Med. 2012;33:824–8.
23. Edouard P, Junge A, Kiss-Polauf M, Ramirez C, Sousa M, Timpka T, et al. Interrater reliability of the injury reporting of the injury surveillance system used in international athletics championships. J Sci Med Sport. 2018;21(9):894–8. pii: S1440–2440(18)30038–0
24. Bennell KL, Crossley K. Musculoskeletal injuries in track and field: incidence, distribution and risk factors. Aust J Sci Med Sport. 1996;28:69–75.
25. Watson MD, Dimartino PP. Incidence of injuries in high school track and field athletes and its relation to performance ability. Am J Sports Med. 1987;15:251–4.
26. D'Souza D. Track and field athletics injuries - a one-year survey. Br J Sports Med. 1994;28:197–202.
27. Boden BP, Pasquina P, Johnson J, Mueller FO. Catastrophic Injuries in Pole-Vaulters. Am J Sports Med. 2001;29:50–4.
28. Boden BP, Boden MG, Peter RG, Mueller FO, Johnson JE. Catastrophic injuries in pole vaulters: a prospective 9-year follow-up study. Am J Sports Med. 2012;40:1488–94.
29. Edouard P, Kerspern A, Pruvost J, Morin JB. Four-year injury survey in heptathlon and decathlon athletes. Sci Sport. 2012;27:345–50.
30. Jacobsson J, Timpka T, Kowalski J, Nilsson S, Ekberg J, Dahlström Ö, et al. Injury patterns in Swedish elite athletics: annual incidence, injury types and risk factors. Br J Sports Med. 2013;47:941–52.
31. Jacobsson J, Timpka T, Kowalski J, Nilsson S, Ekberg J, Renström P. Prevalence of musculoskeletal injuries in Swedish elite track and field athletes. Am J Sports Med. 2012;40:163–9.
32. Raysmith BP, Drew MK. Performance success or failure is influenced by weeks lost to injury and illness in elite Australian track and field athletes: a 5-year prospective study. J Sci Med Sport. 2016;19:778–83.
33. Timpka T, Jacobsson J, Dahlström Ö, Kowalski J, Bargoria V, Ekberg J, et al. The psychological factor "self-blame" predicts overuse injury among top-level Swedish track and field athletes: a 12-month cohort study. Br J Sport Med. 2015;49:1472–7.
34. Feddermann-Demont N, Junge A, Edouard P, Branco P, Alonso J-M. Injuries in 13 international Athletics championships between 2007–2012. Br J Sports Med. 2014;48:513–22.
35. Edouard P, Feddermann-Demont N, Alonso JM, Branco P, Junge A. Sex differences in injury during top-level international athletics championships: surveillance data from 14 championships between 2007 and 2014. Br J Sports Med. 2015;49:472–7.
36. Edouard P, Branco P, Alonso J-M. Muscle injury is the principal injury type and hamstring muscle injury is the first injury diagnosis during top-level international athletics championships between 2007 and 2015. Br J Sports Med. 2016;50:619–30.
37. Edouard P, Navarro L, Branco P, Gremeaux V, Timpka T, Junge A. Injury frequency and characteristics (location, type, cause and severity) differed signifi-

cantly among athletics ('track and field') disciplines during 14 international championships (2007–2018): implications for medical service planning. Br J Sports Med. 2020;54:159–67.
38. Edouard P, Richardson A, Navarro L, Gremeaux V, Branco P, Junge A. Relation of team size and success with injuries and illnesses during eight international outdoor athletics championships. Front Sport Act Living. 2019;1:8.
39. Opar DA, Drezner J, Shield A, Williams M, Webner D, Sennett B, et al. Acute hamstring strain injury in track-and-field athletes: a 3-year observational study at the Penn Relay Carnival. Scand J Med Sci Sport. 2014;24:254–9.
40. Bigouette JP, Owen EC, Greenleaf J, James SL, Strasser NL. Injury surveillance and evaluation of medical services utilized during the 2016 track and field Olympic trials. Orthop J Sport Med. 2018;6:1–9.
41. Rebella GS, Edwards JO, Greene JJ, Husen MT, Brousseau DC. A prospective study of injury patterns in high school pole vaulters. Am J Sports Med. 2008;36:913–20.
42. Rebella G. A prospective study of injury patterns in collegiate pole vaulters. Am J Sports Med. 2015;43:808–15.
43. Carragher P, Rankin A, Edouard P. A one-season prospective study of illnesses, acute, and overuse injuries in elite youth and junior track and field athletes. Front Sport Act Living. 2019;1:1–12.
44. Lysholm J, Wiklander J. Injuries in runners. Am J Sports Med. 1987;15:168–71.
45. Ahuja A, Ghosh AK. Pre-Asiad '82 injuries in elite Indian athletes. Br J Sports Med. 1985;19:24–6.

Sprinter Muscle. Anatomy and Biomechanics

2

George A. Komnos and Jacques Menetrey

2.1 Introduction

The skeletal muscle cell is called muscle fiber or myofiber. The two types of skeletal muscle fibers are the slow-twitch (type I) and the fast-twitch (type II) fibers. Fast-twitch muscles are further divided into two categories: type IIa (moderate fast-twitch) and type IIb or type IIx. Slow-twitch (ST) muscles are activated in long resistance exercise, while fast-twitch (FT) muscles are used in forceful breakouts. The proportion between slow-twitch and fast-twitch fibers may vary depending on the exercise. If it comes to sprinting, training decreases the proportion of ST fibers and increases the proportion of FT fibers.

It is proposed that muscle fiber composition is genetically based and thus is difficult to change with training [1]. However, muscle fiber volume can increase with specified training targeting at type II fibers [2]. As a result, sprinters have larger type II than type I fiber areas in their leg extensor muscles because their training mainly includes fast repetitive movements. The proportion of type II fibers in the vastus lateralis muscle is shown to be related to blocking velocity and running velocity in the phases of acceleration and maximum constant speed, and to the final sprint performance (100 m) [3]. It is worth noting that as enhancement in maximal running velocity during sprint training is very limited, discovery of potential talents could be achieved by detecting athletes with a high proportion of type II fibers [2].

2.2 Sprinter's Specificity

Muscle size is strongly related to better performance in the literature, with sprinters appearing to have more developed lower limb muscles [4–6]. Although thigh and leg muscles have been reported to lead to successful sprinting, literature is not so rich regarding the foot muscles. Tanaka et al. [4] hypothesized that sprinters may also have developed foot muscles because of enhancement of the role of MTP joint during sprinting. They found in their study that thicknesses of the foot muscles, in addition to the lower leg muscles, were larger in sprinters than in non-sprinters. Furthermore, they concluded that the foot muscles might be especially developed in sprinters compared to non-sprinters, since the foot muscle thickness difference between the two groups was relatively greater than in the lower leg muscles.

G. A. Komnos
Centre de Medecine du Sport et de l'Exercice, Swiss Olympic Medical Center, Hirslanden Clinique la Colline, Geneva, Switzerland

J. Menetrey (✉)
Centre de Medecine du Sport et de l'Exercice, Swiss Olympic Medical Center, Hirslanden Clinique la Colline, Geneva, Switzerland

Orthopaedic Surgery Service, University Hospital of Geneva, Geneva, Switzerland
e-mail: jacques.menetrey@hirslanden.ch

G. L. Canata et al. (eds.), *Management of Track and Field Injuries*,
https://doi.org/10.1007/978-3-030-60216-1_2

Another interesting point of their study is that although sprinters appear to have a unique foot structure with greater foot muscularity, this foot muscularity may not always contribute to superior sprint performance. More specifically, they found that despite the desirable increased thickness of the other foot muscles, higher thickness in the abductor hallucis muscle (ABH) could be a negative prognostic factor for sprint performance. This said that it makes no doubt that a strong and quick foot is a key element to be performant in sprinting.

2.3 Essential Elements

Essential elements of a high sprint performance are the ability to accelerate rapidly, the size of maximal velocity, and the ability to maintain this velocity [7]. Even more significant is the ability to accelerate rapidly in the first steps of a sprint, which can distinguish an elite sprinter from a good one [8]. At the muscle level, force, velocity, and power are mainly influenced by fiber type distribution and architecture. So, fast contracting fibers can shorten up to 2–3 times faster than slow ones, muscles with larger cross-sectional area (CSA) generate larger tensions, and muscles with longer fibers can contract more rapidly and generate peak power at a higher velocity [9–11].

Sprint performance and muscle architecture have been thoroughly investigated in the literature. A worth mentioning parameter of muscle architecture concerning sprint running performance, besides muscle thickness, is muscle length. As proposed by Abe et al. [12], a greater fascicle length would confer greater velocity capacity in the sprint acceleration phase. This applies due to the fact that a fiber that contains more sarcomeres in series would contract at a greater velocity than a fiber containing less sarcomeres in series; consequently, power production is greater and sprint performance as well [10]. Monte et al. [13] also support this theoretical background, observing a strong positive correlation between (relative) fascicle length and mechanical power production.

Kubo et al. [16] demonstrated a significant positive relationship between 100 m best sprint time and muscle thickness of knee extensors, but no relationship with tendon stiffness, and elongation, of the knee extensor muscles. In another study, Kumagai et al. [14] reported a significant negative relationship between 100 m best sprint time and fascicle length of vastus lateralis (VL), gastrocnemius medialis (GM), and lateralis (GL). In accordance with this, Abe et al. [15] found a significant negative relationship between 100 m best sprint time and fascicle length of VL and GL but not with GM. Moreover, a negative relationship between maximal elongation of VL tendon and aponeurosis with 100 m sprint times has also been reported [17].

Monte et al. [13] suggested that muscle thickness is positively correlated with power production during sprint running. An increase in muscle thickness (e.g., as a result of a strength training protocol) leads to a greater force production capacity of the muscle with a subsequent expectation of improved acceleration ability of the athlete, due to the positive relationship between force production and acceleration performance [10, 18].

Investigation of possible differences between male and female sprinters has been also of interest in the literature. The sex difference in 100 m sprint performance between the world's best athletes is approximately 10%. This difference is hypothesized to depend on the skeletal muscle mass (SM) relative to body mass, which differs between the two genders. Nevertheless, studies have demonstrated that the muscle fiber type composition and muscle fascicle length are similar between male and female elite sprinters [12, 19]. On the contrary, marked sex differences have been reported in muscle fiber size in athletes, especially fast-twitch fiber cross-sectional area, but not especially in elite sprinters [19]. Besides, sex differences in musculotendinous stiffness and greater structural compliance in females have been also reported [20]. Thus, it is generally considered that males are faster than females because males have more muscle mass [21]. Interestingly, Abe et al. [22] found that even though female sprinters had lower absolute and relative muscle thickness and muscle mass and a higher percentage of body fat compared with male sprinters, differences in muscle mass may not play such a large role in determining successful performance in elite male and female sprinters.

2.4 Measurement of Muscle Size

Muscle size measurement is performed through radiological means. Magnetic resonance imaging (MRI) is the gold standard for muscle size measurement. However, this procedure is not always convenient due to its inherent drawbacks (claustrophobia, not always easily performed, considerable cost). In the clinical setting, ultrasonography (US) is widely used because of its non-invasive nature, lower cost, higher portability, and faster feedback than MRI.

2.5 Biomechanics of Sprint

The biomechanics of sprint running has always been of interest in the scientific literature. The first studies investigating the mechanics of running were published back in the 1920s [23, 24]. The mechanical principles of sprint running have many similarities with running in general. Thus, a major difference is the large acceleration at the start [25]. As a sprint begins, the generation of forward (horizontal) acceleration is most likely the most significant factor that determines the performance (Fig. 2.1). High mean horizontal forces lead to better performance [26, 27] (Fig. 2.2). Another essential factor for achieving the best performance, besides net horizontal force, is minimizing the braking forces.

Accelerating is a key point of performance in many sports and especially in sprint. It is reported that in the 100-m run, the full acceleration phase (the phase from the start to the maximal running velocity) is directly correlated with performance [28, 29]. Literature highlights the significance of horizontal ground reaction force (GRF) production for sprint acceleration performance [8, 30] (Fig. 2.3). According to experimental and clinical studies, hip extensors contribute to sprint acceleration performance. In an attempt to explain the muscular origin of this efficient horizontally oriented GRF production, previous researchers have investigated the important role of the hip extensors (gluteal and especially hamstring muscles) in running performance [31, 32]. They reported that the hip extensor/knee flexor muscle actions played a predominant role as running speed increased and reached maximal sprint speeds. In most of these studies, this predominance was shown to occur during both swing and contact phases [33, 34]. Due to the overall fast motion of the lower limb, the transition between swing phase and stance phase is too short. Clark and Weyand [35] aimed to evaluate the interaction between these two phases in order to maximize running speed. They demonstrated that the amount of knee elevation sprinters achieved late in the swing phase, i.e., when hamstrings are actively lengthened, appears to contribute to the subsequent early stance GRF application through a reduced deceleration time. Therefore, as great limb velocities prior to foot ground impact occur during sprinting, this swing-stance transition moment is of crucial significance for hamstrings, which counteract both external hip flexion and knee extension moments and support forces as high as eight times of body weight [36].

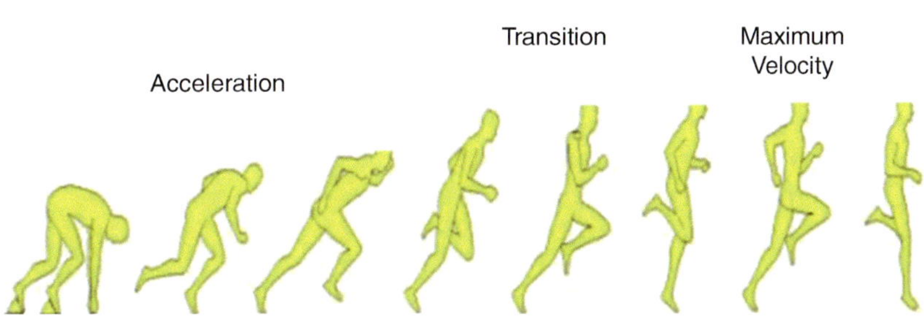

Fig. 2.1 Phases of sprinting

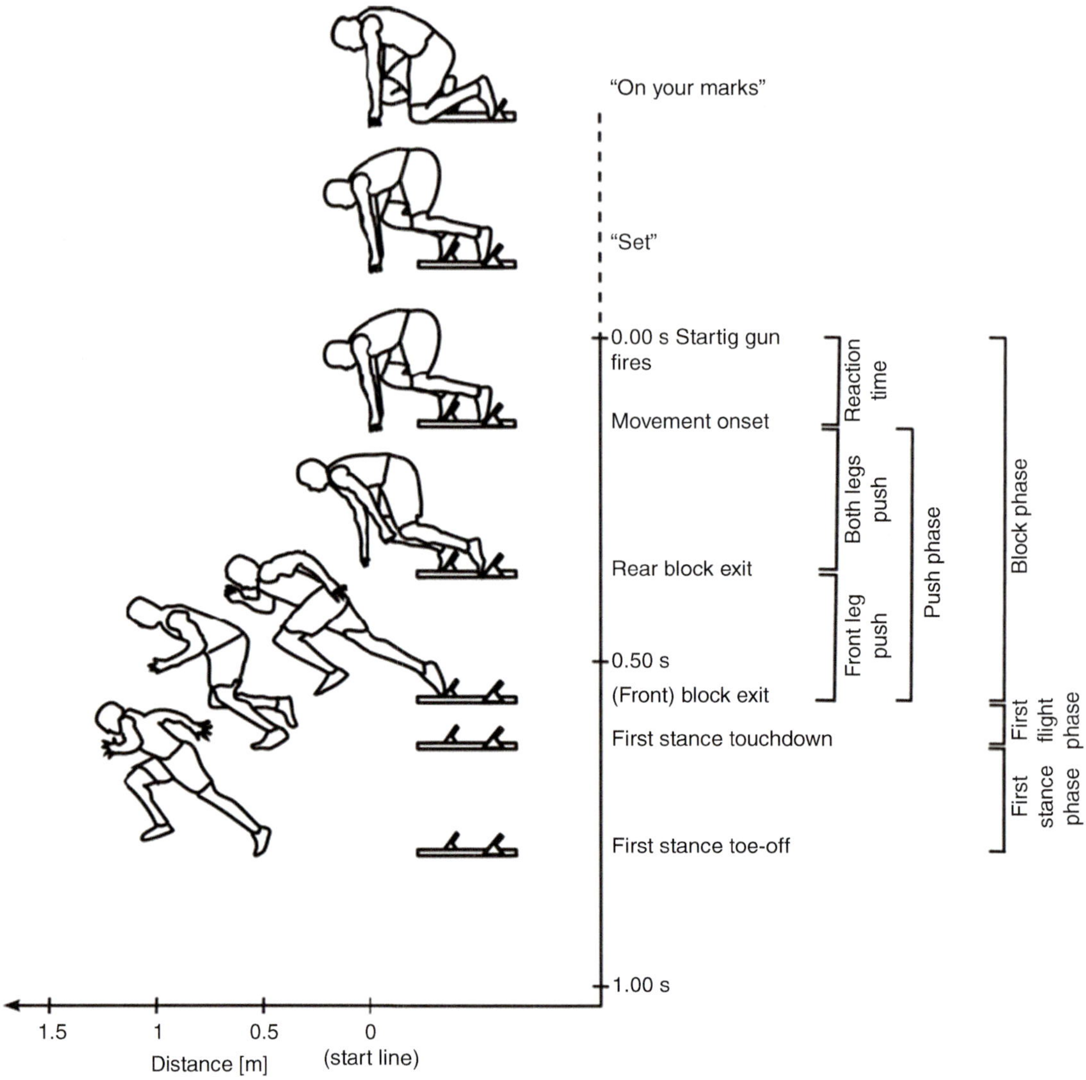

Fig. 2.2 Events and phases during the initial phase (sprint start)

Of major importance for achieving a perfect start is the coordination preparation, which develops the ability to harmoniously motor activities, and enhances the maximum utilization of sprinter's potential [37]. If a disruption in the coordination of a single movement in the sprinting stride cycle occurs, this will result in the delay of the start, the stance, and the swing phases [38]. Sprint start movement patterns demonstrate that biceps femoris and semitendinosus coordinate during a post-start phase, from lifting the front foot to the completion of the first two strides. Sciatic tibial muscles are responsible for knee flexion and thus for prolonging the midflight phase of the back foot during a sprint run. In addition, the gastrocnemius medialis muscles display similar correlations after the start phase. These activate in the support phase and remain active during the run until the next stance [39, 40]. The vastus lateralis is activated during a quick start reaction and the rectus femoris at 10 m of the running distance. These muscles participate in movement between the commands and are responsible for extension of the leg. In conclusion, an ideal sprint start depends on the muscle strength of the legs and the appropriate motor coordination, which

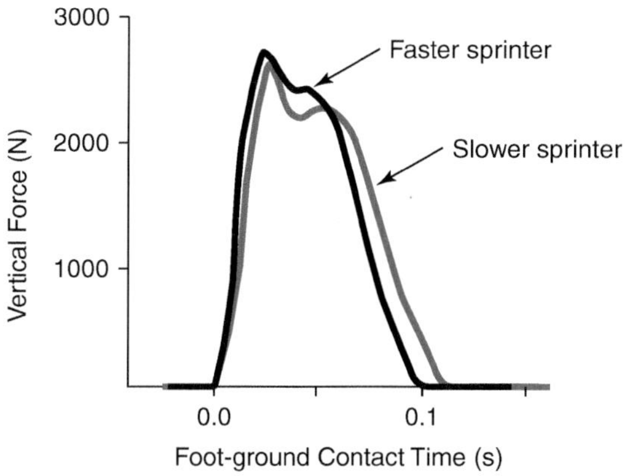

Fig. 2.3 Ground forces applied during sprint running. Elite sprinters can generate greater forces at a shorter time

greatly affects the generation of power in the legs at the right time and optimal duration [40].

In sprinting, high running speeds can be achieved by generating hip extension torque during the terminal swing through the stance phase [41]. The hamstrings and gluteal muscles are agonist muscles of hip extension, and neuromuscular coordination of these muscles contributes to the stabilization of the pelvis during sprint performance [42] As a result, functional imbalances between these muscles can result in an increase in the functional load on the hamstring muscles ending up in injuries [43]. From the late-swing phase to the early-contact phase during running at full speed, the hamstring is required to switch rapidly from eccentric to concentric contraction in the stretch-shortening cycle, while under the influence of the contractile activity of the quadriceps femoris muscle [44]. Therefore, neuromuscular coordination plays an important role in this activity. A hamstring muscle injury is conjectured to occur when there is muscular dyssynergia, such as a disorder in the timing of the contraction from the late-swing phase to the early-contact phase [45].

Regarding stiffness, the muscles of sprinters exhibit characteristic stiffness that can be beneficial to their performance. Passive and active muscle stiffness may play different roles in human locomotion, depending on locomotion speeds. Miyamoto et al. [46] found that higher passive muscle shear wave speed was weakly, but significantly, related to superior sprint performance. High passive muscle stiffness can help in quickly repositioning the limb during the aerial (swing) phase in sprinting. More particularly, the VL muscle is stiffer in long-distance runners than in sprinters under both passive and active conditions. Therefore, a high passive VL shear wave speed is associated with superior sprint performance.

In terms of foot muscle biomechanics, it is demonstrated that the flexor digitorum longus muscle (FDL) and flexor hallucis longus muscle (FHL) activate during the push-off phase and contribute to enhancing the plantar flexor moment [47]. Extrinsic muscles activate during the late stance phase while running contributing to ankle stability [48]. Regarding the intrinsic muscles, the abductor hallucis muscle (AbH) contributes to ankle stability during the late stance phase, and flexor digitorum brevis (FDB) and flexor hallucis brevis (FHB) muscles play important roles in toe flexion [49].

2.6 Sprint Training

Morphological adaptations to sprint training include changes in muscle fiber type, sarcoplasmic reticulum, and fiber cross-sectional area [50]. Therefore, an appropriate sprint training program could be expected to induce a shift toward type IIa muscle, increase muscle cross-sectional area, and increase the sarcoplasmic reticulum volume. Adaptations of the contractile apparatus to a variety of training types have been reviewed. As mentioned before, sprint runners have a larger percentage of type II fibers than other athletes and sprint performance has been strongly correlated with the percentage of histochemically typed type II fibers [3, 51]. Additionally, examination of the contractile nature of whole muscle using stimulated contractions in cross-sectional studies demonstrates that sprint athletes have greater rates of both force development and relaxation than untrained or endurance-trained individuals [52]. Changes to muscle contractile characteristics may also depend on the frequency of sprint training. It has been shown that 6 weeks of sprint training performed three times a week induces a significant increase in the percentage of type IIa muscle fibers in contrast to training twice daily for an additional week, which leads to an increase in the percentage of type I muscle fiber [53]. So, training should aim at developing muscle power, muscle coordination, core stability, and sprinting technique.

2.7 Differences between Young and Old Sprinters

It is worth noting how several biomechanical parameters differ between young and adult sprinters. Aeles et al. [54] aimed to compare the biomechanics of well-trained young and adult sprinters during the first stance phase of sprint running, with a specific emphasis on muscle–tendon unit (MTU) behavior. They found no difference in some of the highlighted performance-related parameters, such as ankle joint stiffness, the range of dorsiflexion, and plantar flexor moment. The young sprinters showed a greater maximal and mean ratio of horizontal to total ground reaction force (GRF), which resulted in a greater change in horizontal center of mass (COM) velocity during the stance phase. Results from the muscle–tendon unit (MTU) length analyses showed that adult sprinters had more MTU shortening and higher maximal MTU shortening velocities in all plantar flexors and the rectus femoris. The pattern of length changes in these MTUs provides ideal conditions for the use of elastic energy storage and release for power enhancement. In other words, a top sprinter needs to train his musculoskeletal system for a while and mature before reaching his best performance.

References

1. Komi PV, Viitasalo JHT, Havu M, Thorstensson A, Sjödin B, Karlsson J. Skeletal muscle fibres and muscle enzyme activities in monozygous and dizygous twins of both sexes. Acta Physiol Scand. 1977;100(4):385–92.
2. Mero A, Komi PV, Gregor RJ. Biomechanics of Sprint running: a review. Sports Med. 1992;13(6):376–92.
3. Mero A, Luhtanen P, Viitasalo JT, Komi PV. Relationships between the maximal running velocity, muscle fiber characteristics, force production and force relaxation of sprinters. Scand J Sport Sci. 1981;1:16–22.
4. Tanaka T, Suga T, Imai Y, Ueno H, Misaki J, Miyake Y, et al. Characteristics of lower leg and foot muscle thicknesses in sprinters: does greater foot muscles contribute to sprint performance? Eur J Sport Sci. 2019;19(4):442–50.
5. Fukunaga T, Miyatani M, Tachi M, Kouzaki M, Kawakami Y, Kanehisa H. Muscle volume is a major determinant of joint torque in humans. Acta Physiol Scand. 2001;172(4):249–55.
6. Hoshikawa Y, Muramatsu M, Iida T, Ii N, Nakajima Y, Kanehisa H. Sex differences in the cross-sectional areas of psoas major and thigh muscles in high school track and field athletes and nonathletes. J Physiol Anthropol. 2011;30(2):47–53.
7. Ross A, Leveritt M, Riek S. Neural influences on sprint running training adaptations and acute responses. Sports Med. 2001;31(6):409–25.
8. Hunter JP, Marshall RN, McNair PJ. Relationships between ground reaction force impulse and kinematics of sprint-running acceleration. J Appl Biomech. 2005;21(1):31–43.
9. Cormie P, McGuigan MR, Newton RU. Developing maximal neuromuscular power. Sport Med. 2011;41(2):125–46.
10. Lieber RL. Skeletal muscle structure, function, and plasticity. Physiotherapy. 2011;89(9):P565.
11. Lee SSM, Piazza SJ. Built for speed: musculoskeletal structure and sprinting ability. J Exp Biol. 2009;212(Pt 22):3700–7.
12. Abe T, Fukashiro S, Harada Y, Kawamoto K. Relationship between sprint performance and muscle fascicle length in female sprinters. J Physiol Anthropol Appl Hum Sci. 2001;20(2):141–7.
13. Montei A, Zamparoi P. Correlations between muscle-tendon parameters and acceleration ability in 20m sprints. PLoS One. 2019;14(3):e0213347.
14. Kumagai K, Abe T, Brechue WF, Ryushi T, Takano S, Mizuno M. Sprint performance is related to muscle fascicle length in male 100-m sprinters. J Appl Physiol. 2000;88(3):811–6.
15. Abe T, Kumagai K, Brechue WF. Fascicle length of leg muscles is greater in sprinters than distance runners. Med Sci Sports Exerc. 2000;32(6):1125–9.
16. Kubo K, Ikebukuro T, Yata H, Tomita M, Okada M. Morphological and mechanical properties of muscle and tendon in highly trained sprinters. J Appl Biomech. 2011;27(4):336–44.
17. Stafilidis S, Arampatzis A. Muscle-tendon unit mechanical and morphological properties and sprint performance. J Sports Sci. 2007;25(9):1035–46.
18. Morin JB, Bourdin M, Edouard P, Peyrot N, Samozino P, Lacour JR. Mechanical determinants of 100-m

sprint running performance. Eur J Appl Physiol. 2012;112(11):3921–30.
19. Costill DL, Daniels J, Evans W, Fink W, Krahenbuhl G, Saltin B. Skeletal muscle enzymes and fiber composition in male and female track athletes. J Appl Physiol. 1976;40(2):149–54.
20. Blackburn JT, Riemann BL, Padua DA, Guskiewicz KM. Sex comparison of extensibility, passive, and active stiffness of the knee flexors. Clin Biomech. 2004;9(1):36–43.
21. Haugen T, Tønnessen E, Seiler S. 9.58 and 10.49: nearing the citius end for 100 m? Int J Sports Physiol Perform. 2015;10(2):269–72.
22. Abe T, Dankel SJ, Buckner SL, Jessee MB, Mattocks KT, Mouser JG, et al. Differences in 100-m sprint performance and skeletal muscle mass between elite male and female sprinters. J Sports Med Phys Fitness. 2019;59(2):304–9.
23. Furusawa K, Hill AV, Parkinson JL. The dynamics of "sprint" running. Proc R Soc London Ser B, Contain Pap a Biol Character. 1927;102:713.
24. Best CH, Partridge RC. The equation of motion of a runner, exerting a maximal effort. Proc R Soc London Ser B, Contain Pap a Biol Character. 1928;103:724.
25. Haugen T, McGhie D, Ettema G. Sprint running: from fundamental mechanics to practice—a review. Eur J Appl Physiol. 2019;119(6):1273–87.
26. Rabita G, Dorel S, Slawinski J, Sàez-de-Villarreal E, Couturier A, Samozino P, et al. Sprint mechanics in world-class athletes: a new insight into the limits of human locomotion. Scand J Med Sci Sport. 2015;25(5):583–94.
27. Nagahara R, Mizutani M, Matsuo A, Kanehisa H, Fukunaga T. Association of sprint performance with ground reaction forces during acceleration and maximal speed phases in a single sprint. J Appl Biomech. 2018;34(2):104–10.
28. Morin JB, Slawinski J, Dorel S, de Villareal ES, Couturier A, Samozino P, et al. Acceleration capability in elite sprinters and ground impulse: push more, brake less? J Biomech. 2015;48(12):3149–54.
29. Delecluse C. Influence of strength training on sprint running performance. Current findings and implications for training. Sports Med. 1997;24(3):147–56.
30. Otsuka M, Shim JK, Kurihara T, Yoshioka S, Nokata M, Isaka T. Effect of expertise on 3D force application during the starting block phase and subsequent steps in sprint running. J Appl Biomech. 2014;30(3):390–400.
31. Schache AG, Dorn TW, Williams GP, Brown NAT, Pandy MG. Lower-limb muscular strategies for increasing running speed. J Orthop Sports Phys Ther. 2014;44(10):813–24.
32. Bartlett JL, Sumner B, Ellis RG, Kram R. Activity and functions of the human gluteal muscles in walking, running, sprinting, and climbing. Am J Phys Anthropol. 2014;153(1):124–31.
33. Kyröläinen H, Avela J, Komi PV. Changes in muscle activity with increasing running speed. J Sports Sci. 2005;23(10):1101–9.
34. Bezodis IN, Kerwin DG, Salo AIT. Lower-limb mechanics during the support phase of maximum-velocity sprint running. Med Sci Sports Exerc. 2008;40(4):707–15.
35. Clark KP, Weyand PG. Are running speeds maximized with simple-spring stance mechanics? J Appl Physiol. 2014;117(6):604–15.
36. Morin JB, Gimenez P, Edouard P, Arnal P, Jiménez-Reyes P, Samozino P, et al. Sprint acceleration mechanics: the major role of hamstrings in horizontal force production. Front Physiol. 2015;6:404.
37. Rimmer E, Sleivert G. Effects of a plyometrics intervention program on sprint performance. J Strength Cond Res. 2000;14(3):295–301.
38. Sessa F, Messina G, Valenzano A, Messina A, Salerno M, Marsala G, et al. Sports training and adaptive changes. Sport Sci Health. 2018;14:2.
39. Wiemann K, Tidow G. Relative activity of hip and knee extensors in sprinting - implications for training. New Stud Athl. 1995;10(1):29–49.
40. Borysiuk Z, Waśkiewicz Z, Piechota K, Pakosz P, Konieczny M, Blaszczyszyn M, et al. Coordination aspects of an effective sprint start. Front Physiol. 2018;9:1138.
41. Hunter JP, Marshall RN, McNair P. Reliability of biomechanical variables of Sprint running. Med Sci Sports Exerc. 2004;36(5):850–61.
42. Chumanov ES, Heiderscheit BC, Thelen DG. The effect of speed and influence of individual muscles on hamstring mechanics during the swing phase of sprinting. J Biomech. 2007;40(16):3555–62.
43. Panayi S. The need for lumbar-pelvic assessment in the resolution of chronic hamstring strain. J Bodyw Mov Ther. 2010;14(3):294–8.
44. Simonsen EB, Thomsen L, Klausen K. Activity of mono- and biarticular leg muscles during sprint running. Eur J Appl Physiol Occup Physiol. 1985;54(5):524–32.
45. Sugiura Y, Sakuma K, Sakuraba K, Sato Y. Prevention of hamstring injuries in collegiate sprinters. Orthop J Sport Med. 2017;5(1):2325967116681524.
46. Miyamoto N, Hirata K, Inoue K, Hashimoto T. Muscle stiffness of the vastus lateralis in sprinters and long-distance runners. Med Sci Sports Exerc. 2019;51(10):2080–7.
47. Péter A, Hegyi A, Stenroth L, Finni T, Cronin NJ. EMG and force production of the flexor hallucis longus muscle in isometric plantarflexion and the push-off phase of walking. J Biomech. 2015;48(12):3413–9.
48. O'Connor KM, Price TB, Hamill J. Examination of extrinsic foot muscles during running using mfMRI and EMG. J Electromyogr Kinesiol. 2006;16(5):522–30.
49. Gooding TM, Feger MA, Hart JM, Hertel J. Intrinsic foot muscle activation during specific exercises: a T2 time magnetic resonance imaging study. J Athl Train. 2016;51(8):644–50.
50. Ross A, Leveritt M. Long-term metabolic and skeletal muscle adaptations to short-sprint training:

implications for sprint training and tapering. Sports Med. 2001;31(15):1063–82.
51. Denis C, Linossler MT, Dormols D, Padilla S, Geyssant A, Lacour JR, et al. Power and metabolic responses during supramaximal exercise in 100-m and 800-m runners. Scand J Med Sci Sports. 1992;2(2):62–9.
52. Carrington CA, Fisher W, White MJ. The effects of athletic training and muscle contractile character on the pressor response to isometric exercise of the human triceps surae. Eur J Appl Physiol Occup Physiol. 1999;80:337–43.
53. Esbjörnsson M, Hellsten-Westing Y, Balsom PD, Sjödin B, Jansson E. Muscle fibre type changes with sprint training: effect of training pattern. Acta Physiol Scand. 1993;149(2):245–6.
54. Aeles J, Jonkers I, Debaere S, Delecluse C, Vanwanseele B. Muscle–tendon unit length changes differ between young and adult sprinters in the first stance phase of sprint running. R Soc Open Sci. 2018;5(6):180332.

Tendons and Jumping: Anatomy and Pathomechanics of Tendon Injuries

3

Lukas Weisskopf, Thomas Hesse, Marc Sokolowski, and Anja Hirschmüller

3.1 Biomechanics

Athletes competing in track and fields sustain huge impact forces, which need to be transported through the body. The basic function of the tendons is to transmit the force created in the muscle to the bone, thus making joint and limb movement possible [4]. To do this effectively, tendons must be capable of resisting high tensile forces with limited elongation [8]. Tendons tolerate extreme tensile forces during sprinting and jumping. Already during normal walking forces of about 2 times, body weight is acting on the Achilles tendon. With increased speed, these forces increase up to 12.5 times the body weight [5].

The top values reach 1.4 tons (calculated for the high jump world record of 2.45 m). Reference values of in vivo loads on the Achilles tendon during various sports activities are shown in Table 3.1.

Comparably, forces acting on the patellar tendon sometimes reach extremely high values (Fig. 3.1). For example, maximum forces of up to 17.5 times body weight were calculated during weightlifting [7]. In jumping, forces on the patellar tendon are about 2000 Nm in take-off and 3000 Nm in landing, which corresponds to a force equivalent of approx. 200 kg and 300 kg, respectively. In general, these huge stresses can be well compensated by the special microarchitecture of the tendons and by their enormous adaptability with a gradual increase in stress. However, in the presence of specific risk factors (Fig. 3.2) or pre-damage of the tendon, the risk of structural damage increases even without the maximum force values having to be achieved. The primary tear force of the tendon is described for the Achilles tendon with 1.8 tons or 25 times the body weight. Basically, all tendons are subject to a so-called stress–strain mechanism, whereby their elongation capacity up to structural injury is about 8%. Crucially, their stiffness/softness is determined by tendon quality, training condition, and various other influencing factors.

3.2 Anatomy

Tendons are composed of collagen fibers and tenocytes, which lie in parallel rows in the extracellular matrix that contains proteoglycans. It forms a dense connective tissue whose purpose usually is to connect muscle and bone and consecutively stabilizes joints and allows for movement through storage and release of energy.

L. Weisskopf (✉) · T. Hesse · M. Sokolowski
A. Hirschmüller
ALTIUS Swiss Sportmed Center AG, Rheinfelden, Switzerland
e-mail: lukas.weisskopf@altius.ag; thomas.hesse@altius.ag; marc.sokolowski@altius.ag; anja.hirschmueller@altius.ag

G. L. Canata et al. (eds.), *Management of Track and Field Injuries*,
https://doi.org/10.1007/978-3-030-60216-1_3

Collagen fibers provide resistance to tensional stress, whereas proteoglycans add viscoelasticity to the tendon. From smallest to largest, the units forming the tendon are tropocollagen < collagen < fibril < fiber < fascicle. Multiple fascicles are surrounded by endotenon, which connects them to form the tendon. It allows for gliding of the fascicles to each other. Epitenon encircles the entire tendon and prevents adhesion to surrounding tissue.

The paratenon finally is the outermost layer further reducing friction between tendon and surrounding tissue [9, 10]. Endotenon and epitenon allow blood vessels and nerves to reach the deeper structures within the unit and prevent separation of the fascicles under stress. At the junction of tendon to bone, the enthesis represents a complex structure with different tissue properties including chondrocytes [11], vulnerable to asymmetrical load and potential of building heterotopic/intratendinous ossification (Fig. 3.3).

Table 3.1 In vivo forces acting on the Achilles tendon during different activities and track and fields [6]

Activity	Force (kN)	Author
Walking	1.3–1.5	Finni et al. (1998)
Counter movement jump	1.9–2.0	Fukashiro et al. (1995)
Squat jump	1.9–2.2	Fukashiro et al. (1995)
Drop jump	3.5–5.0	Brüggemeann et al. (2000)
Running	3.7–3.9	Komi et al. (1990)
Hopping	3.7–4.0	Fukashiro et al. (1995)
Sprint	Up to 9.0	Komi et al. (1990)

3.3 Mechanobiology

Adaptation of tendons to repetitive loading has been increasingly understood in recent years, especially the fact that load is important for remodeling and/or healing of tendons. While tendons in the past were primarily considered poorly vascularized, bradytrophic tissue, their high adaptability to physical stress and their outstanding mechanical properties have recently been increasingly recognized. The latter make them at the same time highly resistant and elastic [13]. The interaction between the mechanical stresses and the responses at the cellular level takes place via a complex homeostatic, mechanobiological feedback [14]. It has long been assumed that the adaptability of tendons in terms of blood circulation and implementation of the extracellular matrix under load are very low. Today, however, it is known that the metabolism of the collagen and the remaining connective tissue adapts to the load and the metabolic activity changes according to the physical activity. Various clinical–experimental works have shown that the oxygen and glucose uptake of the tendon increases under

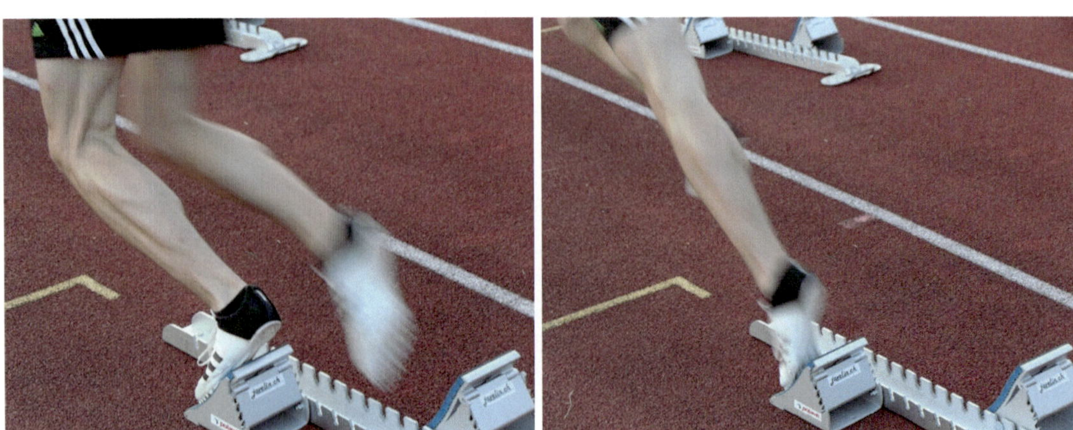

Fig. 3.1 Sprint starts with visualization of the enormous soleus activity (40% push-off capacity) during knee flexion positions going over to gastrocnemius action (33% push-off capacity [5]) while stretching the knee

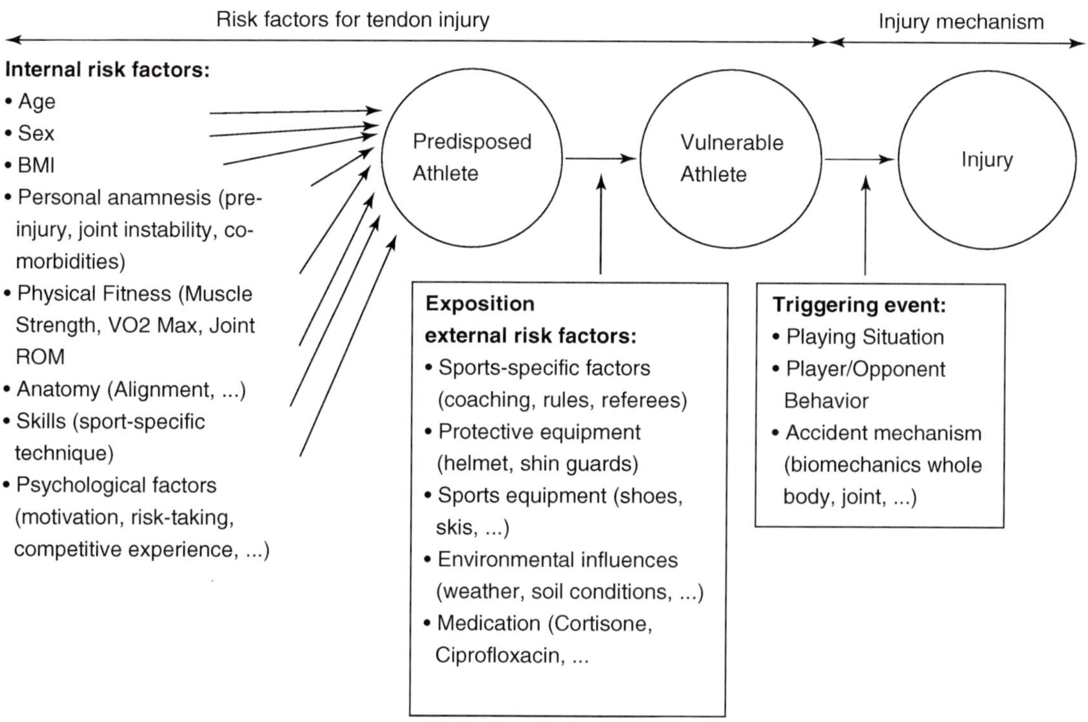

Fig. 3.2 Risk factors to be taken into account in the occurrence of sports/tendon injuries (modified according to Meeuwisse [6, 8]). A distinction is made between external and internal risk factors

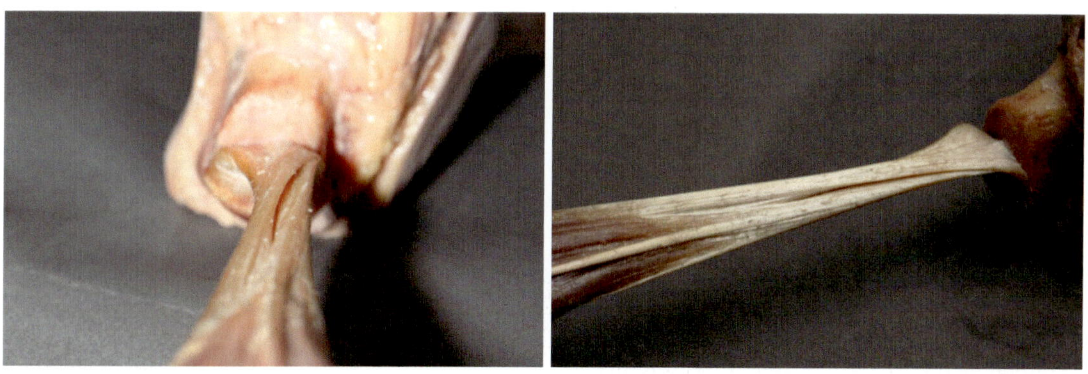

Fig. 3.3 Unique anatomical structure of the Achilles tendon with a twist of 90°, from a frontal and a lateral point of view. Soleus fibers aim to the medial calcaneus, which is important to recognize for the biomechanical understanding of the Achilles tendon function (with kind permission by Robert Smigielski, Poland) from Ref. [12]

mechanical stress [13]. A recently published meta-analysis impressively summarized how enormously adaptable the tendon is in terms of its mechanical, morphological, and structural properties [15]. Sustainable adaptation can be achieved in particular through high-load training and high intensities over a longer period of time (>12 weeks). On the other hand, the training or contraction form of the muscles (isometric/concentric/eccentric) seems to play a subordinate role.

Current data demonstrate that chronic exposure of the AT to elevated jumping loads results in adaptation of its mechanical and material properties. The Achilles tendon in the jump leg of male collegiate-level jumping athletes had 17.8%

greater stiffness and a 24.4% greater Young's modulus (compared to the contralateral lead (non-jump) leg, respectively). The side-to-side differences in jumpers were greater than observed in a cohort of athletic controls, suggesting that they are not simply due to limb dominance [16].

Jumpers also had 35.3% and 76.7% greater tendon stiffness and Young's modulus in their jump leg compared to that in the jump leg of athletic (non-jumping) controls [17].

The greater AT Young's modulus and stiffness in the jump leg of jumpers represent a favorable adaptation. During jumping, a structurally and materially stiffer tendon enables an improved ability to transmit muscle-generated forces, which improved explosive activity performance (jump force and height).

From a pathological standpoint, a stiffer tendon is exposed to greater stress, which may be considered potentially dangerous. However, tendon ultimate stress (i.e., the stress at which a tendon fails) is directly correlated with the tendon Young's modulus [18]. Thus, the increase in Young's modulus observed in jump leg of jumpers would be associated with the tendon being able to tolerate more stress before failure.

3.4 Pathophysiology

Pathophysiological processes have to be divided into different subgroups: "Tendinosis" a group of chronic-degenerative conditions usually of the midsubstance tendon caused by repetitive microtrauma. No inflammatory process can be made accountable for this condition. "Tendinitis" on the other hand is a painful inflammatory process mediated by cytokines and matrix metalloproteinases (MMPs). "Tenosynovitis" is a term describing inflammation of the paratenon with or without additional tendinosis. Lastly, a "rupture" or "tendon tear" is the loss of continuum of the tendon resulting in significant loss of function [19]. As already mentioned with the term "tendinosis," classic inflammatory changes can rarely be histologically detected. Terms such as "epicondylitis humeri lateralis" or "patellar tendonitis" should therefore be abandoned and named "-tendinosis" or "-tendinopathy" instead [20].

Significant changes can, however, be found histologically, indicating a dysfunctional healing response after microtrauma: thinning, disrupted collagen fibers, neoangiogenesis resulting in increased vascularity and cellularity, granulation tissue, and increased proteoglycan content [21]. Adams et al. already demonstrated in 1974 that age-related changes like tenocyte degeneration, accumulation of lipid amorphous extracellular matrix, and hydroxyapatite deposits could be found in early age affecting different tendons throughout the human body [22]. In comparison with normal tendon with well-aligned parallel and compact collagen fibers with adjacent tenocytes, the most prominent changes occur in the disorganization of the tendon matrix represented by discontinuous, crimped, and thinned collagen fibers with loss of their typical organized structure. Pathological tendons reveal loss of matrix integrity by reduction in total collagen content and increased production of extracellular matrix components that result in tendon stiffening [23].

Sonographic evaluation can reveal intra- and peritendinous changes including collagen disorganization and hypoechogenicity. Neovascularization can be found in combination with these degenerative changes, which are accompanied by nerve sprouting and hypersensitivity [24]. Jumpers are at high risk to be affected by tendinopathy of the patellar tendon as shown above. That is why it is also termed "jumper's knee." It can be classified depending on the location: The inferior pole of the patella is predisposed to injury due to maximum tensional stress during loading [25]. Less often but still relevant are the midportion and insertion at the tibial tuberosity [26]. It is important to detect coexisting changes in the Hoffa fat pad to initiate the correct therapy [27].

High levels of tendon strain are associated with a micromorphological deterioration of the collagenous network in the proximal patellar tendon of adolescent jumping athletes. Further, athletes suffering from or developing tendinopathy demonstrated both greater levels of tendon strain and lower levels of fascicle packing and alignment, which lends support to the idea that mechanical strain is the primary mechanical factor for tendon damage accumulation and the progression of overuse [28]. Finally, tendon rupture is associated with degenerative changes and also linked to the impairment of native repair mechanisms to defend the tendon from degeneration and ultimately rupture [29].

3.5 Pathomechanics

When classifying tendon injury mechanisms, acute injuries have to be distinguished from overload-associated damage and chronic-degenerative injuries (tendinopathy).

3.5.1 Acute Injury Patterns

Acute injuries mainly occur when large eccentric force acts on the tendon. A tendon is a remarkably strong tissue. Its in vitro tensile strength is about 50–100 N/mm^2. The cross-sectional area and the length of the tendon affect their mechanical behavior. The greater the tendon cross-sectional area, the larger loads can be applied prior to failure (increased tendon strength and stiffness). A tendon with a cross-sectional area of 1 cm^2 is capable of supporting a weight of 500–1000 kg. Athletes who subject their Achilles tendon to repetitive loads as habitual runners have shown larger Achilles tendon cross-sectional area than control subjects [12, 13]. An increased tendon cross-sectional area would reduce the average stress of the tendon, thereby decreasing the risk of acute tensile tendon rupture.

The breaking force of the Achilles tendon in vivo is as high at 18`000 Newton (equivalent to about 1.8 tons) or 25 times body weight. However, this only applies to axial load on the tendon. Brüggemann and Segesser were able to demonstrate a different tensile behavior with nonaxial strains act on the tendon and postulate it as a risk factor for Achilles tendon ruptures [7] and possible risk factor for overuse (Fig. 3.4).

This can be illustrated by the example of the tensile load of a sheet of paper. As long as the paper is pulled straight, it is very resistant. If the tension is applied asymmetrically, side-

Fig. 3.4 Asymmetrical load on the Achilles tendon during high jump push-off and pronation position of the foot

ways, the paper tears with significantly less effort. Thus, it is also easy to understand that pre-damaged tendons are particularly at risk for acute (partial) ruptures in the case of asymmetric tensile loads. The decrease in the tear force of symptomatic Achilles tendons with detectable pathological structural changes was illustrated in a prospective study by Nehrer et al. 28% of patients with sonographically detectable degenerations showed spontaneous ruptures within the following 4 years. Achilles tendon tears, however, are not limited to the structure of the tendon, but also extend to adjacent structures. Thus, more than 80% of the acute Achilles tendon ruptures also have lesions of the M. soleus usually found at the level of the soleus insertion. This could be caused by an asymmetric tensile load between gastrocnemius and soleus muscles.

Thus, the injury of the medial musculotendinous junction of the gastrocnemius head is easily comprehensible when there is an unnatural position of the calcaneus. If an increased varus position of the calcaneus is added with activated gastrocnemius muscles, the medial gastrocnemius fibers are maximally stressed. Because the corresponding tendon fibers of the medial gastrocnemius portion insert distally and laterally at the calcaneus, they have the largest lever arm of the triceps surae muscles and the longest fibers. The eccentric braking movement in this strain position and the maximum, partly asymmetric pull lead to an increased risk of injury. Very high forces associated with a high risk of complete tendon tearing act on the patellar tendon in knee dislocations, whereby chronic-degenerative overload damage at the patellar tendon is still much more common than ruptures. Typical tendon ruptures still occur as an injury to the most eccentrically loaded tendons and as bony apophysis tears in adolescence. Other acute forms of injury are tendon dislocations, mainly on the peroneal tendons (in ankle sprain) and biceps femoris (in knee dislocation) and tibialis posterior (in pronation trauma).

3.5.2 Chronic Injury Patterns

Chronic tendinopathies belong to the category of overload damage, which is very common in sports, but is often perceived only poorly or even belatedly [30]. The tendons are often subject to a disproportion of high loads with too low regeneration times. Depending on the type and quantity of the load acting on the tissue, dye distinguished a zone of homeostasis, a zone of supraphysiological overload, and a zone of overload, which can cause structural tissue damage. Repetitive loads are associated with immense force values. For example, in a marathon run in world record time, the Achilles tendon is charged at an average speed of more than 20 km/h at each step with approx. 9000 Nm (900 kg) [31], which in total at approx. 800 steps per km, to 42 km (approx. 33,000 steps), corresponds to an equivalent weight force of about 33,000 times the weight of a small car (900 kg) acting on the Achilles tendons. Another example is the total load of approx. 150 tons per patellar tendon during volleyball training with approx. 300 jumps.

It should be noted that an optimally dosed and axis appropriate training can lead to a structural adaptation of the tendon (mechanobiology) and to an enlarged tendon cross section, as Couppe et al. (2009) have shown for the patellar tendon [32]. Here, 30% larger tendon cross sections in the jumping leg of female athletes compared to the nondominant leg and 20% larger tendon cross sections in male athletes in jumping and running sports compared to nonstressing sports (kayak) [33]. In order to prevent tendon injuries, therefore, in sports medicine and sports science, monitoring of stress or symptoms is increasingly being used [34–36]. The risk constellation for tendon injuries in old age is controversially discussed. Although the tendons of older persons have histopathologically 30% lower collagen concentrations, they nevertheless show the same mechanical strength due to compensatory increase in the collagen cross-connections ("crosslinks") [37]. It should also be noted that tendon adaptation can work through ideal training even in old age. Accordingly, one cannot assume an aging process alone, but rather an inactivity process. This is naturally due to comorbidities, such as movement-limiting cardiovascular diseases, arthrosis of the joints, gout, diabetes mellitus or other metabolic pathologies, and increased drug requirements (Fig. 3.2). A very plausible explanation of pathomechanisms was recently postulated by Kjaer et al. [38]. They showed the expression of growth factors and inflammatory mediators that affect collagen synthesis and proteoglycan activity in the peritendineum. The tendons most commonly affected by overload damage (tendinopathies) are the Achilles tendon (usually called midportion tendinopathies in the middle of the tendon, rarer than insertion tendinopathies at the calcaneus approach), the patellar tendon ("jumper's knee"), the quadriceps tendon, the plantar fascia, and the proximal tendons of the ischiocrural muscles. On average, 36% of volleyball players complain of knee pain over the course of a season, most often due to tendinopathy of the patellar tendon [35]. For a sustainable treatment of affected athletes, a good biomechanical understanding of the load and the knowledge of the intrinsic and extrinsic risk factors are of enormous importance. For the tendinopathy of the patellar tendon, nine specific factors could be identified. These include male sex [39], high weight [40], high training volume [41, 42], high muscle strength of the quadriceps [43, 44], high bounce [36] and training on asphalt [42], sports specialization [45], and reduced flexibility of quadriceps [46] and hamstrings [46, 47]. Frequently attempts have been made to prove a link between axis abnormalities and disturbances of the kinematic chain and the occurrence of patellar tendon tendinopathies (PTs), as it is obvious that these can lead to asymmetric tensile forces and thus increased loads of the patellar tendon and cause damage similar to the finding of Segesser and Brüggemann in the Achilles tendon. While no clear link has been established for the often accused pathological Q-angle (e.g., [42]), there is evidence that both leg length differences, a flattened arch [48, 49], a patella alta [50], and a disturbed patella tracking [51] can be accompanied by patellar tendon tendinopathies. Van der Worp et al. (2014) [49] also showed that the horizontal landing phase of jumps forward is crucial for the development of patellar tendinopathy. Patients with patellar tendon tendinopathies often end up with more bent knee and hip joints, so that further hip and knee flexion and thus a cushioning of the eccentric forces are less possible. The landing is therefore "harder" and is coined with higher peak forces in the patellar tendon. In summary, it can be said that both a thorough orthopedic examination and a biomechanical analysis of the movement patterns in chronic patellar tendon tendinopathies are of great importance (Table 3.2).

Table 3.2 Specific intrinsic risk factors for tendinopathies [27]

• Male sex
• Diabetes mellitus
• Metabolic disorders (e.g., hypercholesterolemia)
• Cortisone medication (local or oral)
• Quinolone antibiotics (e.g., Cipro and levofloxacin)
• Blood group 0
• >6000 km of running, > 10 years of running experience, training range > 60 km/week
• Increased tendon stiffness
• Expression of interleukins and metalloproteinases
• Decorin reduction
• Degeneration of tendon in old age
• Movement pattern
• Soleus lesion

3.6 Biomechanical Diagnostics and Therapy

According to these explanations, which are similarly transferable to other tendinopathies, both the stress pattern and the therapy should be biomechanically analyzed or verified and causal malfunctions should be eliminated. The latter can be done, for example, by adaptation of the movement patterns, e.g., by shoe insert supply or specific muscular stabilization forms. The biomechanical diagnosis of tendon injuries should be one of the standard examination methods today, as should imaging methods. It includes stabilometry, isokinetic force measurements (maximum force and rate of force development), running analysis, gait analysis, jump measuring plate, and isokinetic video analysis for the lower extremity. For the upper limb, these are isokinetic force measurements, video analysis, physiotherapeutic verification of scapula coordination, and muscle length measurement. Due to the explained biomechanical and mechanobiological aspects, the individualized therapy of the causal disorder should also be adapted and include various treatment approaches such as training adaptation, technology optimization, material equipment (shoe equipment, inserts), axis training, elimination of disruptive influencing factors (where possible), and "heavy slow resistance" training, "heavy load eccentric training," and "tendon neuroplastic training" (TNT) [52–56]. Summary tendons are subject to extremely large force effects, which are well-tolerated under normal conditions and even lead to tendon adaptation with improvement of the mechanical properties of the tendon during ideal training. Various influencing factors of an intrinsic and extrinsic nature can make the tendon susceptible for overload damage (tendinopathies). An asymmetrically eccentric load is particularly dangerous for the tendons. These pathomechanical aspects in the development of tendon pathologies must be diagnosed and eliminated in order to ensure a sustainable freedom of complaint for the patients concerned.

References

1. Cassel M, Muller J, Moser O, et al. Orthopedic injury profiles in adolescent elite athletes: a retrospective analysis from a sports medicine department. Front Physiol. 2019;10:544.
2. Vosseller JT, Ellis SJ, Levine DS, et al. Achilles tendon rupture in women. Foot Ankle Int. 2013;34(1):49–53.
3. Lundberg Zachrisson A, Ivarsson A, Desai P, et al. Athlete availability and incidence of overuse injuries over an athletics season in a cohort of elite Swedish athletics athletes - a prospective study. Inj Epidemiol. 2020;7(1):16.
4. Maquirriain J. Achilles tendon rupture: avoiding tendon lengthening during surgical repair and rehabilitation. Yale J Biol Med. 2011;84(3):289–300.
5. Komi PV, Fukashiro S, Jarvinen M. Biomechanical loading of Achilles tendon during normal locomotion. Clin Sports Med. 1992;11(3):521–31.
6. Weisskopf L, Drews B, Hirschmüller A, et al. In: Engelhardt M, Mauch F, editors. Pathomechanik und Verletzungsmuster – Teil 2: Sehnen, in Muskel- und Sehnenverletzungen. Verlag VOPELIUS, Jena; 2017. p. 49–58.
7. Zernicke RF, Garhammer J, Jobe FW. Human patellar-tendon rupture. J Bone Joint Surg Am. 1977;59(2):179–83.
8. Meeuwisse WH. Predictability of sports injuries. What is the epidemiological evidence? Sports Med. 1991;12(1):8–15.
9. Docheva D, Muller SA, Majewski M, et al. Biologics for tendon repair. Adv Drug Deliv Rev. 2015;84:222–39.
10. Schneider M, Docheva D. Mysteries behind the cellular content of tendon tissues. J Am Acad Orthop Surg. 2017;25(12):e289–90.
11. Yan Z, Yin H, Nerlich M, et al. Boosting tendon repair: interplay of cells, growth factors and scaffold-free and gel-based carriers. J Exp Orthop. 2018;5(1):1.
12. Freedman BR, Gordon JA, Soslowsky LJ. The Achilles tendon: fundamental properties and mechanisms governing healing. Muscles Ligaments Tendons J. 2014;4(2):245–55.
13. Kjaer M, Langberg H, Heinemeier K, et al. From mechanical loading to collagen synthesis, structural changes and function in human tendon. Scand J Med Sci Sports. 2009;19(4):500–10.
14. Thompson MS. Tendon mechanobiology: experimental models require mathematical underpinning. Bull Math Biol. 2013;75(8):1238–54.
15. Bohm S, Mersmann F, Arampatzis A. Human tendon adaptation in response to mechanical loading: a systematic review and meta-analysis of exercise intervention studies on healthy adults. Sports Med Open. 2015;1(1):7.
16. Bayliss AJ, Weatherholt AM, Crandall TT, et al. Achilles tendon material properties are greater in the jump leg of jumping athletes. J Musculoskelet Neuronal Interact. 2016;16(2):105–12.
17. Bojsen-Moller J, Magnusson SP, Rasmussen LR, et al. Muscle performance during maximal isometric and dynamic contractions is influenced by the stiffness of the tendinous structures. J Appl Physiol. 1985;99(3):986–94.
18. LaCroix AS, Duenwald-Kuehl SE, Lakes RS, et al. Relationship between tendon stiffness and failure: a metaanalysis. J Appl Physiol. 1985;115(1):43–51.
19. Dakin SG, Dudhia J, Smith RK. Resolving an inflammatory concept: the importance of inflammation and resolution in tendinopathy. Vet Immunol Immunopathol. 2014;158(3-4):121–7.
20. Khan KM, Cook JL, Kannus P, et al. Time to abandon the "tendinitis" myth. BMJ. 2002;324(7338):626–7.
21. Khan KM, Cook JL, Bonar F, et al. Histopathology of common tendinopathies. Update and implications for clinical management. Sports Med. 1999;27(6):393–408.
22. Adams CW, Bayliss OB, Baker RW, et al. Lipid deposits in ageing human arteries, tendons and fascia. Atherosclerosis. 1974;19(3):429–40.
23. Jarvinen M, Jozsa L, Kannus P, et al. Histopathological findings in chronic tendon disorders. Scand J Med Sci Sports. 1997;7(2):86–95.
24. Comin J, Cook JL, Malliaras P, et al. The prevalence and clinical significance of sonographic tendon abnormalities in asymptomatic ballet dancers:

a 24-month longitudinal study. Br J Sports Med. 2013;47(2):89–92.
25. Malliaras P, Cook J, Purdam C, et al. Patellar tendinopathy: clinical diagnosis, load management, and advice for challenging case presentations. J Orthop Sports Phys Ther. 2015;45(11):887–98.
26. Sarimo J, Sarin J, Orava S, et al. Distal patellar tendinosis: an unusual form of jumper's knee. Knee Surg Sports Traumatol Arthrosc. 2007;15(1):54–7.
27. Ward ER, Andersson G, Backman LJ, et al. Fat pads adjacent to tendinopathy: more than a coincidence? Br J Sports Med. 2016;50(24):1491–2.
28. Schechtman H, Bader DL. In vitro fatigue of human tendons. J Biomech. 1997;30(8):829–35.
29. Carmont MR, Silbernagel KG, Nilsson-Helander K, et al. Cross cultural adaptation of the Achilles tendon Total rupture score with reliability, validity and responsiveness evaluation. Knee Surg Sports Traumatol Arthrosc. 2013;21(6):1356–60.
30. Bahr R. No injuries, but plenty of pain? On the methodology for recording overuse symptoms in sports. Br J Sports Med. 2009;43(13):966–72.
31. Arndt AN, Komi PV, Bruggemann GP, et al. Individual muscle contributions to the in vivo achilles tendon force. Clin Biomech (Bristol, Avon). 1998;13(7):532–41.
32. Couppe C, Hansen P, Kongsgaard M, et al. Mechanical properties and collagen cross-linking of the patellar tendon in old and young men. J Appl Physiol (1985). 2009;107(3):880–6.
33. Kongsgaard M, Aagaard P, Kjaer M, et al. Structural Achilles tendon properties in athletes subjected to different exercise modes and in Achilles tendon rupture patients. J Appl Physiol (1985). 2005;99(5):1965–71.
34. Clarsen B, Bahr R, Heymans MW, et al. The prevalence and impact of overuse injuries in five Norwegian sports: application of a new surveillance method. Scand J Med Sci Sports. 2015;25(3):323–30.
35. Clarsen B, Ronsen O, Myklebust G, et al. The Oslo sports trauma research center questionnaire on health problems: a new approach to prospective monitoring of illness and injury in elite athletes. Br J Sports Med. 2014;48(9):754–60.
36. Visnes H, Aandahl HA, Bahr R. Jumper's knee paradox--jumping ability is a risk factor for developing jumper's knee: a 5-year prospective study. Br J Sports Med. 2013;47(8):503–7.
37. Couppe C, Kongsgaard M, Aagaard P, et al. Differences in tendon properties in elite badminton players with or without patellar tendinopathy. Scand J Med Sci Sports. 2013;23(2):e89–95.
38. Kjaer M, Bayer ML, Eliasson P, et al. What is the impact of inflammation on the critical interplay between mechanical signaling and biochemical changes in tendon matrix? J Appl Physiol (1985). 2013;115(6):879–83.
39. Lian O, Dahl J, Ackermann PW, et al. Pronociceptive and antinociceptive neuromediators in patellar tendinopathy. Am J Sports Med. 2006;34(11):1801–8.
40. Lian O, Refsnes PE, Engebretsen L, et al. Performance characteristics of volleyball players with patellar tendinopathy. Am J Sports Med. 2003;31(3):408–13.
41. Visnes H, Bahr R. Training volume and body composition as risk factors for developing jumper's knee among young elite volleyball players. Scand J Med Sci Sports. 2013;23(5):607–13.
42. Ferretti A. Epidemiology of jumper's knee. Sports Med. 1986;3(4):289–95.
43. Janssen I, Steele JR, Munro BJ, et al. Sex differences in neuromuscular recruitment are not related to patellar tendon load. Med Sci Sports Exerc. 2014;46(7):1410–6.
44. Janssen I, Steele JR, Munro BJ, et al. Predicting the patellar tendon force generated when landing from a jump. Med Sci Sports Exerc. 2013;45(5):927–34.
45. Cerullo G, Puddu G, Conteduca F, et al. Evaluation of the results of extensor mechanism reconstruction. Am J Sports Med. 1988;16(2):93–6.
46. Witvrouw E, Bellemans J, Lysens R, et al. Intrinsic risk factors for the development of patellar tendinitis in an athletic population. A two-year prospective study. Am J Sports Med. 2001;29(2):190–5.
47. Witvrouw E, Lysens R, Bellemans J, et al. Intrinsic risk factors for the development of anterior knee pain in an athletic population. A two-year prospective study. Am J Sports Med. 2000;28(4):480–9.
48. Van der Worp H, de Poel HJ, Diercks RL, et al. Jumper's knee or lander's knee? A systematic review of the relation between jump biomechanics and patellar tendinopathy. Int J Sports Med. 2014;35(8):714–22.
49. van der Worp H, van Ark M, Roerink S, et al. Risk factors for patellar tendinopathy: a systematic review of the literature. Br J Sports Med. 2011;45(5):446–52.
50. Kujala UM, Friberg O, Aalto T, et al. Lower limb asymmetry and patellofemoral joint incongruence in the etiology of knee exertion injuries in athletes. Int J Sports Med. 1987;8(3):214–20.
51. Allen GM, Tauro PG, Ostlere SJ. Proximal patellar tendinosis and abnormalities of patellar tracking. Skelet Radiol. 1999;28(4):220–3.
52. Alfredson H, Pietila T, Jonsson P, et al. Heavy-load eccentric calf muscle training for the treatment of chronic Achilles tendinosis. Am J Sports Med. 1998;26(3):360–6.
53. Beyer R, Kongsgaard M, Hougs Kjaer B, et al. Heavy slow resistance versus eccentric training as treatment for Achilles tendinopathy: a randomized controlled trial. Am J Sports Med. 2015;43(7):1704–11.
54. Kongsgaard M, Aagaard P, Roikjaer S, et al. Decline eccentric squats increases patellar tendon loading compared to standard eccentric squats. Clin Biomech (Bristol, Avon). 2006;21(7):748–54.
55. Kongsgaard M, Kovanen V, Aagaard P, et al. Corticosteroid injections, eccentric decline squat training and heavy slow resistance training in patellar tendinopathy. Scand J Med Sci Sports. 2009;19(6):790–802.
56. Rio E, Kidgell D, Moseley GL, et al. Tendon neuroplastic training: changing the way we think about tendon rehabilitation: a narrative review. Br J Sports Med. 2016;50(4):209–15.

Ligament Function and Pathoanatomy of Injury and Healing

4

Gabrielle C. Ma, James M. Friedman, Jae S. You, and Chunbong B. Ma

4.1 Structure and Function

Ligaments are fibrous connective tissues that span between bony surfaces acting to stabilize joints. They vary in size, location, shape, and orientation. Ligaments are fairly similar to tendons in both structure and physiology; however, ligaments and tendons differ in function. Tendons connect bone to muscle, whereas ligaments connect bone to bone [1]. Ligaments are responsible for allowing the body to perform specific movements by providing stabilization, guiding joints through a normal range of motion, and distributing tensile loads. For example, the medial collateral ligament spans the medial knee joint preventing valgus opening as the tibia swings in the sagittal plane [2].

Ligaments consist of bundles of collagen fibrils forming a wave crimp pattern [3]. This pattern gives ligaments an elastic property allowing elongation without damage. Depending on the type of ligament, there can be differing numbers of collagen fibril bundles allowing for different levels of elasticity [1, 3]. The alignment of collagen fibrils follows where the tension is applied to the ligament. Within the ligament substance are blood vessels that are parallel to the collagen fibrils [4]. Tissue fluid makes up 60% of the ligament weight, which allows for nutrient and metabolite diffusion to the embedded cells [1]. The solid components of the ligament consist mostly of type I collagen (90%) and type III collagen (10%) [5]. Collagen contributes to the ligament's strength and form, accounting for most of the dry weight [1]. The remainder consists of elastin, proteoglycans, and proteins. In the ligament, elastin is located near the collagen fibrils in the matrix [3, 6, 7]. While the ligament is minimally composed of elastin, elastin plays a large role in reducing tensile stress [8]. The coil conformation of the protein fibrils that make up the elastin allows the ligament to deform without rupturing or tearing [1].

Due to the amount of tension applied in activities, extracellular matrix varies between ligaments [9]. Compared to tendons, ligaments have lower percentage of collagen, less organized fibers, and higher percentage of proteoglycans and water in the extracellular matrix [2]. Proteoglycans can be classified into two main divisions of proteoglycans that play a role in the organization of the matrix and the ligament's ability to lengthen [7]. The larger articular cartilage-type proteoglycans fill the regions between the collagen fibrils by exerting pressure and maintaining water within the tissue. The small leucine-rich proteoglycans are involved with the formation and stability of the extracellular matrix and activity of growth factors [10,

G. C. Ma · J. M. Friedman · J. S. You
C. B. Ma (✉)
Department of Orthopaedic Surgery, University of California– San Francisco, San Francisco, CA, USA
e-mail: Maben@ucsf.edu

G. L. Canata et al. (eds.), *Management of Track and Field Injuries*,
https://doi.org/10.1007/978-3-030-60216-1_4

11]. The extracellular matrix dominant cell type is fibroblasts, which are located between collagen fibers [2]. They help maintain the matrix, and recently, they have been shown to be capable of cell-to-cell communication [2, 10]. Lastly, non-collagenous proteins, like monosaccharides and oligosaccharides, make up little tissues, but have been shown to help maintain the extracellular matrix and influence cell function [3, 12] . Fibronectin, a non-collagenous protein, was found to be associated with the molecules and blood vessels in the ligament matrix [3, 10].

Ligament insertions are sites where ligaments attach to the bones, which vary based on the angle between collagen fibers and proportion of collagen fibers [1, 6] . Ligament insertions are small and contribute little to the ligament's volume and length [6]. However, they contribute to transfer of blood supply to ligaments [12]. There are two main classifications of ligament insertions: indirect and direct [1, 13]. Indirect insertions are the more common kind of insertions [1, 12]. Superficial fibers like Sharpey fibers are inserted into the periosteum of the bone, which is connective tissue around the bone that plays a role in bone growth and repair, e.g., the MCL insertion on the tibial side. In contrast, direct insertions pass directly into bone through fibrocartilage and surrounding periosteum, transitioning from tendon to uncalcified fibrocartilage, calcified fibrocartilage to bone [2], e.g., the MCL insertion on the femoral side.

4.2 Injury

Ligament injuries represent some of the most common musculoskeletal injuries. Shoulder, knee, ankle, and wrist joints are most commonly affected by ligament injuries [2].

Injury to ligaments is caused by disruptions in joint mobility and stability, which can damage other surrounding structures [6, 14]. Ligament injuries tend to occur during strenuous physical activity, overuse, repetitious movements, or cutting motions [6]. During these activities, the ligament's ability to deform under stress is overwhelmed, leading to strain or tear (Fig. 4.1) [2]. Strains and tears ultimately disrupt the load-bearing collagenous matrix, disrupt nutrient-delivering blood vessels, and kill matrix-building cells [1].

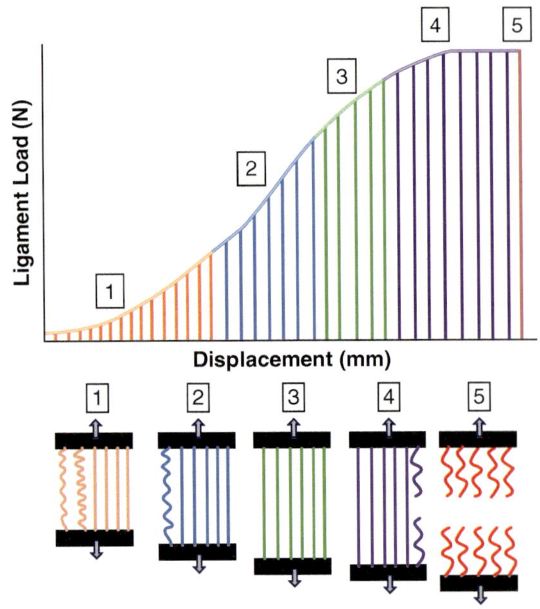

Fig. 4.1 Graph showing change in ligament length (displacement) with increasing load. As load increases, number of engaged ligament fibres and ligament length increases. 5 represents ultimate ligament tear and failure to withstand load

There are two main classifications of ligament injuries: intrinsic or extrinsic [2]. Intrinsic ligament injuries are caused by improper motion of the joint, whereas extrinsic ligament injuries are due to external factors such as a direct blow to the joint [2, 15]. At the time of injury, patients characteristically describe a distinct "pop" noise [16]. Symptoms include pain, swelling, instability, and inability to withstand weight [15, 17].

Ligament healing is slow and often the healed tissue is inferior to original ligament, which leads to further joint pathology [18]. When the ligament becomes lax, intra-articular pressure alters, leading to non-physiologic rubbing on articular cartilage [2]. This causes the breakdown and deterioration of the cartilage, ultimately leading to osteoarthritis [2, 18, 19]. The inability for ligaments to properly heal to the appropriate tension

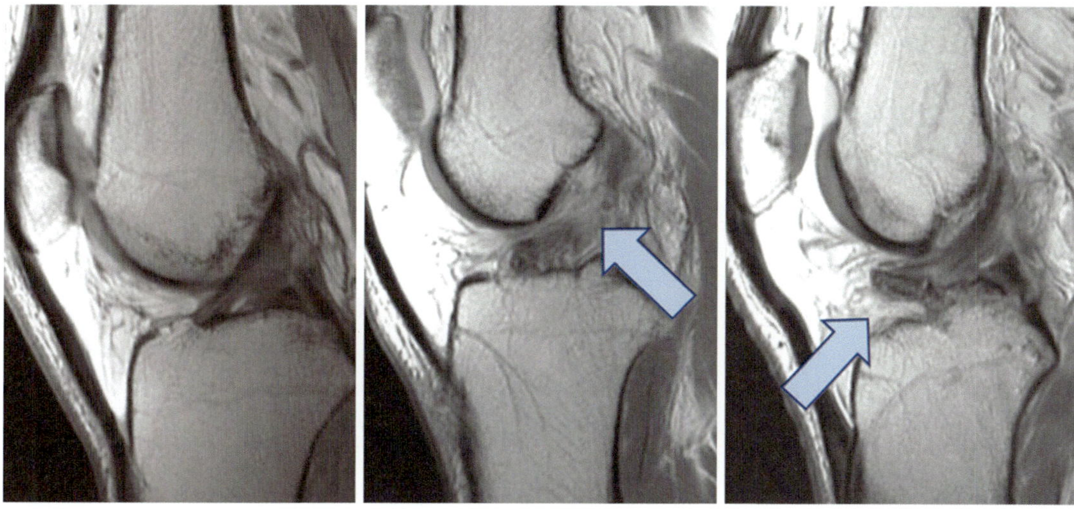

Fig. 4.2 Normal ACL (left) along with midsubstance (middle) and avulsion (right) ligament tears. Midsubstance tears occur in the body of the ligament, avulsion tears occur with the ligament pulling off a bony piece

leads to earlier osteoarthritis [20–22]. Ligament laxity also causes muscle weakness, joint laxity, knee instability, and decreased function [18].

The most common knee ligament tears are the anterior cruciate ligament (ACL) and medial collateral ligament (MCL) [2, 6] with a total annual incidence of 2 per 1000 people [23]. The ACL is an intra-articular ligament and connects from the femur to the tibia, preventing anterior laxity of the tibia on the femur as the knee performs its "hinge-like" motion. The MCL is an extra-articular ligament on the medial side of the knee and attaches the medial femur to the medial tibia, limiting joint valgus laxity. Both help reduce the load on the knee by absorbing the force and providing stability [20]. Despite its proximity in the same joint, the healing potential of these two knee ligaments is very different.

An important distinction between the MCL and ACL is the healing capacity due to different stem cell properties [24] and supporting structures. MCL tends to heal spontaneously, while ACL tends to have limited healing abilities [14]. Because of the differences between healing abilities, the respective ligaments are treated differently. MCL injuries (depending on tear location) do not usually require surgery and tend to heal quicker than ACL injuries. ACL tears, on the other hand, are usually treated with surgical reconstruction. These reconstructions replace the injured ligament with a graft [25].

Overload of tensile forces can cause unbalanced muscular contractions and lead to different locations of a ligament tear [26]. Avulsion tears tend to be associated with better outcomes compared to midsubstance tears. Avulsion tears tend to occur in older patients, while midsubstance tears tend to occur in younger patients. Disruption in the ligament is more common in the midsubstance location (Fig. 4.2) [27].

4.3 Healing

Disruptions and tears in the ligament cause a cascade of events to heal and recover the injury site. Ligament injuries lead to the initiation of the healing process, which consists of three phases: inflammatory, proliferation, and remodeling phases (Fig. 4.3) [1].

The inflammatory phase occurs during the first week of the injury incident [1, 28, 29]. During the inflammatory phase, cytokines and growth factors are released to stimulate tissue repair. Some examples include TGFB, IGF-1, and PDGF. Exudation of fluid from vessels in the injured region occurs due to vascular dilation and vascular permeability, causing the tissue to

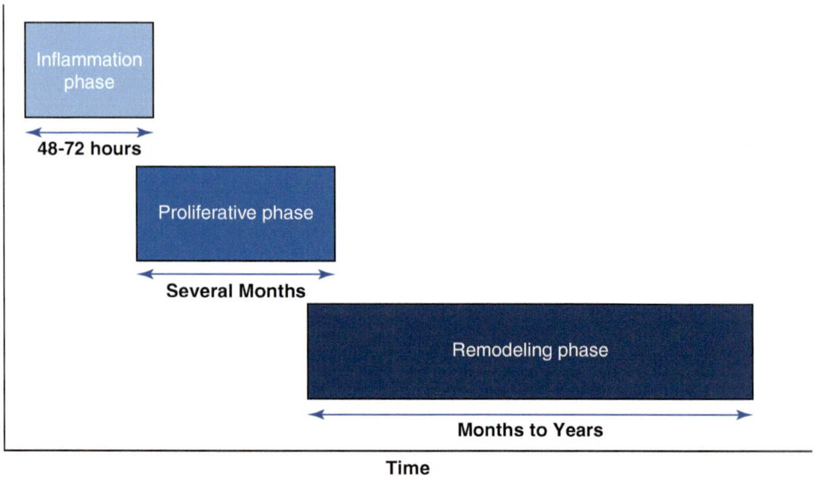

Fig. 4.3 Duration of the three different healing phases in the injured ligament: inflammatory, proliferative, and remodeling

become swollen [1]. Blood from damaged vessels accumulates within damaged tissue forming clots that are made up of fibrin, platelets, red blood cells, and cell and matrix debris [28]. These clots act as scaffolds that healing cells and related growth factors can anchor to. Polymorphonuclear leukocytes appear in the damaged tissue and clot [1]. Growth factors that are released from platelets and cells recruit neutrophils, which, in turn, recruit macrophages [9]. Monocytes become the dominant cell type at the injury site and phagocytose the necrotic tissue along with enzymes [2]. Endothelial cells in the blood vessels begin to proliferate, allowing tissue growth. The production of type III collagen also increases [28]. The release of the inflammatory cells in the inflammation phase recruits fibroblasts, which allows healing to enter the proliferation phase when the repair process begins [1].

During the proliferation phase, damaged tissue is repaired through cell regeneration and the expansion of the extracellular matrix [1, 9]. Many growth factors are released by immune cells in order to attract fibroblasts and increase ECM production [9, 30]. Soft, loose fibrous matrix is created by the new fibroblasts entering the tissue and clot and replacing the damaged tissue [30, 31]. Vascular buds soon grow into repair tissue and allow blood flow to injured tissue, creating vascular granulation tissue [32]. The type III collagen in the vascular granulation tissue is gradually replaced with type I collagen since type I collagen has more crosslinks and tensile strength [9]. When this happens, collagen fibrils size increases, matrix organization increases, number of blood vessels increases, elastin increases, and the tensile strength increases [1, 2]. The injury site ends up with excessive amounts of highly cellular tissue, explaining why ligaments are the weakest during this phase [1]. The newly deposited collagen in the ligament needs more organization and stability before it is finally healed, which is why the remodeling phase is necessary [30].

The last phase is the remodeling phase occurs within several weeks of injury. Injured ligament structure is first replaced by tissue resembling scar tissue [2]. Tissue is reshaped and strengthened by the removal and reorganization and cells and the matrix [1]. Fibroblasts and macrophages begin to decrease, water and proteoglycan concentrations decrease, and type III collagen decreases. The collagen fibrils of the matrix begin to settle in a more organized appearance [3, 30]. Signs of remodeling disappearing tend to occur within 4–6 months since injury, but the whole process can last for years as the ligament is constantly adapting and improving. Even most vascularized ligaments generally cannot heal [1, 9]. Remodeled tissue is also weaker compared to normal ligament tissue as the remolded matrices may consist of smaller collagen fibrils, failed collagen crosslinks, and alternations to proteoglycans and collagen [1, 3, 30].

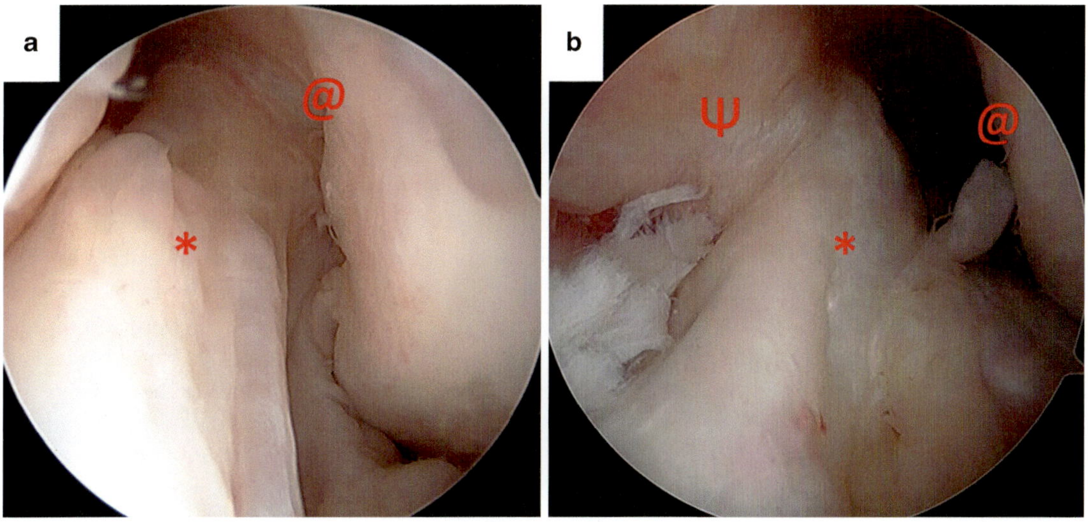

Fig. 4.4 (**a**) Intact ACL (*) inserting into the lateral femoral condyle (@). (**b**) ACL (*) that has torn from the femoral attachment (@) and fallen behind the PCL (Ψ). Without stable apposition to the femoral insertion site, this ACL cannot heal to its native attachment site (@)

4.4 Factors Affecting Healing

Factors affecting ligament healing include the type of the ligament, apposition, and stability of the injured ligament, and the amount of load applied. Intra-articular ligaments such as the ACL have demonstrated a poor healing response compared to extra-articular ligaments such as the MCL. Although the cells and vascularity of the ACL are capable of mounting a functional healing response similar to those found in the MCL [33, 34], the provisional scaffold found in the healing environment of MCL is not found in the ACL (Fig. 4.4). This may be explained by the altered environment between the two ligaments, as the ACL is surrounded by synovial fluid, whereas the MCL and other extracapsular ligaments are not [25].

In addition, apposition and stability of the torn ligament can aid in the healing process by decreasing the amount of collagen tissue and remodeling required to heal the injury. Therefore, treatment options that maintain some stability at the site of injury and close apposition of the ligament ends are favorable during the initial stages of healing.

Early controlled loading of the ligament can promote healing and improvement in biomechanical properties. Studies have shown that decrease in joint loading decreases the tensile strength of the bone–ligament interface and results in matrix degradation and decrease in the mass and strength of the ligament [35]. However, excessive and uncontrolled loading can disrupt tissue repair and alter healing [36–38].

The biological effect of immobilization on ligament injury has been widely studied. In superficial medial collateral ligament models, increased collagen degradation after 12 weeks of immobilization was observed in rabbit models [5]. In addition, detrimental effects of immobilization were seen in collagen, with increase in collagen degradation, decrease in synthesis, and a greater percentage of disorganized collagen fibrils in healing ligaments [39–42]. In another study using dog models, enhanced healing and improved biomechanical properties of the MCL were seen in early motion protocols [13]. Furthermore, according to two recent systematic reviews, there have been no controlled studies favoring prolonged immobilization for the treatment of ligament injuries [43, 44].

In contrast, early controlled resumption of activity including repetitive loading of the soft tissue has shown beneficial effects on the recovery of injured ligaments with enhancement of cellular activity resulting in increased tissue mass, strength, and improvement in matrix organization and organized collagen formation [35]. Controlled motion and exercise have been shown to increase blood flow to the affected joint and ligament, aiding in increased delivery of metabolites necessary for repair and healing.

4.4.1 Nonsteroidal Anti-Inflammatory Drugs (NSAIDs)

NSAIDs have been a mainstay in the treatment of ligament injuries; however, there is recent research to suggest that these drugs are only mildly effective in relieving symptoms while having a potentially harmful effect on soft tissue healing [45, 46]. NSAIDs are known to inhibit key steps of the inflammatory cascade including the recruitment of cells responsible for the initiation of the healing process [47].

In a rat model study, investigators studied the effects of a nonselective anti-inflammatory drug and a cyclooxygenase-2-specific anti-inflammatory on bone–tendon healing. The authors concluded that the inhibition of cyclooxygenase-2 in the inflammatory phase of healing resulted in adverse effects of bone-tendon healing [48]. A randomized control study looked at the use of NSAIDs in the treatment of acute ankle sprains in recruits in the Australian military. Investigators found that recruits treated with NSAIDs had a shorter time from injury to return to training; however, they also experienced increased ankle instability over the long term [49]. In addition, numerous other studies have concluded that the use of NSAIDs inhibits ligament healing and leads to impaired mechanical properties of the ligament [50, 51]. Therefore, NSAIDs are no longer recommended in the treatment of chronic ligament injuries and the use of these drugs is cautioned in the treatment of athletes with acute ligament injuries.

4.4.2 Cortisone Injections

Cortisone injections have shown a short-term benefit in decreasing pain and inflammation in ligament injuries. However, there is increasing evidence to suggest that cortisone injections into ligaments have a deleterious effect on the histological and biomechanical properties of ligament healing. On a cellular level, cortisone injections inhibit fibroblast function, which interferes with collagen synthesis [52–54]. In addition, the anti-inflammatory properties of corticosteroids disrupt the cascade of inflammatory cytokines and mediators essential in the healing process of ligaments [55]. Biomechanically, steroid-injected ligaments have been found to be smaller in cross-sectional areas with decrease in tensile strength and load to failure [56–59]. Therefore, the use of cortisone injections in the treatment of ligament injuries is discouraged, especially in athletes [60, 61].

4.5 Healing Augmentation

As extra-articular ligaments often heal with inferior biomechanics and intra-articular ligaments fail to heal at all, there has been increasing interest in augmentation of ligament healing to ensure a strong repair [1]. Healing augmentation strategies under investigation are based upon our understanding of staged ligament healing as described above. Broadly, healing augmentation research can be separated into cell-based therapy, growth factors, and scaffolds. Cell-based therapies provide cells that create the extracellular collagen matrix of ligament to the injury site [62]. Of particular interest is mesenchymal stem cells (MSCs), which can be isolated from bone marrow, adipose, or even tendon and ligament [63, 64]. MSCs can replicate and are associated with ligament healing. Replication allows for in vitro expansion prior to in vivo implantation. Ligament healing is related to the cells' ability to differentiate into multiple matrix-producing cells and MSC ability to secrete cytokines that activate surrounding cells and modulate the immune response [65]. Delivery to the injury site has been attempted with injection of MSCs in solution, in a fibrin or collagen carrier, and attached to a scaffold [9]. To date, a majority of outcome data have come from animal studies, which do show promising results; however, only preliminary data have emerged from human trials [64]. Although it is thought that cell-based therapy will ultimately play a role in ligament-healing augmentation, the best cell type and delivery methods are still under investigation.

Growth factors represent the small molecules, or cytokines, found throughout the ligament repair process that acts through cell differentiation, cell proliferation, chemotaxis, and/or cell-matrix synthesis. Growth factors placed at the site of injury work to stimulate or enhance the early phases of the healing response [1, 9]. Individual factors that have been tested include but are not limited to bFGF, GDF5, GDF6 (BMP13), GDF7 (BMP12), IGF1, PDGF, TGF-β1, TGF-β2, VEGF, and combinations of these growth factors (Fig. 4.5) [9]. Studies with many of these factors have shown some early benefit to tendon healing; however, the long-term outcomes have been mixed [66]. Discovering the best mixture of growth factors is particularly complex and has led to the increasing interest in platelet-rich plasma (PRP). PRP is obtained from the removal of red blood cells from autologous venous blood leaving behind a solution of concentrated platelets and growth factor-rich plasma [67]. PRP includes PDGF, VEGF, TGF-β, EGF, FGF, and IGF at varying concentrations [9]. Unfortunately, definitive evidence of PRP's long-term ability to enhance ligament healing has not yet been produced. Clinical studies are difficult to perform as differences between patients and preparation methods make the concentrations of growth factor within each PRP injection variable [1, 9]. Ongoing work with growth factors will likely focus on standardizing growth factor solutions in addition to continuing early promising work on how growth factors and cell-based therapies can be combined to recreate embryonic-like ligament growth [62, 66].

Scaffolds can act to stabilize an injured ligament, direct ligament growth, and act as an anchor site for cells and growth factors. Much of the current clinical scaffold research is focused on mimicking the natural fibrous scaffolds found in extra-articular ligaments for intra-articular ligaments such as the ACL. As described, these natural scaffolds are thought to play an important role in allowing extra-articular ligament healing

Inflammatory Phase	Function
Platelet-Derived Growth Factor (PDGF)	Influx of mononuclear cells and fibroblasts, enhanced angiogenesis and collagen deposition
Insulin-like Growth Factor-I (IGF-I)	Proliferation of fibroblasts, enhanced collagen deposition
Transforming Growth Factor-B (TGF-B)	Influx of mononuclear cells and fibroblasts, enhances collagen deposition
Proliferative Phase	
Insulin-like Growth Factor-I (IGF-I)	Proliferation of fibroblasts, enhanced collagen deposition
Transforming Growth Factor-B (TGF-B)	Influx of mononuclear cells and fibroblasts, enhances collagen deposition
Vascular Endothelial Growth Factor (VEGF)	Enhanced angiogenesis and collagen deposition
Basic Fibroblast Growth Factor (bFGF)	Proliferation of fibroblasts, enhanced collagen deposition

Fig. 4.5 Examples of some of the most studied growth factors and their functions during healing stages of soft tissue repair

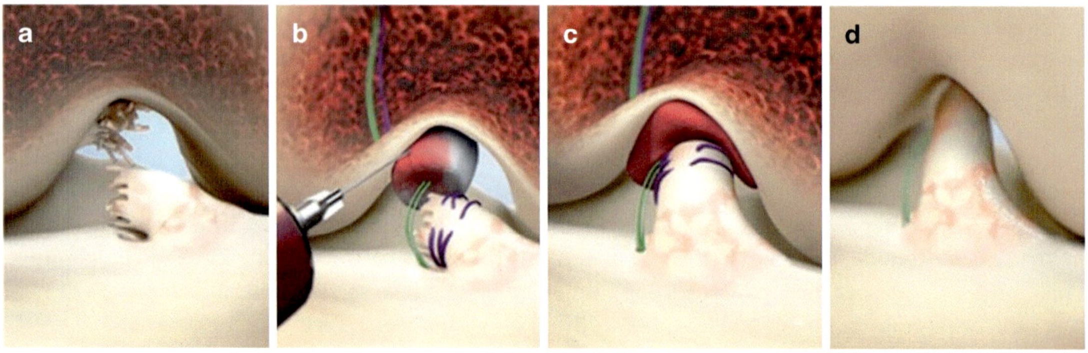

Fig. 4.6 Steps of Bridge-Enhanced ACL Repair (BEAR) technique using a collagen-based scaffold. (**a**) represents torn ACL tissue. (**b**). pictures the implantation of blood saturated collagen-based scaffold. (**c**) shows the tibial stump pulled into scaffold and secured with stiches. (**d**) depicts torn parts of ACL growing into the collagen-based scaffold. ACT tissue replaces BEAR implant and ligament is reunited. Reprinted with permission from "Bench-to-bedside: bridge-enhanced anterior cruciate ligament repair" by G. Perrone et al., Journal of Orthopaedic Research, 2017

[25, 68]. Early attempts at scaffold implantation into intra-articular ligaments suffered from overreactive inflammatory responses and poor healing. However, new low-DNA collagen-based scaffolds, when combined with autologous blood, have shown early potential to successfully heal ACLs in vivo without harmful inflammatory responses [69, 70]. Murray et al. have recently reported on human clinical trials in the Bridge-Enhanced Anterior Cruciate Ligament Repair (BEAR) study with promising, albeit preliminary, outcomes (Fig. 4.6) [71]. Further research is also being conducted into different materials to control the biomechanical properties of scaffolds, such as elasticity, to match the healing ligament and to enhance healing strength through mechanical stimulation [72]. Growth factor and cell gradients can also be created, which may allow the recreation of complex ligament structures such as the bone–ligament attachment [73]. As our understanding of ligament-healing cells, growth factors, and scaffold material improves, it is most likely that a combination of all three categories of healing augmentation will play a role in stronger and more predictable ligament healing.

In this chapter, we described the complex structural organization of ligaments and its important role in joint stability and function. Ligament injury and healing remain an active area of research where focus has been on enhancing healing with better biomechanical properties of the healed ligament and improving healing of the healing response. The future aspiration is to have a fast and reliable recovery from these common ligament injuries.

References

1. Nakamura N, Rodeo SA, Alini M, et al. Physiology and pathophysiology of musculoskeletal tissues. Philadelphia: Elsevier; 2015. Epub ahead of print 2015. https://doi.org/10.1016/B978-1-4557-4376-6.00001-9.
2. Hauser RA, Dolan EE, Phillips HJ, et al. Ligament injury and healing: a review of current clinical diagnostics and therapeutics. Open Rehabil J. 2013;6:1–20.
3. Buckwalter J, Maynard J, Vailas A. Skeletal fibrous tissue: tendon, joint capsule, and ligament. In: Albright JA, Brand RA, editors. The scientific basis of orthopedics. CT: Norwalk; 1987. p. 387–405.
4. Barrack R, Skinner H. The sensory function of knee ligaments. In: Akeson D, O'Connor D, editors. Knee ligaments: structure, function, injury, and repair. New York: Raven Press; 1990.
5. Amiel D, Akeson WH, Harwood FL, et al. Stress deprivation effect on metabolic turnover of the medial collateral ligament collagen. A comparison between 9- and 12-week immobilization. Clin Orthop Relat Res. 1983;172:265–70.
6. Frank CB. Normal ligament structure and physiology. J Musculoskelet Neuronal Interact. 2004;4:199–201.

7. Hardingham T. Proteoglycans: their structure, interactions and molecular organization in cartilage. Biochem Soc Trans. 1981;9:489–97.
8. Henninger HB, Valdez WR, Scott SA, et al. Elastin governs the mechanical response of medial collateral ligament under shear and transverse tensile loading. Acta Biomater. 2015;25:304–12.
9. Leong NL, Kator JL, Clemens TL, et al. Tendon and ligament healing and current approaches to tendon and ligament regeneration. J Orthop Res. 2020;38:7–12.
10. Frank C, Shrive N, Hiraoka H, et al. Optimisation of the biology of soft tissue repair. J Sci Med Sport. 1999;2:190–210.
11. Schaefer L, Iozzo RV. Biological functions of the small leucine-rich proteoglycans: from genetics to signal transduction. J Biol Chem. 2008;283:21305–9.
12. Cooper RR, Misol S. Tendon and ligament insertion. A light and electron microscopic study. J Bone Jt Surg A. 1970;52:1–20.
13. Woo SL, Gomez MA, Seguchi Y, et al. Measurement of mechanical properties of ligament substance from a bone-ligament-bone preparation. J Orthop Res. 1983;1:22–9.
14. Jung H-J, Fisher MB, Woo SL-Y. Role of biomechanics in the understanding of normal, injured, and healing ligaments and tendons. BMC Sports Sci Med Rehabil. 2009;1:9. Epub ahead of print December. https://doi.org/10.1186/1758-2555-1-9.
15. Schulz MS, Russe K, Weiler A, et al. Epidemiology of posterior cruciate ligament injuries. Arch Orthop Trauma Surg. 2003;123:186–91.
16. Gianotti SM, Marshall SW, Hume PA, et al. Incidence of anterior cruciate ligament injury and other knee ligament injuries: a national population-based study. J Sci Med Sport. 2009;12:622–7.
17. Agel J, Arendt EA, Bershadsky B. Anterior cruciate ligament injury in National Collegiate Athletic Association basketball and soccer: a 13-year review. Am J Sports Med. 2005;33:524–30.
18. Fleming BC, Hulstyn MJ, Oksendahl HL, et al. Ligament injury, reconstruction and osteoarthritis. Curr Opin Orthop. 2005;16:354–62.
19. Koh J, Dietz J. Osteoarthritis in other joints (hip, elbow, foot, ankle, toes, wrist) after sports injuries. Clin Sports Med. 2005;24:57–70.
20. Elkin JL, Zamora E, Gallo RA. Combined anterior cruciate ligament and medial collateral ligament knee injuries: anatomy, diagnosis, management recommendations, and return to sport. Curr Rev Musculoskelet Med. 2019;12:239–44. https://doi.org/10.1007/s12178-019-09549-3.
21. Panjabi MM. A hypothesis of chronic back pain: ligament subfailure injuries lead to muscle control dysfunction. Eur Spine J. 2006;15:668–76.
22. Øiestad BE, Engebretsen L, Storheim K, et al. Knee osteoarthritis after anterior cruciate ligament injury: a systematic review. Am J Sports Med. 2009;37:1434–43.
23. Papalia R, Vasta S, Denaro V, et al. Operative vs. nonoperative treatment of combined anterior cruciate ligament and medial collateral ligament injuries. In: Bhandari M, editor. Evidence-based orthopedics. Hoboken, NJ: Blackwell; 2011. p. 832.
24. Zhang AL, Feeley BT, Schwartz BS, et al. Management of deep postoperative shoulder infections: is there a role for open biopsy during staged treatment? J Shoulder Elb Surg. 2015;24:e15–20.
25. Murray MM, Fleming BC. Biology of anterior cruciate ligament injury and repair: kappa delta ann doner Vaughn award paper 2013. J Orthop Res. 2013;31:1501–6.
26. Stevens MA, El-Khoury GY, Kathol MH, et al. Imaging features of avulsion injuries. Radiographics. 1999;19:655–72.
27. van der List JP, Mintz DN, DiFelice GS. The location of anterior cruciate ligament tears: a prevalence study using magnetic resonance imaging. Orthop J Sport Med. 2017;5:2325967117709966. Epub ahead of print 21 June 2017. https://doi.org/10.1177/2325967117709966.
28. Buckwalter J, Cruess R. Healing of musculoskeletal tissues. In: Green D, Rockwood C, editors. Fractures. Philadelphia: JB Lippincott; 1991.
29. Woo S, Horibe S, KJ O. The response of ligaments to injury: healing of the collateral ligaments. In: Daniel D, Akeson W, O'Connor J, editors. Knee ligaments: structure, function, injury, and repair. New York: Raven Press; 1990.
30. Frank C, Schachar N, Dittrich D. Natural history of healing in the repaired medial collateral ligament. J Orthop Res. 1983;1:179–88.
31. Luiza M, Mello S, Godo C, et al. Changes in macromolecular orientation on collagen fibers during the process of tendon repair in the rat. Ann Histochim. 1975;20:145–52.
32. Clayton ML, Weir GJ. Experimental investigations of ligamentous healing. Am J Surg. 1959;98:373–8.
33. Murray MM, Martin SD, Martin TL, et al. Histological changes in the human anterior cruciate ligament after rupture. J Bone Jt Surg A. 2000;82:1387–97.
34. Murray MM, Spindler KP, Ballard P, et al. Enhanced histologic repair in a central wound in the anterior cruciate ligament with a collagen-platelet-rich plasma scaffold. J Orthop Res. 2007;25:1007–17.
35. Buckwalter JA. Activity vs. rest in the treatment of bone, soft tissue and joint injuries. Iowa Orthop J. 1995;15:29–42.
36. Fronek J, Frank C, Amiel D. The effects of intermittent passive motion (IPM) in the healing of medial collateral ligament. Trans Orthop Res Soc. 2017;8:31.
37. Andriacchi T, Sabiston P, Dehaven K, et al. Ligament injury and repair. In: Woo S, Buckwalter JA, editors. Injury and repair of the musculoskeletal soft tissues. Park Ridge, IL: American Academy of Orthopaedic Surgeons; 1988.
38. Vailas AC, Tipton CM, Matthes RD, et al. Physical activity and its influence on the repair process of medial collateral ligaments. Connect Tissue Res. 1981;9:25–31.

39. Padgett LR, Dahners LE. Rigid immobilization alters matrix organization in the injured rat medial collateral ligament. J Orthop Res. 1992;10:895–900.
40. West RV, Fu FH. Soft-tissue physiology and repair. In: Vaccaro AR, editor. Orthopaedic knowledge update. Rosemont, IL: Am Academy of Orthopaedic Surgeons; 2005. p. 15–27.
41. Woo SLY, Abramowitch SD, Kilger R, et al. Biomechanics of knee ligaments: injury, healing, and repair. J Biomech. 2006;39:1–20.
42. Woo SL, Gomez MA, Sites TJ, et al. The biomechanical and morphological changes in the medial collateral ligament of the rabbit after immobilization and remobilization. J Bone Jt Surg Am. 1987;69:1200–11.
43. Kerkhoffs GM, Rowe BH, Assendelft WJ, et al. Immobilisation and functional treatment for acute lateral ankle ligament injuries in adults. In: Cochrane database of systematic reviews. Hoboken, NJ: John Wiley & Sons, Ltd.. Epub ahead of print 22 July 2002; 2002. https://doi.org/10.1002/14651858.cd003762.
44. Nash CE, Mickan SM, Del Mar CB, et al. Resting injured limbs delays recovery: A systematic review, www.jfponline.com (1 September 2004, accessed 5 July 2020).
45. Dahners LE, Mullis BH. Effects of nonsteroidal antiinflammatory drugs on bone formation and soft-tissue healing. J Am Acad Orthop Surg. 2004;12:139–43.
46. Mehallo CJ, Drezner JA, Bytomski JR. Practical management: nonsteroidal antiinflammatory drug (NSAID) use in athletic injuries. Clin J Sport Med. 2006;16:170–4.
47. Radi ZA, Khan NK. Effects of cyclooxygenase inhibition on bone, tendon, and ligament healing. Inflamm Res. 2005;54:358–66.
48. Cohen DB, Kawamura S, Ehteshami JR, et al. Indomethacin and celecoxib impair rotator cuff tendon-to-bone healing. Am J Sports Med. 2006;34:362–9.
49. Slatyer MA, Hensley MJ, Lopert R. A randomized controlled trial of piroxicam in the management of acute ankle sprain in Australian regular Army recruits: the kapooka ankle sprain study. Am J Sports Med. 1997;25:544–53.
50. Warden SJ. Cyclo-oxygenase-2 inhibitors: beneficial or detrimental for athletes with acute musculoskeletal injuries? Sports Med. 2005;35:271–83.
51. Warden SJ, Avin KG, Beck EM, et al. Low-intensity pulsed ultrasound accelerates and a nonsteroidal antiinflammatory drug delays knee ligament healing. Am J Sports Med. 2006;34:1094–102.
52. Berliner D, Nabors C. Effects of corticosteroids on fibroblast functions. Res J Reticuloendothel Soc. 1967;4:284–313.
53. Kapetanos G. The effect of the local corticosteroids on the healing and biomechanical properties of the partially injured tendon. Clin Orthop Relat Res. 1982;163:170–9.
54. Oxlund H. The influence of a local injection of cortisol on the mechanical properties of tendons and ligaments and the indirect effect on skin. Acta Orthop. 1980;51:231–8.
55. Shapiro PS, Rohde RS, Froimson MI, et al. The effect of local corticosteroid or ketorolac exposure on histologic and biomechanical properties of rabbit tendon and cartilage. Hand. 2007;2:165–72.
56. Noyes F, Grood E, Nussbaum N, et al. Effect of intraarticular corticosteroids on ligament properties. A biomechanical and histological study in rhesus knees. Clin Orthop Relat Res. 1977;123:197–209.
57. Nichols AW. Complications associated with the use of corticosteroids in the treatment of athletic injuries. Clin J Sport Med. 2005;15:370–5.
58. Wiggins ME, Fadale PD, Ehrlich MG, et al. Effects of local injection of corticosteroids on the healing of ligaments: a follow-up report. J Bone Jt Surg A. 1995;77:1682–91.
59. Wiggins ME, Fadale PD, Barrach H, et al. Healing characteristics of a type I collagenous structure treated with corticosteroids. Am J Sports Med. 1994;22:279–88.
60. Fadale PD, Wiggins ME. Corticosteroid injections: their use and abuse. J Am Acad Orthop Surg. 1994;2:133–40.
61. Fredberg U. Local corticosteroid injection in sport: review of literature and guidelines for treatment. Scand J Med Sci Sports. 2007;7:131–9.
62. Gulotta LV, Kovacevic D, Ehteshami JR, et al. Application of bone marrow-derived mesenchymal stem cells in a rotator cuff repair model. Am J Sports Med. 2009;37:2126–33.
63. Walia B, Huang AH. Tendon stem progenitor cells: understanding the biology to inform therapeutic strategies for tendon repair. J Orthop Res. 2019;37:1270–80.
64. Hevesi M, LaPrade M, Saris DBF, et al. Stem cell treatment for ligament repair and reconstruction. Curr Rev Musculoskelet Med. 2019;12:446–50.
65. Fu Y, Karbaat L, Wu L, et al. Trophic effects of mesenchymal stem cells in tissue regeneration. Tis Eng B Rev. 2017;23:515–28.
66. Liu CF, Aschbacher-Smith L, Barthelery NJ, et al. What we should know before using tissue engineering techniques to repair injured tendons: a developmental biology perspective. Tis Eng B Rev. 2011;17:165–76.
67. Southworth TM, Naveen NB, Tauro TM, et al. The use of platelet-rich plasma in symptomatic knee osteoarthritis. J Knee Surg. 2019;32(1):37–45. https://doi.org/10.1055/s-0038-1675170.
68. Spindler KP, Murray MM, Devin C, et al. The central ACL defect as a model for failure of intra-articular healing. J Orthop Res. 2006;24:401–6.
69. Iannotti JP, Codsi MJ, Kwon YW, et al. Porcine small intestine submucosa augmentation of surgical repair of chronic two-tendon rotator cuff tears: a randomized, controlled trial. J Bone Jt Surg Ser A. 2006;88:1238–44.
70. Zheng MH, Chen J, Kirilak Y, et al. Porcine small intestine submucosa (SIS) is not an acellular collag-

enous matrix and contains porcine DNA: possible implications in human implantation. J Biomed Mater Res B Appl Biomater. 2005;73:61–7.
71. Murray MM, Fleming BC, Badger GJ, et al. Bridge-enhanced anterior cruciate ligament repair is not inferior to autograft anterior cruciate ligament reconstruction at 2 years: results of a prospective randomized clinical trial. Am J Sports Med. 2020;48:1305–15.
72. Chokalingam K, Juncosa-Melvin N, Hunter SA, et al. Tensile stimulation of murine stem cell-collagen sponge constructs increases collagen type I gene expression and linear stiffness. Tissue Eng Part A. 2009;15:2561–70.
73. Spalazzi JP, Dagher E, Doty SB, et al. In vivo evaluation of a tri-phasic composite scaffold for anterior cruciate ligament-to-bone integration. In: Annual International Conference of the IEEE Engineering in Medicine and Biology – Proceedings. Conf Proc IEEE Eng Med Biol Soc, 2006. pp. 525–8.

Anatomy and Function of Articular Cartilage

5

Alberto Gobbi, Eleonora Irlandini, and Alex P. Moorhead

5.1 Introduction

Cartilage tissue is a nonlinear, anisotropic, visco-elastic, and multiphasic complex with a low coefficient of friction, which distributes loads across the knee joint, protecting the subchondral bone and allowing for numerous cycles of joint loading before wearing [1, 2].

5.1.1 Chondrogenesis

Cartilage starts as undifferentiated mesenchyme, which changes into three different stratified layers as the mesoblast differentiates into chondrogenic structures. The top and bottom layers begin to join and grow eccentrically, integrating with the bone ends and acquiring chondrogenic features. The intermediate layer, which is less dense than the other layers, contains small lacunae, which grow and coalesce to give rise to the future joint cavity [3].

As a consequence of increased proliferating activity, nuclei of blastemic condensation are seen in 41-day-old embryos beginning pre-cartilaginous areas at the distal end of the femur and the proximal end of the tibia, in a continuous arrangement bound by undifferentiated mesoblastic tissue (Fig. 5.1). Three tissue areas or levels are formed in the mesenchyme located between the pre-cartilaginous folds. A centrally located undifferentiated area and two eccentric ones undergo chondral predetermination [3].

In the 48-day-old embryo, patellar mesenchyme condensation happens and wide organized chondral areas start to appear at the femoral condyles and tibial platform as seen in Fig. 5.2 [3].

After this, the femoral condyles and the tibial platform are now at the cartilaginous stage. The chondrification areas of the patellar mesoblastic aggregate increase and group together (Fig. 5.3). In the patella, the cartilaginous modeling is characterized by the growth of the cartilaginous mold through subperichondrial apposition and cell division [3].

Type

5.2 Types of Cartilage

There are three types of cartilage: elastic, fibro-elastic, and hyaline/articular cartilage. Elastic cartilage is found in the ear and in the larynx (4), and fibro-elastic cartilage is found in inter-

A. Gobbi (✉) · E. Irlandini
Orthopaedic Arthroscopic Surgery International (O.A.S.I.) Bioresearch Foundation, Gobbi N.P.O., Milan, Italy
e-mail: gobbi@cartilagedoctor.it; fellow@oasiortopedia.it

A. P. Moorhead
C.T.O., Milan, Italy
e-mail: info@oasiortopedia.it

G. L. Canata et al. (eds.), *Management of Track and Field Injuries*,
https://doi.org/10.1007/978-3-030-60216-1_5

vertebral disks and knee menisci (8). Hyaline/articular cartilage is the most widespread cartilage, which is a thin, connective tissue of diarthrodial (synovial) joints and is highly specialized with unique characteristics [1, 4, 5]. It contains no blood vessels, lymphatics, or nerves, which results in a limited capacity for healing and repair [6].

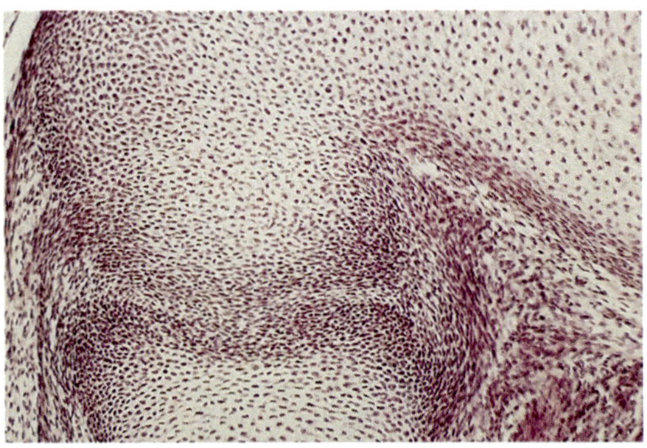

Fig. 5.1 Precartilaginous mesenchyme of femur and tibia bound together by undifferentiated mesenchymal cells. By courtesy of Collado JJ, Garcia PG et al. [3]

Hyaline cartilage is present in the embryo during endochondral ossification and in adults at the costal cartilages, in the respiratory system in the trachea, and in the growth plate of bones [4, 5]. Immature cartilage has a bluish color, but with maturation, becomes shiny, smooth, and white in young healthy adult mammals, and then becomes yellowish in older animals (Fig. 5.4) [5].

The main function of the articular cartilage is to maintain smooth movement and facilitate load transmission to the underlying subchondral bone. The main function of articular cartilage is to maintain smooth movement, facilitating load transmission to the underlying bone, and offering through a complex lubrification mechanism low shear stresses. It also protects the subchondral bone from compressive loading and mechanical trauma [5].

Articular cartilage consists of a liquid and a solid component. The liquid component is primarily water, and the solid component is mainly comprised of extracellular matrix [5].

The growth plate is an area that maintains cellular organization for long bone elongation [7].

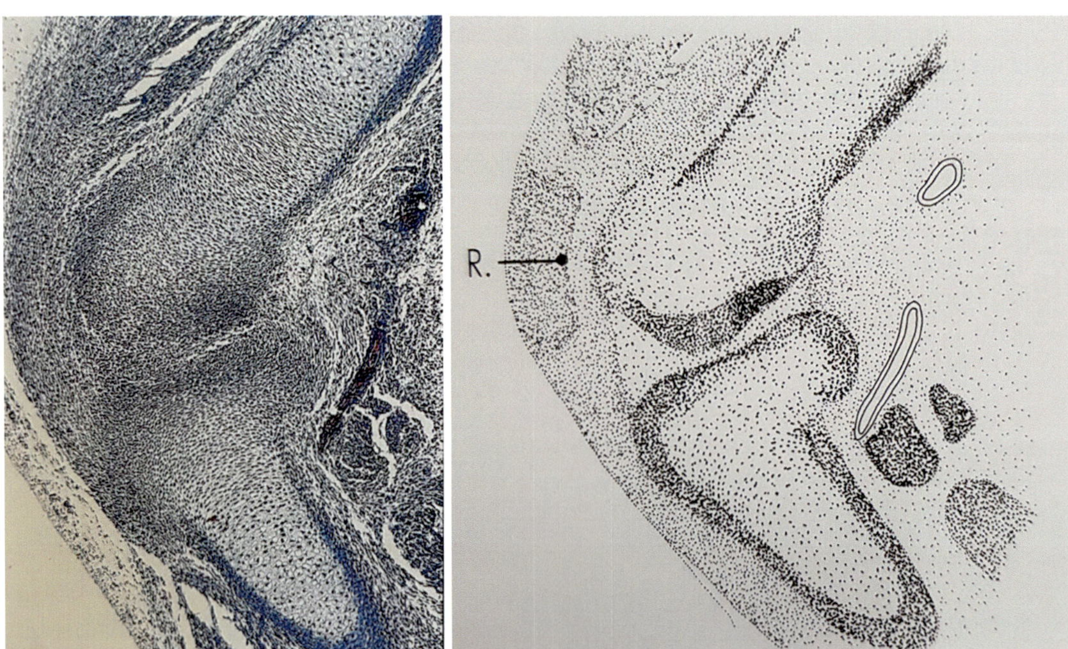

Fig. 5.2 Beginning of patellar mesenchymal condensation (left) and patellar primordium (R). Collado JJ, Garcia PG et al. [3]

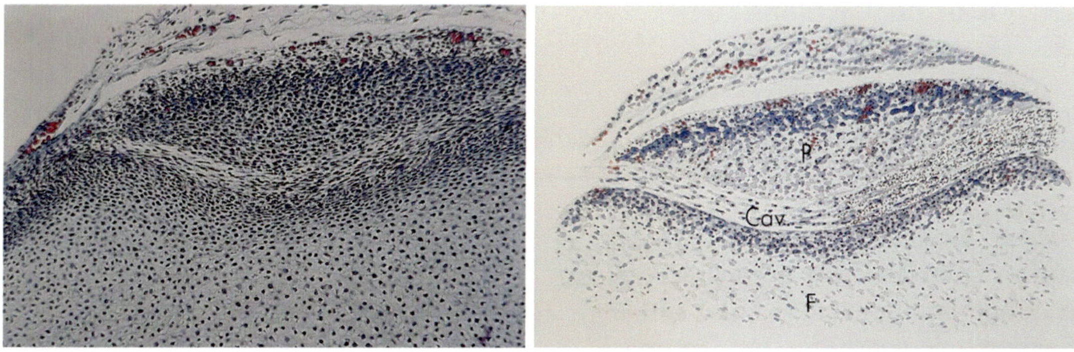

Fig. 5.3 Cartilaginous state with beginning of patellofemoral cavitation (*P* Patella, *Cav* Cavity, *F* Femur). Collado JJ, Garcia PG et al. [3]

Fig. 5.4 Normal adult articular cartilage with its chondrocyte and collagen fibril orientation. By courtesy of March et al. [45]

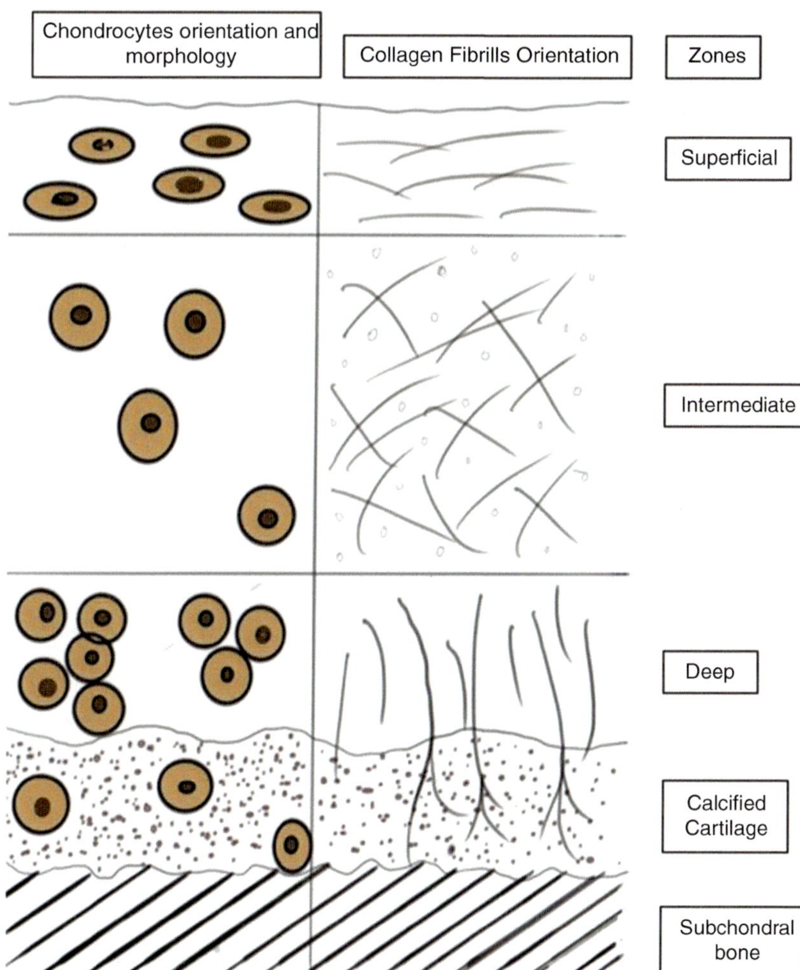

5.3 Articular Cartilage Components

Articular cartilage of the knee is approximately 2–4 mm thick composed of an extracellular matrix and highly specialized cells known as chondrocytes [6]. A network of collagen fibers begins as parallel to the surface and becomes perpendicular as it goes deeper as seen in Fig. 5.4 [8, 9].

5.3.1 Extracellular Matrix

The extracellular matrix is approximately 70–80% water and contains collagen, proteoglycans, and other glycoproteins [6]. Generally, these components maintain the water within the extracellular matrix, which is permeable and porous [5].

Cartilage is comprised of many types of collagen but primarily type II, which is responsible for approximately 60% of the dry weight of articular cartilage [5, 8, 10, 11]. Collagen fibers are composed of 4 polypeptide α-chains, which are twisted into a right-handed helix forming a rope structure stabilized by hydrogen bonds [8].

Collagen precursors, or the procollagens, are synthesized with C- and N-terminals. They are used for chain assembly prior to triple-helix formation. These will be cleaved by specific procollagen peptidases prior to fibril formation. Then, these fibrils will be stabilized further by making crosslinks with lysine residues. The biological functional form is the fibrillar collagen. Proper formation of fibril is needed for proper development of cartilage [8].

Type II collagen is a marker for chondrocyte differentiation, which is a homotrimer composed of an α1 (II) chain. It is the most abundant collagen present in the body representing 80% of all the collagen [12]. Fibrils are thinner than type I collagen found on other tissues. Type II collagen also forms crosslinks with type IX collagen. Antiparallel orientation of the molecules permits the necessary deformation under compression as observed from wet cartilage compression [8, 13]. It becomes parallel on the surface. Aside from type II collagen, hyaline cartilage also has type III (10%), type XI (3%), type IX (1%), and type VI (<1%). Type X collagen is in a calcified layer representing hypertrophic cartilage [5].

5.3.2 Non-collagenous Proteins

5.3.2.1 Proteoglycan

Proteoglycans are 20–30% of the dry weight [5, 10, 11]. Proteoglycan are needed to function normally. It has numerous functions depending on its core proteins and glycosaminoglycan chains [4].

Proteoglycan aggrecan, in the form of proteoglycan aggregates as hyaluronan and link protein, is responsible for its turgidity and osmotic properties [7]. This will now provide flexibility and viscoelasticity to the musculoskeletal system [12]. Aggrecan is the largest and produces multimolecular complex with hyaluronan where the glycosaminoglycan keratin sulfate and chondroitin sulfate attach furtherly stabilized by link proteins (Fig. 5.5) [5].

Proteoglycan aggregate and the interstitial fluid together maintain the compressive resilience through negative electrostatic repulsion forces (Fig. 5.6) [5]. Small amounts of leucine-rich repeat proteoglycans (SLRPs) are also present to maintain the tissue integrity and control metabolism. Examples of SLRPs include biglycan and decorin, which contain the dermatan sulfate, while the fibromodulin and lumican contain the keratan sulfate [5].

5.3.2.2 Glycosaminoglycans (GAGs)

These are carbohydrates with six major subunits in articular cartilage made from repeating disaccharide units. These major subunits are negatively charged, attracting water, calcium, and sodium but repel each other [5, 14, 15]. Their main function is to absorb water and maintain mechanical properties of the extracellular matrix (9). The synthesis of GAGs needs glucose, which diffuses from synovial fluid into the chondrocyte through glucose transporters (GLUT) [12, 16].

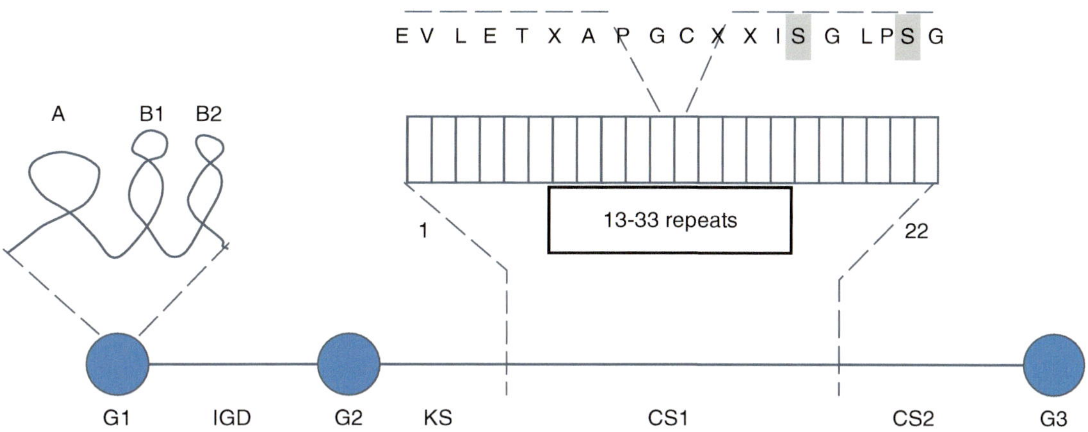

Fig. 5.5 Structure of aggrecan that consists of three disulphide-bonded globular domains (G1–3), an interglobular domain (IGD), and attachment regions for keratan sulfate (KS) and chondroitin sulfate (CS1 and CS2). By courtesy March, Lyn et al. [45]

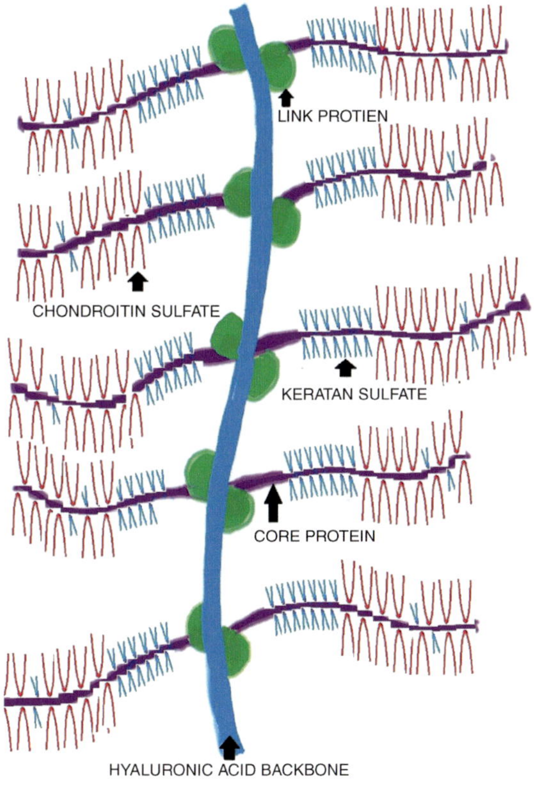

Fig. 5.6 Proteoglycan aggregation. Schematic diagram involving the interaction of proteoglycan monomers and link protein. By courtesy of King, Michael [46]

5.3.2.3 Structural Proteins

These proteins include cartilage matrix protein (matrilin-1 and matrilin-3), cartilage oligomeric protein (thrombospondin-5), cartilage intermediate layer protein, fibronectin, and tenascin-C [5].

5.3.2.4 Regulatory Proteins

These proteins include growth factors such as transforming growth factor-β (TGF-β), bone morphogenic proteins (BMPs), cartilage-derived retinoic acid-sensitive proteins, gp-39/YKL-40, matrix Gla protein, chondromodulin I, and chondromodulin II. This group of proteins affects cell metabolism with no structural role in the matrix [5] (Fig. 5.7).

5.3.3 Chondrocytes

Chondrocytes are cells that produce and maintain extracellular matrix of cartilage. It occupies only 2% of the total volume of the articular cartilage [5, 17]. It resists very high compressive loads. They are responsible for the maintenance of cartilage homeostasis by producing growth factors, enzymes, and inflammatory mediators [5]. Different pathways regulate chondrocyte func-

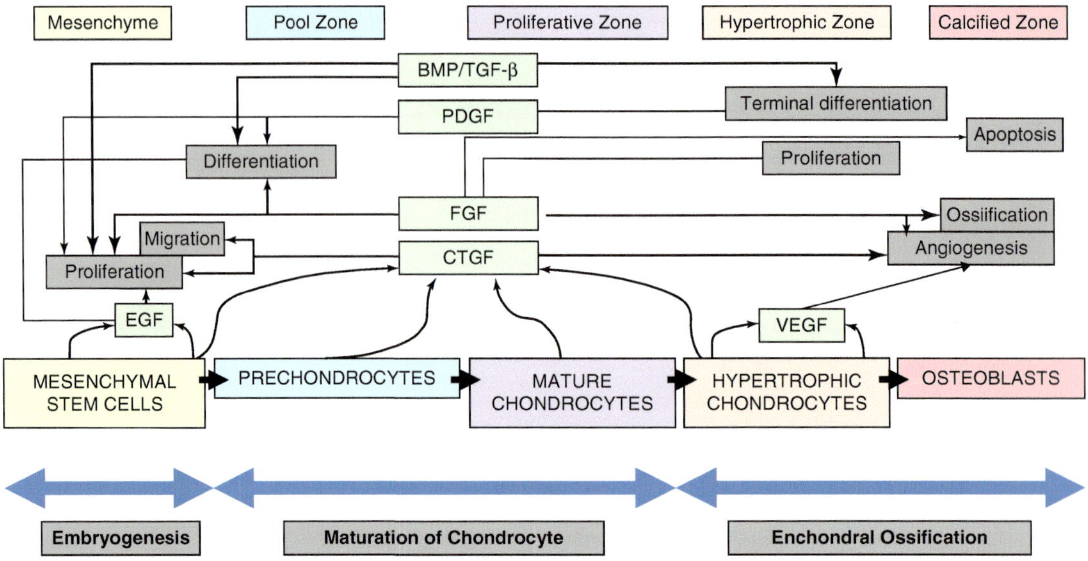

Fig. 5.7 Schematic diagram of the role of regulatory proteins at different stages of the chondrogenesis. By courtesy of Demoor, M. et al. 2014

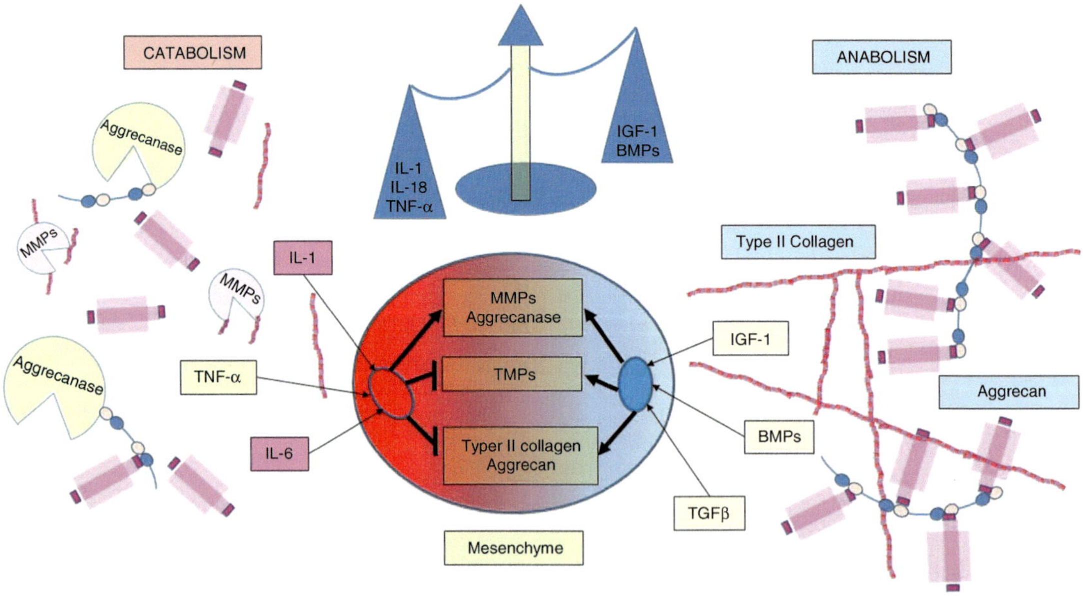

Fig. 5.8 Chondrocyte property of balancing anabolism versus catabolism. By courtesy of Demoor, M. et al. [12]

tion, regulate cartilage and bone formation, and maintain homeostasis of mature articular cartilage in adults [5, 18, 19]. It differs from other mesenchymal cells in terms of its properties and capabilities. It does not divide, and its apoptotic activity is low [12, 20, 21].

Chondrocytes are subjected to different mechanical and environmental factors that affect their metabolic activity and phenotype. Thus, according to the signals that they perceive, chondrocytes are now accountable for the production, organization, and maintenance of the integrity of the extracellular matrix [12]. They maintain the matrix by moderating the balance between anabolism and catabolism (Fig. 5.8). It is controlled by relative amount of growth factors and cytokines in synovial fluid. The result of this balance regulates cartilage homeostasis [12].

The number of chondrocytes formed by proliferating monolayer cultures is low, which is why it is not easily characterized. There are no cell surface markers, but the accepted indicator of the chondrocyte phenotype is type II collagen [5].

Adult cartilage chondrocytes rarely divide but live for a long time and maintain the capacity to replicate [5].

5.4 Zones of Articular Cartilage

Articular cartilage has four different zones, which are highly organized. Each zone has its own characteristics (Fig. 5.9) [5, 22, 23].

5.4.1 Superficial/Tangential Zone

This zone is a thin layer that protects other layers from shear stress. It is approximately 10–20% of the entire articular cartilage thickness. Collagen content is highest, while the proteoglycan content is lowest in this zone [8].

Most collagen fibers are type II and type IX collagen, which are parallel and tightly packed. It has numerous flattened chondrocytes and has the integrity to protect deeper layers. Since this zone is in contact with synovial fluid, it has most of the tensile properties. This layer generally prevents shear, tensile, and compressive forces during articulation [6].

This stains for fast green but not for safranin-O. Lamina splendens or the fine collagens at the surface can be seen. These cells are elongated but arranged tangentially [5].

5.4.2 Middle/Transitional Zone

This zone bridges the superficial and the deep zones. It comprises approximately 40–60% of the articular cartilage. Collagen is 20% less than the superficial zone, while proteoglycan content is 50% more compared to superficial zone. Collagens are arranged obliquely, while the chondrocytes have low density and are spherical. Compressive forces are first resisted by this zone [5, 10, 11].

Safranin-O staining first appears in this zone where cells are round or ovoid but with random distribution [5].

5.4.3 Deep/Basal Zone

Collagen and chondrocyte distribution is approximately equal with the middle or transition zone [5, 10, 11]. Cells in this zone are seen as short columns [5, 24].

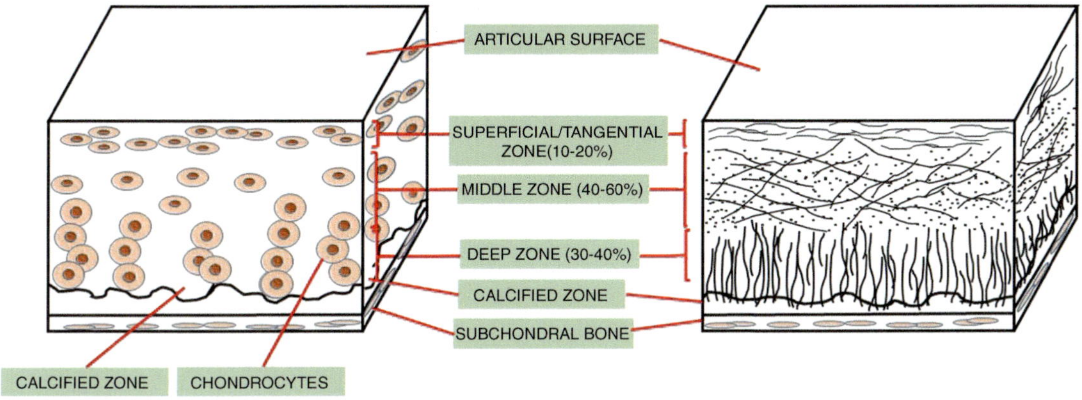

Fig. 5.9 Cross section of adult articular cartilage. By courtesy of Newman [47]

5.4.4 Tidemark and Calcified Zone

The tidemark represents the border between mineralized and unmineralized regions that separate the deep from the calcified zone. It is a thin basophilic area seen through eosin and hematoxylin stains [5].

5.5 Main Function of Articular Cartilage

The main function of the articular cartilage is to provide a smooth movement and facilitate load transmission with low friction [6]. An increase in local pressure causes the fluid to flow out of the extracellular matrix, but when the pressure or the compression load is removed interstitial fluid gets back to cartilage [5, 25–27]. Since the articular cartilage has a low permeability, fluid is prevented from being easily squeezed out of the matrix. Mechanical deformation is restricted by confining the cartilage under contact surface between the two opposing bones [27].

Synovial fluid also has a role in lubrication and nutrition of the articular cartilage. It is the major source of nutrients since it is avascular. It is also a reservoir of proteins originating from the cartilage and synovial tissues. With this, it could serve as biomarker reflecting the condition of the joint. Three of the most important components of synovial fluid are the hyaluronic fluid, lubricin, and the phospholipids, which help in effective boundary friction in cartilage [27].

5.6 Aging in Articular Cartilage

Degeneration of articular cartilage leads to mechanical and inflammatory responses that activate signal transduction pathways on all joint tissues [5]. Osteoarthritic cartilage decreases tensile stiffness, which increases water content and softens cartilage [8, 28]. Aging also showed separation of collagen fibers [8, 9].

In osteoarthritis, adult cartilage chondrocytes reappear when their collagenous network of local matrix is damaged. Responses include increased type II collagen and matrix protein synthesis but with inferior biomechanical properties. It is this progressive deterioration that signifies the early stages of the osteoarthritis [5]. Chondrocyte dedifferentiation is characterized by increased synthesis of type I collagen [12].

Early event in osteoarthritis signifies the attachment of stromelysin (MMP-3) and alteration of TGF-β signals with high concentration of TGF-β1 [8, 12]. Inhibition of TGF-β1 lessens cartilage degeneration [5, 29]. TGF-β has a role in both cartilage health and disease [5].

Pro-inflammatory cytokines like tumor necrosis-alpha and interleukin-1 beta promote expression of prostaglandin, matrix metalloproteinase (MMP), cyclooxygenase, and nitric oxide and may promote other pro-inflammatory cytokines such as interleukins 6, 8, 17, and 18. MMP-13 has the highest count in any proteinase in osteoarthritis. These catabolic molecules interrupt the integrity of the extracellular matrix and decrease the response of chondrocytes to external anabolic signals [5, 12].

MMP-13 also degrades collagen II and aggrecan. This is why the MMP-13 seems to be the target in preventing osteoarthritis. Aggrecanases ADAMTS-5 and ADAMTS-4 are both responsible as the primary mediators of aggrecan cleavage [5, 30, 31]. Upregulation of the transcriptional regulator cAMP-responsive element-binding protein (CITED2) coincided with the downregulated expression of MMP-1 and MMP-13. A pro-catabolic factor is identified as contributory to cartilage remodeling and degradation by regulating MMP-13 gene transcription. Recently, it was identified that a serum proteases inhibitor, alpha 2 macroglobulin, is an inhibitor of many types of cartilage-degrading enzymes by decreasing gene expression and protein levels in posttraumatic joint osteoarthritis. Discoidin domain receptor (DDR2) is associated with induction and upregulation of MMP-13 and disruption of pericellular matrix [5, 32].

5.7 Healing in Articular Cartilage

In general, articular cartilage self-repair is significantly diminished because of the inherent poor vascularity and reduced regenerative capacity of hyaline articular cartilage in adult life [33].

An injury that disturbs the homeostatic balance in maintaining smooth articulation will result in the release and activation of chondrocytes as well as the expression of catabolic and pro-inflammatory genes (Fig. 5.10) [5, 34].

Articular cartilage injuries have a limited capacity for repair and limited ability of chondrocytes to yield a sufficient amount of extracellular matrix. Therefore, osteoarthritis develops when injury to the cartilage is left untreated (Fig. 5.8) Since articular cartilage is avascular, there is little ability for clot formation, which is a much-needed step in the healing cascade [5, 35]. Injuries, if left untreated, have little or no potential to heal spontaneously with normal hyaline cartilage [2]. However, lesions that reach the subchondral bone can undergo some amount of repair because of fibrin clot formation [36–38].

Adult chondrocytes have limited potential to proliferate enough extracellular matrix to fill a defect. Defects can be characterized as partial-thickness defect, which does not traverse the subchondral bone or full-thickness defect, which penetrates the subchondral bone. Partial-thickness defect has no ability to repair spontaneously, while full-thickness defect has the potential to repair due to the local influx of blood-forming fibrin clot and mesenchymal stem cells [5].

Recent analysis of synovial fluid after a knee injury or in osteoarthritis shows a larger number of mesenchymal stems cells compared to normal knees [5, 39, 40]. We know that the MSCs have

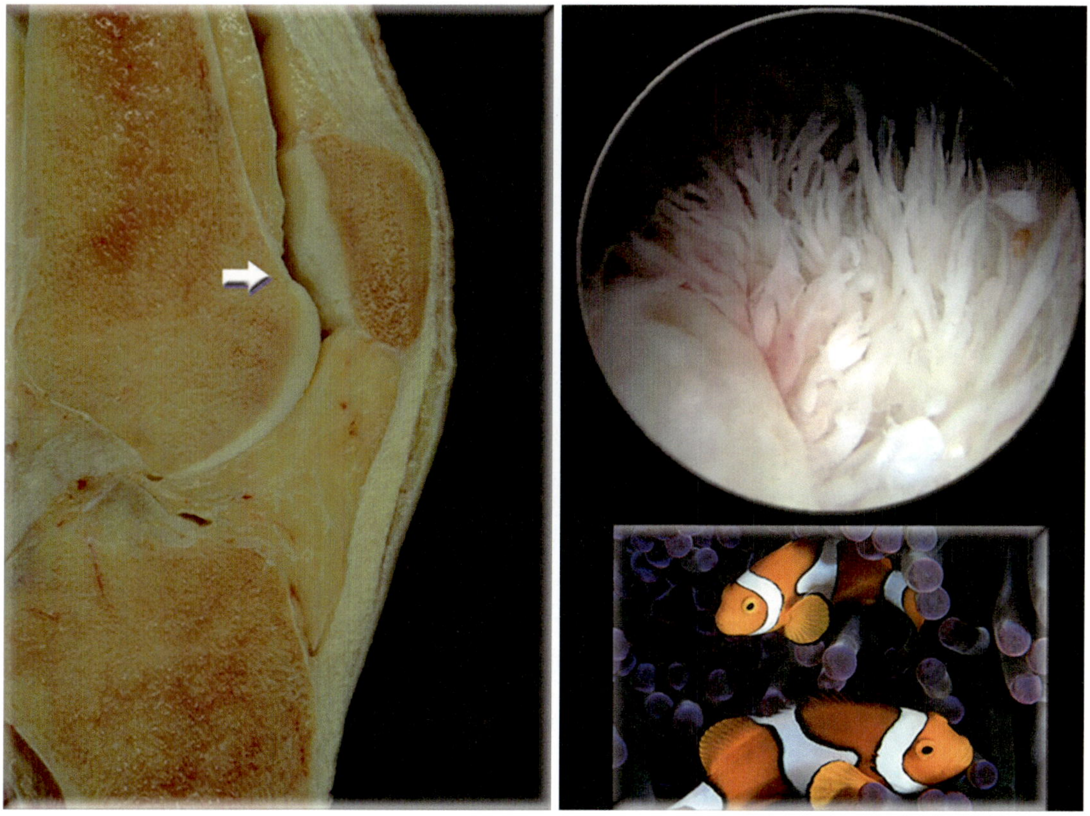

Fig. 5.10 Synovial inflammatory response of the knee that could lead to osteoarthritis if left untreated. (*white* arrow—supratrochlear fossa). By courtesy of Pau Golano

the capacity to differentiate into mature articular chondrocytes and thus contribute to the repair of lesion in articular cartilage [12, 41].

There are a lot of challenges in maintaining good joint articulation. A combination of different factors may be able to inhibit cartilage degeneration. Different culture systems maintain the chondrocyte phenotype like the high cell seeding density in pellet culture or micromass culture, suspension cultures, culture on different biomaterials [5, 42], and scaffolds [5, 43, 44]. New tissue engineering approaches and cell-based tissue engineering are still needed to continue to be evaluated to optimize cartilage regeneration [5].

References

1. Correia CR, Reis RL, Mano JF. Multiphasic, multi-structured and hierarchical strategies for cartilage regeneration. In: Adv. Exp. Med. Biol. New York: Springer New York LLC; 2015. p. 143–60.
2. Gobbi A, Nunag P, Malinowski K. Treatment of full thickness chondral lesions of the knee with microfracture in a group of athletes. Knee Surgery, Sport Traumatol Arthrosc. 2005;13:213–21.
3. Collado JJ, Garcia PG, Perez JS. In: Mafpre SA, editor. Rodilla: Morfogenesis, Anatomica Aplicada, Vias De Accesso - The Knee: Morphogenesis, Applied Anatomy, Approach Routes. Madrid: Paseo de Recoletos; 1994.
4. Roughley PJ. The structure and function of cartilage proteoglycans. Eur Cells Mater. 2006;12:92–101.
5. Carballo CB, Nakagawa Y, Sekiya I, Rodeo SA. Basic science of articular cartilage. Clin Sports Med. 2017;36:413–25.
6. Sophia Fox AJ, Bedi A, Rodeo SA. The basic science of articular cartilage: structure, composition, and function. Sports Health. 2009;1:461–8.
7. Melrose J, Fuller ES, Roughley PJ, Smith MM, Kerr B, Hughes CE, Caterson B, Little CB. Fragmentation of decorin, biglycan, lumican and keratocan is elevated in degenerate human meniscus, knee and hip articular cartilages compared with age-matched macroscopically normal and control tissues. Arthritis Res Ther. 2008;10:R79.
8. Muir H. The chondrocyte, architect of cartilage. Biomechanics, structure, function and molecular biology of cartilage matrix macromolecules. BioEssays. 1995;17:1039–48.
9. Stockwell RA, Billinghamt MEJ, Muir H. Ultrastructural changes in articular cartilage after experimental section of the anterior cruciate ligament of the dog knee. J Anat. 1983;136(Pt 2):425–39.
10. Mow VC, Guo XE. Mechano-electrochemical properties of articular cartilage: their Inhomogeneities and anisotropies. Annu Rev Biomed Eng. 2002;4:175–209.
11. Brocklehurst R, Bayliss MT, Maroudas A, Coysh HL, Freeman MA, Revell PA, Ali SY. The composition of normal and osteoarthritic articular cartilage from human knee joints. With special reference to unicompartmental replacement and osteotomy of the knee. J Bone Jt Surg Ser A. 1984;66:95–106.
12. Demoor M, Ollitrault D, Gomez-Leduc T, et al. Cartilage tissue engineering: molecular control of chondrocyte differentiation for proper cartilage matrix reconstruction. Biochim Biophys Acta - Gen Subj. 2014;1840:2414–40.
13. Broom ND. Abnormal softening in articular cartilage. Its relationship to the collagen framework. Arthritis Rheum. 1982;25:1209–16.
14. Culav EM, Clark CH, Merrilees MJ. Connective tissues: matrix composition and its relevance to physical therapy. Phys Ther. 1999;79:308–19.
15. Brody LT. Knee osteoarthritis: clinical connections to articular cartilage structure and function. Phys Ther Sport. 2015;16:301–16.
16. Mobasheri A, Vannucci SJ, Bondy CA, Carter SD, Innes JF, Arteaga MF, Trujillo E, Ferraz I, Shakibaei M, Martín-Vasallo P. Glucose transport and metabolism in chondrocytes: a key to understanding chondrogenesis, skeletal development and cartilage degradation in osteoarthritis. Histol Histopathol. 2002;17:1239–67.
17. Ulrich-Vinther M, Maloney MD, Schwarz EM, Rosier R, O'Keefe RJ. Articular cartilage biology. J Am Acad Orthop Surg. 2003;11:421–30.
18. Kozhemyakina E, Lassar AB, Zelzer E. A pathway to bone: signaling molecules and transcription factors involved in chondrocyte development and maturation. Development. 2015;142:817–31.
19. Goldring MB, Tsuchimochi K, Ijiri K. The control of chondrogenesis. J Cell Biochem. 2006;97:33–44.
20. Nuttall ME, Nadeau DP, Fisher PW, et al. Inhibition of caspase-3-like activity prevents apoptosis while retaining functionality of human chondrocytes in vitro. J Orthop Res. 2000;18:356–63.
21. Poole CA. Review. Articular cartilage chondrons: form, function and failure. J Anat. 1997;191:1–13.
22. Poole AR. What type of cartilage repair are we attempting to attain? J Bone Joint Surg Am. Journal of Bone and Joint Surgery Inc. 2003;85-A:40–4.
23. Mainil-Varlet P, Aigner T, Brittberg M, et al. Histological assessment of cartilage repair: a report by the Histology Endpoint Committee of the International Cartilage Repair Society (ICRS). J Bone Joint Surg Am. Journal of Bone and Joint Surgery Inc. 2003;85-A:45–57.
24. Nakagawa Y, Muneta T, Otabe K, et al. Cartilage derived from bone marrow mesenchymal stem cells expresses lubricin in vitro and in vivo. PLoS One.

2016;11(2):e0148777. https://doi.org/10.1371/journal.pone.0148777.
25. Tandon PN, Agarwal R. A study on nutritional transport in a synovial joint. Comput Math Appl. 1989;17(7):1131–41.
26. Ateshian GA, Warden WH, Kim JJ, Grelsamer RP, Mow VC. Finite deformation biphasic material properties of bovine articular cartilage from confined compression experiments. J Biomech. 1997;30:1157–64.
27. Frank EH, Grodzinsky AJ. Cartilage electromechanics-I. Electrokinetic transduction and the effects of electrolyte pH and ionic strength. J Biomech. 1987;20:615–27.
28. Venn M, Maroudas A. Chemical composition and swelling of normal and osteoarthrotic femoral head cartilage I. Chemical composition. Ann Rheum Dis. 1977;36(2):121–9.
29. Zhen G, Wen C, Jia X, et al. Inhibition of TGF-β signaling in mesenchymal stem cells of subchondral bone attenuates osteoarthritis. Nat Med. 2013;19:704–12.
30. Rengel Y, Ospelt C, Gay S. Proteinases in the joint: clinical relevance of proteinases in joint destruction. Arthritis Res Ther. 2007;9:221.
31. Little CB, Barai A, Burkhardt D, Smith SM, Fosang AJ, Werb Z, Shah M, Thompson EW. Matrix metalloproteinase 13-deficient mice are resistant to osteoarthritic cartilage erosion but not chondrocyte hypertrophy or osteophyte development. Arthritis Rheum. 2009;60:3723–33.
32. Xu L, Servais J, Polur I, Kim D, Lee PL, Chung K, Li Y. Attenuation of osteoarthritis progression by reduction of discoidin domain receptor 2 in mice. Arthritis Rheum. 2010;62:2736–44.
33. Gobbi A, Berruto M, Kon E, Francisco R, Filardo G. Hyaluronan based scaffold as treatment option for the management of patellofemoral chondral defects. Elizavetakon.it. 2006;1:470–4.
34. Goldring MB, Otero M. Inflammation in osteoarthritis. Curr Opin Rheumatol. 2011;23:471–8.
35. Mow VC, Rosewasser M. Articular cartilage: biomechanics. Park Ridge, IL: American Academy of Orthopaedic Surgeons. [Am Acad Orthop Surg]; 1988. p. 427–46.
36. Knutsen G, Isaksen V, Johansen O, Engebretsen L, Ludvigsen TC, Drogset JO, Grøntvedt T, Solheim E, Strand T, Roberts S. Autologous chondrocyte implantation compared with microfracture in the knee: a randomized trial. J Bone Jt Surg - Ser A. 2004;86:455–64.
37. Lorentzon R, Alfredson H, Hildingsson C. Treatment of deep cartilage defects of the patella with periosteal transplantation. Knee Surgery, Sport Traumatol Arthrosc. 1998;6:202–8.
38. Marder RA, Hopkins G, Timmerman LA. Arthroscopic microfracture of chondral defects of the knee: a comparison of two postoperative treatments. Arthrosc - J Arthrosc Relat Surg. 2005;21:152–8.
39. Morito T, Muneta T, Hara K, Ju YJ, Mochizuki T, Makino H, Umezawa A, Sekiya I. Synovial fluid-derived mesenchymal stem cells increase after intra-articular ligament injury in humans. Rheumatology. 2008;47:1137–43.
40. Sekiya I, Ojima M, Suzuki S, et al. Human mesenchymal stem cells in synovial fluid increase in the knee with degenerated cartilage and osteoarthritis. J Orthop Res. 2012;30:943–9.
41. Hattori S, Oxford C, Reddi AH. Identification of superficial zone articular chondrocyte stem/progenitor cells. Biochem Biophys Res Commun. 2007;358:99–103.
42. Berruto M, Delcogliano M, De Caro F, Carimati G, Uboldi F, Ferrua P, Ziveri G, De Biase CF. Treatment of large knee osteochondral lesions with a biomimetic scaffold: results of a multicenter study of 49 patients at 2-year follow-up. Am J Sports Med. 2014;42:1607–17.
43. Samaroo KJ, Tan M, Putnam D, Bonassar LJ. Binding and lubrication of biomimetic boundary lubricants on articular cartilage. J Orthop Res. 2017;35:548–57.
44. Watt FM. Effect of seeding density on stability of the differentiated phenotype of pig articular chondrocytes in culture. J Cell Sci. 1988;89(Pt 3):373–8.
45. March L. Articular cartilage in health and disease. Musculoskelet. Med.; 2016.
46. King M. Glycosaminoglycans and proteoglycans. Med Biochem Page. 2017;11:27.
47. Newman AP. Articular cartilage repair. Am J Sports Med. 1998;26:309–24.

Bone Structure and Function in the Distance Runner

6

Giuseppe M. Peretti and Marco Domenicucci

6.1 Bone Structure and Functions

Bone is a connective tissue characterized by a remarkable strength and mechanical resistance. These properties are guaranteed by an abundant extracellular matrix, composed of an organic and an inorganic portion. The organic portion, responsible for 20–25% of the wet weight of bone tissue, is constituted for more than 90% of type I collagen, organized in fibers; to a lesser extent, type V and type III collagen, proteoglycans, proteins, growth factors, and cytokines are also present. The inorganic portion accounts for 60–70% of the wet weight of bone tissue and is composed of mineral crystals, mainly calcium combined with oxygen, phosphorus, and hydrogen to form a molecule called hydroxyapatite; the high level of mineralization makes this matrix extremely resistant.

The different components of the matrix (organic and inorganic) confer different and interdependent properties to the tissue: The calcified fraction is responsible for the hardness of the bone, while the fibrillary organic fraction is responsible for the flexibility and, therefore, the resistance to traction.

Bone functions include supporting the body, protecting vital organs (for example, in the case of the ribcage) and movement (through the action of the muscles); the bone tissue also constitutes a vast reserve of calcium and phosphate, which are available to the body through the regulation of certain hormones (PTH, calcitonin, vitamin D, etc.).

Based on their macroscopic shape, bones can be classified as long bones (e.g., femur, tibia), short bones (e.g., carpal bones), flat bones (e.g., in the skull), and irregular bones (e.g., vertebrae).

6.2 Bone Cells

Bone cells include osteoprogenitor cells, osteoblasts, osteocytes, and osteoclasts. Osteoblasts, deriving from osteoprogenitor cells, are cuboidal mononucleate cells, with highly developed rough endoplasmic reticulum and Golgi apparatus. They are responsible for the deposition of extracellular matrix. In fact, they synthesize and secrete a large amount of matrix until they are incorporated within it; consequently, they change shape and are transformed into osteocytes that remain within the bone gaps, called lacunae. Osteoclasts—being derived from

G. M. Peretti (✉)
Department of Biomedical Sciences for Health, University of Milan, Milan, Italy

IRCCS Istituto Ortopedico Galeazzi, Milan, Italy
e-mail: giuseppe.peretti@unimi.it

M. Domenicucci
ASST degli Spedali Civili di Brescia, Brescia, Italy

G. L. Canata et al. (eds.), *Management of Track and Field Injuries*,
https://doi.org/10.1007/978-3-030-60216-1_6

hematopoietic cells—are multinucleated and present an external area called "ruffled border" where the resorption of bone tissue takes place; they are able to break down bone mineral and, at the same time, degrade the constituents of the organic matrix.

6.3 Microscopic Structure

There are different types of bone tissue: lamellar (Fig. 6.1), which includes cortical and cancellous bone, and woven (or non-lamellar), which is mechanically weaker and can present with inter-

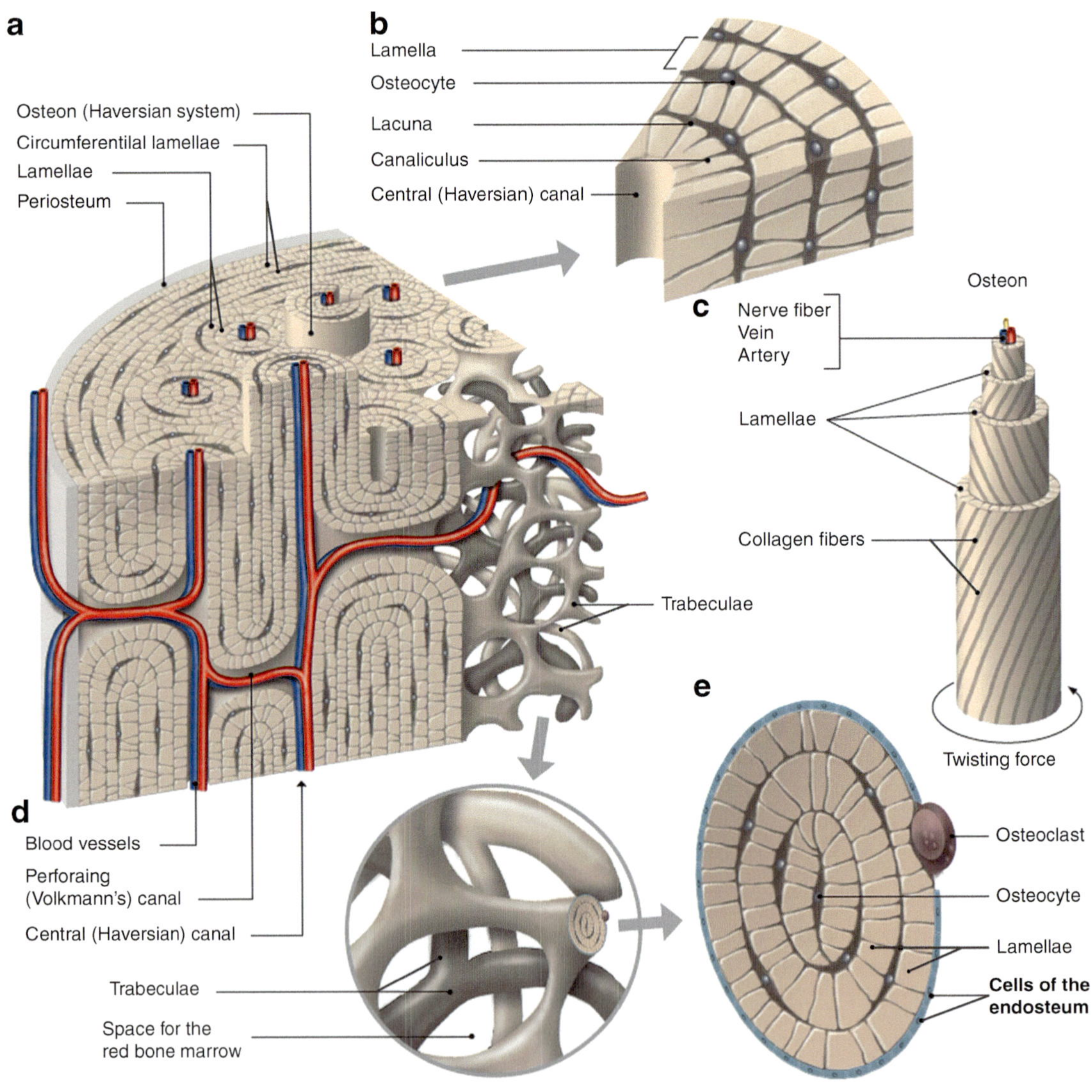

Fig. 6.1 Structure of the cortical (compact) and the trabecular (spongy) bone. The Haversian system of the cortical bone (**a**); a detail of an osteon with lamellae and osteocytes (**b**) and the lamellar organization of an osteon with the peculiar orientation of the collagen fibers (**c**). A detail of the trabeculae (**d**) and a cross section with the lamellar organization (**e**)

woven or parallel fibers. Lamellar cortical bone consists of multiple microscopic columns, called osteons: These structures, with a diameter of about 0.2 mm, are formed by many concentric lamellae and present bone lacunae (containing osteocytes) between the individual lamellae. In the center of each osteon, there is the Haversian canal, containing blood vessels, nerves, and lymphatic vessels; smaller canals (called Volkmann's canal) connect different Haversian canals. In the osteon, the youngest lamella is the one located deeper, closest to the Haversian canal. About 80% of adult skeletal mass is composed by cortical bone tissue.

Cancellous bone (also called trabecular bone) consists of trabeculae formed by lamellae, osteocytes, and a layer of endosteum that covers each trabecula. In the spaces between the trabeculae, bone marrow is present. Cancellous bone has a much greater surface area, compared to its mass, than cortical bone.

In an adult long bone, the central part (diaphysis) is composed of an external region of cortical bone, which mainly performs a mechanical function, and an internal cavity of cancellous bone and bone marrow. The two ends of the long bones (epiphysis), on the other hand, are composed of cancellous bone, with trabecular architecture developed along the main load vectors.

The bone is externally entirely covered by a dense elastic connective tissue membrane called periosteum, with the exception of joint surfaces being covered by hyaline cartilage. The inner layer of the periosteum, called cambium layer, contains osteoprogenitor cells, which can be activated when new bone formation is needed. Periosteum and bone are connected by Sharpey's fibers, mainly composed of type I collagen.

Internally, bone is occupied by a different connective tissue, called bone marrow stroma; it contains a large number of mesenchymal stem cells, which are able to differentiate into osteoblasts, chondrocytes, myocytes, and other types of cells.

6.4 Bone Formation

There are two different pathways by which the bone tissue is formed: endochondral ossification and intramembranous ossification.

Long bones, vertebral bodies, and most small bones are formed through endochondral ossification. In this process, beginning in the first trimester of development and continuing until the end of skeletal growth, bone tissue starts developing from a cartilaginous tissue, with mesenchymal cells differentiating into osteoblasts that produce bone extracellular matrix; then, cartilage and bone continue growing together until the final shape of the bone is reached.

Other bones, like clavicles and cranial flat bones, are formed by intramembranous ossification. In this process, clusters of osteoblasts form within the embryonic mesenchyme and start producing bone extracellular matrix; these small regions then merge to form the mature bone.

6.5 Bone Remodeling

Remodeling is the biological process that allows bone tissue to continuously renew itself. It involves a modification of the composition of the tissue, especially where the bone is damaged, fractured, or aged. This process is the main metabolic activity of the skeleton in the adult life and continues uninterruptedly until death; it has been calculated that the total skeletal mass of an average adult is completely replaced every 15–20 years.

Specifically, bone remodeling is performed by specialized groups of cells called basic multicellular units (BMUs). Their work is divided into four phases: activation, resorption, reversal, and formation.

In the activation phase, osteoclasts are formed in the needed site by the fusion of their progenitor cells. The following resorption phase, in which the osteoclasts break down bone matrix, lasts for 2–4 weeks. The reversal phase represents the overlapping of the end of the resorption and the beginning of the following formation phase. Consequently, in the formation phase, osteoblasts deposit osteoid, which is then mineralized to create the mature bone extracellular matrix; some osteoblasts remain buried within the newly formed matrix and become osteocytes.

In cortical bone tissue, all these phases of the BMUs can be observed in a tunnel-like structure:

A group of newly formed osteoclasts forms a cylindrical resorption cavity, the tip of which is called the cutting cone. Behind these osteoclasts, a reversed area is present, and subsequently, osteoblasts are depositing new matrix.

6.6 Fracture Healing

When a fracture occurs, the body is able to repair bone tissue, under certain mechanical and biological conditions, by two different processes: primary or secondary healing.

Primary (also known as direct) healing requires absolute stability of the bone fragments; it is characterized by direct osteonal remodeling with the combined action of osteoclasts (which create microscopic cavities in the fragments) and osteoblasts (which fill these cavities with new bone matrix).

Secondary (also known as indirect) healing, instead, consists of four phases:

- inflammation (1–7 days): Hematoma forms and cells reach the fracture site,
- soft callus formation (2–3 weeks): Fibroblasts produce collagen fibers, and fibrocartilage replaces the hematoma,
- hard callus formation (3–12 weeks): The soft callus is converted into woven bone tissue, mainly through endochondral ossification,
- remodeling (months–years): Woven bone is converted into lamellar bone.

6.7 Bone Response to Mechanical Stimuli and Stress Fractures

Bone remodeling is stimulated by mechanical stress: According to Wolff's law, bone shape and density depend on the forces acting on the bone; more specifically, the number and frequency of loading cycles directly affect the rate and amount of remodeling [1].

Bone response to repetitive stress is an increase in the osteoclastic activity over the new bone formation, resulting in temporary bone weakening; this is normally followed by new bone formation, providing reinforcement. However, during prolonged periods of intense training without adequate rest, bone tissue deposition is slowed down with respect to resorption; this may result initially in microscopic injuries (microfractures), which in the early stages are typically asymptomatic but trigger a reparative response detectable in magnetic resonance imaging (MRI) by the presence of bone marrow edema. If the intense load is not reduced, microfractures may propagate and eventually create true cortical breaks (stress fractures), with the development of clinical symptoms. Therefore, stress fractures represent only one phase of a broad spectrum of overuse bone lesions.

A sudden increase in physical activity intensity, frequency, or duration without adequate rest periods can therefore induce an imbalance between bone resorption and formation, eventually leading to pathologic changes.

Within the category of stress fractures, it is necessary to distinguish between insufficiency fractures and fatigue fractures: The former result from the application of normal strain in a subject with low bone mineral density, while fatigue fractures originate from excessive or abnormal strain applied on normal bone tissue.

Calcium and vitamin D are extremely important for bone health and the prevention of fractures. Serum vitamin D deficiency is significantly correlated with the incidence of stress fractures [2, 3]. Similarly, the correlation between lower calcium intake and an increased incidence of stress fractures has been demonstrated, especially in female athletes [4]. As expected, lower bone mineral density, as assessed by dual-energy X-ray absorptiometry (DEXA) scans, is correlated with higher incidence of stress fractures [4, 5].

In addition to the microscopic structure and metabolism, the shape of the bones also contributes to increase or decrease the risk of overuse injuries. The tibia, for example, is one of the bones most affected by stress fractures in distance runners. According to a case–control study [6], the risk for tibial stress injury is increased by a combination of factors, which include the pres-

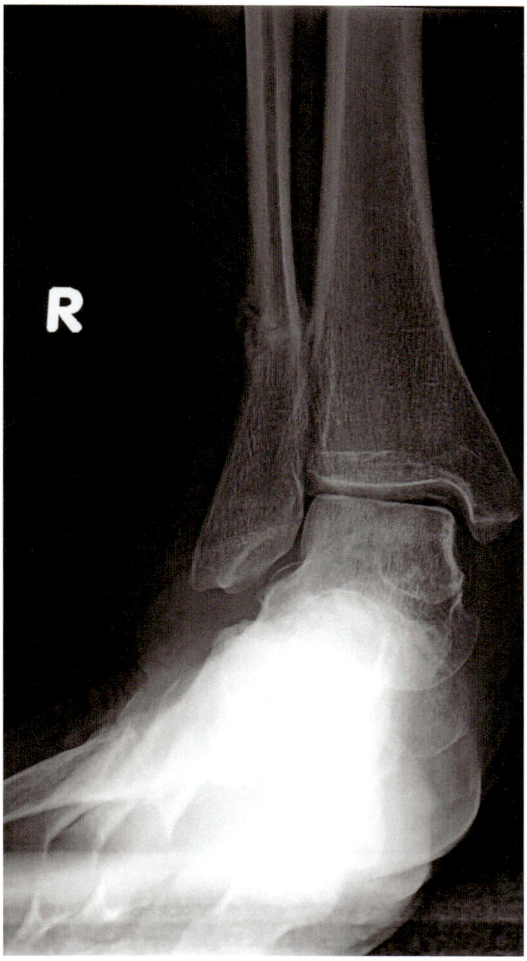

Fig. 6.2 Stress fracture located in the right fibula (by courtesy of Dr. Maria Palmucci)

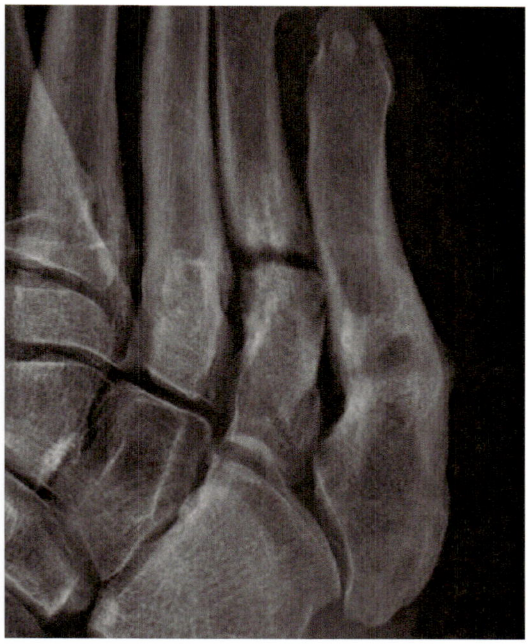

Fig. 6.3 Fracture of the fourth metatarsal bone. A fracture outcome at the level of the fifth metatarsal bone can also be noted (by courtesy of Dr. Maria Palmucci)

ence of thinner and smaller bones (regardless of overall bone density) and foot deformities. Athletes with a history of tibial stress fractures have been observed to have smaller bone geometry and higher bending moments in the medial–lateral axis, with a smaller diameter in the middle diaphyseal third of the tibia compared to athletes without previous stress fractures [7]. However, other sites of the lower limb may be involved in a stress fracture, e.g., the fibula (Fig. 6.2) and the metatarsal bones (Fig. 6.3).

Based on the evidence in the literature, we can affirm that long-distance running, if practiced respecting recovery times and avoiding prolonged periods of extremely intense training, leads to an increase in bone mineral density and mechanical strength of the bones in the lower limbs. This happens because mechanical strains directly stimulate osteoblastic activity and increase the release of hormones involved in bone remodeling (e.g., calcitonin [8]).

However, excessive running (with nonprogressive increase in distances and without adequate recovery times) causes mechanical damage and inflammatory states in bones subjected to the greatest stress, leading to a decrease in bone mineral density over time [9].

6.8 Stress Fractures in the Lower Limbs

As a consequence of continuous loading, the lower limbs are the body segment most affected by stress fractures. In particular, over a third of stress fractures in the lower limbs are located in the metatarsal bones [10] (Fig. 6.3). They typically present with nonspecific, progressive pain in the midfoot and are often related to alterations in the biomechanics of the foot (such as cavovarus foot in fifth metatarsal base fracture) or to prolonged forced movements (such as plantar flexion of the Lisfranc joint in ballet dancers).

As previously mentioned [6], another site often affected is the tibial diaphysis, especially in long-distance runners. Similarly to metatarsal bones, tibial stress fractures have also been shown to be related to biomechanical alterations, including a rotational torque on the longitudinal tibial axis caused by increases in peak hip adduction and peak rearfoot eversion during running [11].

In the case of stress fractures of the calcaneus, they are sometimes unrecognized due to the often-negative X-rays and similar symptoms with plantar fasciitis; MRI allows, however, reaching a precise diagnosis. Some studies have shown that calcaneal stress fractures are associated not only with osteoporosis, but also with recent hip or knee replacement surgery [12]; this is probably due to a change in the biomechanics of walking, accompanied by a decrease in perceived pain from taking postoperative analgesic drugs.

Another locations where stress fractures are often undiagnosed are the cuboid and the navicular bone: In particular, in more than half of navicular stress fractures, radiographs are false negative [13], so it is essential to perform an MRI or CT scan, in the case of diagnostic suspect. Generally, this statement could always be considered valid in the event of suspected fracture for any lower limb bone in the distance runner.

References

1. Warden SJ, Burr DB, Brukner PD. Stress fractures: pathophysiology, epidemiology, and risk factors. Curr Osteoporos Rep. 2006;4(3):103–9.
2. Dao D, Sodhi S, Tabasinejad R, Peterson D, Ayeni OR, Bhandari M, et al. Serum 25-Hydroxyvitamin D levels and stress fractures in military personnel: a systematic review and meta-analysis. Am J Sports Med. 2015;43(8):2064–72.
3. Burgi AA, Gorham ED, Garland CF, Mohr SB, Garland FC, Zeng K, et al. High serum 25-hydroxyvitamin D is associated with a low incidence of stress fractures. J Bone Miner Res Off J Am Soc Bone Miner Res. 2011 Oct;26(10):2371–7.
4. Myburgh KH, Hutchins J, Fataar AB, Hough SF, Noakes TD. Low bone density is an etiologic factor for stress fractures in athletes. Ann Intern Med. 1990;113(10):754–9.
5. Bennell KL, Malcolm SA, Thomas SA, Reid SJ, Brukner PD, Ebeling PR, et al. Risk factors for stress fractures in track and field athletes. A twelve-month prospective study. Am J Sports Med. 1996;24(6):810–8.
6. Beck BR, Rudolph K, Matheson GO, Bergman AG, Norling TL. Risk factors for tibial stress injuries: a case-control study. Clin J Sport Med Off J Can Acad Sport Med. 2015;25(3):230–6.
7. Popp KL, Frye AC, Stovitz SD, Hughes JM. Bone geometry and lower extremity bone stress injuries in male runners. J Sci Med Sport. 2020;23(2):145–50.
8. Lee JH. The effect of long-distance running on bone strength and bone biochemical markers. J Exerc Rehabil. 2019;15(1):26–30.
9. MacDougall JD, Webber CE, Martin J, Ormerod S, Chesley A, Younglai EV, et al. Relationship among running mileage, bone density, and serum testosterone in male runners. J Appl Physiol. 1992;73(3):1165–70.
10. Wilson ESJ, Katz FN. Stress fractures. An analysis of 250 consecutive cases. Radiology. 1969;92(3):481–6. passim
11. Pohl MB, Mullineaux DR, Milner CE, Hamill J, Davis IS. Biomechanical predictors of retrospective tibial stress fractures in runners. J Biomech. 2008;41(6):1160–5.
12. Miki T, Miki T, Nishiyama A. Calcaneal stress fracture: an adverse event following total hip and total knee arthroplasty: a report of five cases. J Bone Joint Surg Am. 2014;96(2):e9.
13. de Clercq PFG, Bevernage BD, Leemrijse T. Stress fracture of the navicular bone. Acta Orthop Belg. 2008;74(6):725–34.

Diagnostic Imaging in Track and Field Athletes

7

Giuseppe Monetti

7.1 Foot and Ankle

7.1.1 Tendinopathy

The tendon subjected to the most severe strain and the one injured most often in track and field athletes is the Achilles tendon, which besides fracture may present a variety of inflammatory and degenerative conditions. The most informative imaging modalities to investigate tendon lesions are dynamic US with power Doppler and elastography (Fig. 7.1a, b) [1], followed by MRI, which can now be used to acquire dynamic upright scans (Fig. 7.2a, b). In patients with overload tendinopathy, the most frequently affected ankle compartment is the medial tarsal tunnel, especially the posterior tibial and flexor hallucis longus tendons. The conditions involving these structures are effectively examined using dynamic US and MRI (Fig. 7.3a, b). In sprains, which often occur with the ankle in inversion, the peroneal tendons are those involved most often. Conditions range from tenosynovitis to subluxation secondary to laxity to rupture of the retinaculum (Fig. 7.4a, b). The most common enthesitis is plantar fasciitis, which is accurately assessed by dynamic compression elastography and MRI (Fig. 7.5a, b) [2].

G. Monetti (✉)
Dipartimento di Diagnostica per Immagini, Ospedale Privato Accreditato Nigrisoli, Bologna, Italy
e-mail: info@muskultrasound.it

7.1.2 Capsule Ligament Injury

In inversion ankle sprains, the external ligament compartment is the one injured most often, particularly the anterior talofibular and calcaneofibular ligaments, which may exhibit partial or full-thickness rupture (Fig. 7.6a, b). Lesions of the internal compartment (deltoid ligament) are less frequent, and those of the tarsal sinus ligaments are even uncommon (Fig. 7.7a, b) [3].

7.2 Knee

7.2.1 Tendinopathy

In track and field athletes, the proximal insertion of the patellar tendon is particularly prone to injury (jumper's knee) due to repeated jumping stress. The most suitable techniques to investigate these lesions are dynamic US with power Doppler and elastography and dynamic upright MRI (Fig. 7.8a, b) [2].

7.2.2 Capsule Ligament Injury

The knee ligaments injured most frequently are the anterior cruciate and the medial collateral ligaments. Dynamic upright MRI is capable of quantifying the damage and of assessing any residual instability (Fig. 7.9a–d) [4].

G. L. Canata et al. (eds.), *Management of Track and Field Injuries*,
https://doi.org/10.1007/978-3-030-60216-1_7

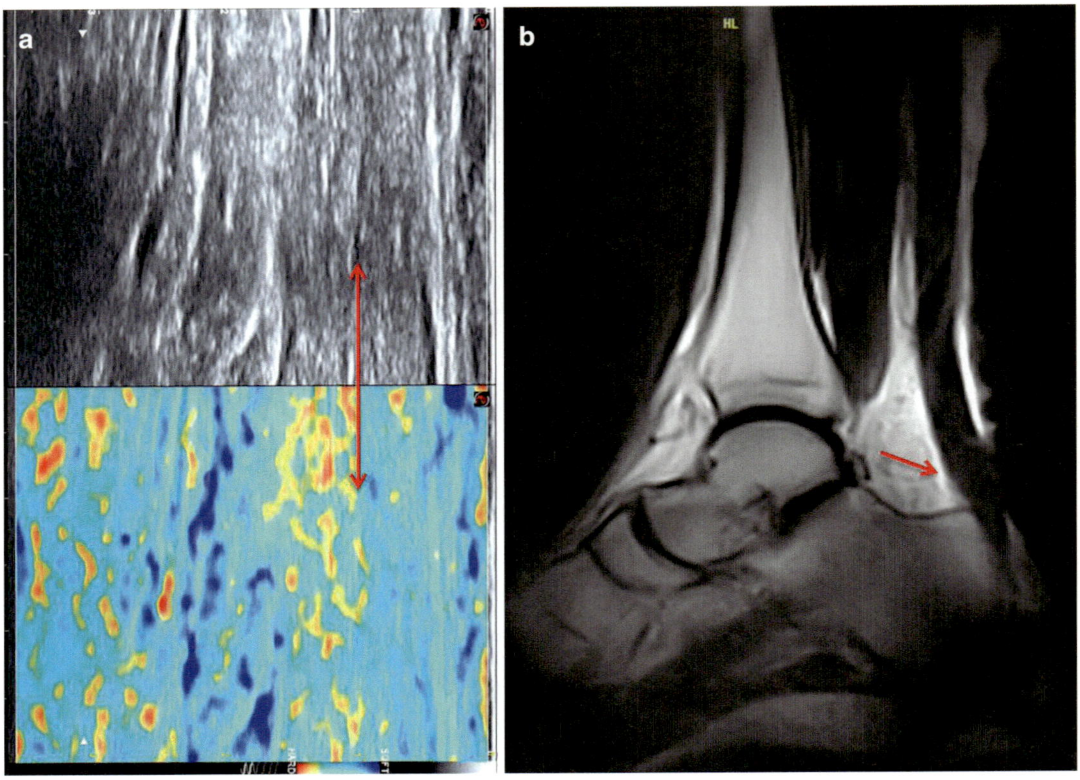

Fig. 7.1 (**a**, **b**) Comparison of dynamic US, elastography and MRI scans demonstrating degeneration secondary to overload tendinopathy of the Achilles tendon

7.2.3 Meniscal Lesions

In these athletes, the meniscal body and the posterior horn of the medial meniscus are the knee structures most prone to degenerative and traumatic lesions and to meniscocapsular separation (Fig. 7.10a, b) [5].

7.3 Pelvis

7.3.1 Pubalgia

The constant loading strain to which the pelvic structures are subjected can induce athletic pubalgia, a common condition that often involves the pubic symphysis. A marked bone marrow oedema extending to neighbouring muscles, especially the obturator internus and the abductor longus, is frequently detected in this area (Fig. 7.11a, b) [6].

7.4 Lumbar Spine

The lumbar spine is the tract most consistently affected by overload conditions like disc herniation and, especially, anterolisthesis with different grades of slippage. Dynamic upright MRI ensures highly accurate evaluation of the diastasis between the vertebral bodies (Fig. 7.12a, b). Sacroiliac joint instability, another common pathology, is also clearly

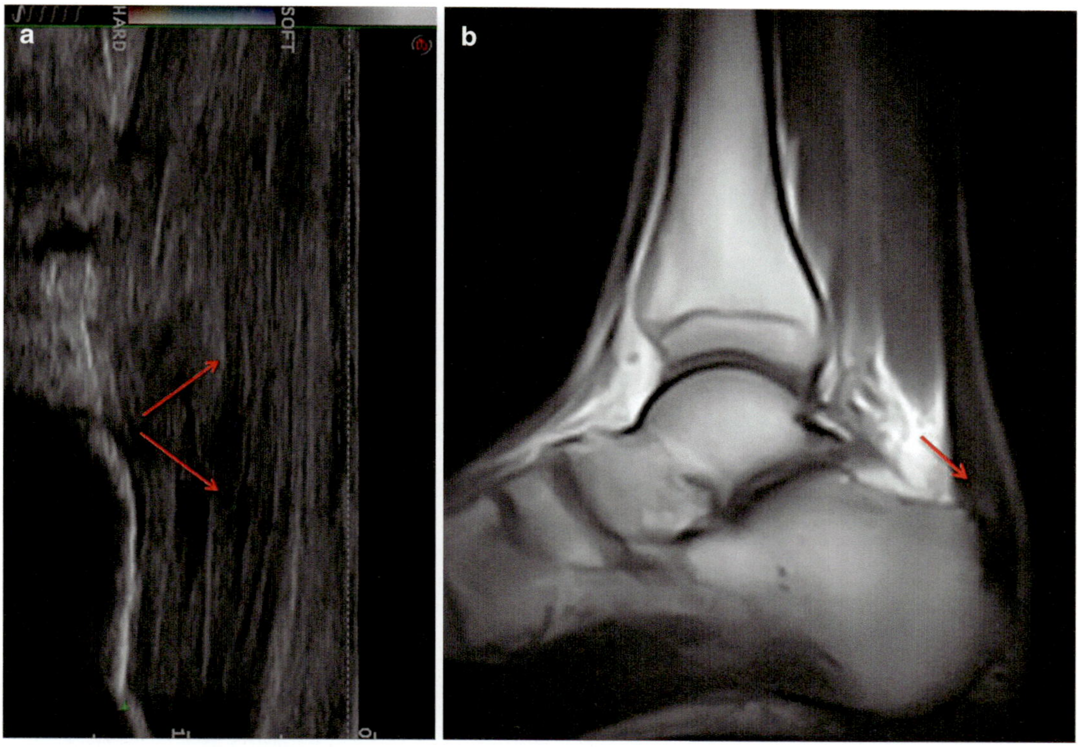

Fig. 7.2 (**a**, **b**) Dynamic US and MRI scans documenting transmural rupture of the Achilles tendon

depicted in dynamic MRI scans acquired with the athlete standing first on each leg and then on both legs (Fig. 7.13a, b) [7].

7.5 Muscle Lesions

The biceps femoris, the semitendinosus and the Gemelli are the most frequently injured muscles in track and field athletes. Like all muscles, they can also suffer distortion and rupture. Again, dynamic US with elastography and MRI is the modality of choice to assess them (Fig. 7.14) [8].

7.6 Stress Fractures

These lesions are more commonly associated with endurance competitions like marathons and typically affect the metatarsals at the level of the foot. The most suitable diagnostic imaging

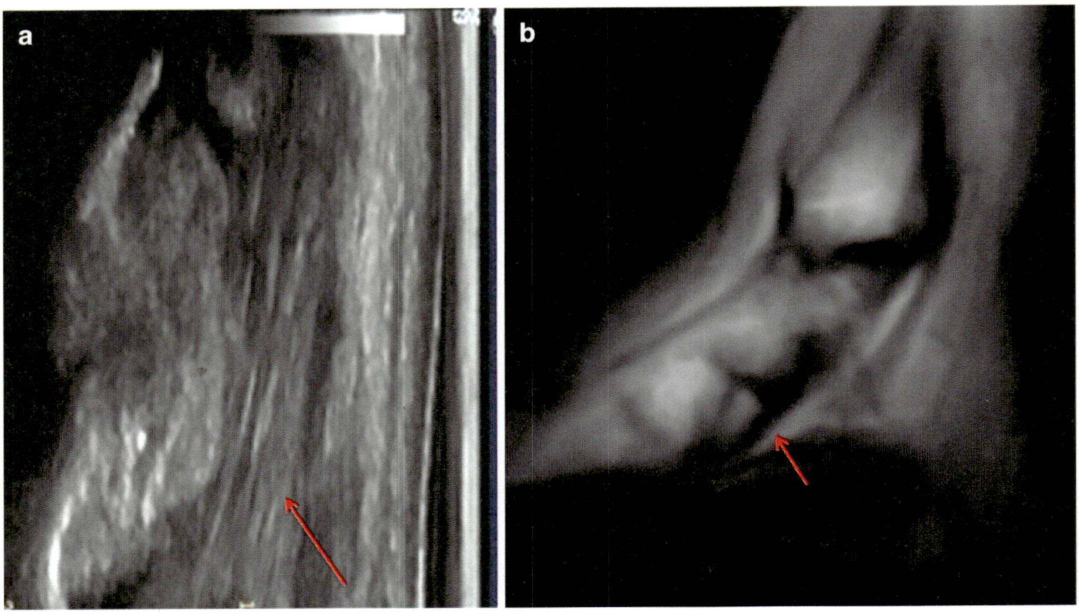

Fig. 7.3 (**a**, **b**) Dynamic US and MRI scan depicting an accessory scaphoid bone and tendinopathy affecting the distal insertion of the posterior tibial muscle

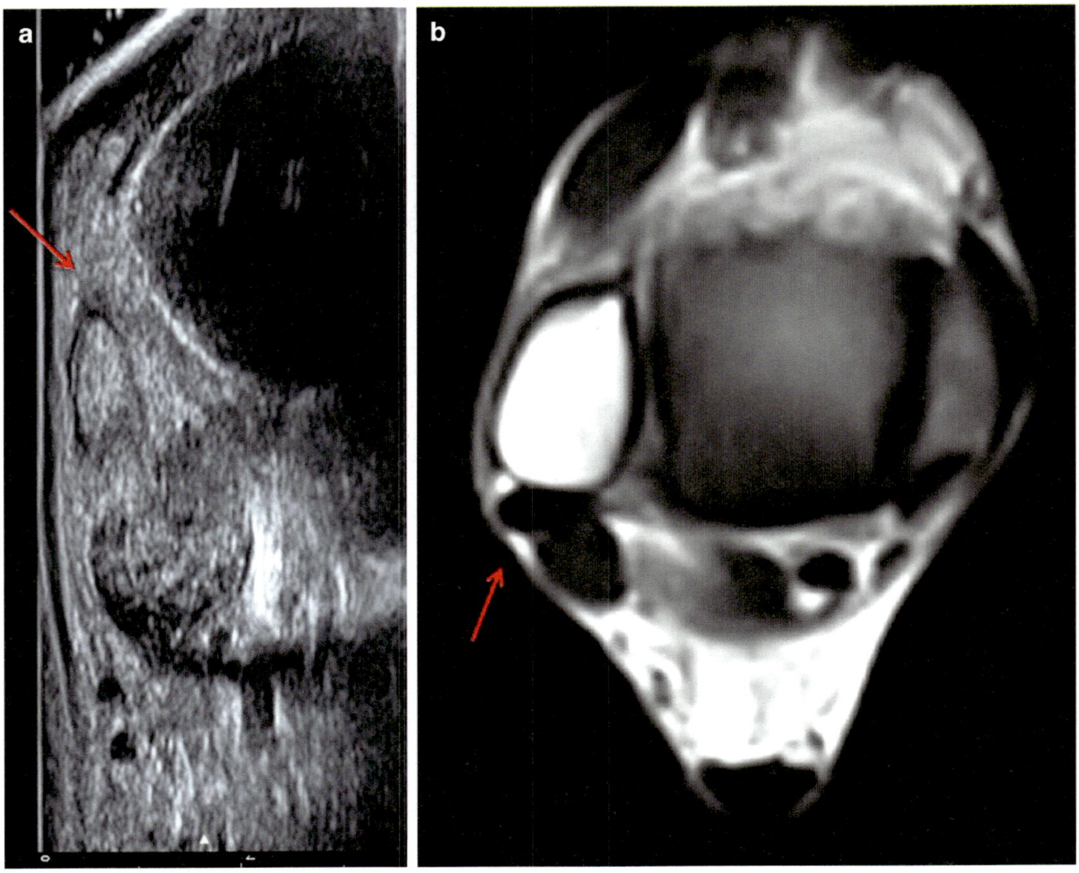

Fig. 7.4 (**a**, **b**) Dynamic US and MRI scans: subluxation and tenosynovitis of the peroneal tendons due to a retinaculum tear secondary to inversion ankle sprain

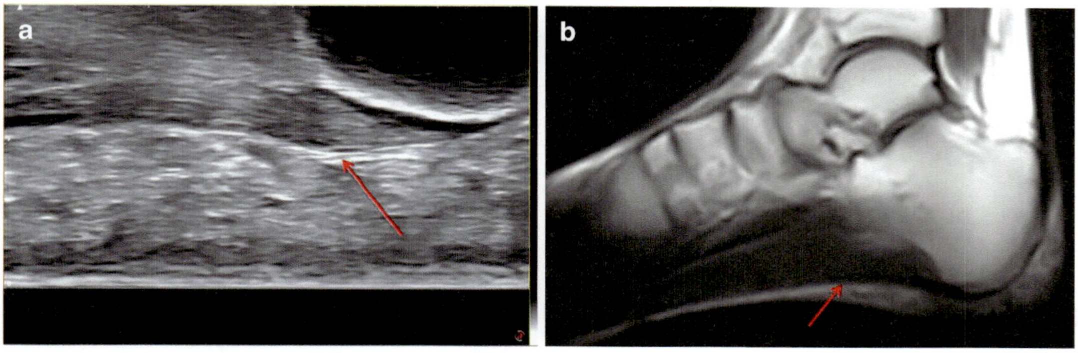

Fig. 7.5 (**a**, **b**) Marked tissue stiffness due to plantar fasciitis depicted by dynamic upright US and MRI

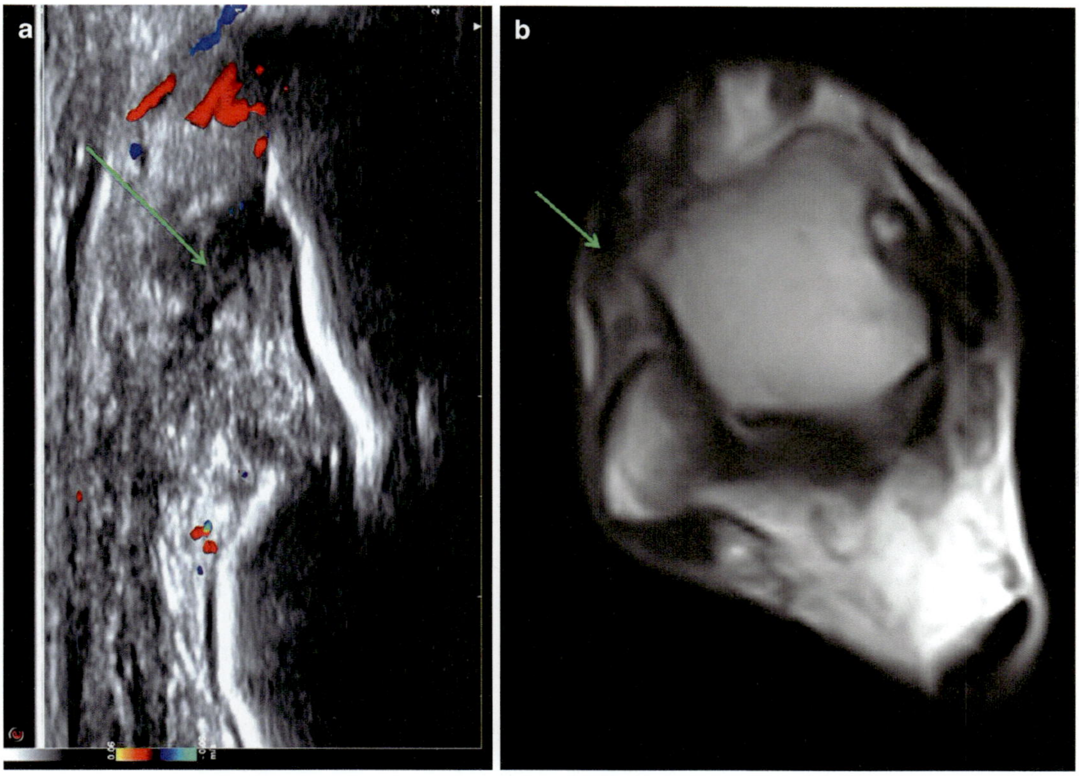

Fig. 7.6 (**a**, **b**) Dynamic US and MRI scans acquired with the ankle inverted demonstrating a full-thickness lesion of the anterior talofibular ligament

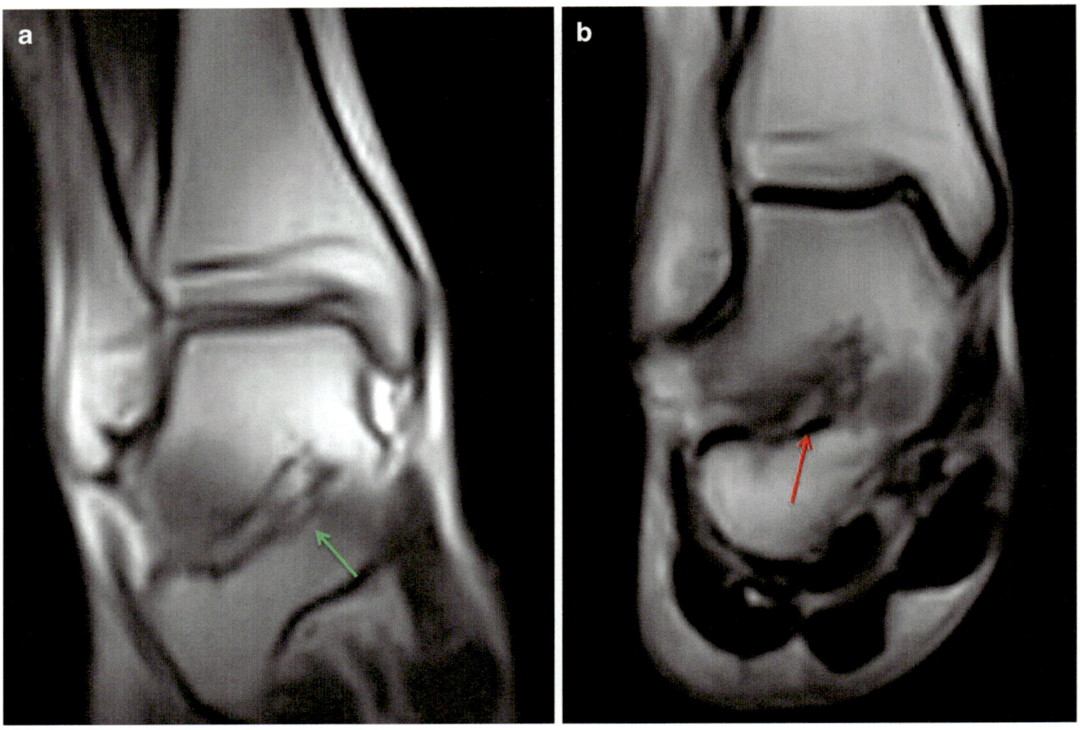

Fig. 7.7 (**a**, **b**) Dynamic MRI acquired with the ankle rotated medially. Left, normal ligament; right, severe distraction of the interosseous ligament at the level of the tarsal sinus

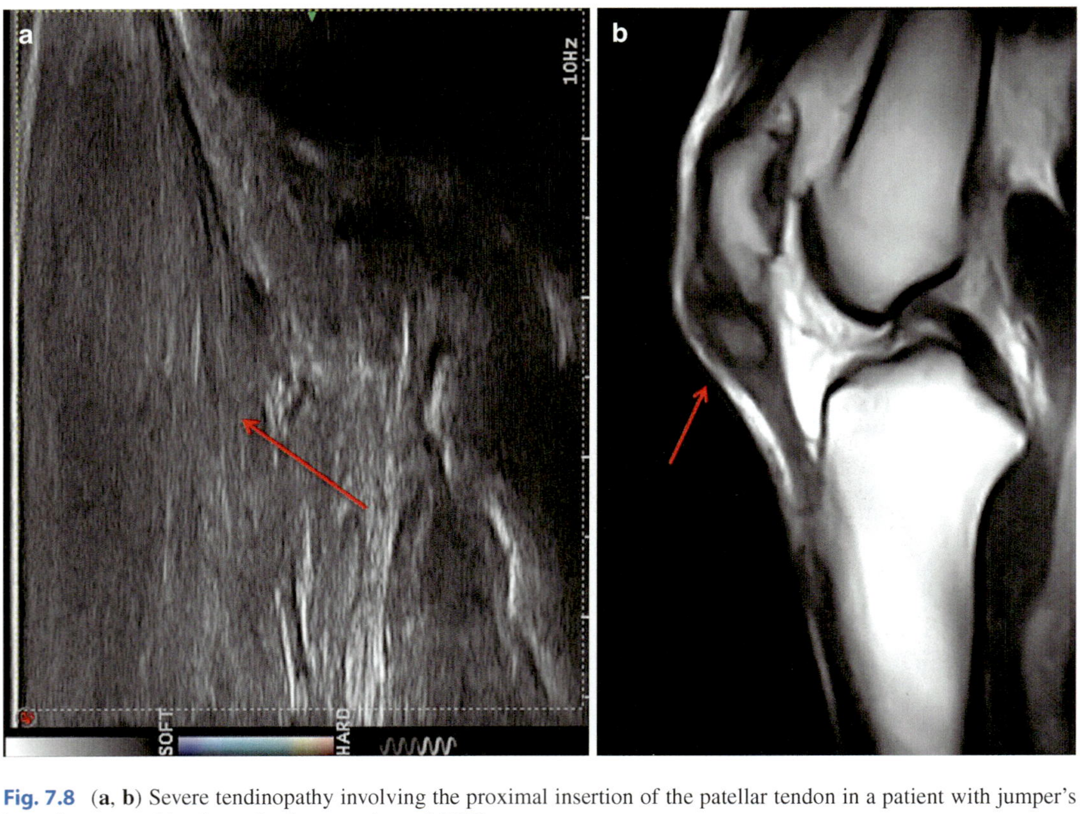

Fig. 7.8 (**a**, **b**) Severe tendinopathy involving the proximal insertion of the patellar tendon in a patient with jumper's knee documented by dynamic elastography and MRI

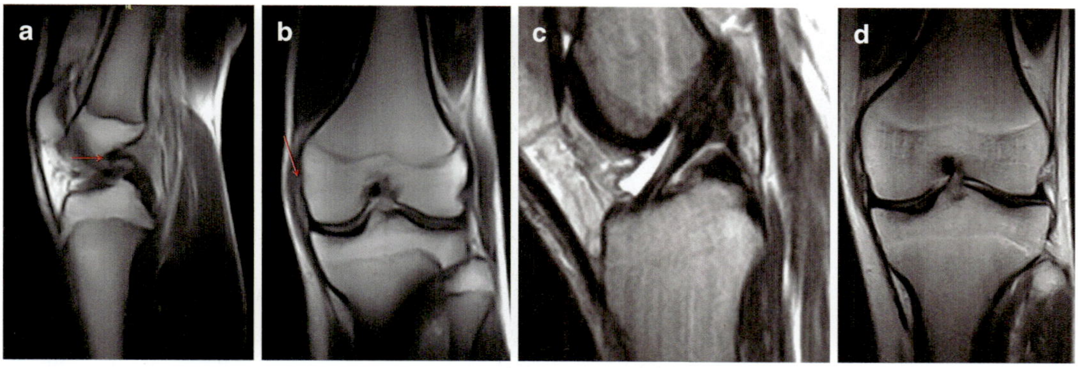

Fig. 7.9 (**a–d**) Static and dynamic MRI scans demonstrating a full-thickness lesion of the anterior cruciate ligament and the medial collateral ligament, which are not clearly depicted in static scans

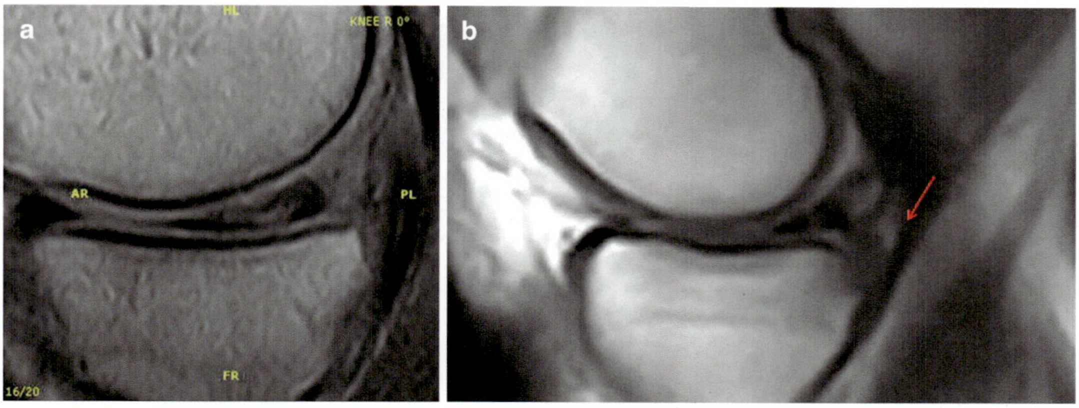

Fig. 7.10 (**a**, **b**) Chronic rupture of the meniscal body and of the posterior horn of the medial meniscus. Whereas the static MRI scan suggests that the lesion is stable, the dynamic scan documents clear meniscocapsular instability

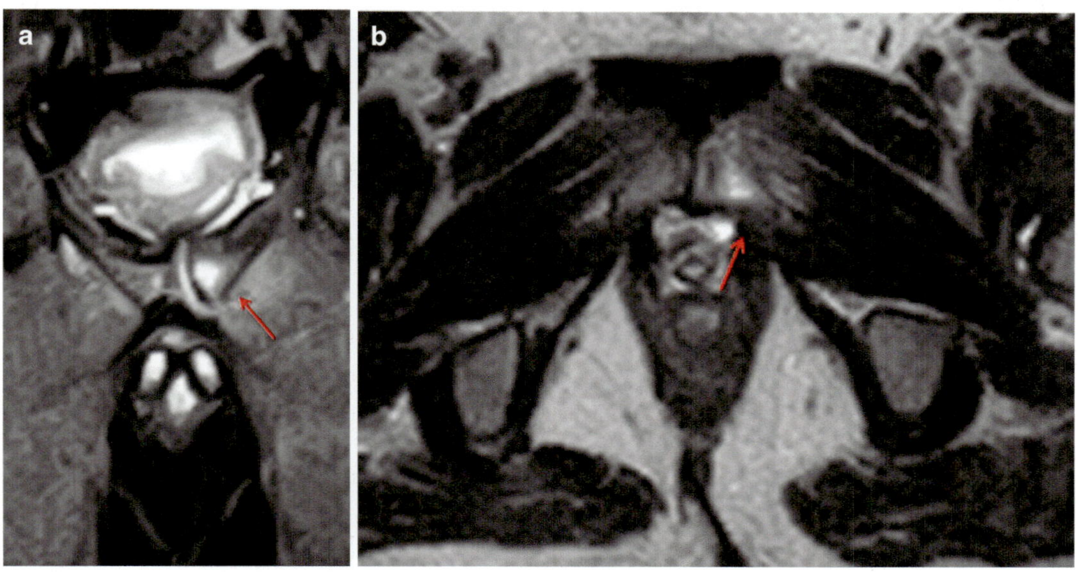

Fig. 7.11 (**a**, **b**) Coronal and axial scans demonstrating a marked bone marrow oedema involving the pubic symphysis (left) and extending to the left abductor longus

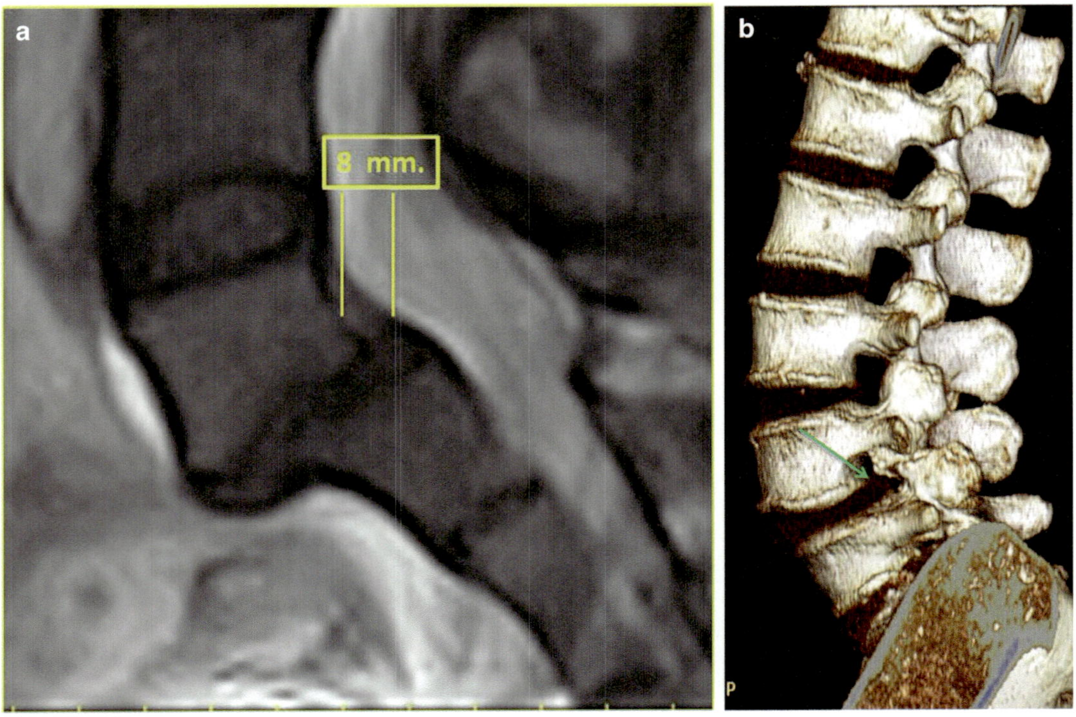

Fig. 7.12 (**a**, **b**) Dynamic upright MRI scans acquired with the spine in flexion and extension. The severe anterolisthesis of L5 on S1 is not depicted by CT

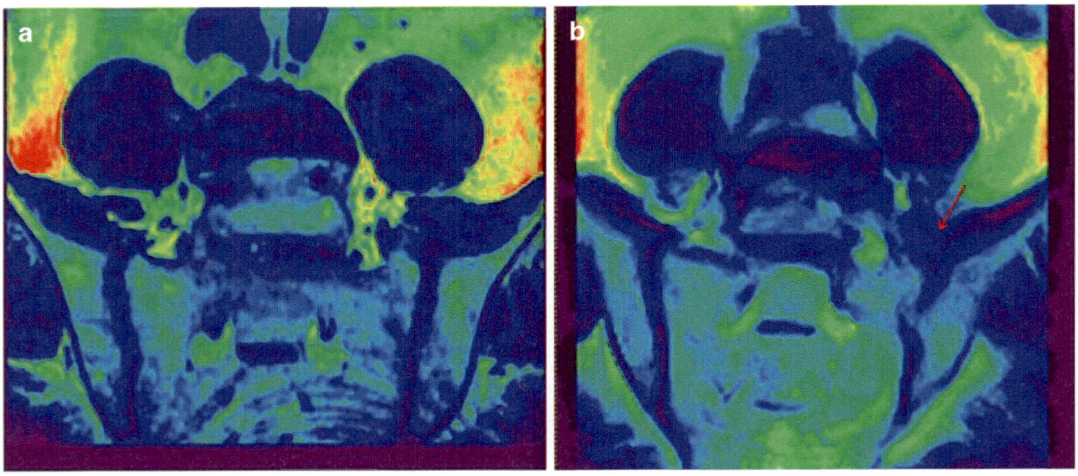

Fig. 7.13 (**a**, **b**) Frank instability of the left iliosacral joint is well documented by the dynamic scans, acquired with the patient standing on the right leg and the left leg, respectively, but is poorly depicted in the static scan

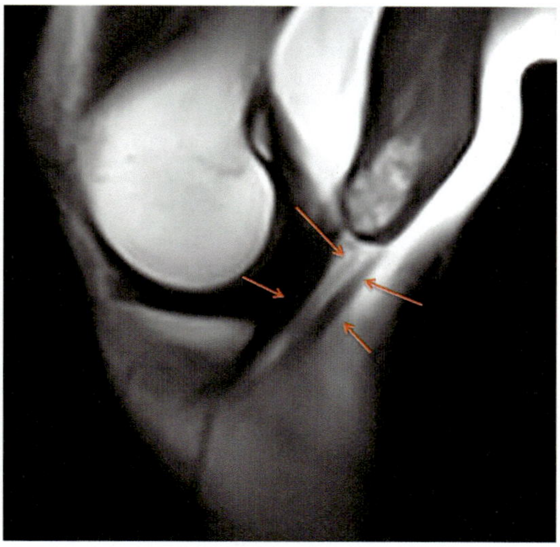

Fig. 7.14 Dynamic MRI demonstrating an extensive haematoma appearing as a cyst at the level of the myotendinous junction of the femoral biceps tendon, without rupture

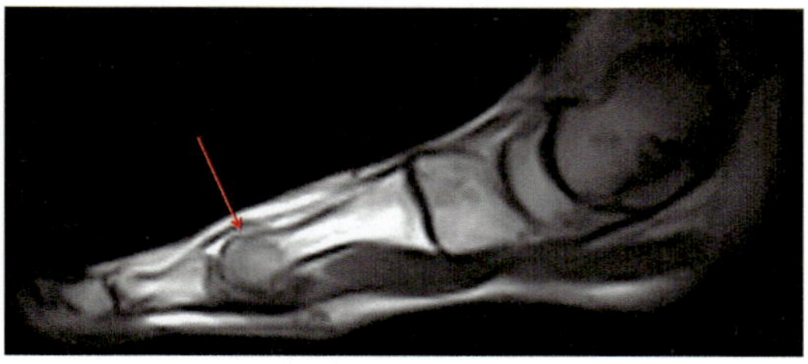

Fig. 7.15 Dynamic upright MRI scan acquired in dorsi-plantar flexion showing a compression oedema due to a stress fracture of the metatarsal head of the second toe

modality to evaluate them is traditional and upright MRI, with dynamic sequences as appropriate (Fig. 7.15) [4].

References

1. Jacobson JA. Musculoskeletal sonography and MR imaging: a role for both imaging methods. Radiol Clin N Am. 1999;37(4):713–35.
2. Monetti G. In: Brasseur JL, editor. L' Elastographie en Ecographie de l'appareil locomoteur – Actualites en ecographie de l'appareil locomoteur. Montpellier: Sauramps Medical; 2012.
3. Peetrons P, Croteur V, Bacq C. Sonography of ankle ligaments. J Clin Ultrasound. 2004;32:491–9.
4. Stoller DW. Stoller's atlas of Orthopaedics and sports medicine. Philadelphia: Lippicott Williams & Wilkins; 2008.
5. Barile A, Masciocchi C. Risonanza magnetica del ginocchio. Naples: Idelson Gnocchi; 1994.
6. Van Holsbeeck MT, Introcaso JH. Musculoskeletal ultrasound. 2nd ed. St. Louis: Mosby; 2001.
7. Jinkins JR, Dworkin JS, Damadian RV. Upright, weight bearing, dynamic-kinetic MRI of the spine: initial results. Eur Radiol. 2005;15(9):1815–25.
8. McNally EG. Practical musculoskeletal ultrasound. Philadelphia: Elsevier Churchill Livingstone; 2005.

Part II

Upper Extremity

Shoulder Instability in Track and Field Athletes

8

Hunter Bohlen and Felix Savoie

8.1 Introduction

Management of the unstable shoulder presents a challenging dilemma for the practicing orthopedic surgeon. The kinetic chain, in which a thrower generates tremendous energy from the legs, translates it through the truck, into the scapula, and ultimately the glenohumeral joint, is foundational to throwing any object overhead with maximal force. Irregularities in the kinetic chain will place undue stress on the athlete, increasing the risk for injury [1–3]. Though the kinetic chain and associated injuries with throwing a baseball have been rigorously studied, the biomechanics and injury profile of many other sports, including track and field events, have received less attention [4]. Here, we will discuss variations in the traditional kinetic chain and subsequent injuries for javelin throw, shot put, discus, hammer throw, and the pole vault.

Shoulder instability can best be understood as a spectrum of disease ranging from traumatic dislocation of the glenohumeral joint on one end to repetitive microtrauma of the capsuloligamentous structures leading to pain and apprehension in the athlete on the other. The latter is also known as multidirectional instability (MDI). For traumatic dislocation, surgical intervention is typically required to repair damaged structures. MDI presents a more complicated clinical entity, as for peak performance in track and field events, the soft tissue stabilizers of the shoulder must possess enough laxity to tolerate the massive forces placed upon them, while also providing enough stability to prevent subluxation and dislocation of the humeral head [5]. These athletes typically require surgical intervention only after they have failed a full course of nonoperative management. In this chapter, we will review the relevant anatomy, biomechanics, and management for track and field athletes with shoulder instability.

8.2 Anatomy

The shoulder joint permits greater degrees of freedom than any other joint in the body, allowing humans to accomplish incredible feats. This mobility necessitates a complex and delicate balance of stabilizers to maintain integrity of the glenohumeral joint. Pain in the overhead athlete can be traced to disruption of these stabilizing mechanisms [6]. Here, we will review the static and dynamic stabilizers of the glenohumeral joint (Fig. 8.1).

The static stabilizers of the shoulder include the bony anatomy, capsuloligamentous structures, and the glenoid labrum [7]. The glenoid is pear- shaped

H. Bohlen · F. Savoie (✉)
Department of Orthopaedic surgery, Tulane University, New Orleans, LA, USA
e-mail: fsavoie@tulane.edu

G. L. Canata et al. (eds.), *Management of Track and Field Injuries*,
https://doi.org/10.1007/978-3-030-60216-1_8

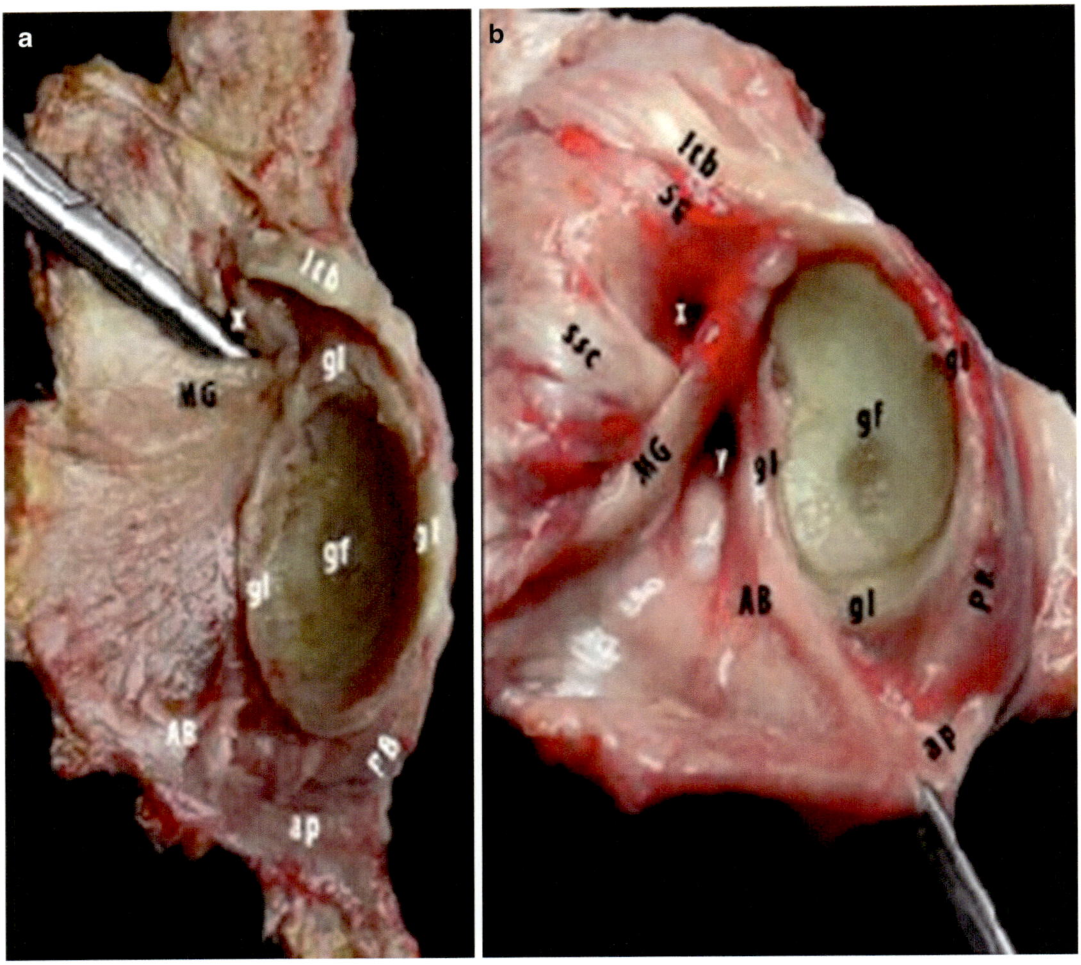

Fig. 8.1 Cadaveric dissection showing the capsular anatomy of the glenohumeral joint

with its width highest inferiorly. Of note, the sphere- shaped humeral head has roughly 3 times the surface area of the glenoid and consequentially only 25–30% of the humeral head is in contact with the glenoid in a given position [8]. This highlights the importance of soft tissue stabilizers in the overall stability of the glenohumeral joint. Additional osseous elements contributing to stability of the shoulder include glenoid retroversion and the coracoacromial arch. Glenoid retroversion can range from 9.5 degrees of anteversion to 10.5 degrees of retroversion, with a mean of 1.23 degrees of retroversion [9]. Excess anteversion or retroversion can be associated with decreased shoulder stability. The coracoacromial arch, which includes the acromion, the coracoid process, and the coracoacromial ligament, acts to prevent anterosuperior migration of the humeral head [7].

Static soft tissue stabilizers are critical to maintaining the glenohumeral joint and will account for the majority of pathology discussed in this chapter. The glenoid labrum provides a rim of fibrocartilaginous tissue that functions to extend the surface area and depth of the bony glenoid. The superior labrum shares an insertion with the long head of the biceps tendon on the supraglenoid tubercle. The capsuloligamentous structures of the glenohumeral joint provide varying degrees of stabilization depending on the position of the shoulder. Of note, the anterior and posterior bands of inferior glenohumeral ligament (IGHL) act as a dynamic sling to support the humeral head [10].

Table 8.1 Functions of the glenohumeral ligaments

Superior glenohumeral ligament	Prevents anterior and inferior displacement when the arm is adducted
Middle glenohumeral ligament	Prevents anterior and inferior displacement when the arm is at 45° of abduction
Anterior band of the inferior glenohumeral ligament	Prevents anterior displacement with the arm abducted to 90° and externally rotated
Posterior band of the inferior glenohumeral ligament	Prevents posterior displacement with the arm abducted to 90° and internally rotated

With the arm in an abducted and externally rotated position, the anterior band of the IGHL prevents anterior translation of the humeral head on the glenoid, whereas in an abducted and internally rotated position, the posterior band of the IGHL prevents posterior translation of the humeral head. The roles of the capsular ligaments are summarized in Table 8.1. The rotator interval is a triangular space constrained by the anterior margin of supraspinatus superiorly, the superior margin of subscapularis inferiorly, and the coracoid process as its base. It contains the coracohumeral ligament (CHL), the superior glenohumeral ligament (SGHL), middle glenohumeral ligament (MGHL), the long head of the biceps, and a thin layer of capsule. It functions to help stabilize the shoulder from posterior inferior translation, and it completes the circular ring of the joint capsule [7].

Dynamic stabilizers of the shoulder joint include the rotator cuff, long head of biceps, and the scapular rotators [7]. The rotator cuff functions to pull the humeral head medially toward the glenoid fossa. Additionally, the tendons prevent superior migration (supraspinatus), posterior migration (infraspinatus, teres minor), and anterior migration (subscapularis) of the humeral head. The scapular rotators, including trapezius, the rhomboids, latissimus dorsi, serratus anterior, and levator scapulae, function to help coordinate movement between the scapula and humerus.

8.3 Biomechanics

8.3.1 Javelin

The javelin throw consists of five steps: [11] first, the approach, in which the athlete runs in the direction of the throw to generate momentum; second, a series of sideway crossover steps, inducing stretch of the trunk and throwing muscles; third, the phase of single support in which the athlete transitions from running to throwing; fourth, an abrupt stop, during which the runner transfers momentum from forward motion into the overhead throw of the javelin, ultimately resulting in release of the javelin; and fifth, a follow-through phase in which the thrower completes the throwing motion and regains balance as he or she decelerates. The biomechanics of the javelin throw closely resemble those of throwing a baseball, with the cocking and acceleration phases taking place during the fourth part of the javelin throw, and the deceleration and following throw phases occurring during the fifth portion [4, 12, 13].

8.3.2 Hammer Throw

The hammer throw is an event in which the athlete generates centrifugal force to throw a 7.3-kg metal ball attached to a 4-ft. steel wire for men, or a 4-kg ball on a 3-ft. 11 in steel wire for women. Through a complex technique, the thrower generates force with initial arm swings followed by 3 to 5 turns before release. The turns are divided into phases of double support, in which both feet are on the ground and the hammer is accelerated, and single support in which one foot is lifted in order to turn [14]. Specific forces on the shoulder for this event have not been studied; however, the large centrifugal forces generated likely require the labrum, rotator cuff, and other secondary stabilizers to activate in order to prevent anterior dislocation of the humerus [14].

8.3.3 Shot Put

In the shot put, the thrower must utilize a 7-ft diameter circle to generate maximal force and throw a 7.26-kg ball (4 kg for women) as far as possible. Two techniques are currently in practice, including the glide technique and the rotation technique.

The glide technique consists of two phases (Fig. 8.2), the approach phase and the delivery phase. The athlete starts the approach phase at the back of the circle, holding the shot put close to the body with the shoulder abducted and elbow flexed. Next, the thrower generates momentum in the lower body by pushing with his or her nondominant leg toward the front of the circle, keeping the upper body passive. Once the thrower reaches the front of the circle, the front leg touches down followed by the back leg, entering the power position of the delivery phase. The delivery of the shot put is achieved by transitioning lower body momentum into a forward strike of the arm, in which the shoulder remains abducted and the elbow moves from a flexed to an extended position [15].

The rotation technique is more complex and requires the thrower to generate rotational inertia as they move forward in the ring with wide sweeping motions of the nondominant leg. Once the athlete reaches the front of the ring, this energy is transferred to the arm for a forward strike in a similar fashion to the glide technique. Of note, activity of the vastus lateralis and pectoralis major during the delivery phase has been correlated with increased performance [16].

8.3.4 Discus

The discus throw requires an athlete to throw a 220-mm-diameter 2-kg disk for men and 181 m 1-kg disk for women as far as possible while utilizing the space of a 2.5-m-diameter circle. The discus throw is broken down into five steps (Fig. 8.3) [17]: First, a preparation double support phase begins with the discus in a backward

Fig. 8.2 Depiction of the shot-put glide technique

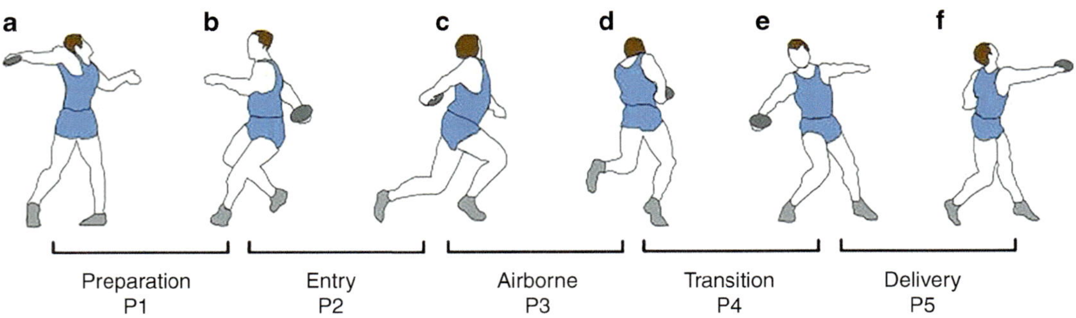

Fig. 8.3 Depiction of the five phases of the discus throw

swing and is completed when the right (front) foot breaks contact with the ground; second, a single leg support phase in which rotational inertia is developed, ending when the left foot leaves the ground; third, an airborne phase, which ends with the right foot touches down; the fourth phase is a transition phase with single leg support and ends when the left leg touches the ground; and the fifth and final phase is delivery, in which the body is perpendicular to the direction of the throw, and generated momentum is released into the discus.

8.3.5 Pole Vault

Though not technically an overhead throwing spot, the pole vault presents a field event in which the athlete must utilize a kinetic chain to channel energy from the legs through the body and into the glenohumeral joint to achieve success. The pole vault can be divided into seven stages, including (i) the run up, (ii) transition with arm elevation, (iii) take-off with pole plant, (iv) swing phase, (v) rock back, (vi) inverted position, and (vii) bar clearance [18]. At the point of take-off, the dominant shoulder must hold the arm above the head and resist the force applied from the ground through the pole, allowing the pole to bend. It is at this stage that maximal force is placed across the glenohumeral joint, which is held in a vulnerable position (Fig. 8.4). Subsequent shoulder instability events are not uncommon [19].

8.4 Management/Examination/Rehabilitation

8.4.1 Presentation

A through patient history is important to help focus the physical examination and make the correct diagnosis. Instability should always be considered in the track and field athlete who presents with shoulder pain. Presenting athletes with fall into two camps. The first includes those who sustained a specific traumatic dislocation event leading to instability. In these patients, it is important to ascertain when the initial event took place, how long they have been out of sport, and if any recurrences have occurred. The second reflects those with chronic microtrauma leading to instability. These patients will often complain

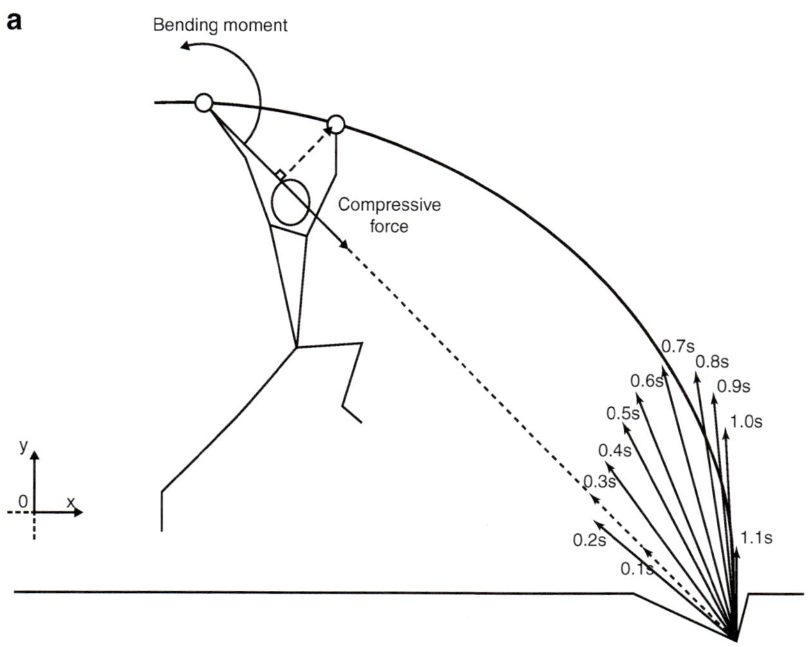

Fig. 8.4 Biomechanics of the take-off phase

of a subjective sense of instability, decreased performance, and weakness after participation in sport [20].

8.4.2 Examination

Examination of the shoulder for a track and field athlete in which instability is suspected must be comprehensive, with an emphasis on evaluation of the relevant anatomy, including the labrum, biceps, and rotator cuff [21]. Differentiation between physiologic laxity and pathologic instability can be difficult to distinguish, necessitating a through physical examination. It is critical to compare the affected and nonaffected sides to appreciate how much laxity is normal in a given patient.

Examination should begin with evaluation of the cervical spine to rule out neck pathology that may manifest as shoulder pain. Limited neck range of motion or pain radiating from the neck into the arm during provocative testing suggests cervical rather than shoulder pathology [22].

Shoulder examination consists of inspection, palpation, motion testing, strength testing, and specialized tests. Inspection should be done by comparing the injured and noninjured shoulders. Visible muscle atrophy, changes in resting position, or squaring of the shoulder girdle could indicate a neurologic cause for shoulder symptoms. Position of the scapula should also be assessed [20]. Tenderness with palpation over the AC joint or biceps tendon suggests pathology in these areas. Tenderness over the anterior or lateral edge of the acromion is common for rotator cuff pathology. Pain over the lateral humerus may be present with a Hill– Sachs lesion or greater tuberosity fracture following a dislocation event.

Active and passive range of motion testing for forward flexion, abduction, and internal and external rotation in adduction and 90° of abduction should be performed. Specific attention should be paid to the total arc of motion in the throwing athlete. Measurements should be done with the patient supine, the arm in 90° of abduction, and the scapula stabilized anteriorly. Internal rotation and external rotation are measured using a goniometer and compared to the unaffected side [23]. Though much attention has been given to glenohumeral internal rotation deficit (GIRD), it is now thought that decrease in total arc of motion is a better measure of an athlete's ability to throw safely. A loss of arc of as little as 10° compared to the unaffected side could increase risk of injury [3]. Strength testing should also be performed to evaluate the rotator cuff and overall shoulder function.

Specialized tests are a critical component of an instability examination to evaluate direction of instability and potential sites of pathology. Differentiating between pathologic instability and physiologic laxity is critical, and thus, a generalized laxity assessment should be performed first. A Beighton score between 0 and 9 can be assigned by taking the patient through a number of tests that access general ligamentous laxity. These tests (done bilaterally) include hyperextension of the small finger metacarpophalangeal joint past 90°, ability to place thumb on the volar forearm, hyperextension of the elbow beyond 10°, and ability to place both palms on the floor with the knees extended. One point is assigned for a positive result, with an additional point if both palms can be placed on the floor. A score of 4 or more is indicative of general ligamentous laxity [24].

Next, directional laxity should be assessed. Inferior laxity can be evaluated using the sulcus test [25]. With the patient sitting, the examiner pulls the humerus inferiorly, recording the amount of displacement. This test is repeated with the arm in maximal external rotation, which tightens the anterior capsule and rotator interval. If the amount of inferior translation does not decrease, an incompetent rotator interval should be suspected.

Anterior instability is the most common type experienced by track and field throwing athletes. The anterior fulcrum test can be used to evaluate anterior instability. This test is performed with the patient supine and the arm in 90° of abduction and external rotation. With one hand stabilizing the arm horizontally at the elbow, an anterior force is applied posteriorly to the humeral head.

The amount of translation and end laxity should be compared to the opposite shoulder. Other tests that can be used include the anterior Lachman and anterior drawer tests [26]. Of note, anterior instability was often thought to be a primary cause of shoulder pain in overhead throwing athletes, but more commonly manifests as pathology to the posterior superior labrum, with transmission of instability to the anterior side of the labral ring. This is known as pseudolaxity [27].

The apprehension–relocation test can also be helpful in analysis of instability. In this test, the affected arm is brought into 90 degrees of external rotation and abduction, and the patient noting apprehension of impending instability is considered a positive test. If the apprehension is relieved with the shoulder manually stabilized with a posteriorly directed force to the humeral head, the relocation part of the test is considered positive. A positive result is indicative of anterior instability.

Testing for posterior instability can be done using the posterior drawer test. With the patient sitting, the examiner stabilizes the scapula with one hand and grasps the humeral head between the thumb and fingers of the other hand. The humeral head is gently translated posteriorly, and displacement is measured as a percentage of the humeral head that can be subluxed posteriorly to the glenoid ring. Comparison to the contralateral side is critical, as up to 50% humeral head displacement can be normal [20].

SLAP lesions can contribute to shoulder instability or occur concomitantly. A number of tests exist to evaluate SLAP lesions, though the most clinically relevant examination maneuvers must reproduce the peel-back mechanism [28]. These tests include the modified dynamic labral shear (DLS), biceps load, biceps load II, pronated load, pain provocation, and resisted supination external rotation tests. The DLS test is the authors' preferred test [29]. This is performed with the examiner standing behind the seated patient, holding the patient's arm at the wrist in 90° abduction and external rotation. The examiner then raises the patient arm from 90° abduction to 150° while applying maximal external rotation. The test is positive with subjective reports of pain or the examiner feeling a click at the posterior joint line between 90° and 120° abduction.

8.4.3 Imaging

Diagnostic imaging is indicated for patients with gross instability events or for those with shoulder pain that does not improve following a period of nonoperative management. Magnetic resonance imaging (MRI) provides a thorough evaluation of the osseous and soft tissue structures that can be affected in the unstable shoulder. Specifically, MR arthrography (MRA) remains the gold standard for preoperative evaluation of soft tissue injury in the unstable athlete [14]. These examinations allow for excellent visualization of the labroligamentous structures, rotator cuff, and articular cartilage. Of note, an MRI/MRA of a throwing athlete must include abduction external rotation (ABER) views to properly evaluate internal impingement of the rotator cuff and superior labrum peel-back changes [29]. MRI for labroligamentous complex injuries is reported to have sensitivities and specificities ranging from 44 to 100% and 66 to 95%, respectively, with higher values for MRA [14, 30]. An MRI or MRA should always be obtained prior to surgical intervention.

Computed tomographic (CT) imaging also plays a role in evaluating instability of the throwing athlete. CT imaging is the preferred modality for visualization of osseous defects that occur in lesions such as the bony Bankart and Hill–Sachs [31]. It is important to note that recurrent instability of the shoulder is often associated with unrecognized bone loss, so the treating surgeon should have a low threshold to include a CT scan in the diagnostic workup of these athletes [62] (Fig. 8.5). Additionally, in patients with contraindications to MR imaging, CT arthrography provides a reliable alternative for evaluating the soft tissue structures in the unstable shoulder [30]. Plain radiographs can also aid in the evaluation of the unstable shoulder, particularly in patients with a dislocation event. In these patients, a complete radiographic set

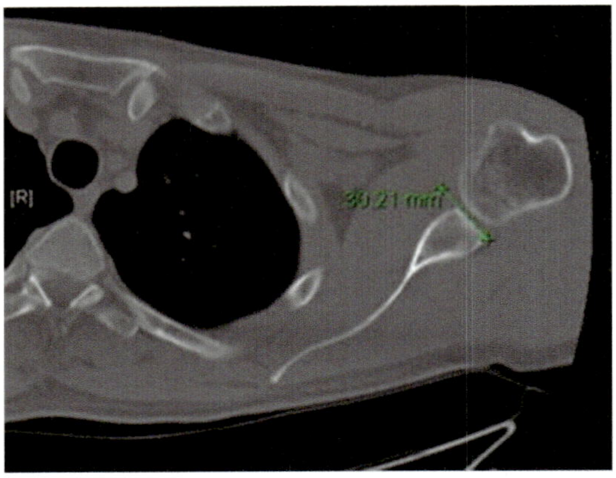

Fig. 8.5 Axillary CT scan image showing measurement of the glenoid to estimate bone loss

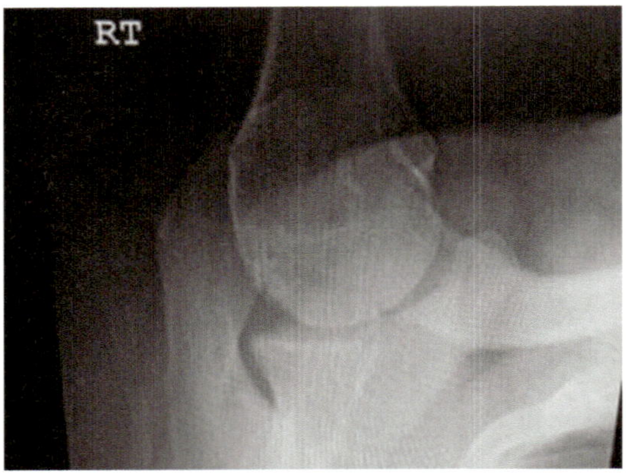

Fig. 8.6 Bernageau view of the shoulder

including a true AP (Grashey), scapular Y, axillary lateral, and Bernageau view should be done. The Bernageau view in particular can help evaluate for anterior glenoid bone loss [32] (Fig. 8.6).

In summary, MRI/ MRA remains the gold standard for the evaluation of the athlete with shoulder instability and must be done prior to surgical intervention. CT arthrography can replace an MRI in patients with contraindications to MR imaging. If osseous lesions are suspected, a CT scan and plain radiographs should be included in the evaluation of a patient. Advanced imaging should be done immediately in patients with dislocation events, and after a period of nonoperative management in patients with subjective instability complaints.

8.4.4 Nonoperative Management

Given that it is difficult to determine by physical examination and imaging how much laxity is too much for a given athlete, nonoperative management should always be attempted prior to surgical intervention in the absence of a traumatic dislocation [33]. Of note, there is little to no literature regarding operative and nonoperative management of specific track and field events. In our professional opinion, instability in the track and field athlete can be managed analogously to how one would manage instability in other athletes. Thus, initial management consists of a trail of 4–6 weeks of rest and rehabilitation. During this time, attention should be given to correcting any abnormalities in the kinetic chain for the athlete's specific sport. Once pain has diminished, physical therapy should begin and focus on shoulder stretching with ER/IR balance, core strengthening, and shoulder/ scapular taping [26]. Following traumatic dislocation, acute surgical intervention can be done without a trial of nonoperative management if the patient's shoulder is grossly unstable on physical examination [33].

8.5 Surgery

Surgery for the unstable shoulder should be undertaken with caution, as "instability" often represents the normal laxity required for throwing in many field sports. For the athlete that has failed nonoperative management or suffered acute traumatic dislocation events, surgery is indicated. The ideal surgery should access stability of the glenohumeral joint in the context of anatomic structures involved, the type of fixation needed, and the potential for healing. The goals of surgery are to perform an anatomic repair of the pathologic tissues and to restore bony anatomy in the case of bone loss. Based on imaging assessment and physical examination, a preoperative plan should be developed to address the relevant areas of instability. The key to obtaining excellent results is creating an anatomic repair of all pathologic structures. Restoring the patient's normal anatomy will yield the best results. The surgical procedure for the unstable shoulder should proceed in the following order: diagnostic arthroscopy, inferior repair, posterior repair, anterior repair, and superior repair. Only indicated procedures should be performed. As such, most unstable shoulders will not require every step described, but structures should be evaluated and repaired in this order, if necessary.

Of note, a dearth of literature currently exists regarding outcomes for track and field athletes following surgical treatment of the unstable shoulder [4]. Thus, studies presented here contain results that are unfortunately not specific to the field events described.

8.5.1 Diagnostic Arthroscopy

With the patient place in the lateral decubitus position (beach chair can also be utilized), a posterior inferior portal is initiated between the infraspinatus and teres minor, roughly 2 cm inferior to the posterolateral corner of the acromion. Under direct visualization, an anterior inferior portal is established adjacent to the subscapularis tendon in the rotator interval. Examination begins with visualization of the glenoid and humeral head, taking note of any osteochondral lesions. The anterior, inferior, posterior, and superior labrum should be visualized and probed. The biceps tendon, middle glenohumeral ligament, superior glenohumeral ligament, and anterior and posterior bands of the inferior glenohumeral ligament should also be visualized and probed. The undersurface of the rotator cuff should be observed, followed by examination of the peel-back mechanism and internal impingement by placing the arm in an abducted and externally rotated position [34]. Following visualization, an anterior superior portal should be developed between the coracoid and acromion, just anterior to but not through the supraspinatus, permitting a view from above to aid in balancing the shoulder [35]. A positive drive through sign, in which the arthroscope is easily passed into the joint at the level of the anterior band of the IGHL, may be a sign of pathologic capsular laxity. Before repair is initiated, preparation for reconstruction begins with debridement of frayed or degenerative tissue from the labrum and undersurface of the rotator cuff, with care given to retain as much normal tissue as possible. The capsule should be released medially and inferiorly from the 1 o'clock to 6 o'clock position, and the glenoid neck lightly abraded to create a large healing bone surface [36].

8.5.2 Inferior Repair

Reconstruction begins by addressing the inferior structures. The goals of inferior repair include restoration of the IGHL complex and creating an inferior capsular shift, which involves superior lateral tensioning of the inferior capsule to recreate a capsule fold to the glenoid neck [37]. This is best accomplished with an initial double-loaded anchor placed at the 6 o'clock position, inferior to any bone lesions that may be present (Fig. 8.7). These sutures should be retrieved via the posterior portal to prepare for inferior repair and shift (Fig. 8.8). The inferior capsule and labrum are grasped below the level anchor and elevated superiorly toward the anchor's insertion, restoring normal capsulolabral complex tension. Multiple passages through the capsule are key to creating a strong and stable capsular shift. Oblique mattress stitches should be used to avoid suture contact with the articular cartilage (Fig. 8.9) [38, 39].

Neer originally published excellent results for open inferior capsular shift to treat capsular redundancy leading to instability [37]. Recently, arthroscopic variants of this technique have been successfully described. Fleega et al. published a minimum 7-year follow-up of 75 patients who received isolated inferior repair for capsular redundancy. Surgical intervention improved ASES and UCLA scores from 70.76 to 97.53 and

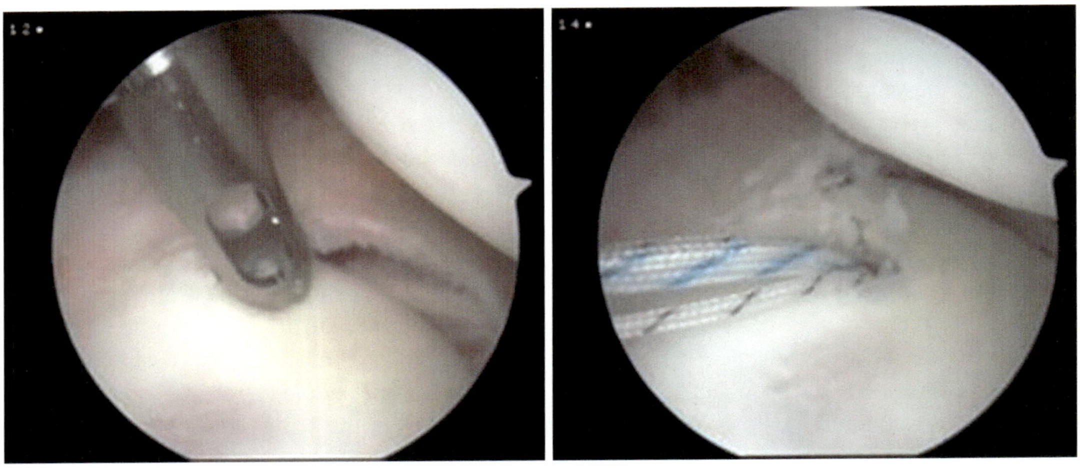

Fig. 8.7 Arthroscopic placement of 6 o'clock suture anchor

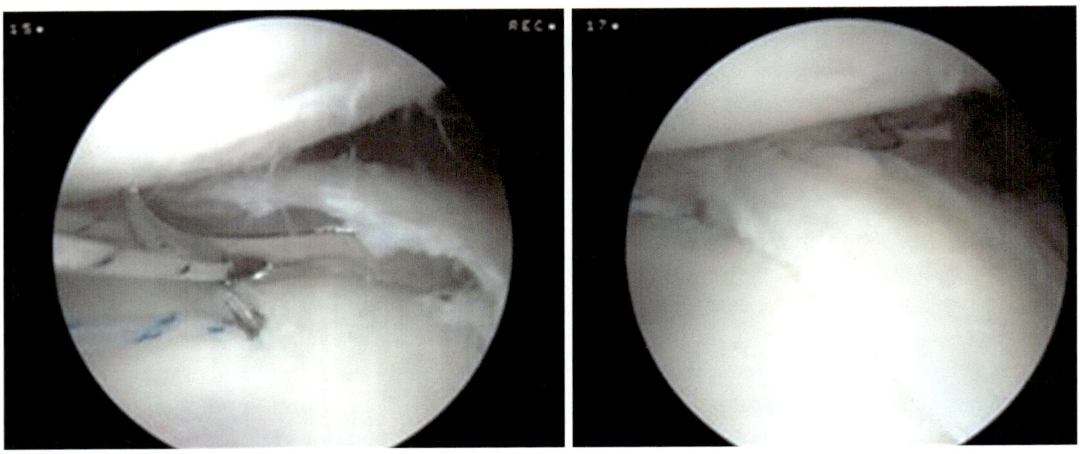

Fig. 8.8 Sutures being retrieved from posterior portal to prepare for inferior repair and shift

21.97 to 33.84, respectively [40]. Uciyama et al. compared isolated Bankart repair to Bankart repair augmented with inferior capsular shift and found lower rates of recurrent instability in the augmented group (0% vs. 26.6%). [56] These studies help demonstrate the importance of addressing the inferior structures in an effort to correct all pathologic anatomy in the unstable shoulder.

8.5.3 Posterior Repair

Following inferior repair, attention should be turned to the posterior shoulder. At this stage, stability of the posterior shoulder should be assessed, including evaluation of the PIGHL and posterior labrum. The presence of any bony abnormalities including a Hill–Sachs lesion should also be assessed. Significant capsulolabral defects, including a deficient PIGHL or tears to the posterior labrum, should be fixed using suture anchors in the glenoid neck. If a humeral avulsion of the glenohumeral ligament (HAGL) lesion is present, it should be repaired with a suture anchor at the PIGHL insertion on the humeral neck [41]. For posterior repair, the glenoid should be prepared with gentle burring of the neck to create a healing face. Suture anchors are utilized as necessary between the 6 and 12 o'clock positions,

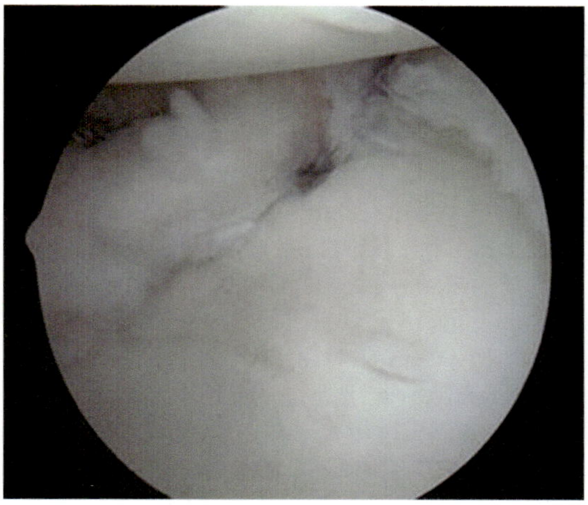

Fig. 8.9 View from anterior superior portal demonstrating completed inferior repair with superior lateral capsular shift from a double-loaded anchor in the 6 o'clock position

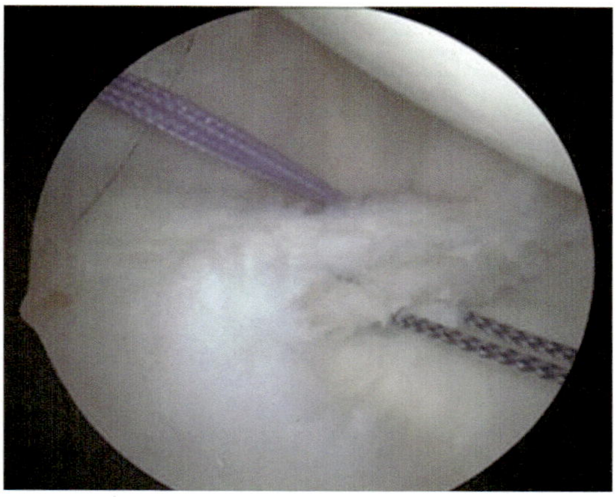

Fig. 8.10 Posterior view of 3 o'clock suture anchor limbs to be used for capsulolabral repair

with care given to utilize mattress sutures or knotless fixation to prevent iatrogenic injury to glenoid or humeral cartilage. Suture plication of the posterior capsule can be used in addition to anchor fixation or on its own to tighten residual posterior laxity [38, 42]. Remplissage can be used as an alternative to suture plication and should be used if a concurrent Hill–Sachs lesion is present. Hill–Sachs lesions will be covered in more detail below. At this stage, the humeral head should be centered on the glenoid. In our experience, javelin throwers are particularly susceptible to posterior instability. Pole vaulters should also be carefully evaluated for damage to the posterior labrum.

Bradley et al. published a series of 297 shoulders in athletes who required posterior capsulolabral repair. 6.4% of patients ended up requiring revision surgery. Those who did not require revision went on to return to sport at the same level 64.3% of the time, with 78.6% returning to sport at some level. This study highlights the importance of a proper initial repair, as revision surgery resulted in significantly diminished ability to return to sport and to return at a pre-injury level. [59].

8.5.4 Anterior Repair

Attention should next be turned to the anterior structures. Goals for anterior repair are to restore the anterior capsulolabral complex. An initial anchor should be placed in the 6 o'clock position if one was not placed during the inferior repair. Subsequent anchors should be placed in the glenoid neck moving superiorly from the 6 o'clock position until adequate stabilization of the capsulolabral complex has been achieved. Special care should be given to any lesion in the 3 o'clock position due to the importance of this area on shoulder biofeedback (Fig. 8.10). If small-to-moderate lesions of the glenoid are present, they can be incorporated into the repair by passing sutures below and through or around the fragments. Glenoid bone loss will be covered in more detail below. Mattress stitches should be used to ensure that the suture and knots do not contact the articular cartilage of the glenoid or humeral head [33].

Allen et al. reported on fifty-eight athletes undergoing anterior capsulolabral repair and found a return to play rate of 87% at 27-month follow-up. 70% of patients returned to pre-injury level of competition [43]. In a review of nine high-quality articles, Donohue et al. found that a cumulative 361 athletes achieved a 73% return to performance at prior level of competition following anterior repair, noting superior outcomes for surgical repair compared to nonoperative management in these patients [44]. In our experience, injuries to the anterior labrum are particularly common in discus and pole vaulters. Returning stability via anterior repair is critical in these athletes.

8.5.5 Superior Repair

Following posterior repair, the arthroscope should be moved to the posterior portal for visualization of the superior structures, including the anterior superior labrum and the rotator interval. Anatomic structures to be addressed here include the middle glenohumeral ligament (MGHL), superior glenohumeral ligament (SGHL), coracohumeral ligament (CHL), anterior superior labrum, and rotator interval [45]. Additional anterior superior stability can be achieved in most patients by tightening the MGHL and SGHL. This is performed by placing a double-loaded suture anchor at the 1 o'clock position after glenoid preparation. Mattress sutures are passed through the MGHL first, followed by the SGHL. This typically provides adequate fixation; however, in high-risk patients, patients with significant intrinsic ligamentous laxity, and patients with observable defects of the rotator interval, proper closure of the rotator interval may be necessary [46]. Our preferred technique for closure of the rotator interval involves suture plication of the supraspinatus tendon to subscapularis (Fig. 8.11). Care should be given to keep sutures lateral on both tendons, with the goal of further tightening the SGHL and CHL. If a SLAP lesion is present, it can be repaired at this time. In shot putters, partial articular- sided supraspinatus avulsion (PASTA) lesions may be present. These can be repaired during this part of the procedure as well, if present.

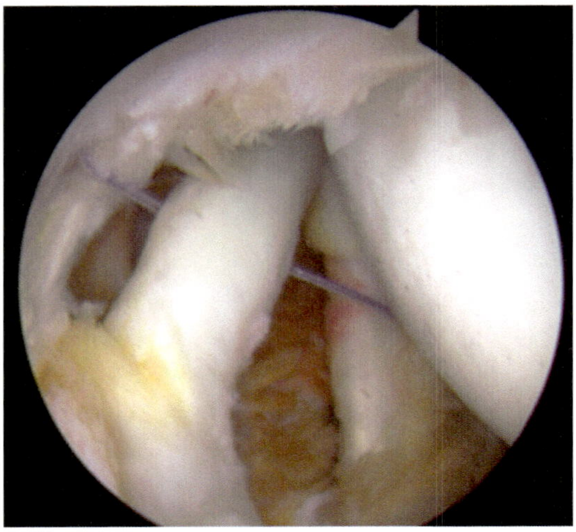

Fig. 8.11 Initial step in closure of the rotator interval showing suture plication of supraspinatus to subscapularis

After Harryman helped demonstrate the importance of the rotator interval in overall stability of the shoulder, several techniques have been developed to address the structures of the rotator interval. [57] A series by Field et al. of patients treated with isolated open rotator interval closure found that all fifteen patients had achieved a good or excellent Rowe score by an average of 3.3-year follow-up. Arthroscopic techniques have since been developed, with our series of 92 shoulders reporting a 97% success rate by Neer–Foster score when rotator interval closure was included in repair for posterior instability. [58].

8.5.6 Bone Loss

Patients with recurrent anterior instability often have anterior glenoid deficiency and/ or a concomitant Hill–Sachs lesion that contributes to instability of the glenohumeral joint. A preoperative evaluation using a CT scan and Bernageau radiograph should be performed to quantify size and shape of bone loss. This allows for strong preoperative decision-making regarding graft choice for repair, if necessary. The glenoid and humeral head should be further evaluated during diagnostic arthroscopy. With a bony Bankart lesion, the shape of the glenoid changes to an "inverted pear" appearance, in which it is wider superiorly. A good estimation of bone loss can be achieved by visualizing the bare spot on the glenoid, which should be equidistant between the anterior and posterior glenoid rims [47].

If a significant Hill–Sachs lesion is present, remplissage should be used to fill the defect and correct instability during the posterior repair part of the procedure. Small Hill–Sachs defects can be ignored, as a well-done posterior repair as described above will obscure it from view. The primary indication for remplissage is an engaging lesion, in which the humeral head lesion traverses the glenoid rim in less than 90 degrees of abduction and external rotation as visualized arthroscopically. [60] Patients are candidates for remplissage if the lesion occupies greater than 10% but less than 50% of humeral articular surface, with associated anterior glenoid bone loss of less than 25% as determined by preoperative CT. The first step in remplissage is preparation of the defect with gentle burring to create a healing face, with care taken to minimize removal of bone. The technique is performed with the use of two suture anchors (one superior and one inferior) placed in the medial aspect of the Hill–Sachs defect. The sutures are then passed through the infraspinatus tendon and posterior capsule on a line straight back from the medial defect, effectively transferring the infraspinatus to fill the osseous defect. The inferior suture should be tied first followed by superior, with knots staying extra-articular [48] (Fig. 8.12). Our series of 30 patients who underwent remplissage with Bankart repair for anterior instability with mild glenoid bone loss demonstrated excellent results for primary repair, with no failures at an average follow-up for 41 months. Results were less satisfactory for this procedure in revision surgery. [61].

A number of techniques and graft choices exist for repairing the anterior glenoid. Sugaya et al. described a technique in which the glenoid fragments were repaired in conjunction with the damaged labrum with the use of suture anchors. In their study, 39/42 (93%) patients achieved good or excellent results by UCLA and Rowe scoring systems, with 32/38 (84%) of athletes returning to play at pre-injury level [49]. Abrams describes a technique in which clavicular autograft is harvested arthroscopically and secured to the anterior glenoid to bolster a Bankart repair with remplissage [50].

Of note, all arthroscopic glenoid reconstruction techniques are technically challenging and carry higher risk of iatrogenic injury to the axillary and musculocutaneous nerves than do open techniques. We caution surgeons to stay within their surgical comfort zone when caring for these athletes.

In athletes with advanced bone loss or prior failed anterior repair, coracoid autograft can be utilized to stabilize the anterior glenoid via the Latarjet procedure (Fig. 8.13). Many variants of the original technique described by Latarjet exist;

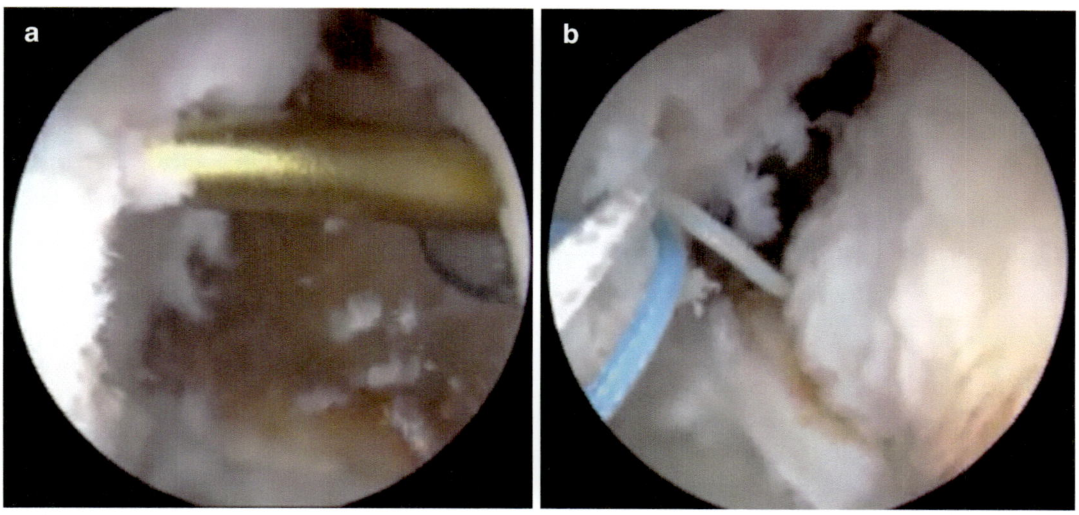

Fig. 8.12 (**a**) preparation of Hill–Sachs defect for anchor placement. (**b**) Tightening of infraspinatus and posterior capsule toward the humeral defect

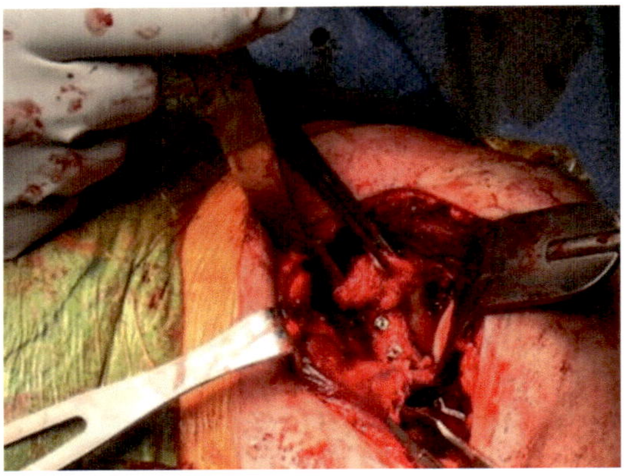

Fig. 8.13 Harvesting of coracoid autograft for Latarjet procedure

however, the preferred technique by the authors is the variant described by Walch and Boileau in which two screws are used to fixate the graft through a deltopectoral approach and subscapularis split [51, 52]. Though no data are available for track and field athletes, return to play for other athletes following Latarjet reconstruction is less than ideal. A MOON study of 65 patients found that over 55% of athletes failed to achieve at least one return to play criteria [53]. Additionally, Higgins et al. reported a 22% recurrence dislocation rate with only a 50% return to play at pre-injury level in their series of contact athletes [54].

8.5.7 Postoperative Rehabilitation

Our preference for rehabilitation after surgical repair of the unstable shoulder is as follows. Immediately after surgery, the patient is immobilized with an adduction brace. During the first week, the shoulder remains immobilized; however, scapular and core exercises can be initiated. During the second week, active and passive range of motion exercises can be initiated and should be limited by patient comfort. If the infraspinatus was transferred during remplissage, external rotation should be limited to protect the infraspinatus. Bracing should continue for at least 4 weeks, with slow weaning until 6 weeks, allowing proper time for the labrum to heal. From weeks 4 to 6, the affected extremity can be used for light activities of daily living as tolerated. From weeks 6 to 16, the athlete should begin integrated rehabilitation consisting of shoulder strengthening, core strengthening, and scapular strengthening and positioning. At approximately 3 months post operatively, the athlete should begin high-speed plyometric training in preparation for return to sport, with most track and field athletes returning between 4 and 6 months after surgery.

8.6 Summary

- Shoulder instability in track and field athletes presents a complicated problem. It is critical to understand the different types of instability and the athlete's history when developing a treatment plan.
- MR arthrogram is the best test to evaluate capsulolabral pathology. CT imaging should be included if bone loss is suspected.
- Based on preoperative findings, a plan should be made to address all pathology in the unstable shoulder. The key to obtaining excellent results is anatomic restoration of the patient's normal anatomy.
- Soft tissue repair must be sure to restore the capsulolabral complex, recreate a PIGHL, eliminate any Hill–Sachs lesion, and close the true rotator interval in high- risk patients.
- Surgery to address bone loss must include careful preoperative measurement of the bony defect to decide on graft choice, if necessary. A number of open and arthroscopic techniques exist to address bony deficiency of the glenoid, and it is important that the practicing surgeon stays within his or her surgical comfort zone for best results.
- A dearth of literature exists regarding the epidemiology and treatment outcomes for shoulder instability in track and field athletes. This presents an area of need in the sports medicine orthopedic literature.

References

1. Frère J, L'Hermette M, Slawinski J, Tourny-Chollet C. Mechanics of pole vaulting: a review. Sports Biomech. 2010;9(2):123–38. https://doi.org/10.1080/14763141.2010.492430.
2. Bird S, Black N, Newton P. Sports injuries. Cheltenham, UK: Stanley Thornes; 1997.
3. Haley COL, Chad A. History and physical examination for shoulder instability. Sports Med Arthrosc Rev. 2017;25(3):150–5.
4. Chu SK, Jayabalan P, Kibler WB, Press J. The Kinetic chain revisited: new concepts on throwing mechanics and injury. PM R. 2016;8(3 Suppl):S69–77.
5. Ben Kibler W, Wilkes T, Sciascia A. Mechanics and pathomechanics in the overhead athlete. Clin Sports Med. 2013;32(4):637–51.
6. O'Brien SJ, Neves MC, Arnoczky SP, et al. The anatomy and histology of the inferior glenohumeral ligament complex of the shoulder. Am J Sports Med. 1990;18:449–56.
7. Meister K. Current concepts injuries to the shoulder in the throwing athlete part one: biomechanics/pathophysiology/classification of injury. Am J Sport Med. 2000;28:265–75. https://doi.org/10.1097/BLO.0b013e3180e79c6a.
8. Murofushi K, Sakurai S, Umegaki K, Takamatsu J. Hammer acceleration due to thrower and hammer movement patterns. Sports Biomech. 2007;6:301–14. https://doi.org/10.1080/14763140701489843.
9. Zatsiorsky VM, Lanka GE, Shalmanov AA. Biomechanical analysis of shot putting technique. Exerc Sport Sci Rev. 1981;9:353–89. https://doi.org/10.1249/00003677-198101000-00009.
10. Terzis G, Karampatsos G, Georgiadis G. Neuromuscular control and performance in shot-put athletes. J Sports Med Phys Fitness. 2007;47:284–90.
11. Weber AE, Kontaxis A, O'Brien SJ, Bedi A. The biomechanics of throwing: simplified and cogent. Sports Med Arthrosc. 2014;22:72–9.
12. Wilk KE, Meister K, Andrews JR. Current concepts in the rehabilitation of the overhead throwing athlete. Am J Sports Med. 2002;30:136–51.
13. Morriss C, Bartlett R. Biomechanical factors critical for performance in the men's javelin throw. Sports Med. 1996;21:438–46. https://doi.org/10.2165/00007256-199621060-00005.
14. Wilk KE, Arrigo C. Current concepts in the rehabilitation of the athletic shoulder. J Orthop Sports Phys Ther. 1993;18(1):365–78.
15. Murray IR, et al. Functional anatomy and biomechanics of shoulder stability in the athlete. Clin Sports Med. 2013;32(4):607–24.
16. Codman E. The shoulder. Boston: Thomas Todd; 1934.
17. Churchill RS, Brems JJ, Kotschi H. Glenoid size, inclination, and version: an anatomic study. J Shoulder Elb Surg. 2001;10(4):327–32.
18. Wilk KE, Macrina LC, Fleisig GS, Porterfield R, Simpson CD, Harker P, Paparesta N, Andrews JR. Correlation of glenohumeral internal rotation deficit and total rotational motion to shoulder injuries in professional baseball pitchers. Am J Sports Med. 2011;39(2):329–35.
19. Cameron KL, Duffey ML, DeBerardino TM, et al. Association of generalized joint hypermobility with a history of glenohumeral joint instability. J Athl Train. 2010;45:253–8.
20. Burkhart SS, Morgan CD, Kibler WB. The disabled throwing shoulder: spectrum of pathology. Part I: pathoanatomy and biomechanics. Arthroscopy. 2003;19:404–20.
21. Owens BD, Burns TC, DeBerardino TM. Examination and classification of instability. Philadelphia, PA: Elsevier Saunders; 2012. p. 20–32.
22. Reinold MM, Gill TJ. Current concepts in the evaluation and treatment of the shoulder in overhead-throwing athletes, part 1: physical characteristics and clinical examination. Sports Health. 2010;2(1):39–50. https://doi.org/10.1177/1941738109338548.
23. Bohlen HL, Savoie FH. ICJR shoulder. 2019.
24. Bois AJ, Walker REA, Kodali P, Miniaci A. Imaging instability in the athlete: the right modality for the right diagnosis. Clin Sports Med. 2013;32(4):653–84.
25. Macmahon PJ, Palmer WE. Magnetic resonance imaging of glenohumeral instability. Magn Reson Imaging Clin N Am. 2012;20(1):295–312.
26. Sanders TG, Zlatkin M, Montgomery J. Imaging of glenohumeral instability. Semin Roentgenol. 2010;45(3):160–79.
27. Dinu D, Houel N, Louis J. Effects of a lighter discus on shoulder muscle activity in elite throwers, implications for injury prevention. Int J Sports Phys Ther. 2019;14(4):592–602.
28. Arciero RA, Parrino A, Bernhardson AS, Diaz-Doran V, Obopilwe E, Cote MP, Provencher MT. The effect of a combined glenoid and hill-sachs defect on glenohumeral stability. Am J Sports Med. 2015;43(6):1422–9. https://doi.org/10.1177/0363546515574677.
29. Sanders TG, Zlatkin M. Imaging of glenohumeral instability. Semin Roentgenol. 2010;45(3):160–79.
30. Bernageau J, Patte D, Debeyre J, et al. Value of the glenoid profile in recurrent luxations of the shoulder. Rev Chir Orthop Reparatrice Appar Mot. 1976;62(Suppl 2):142–7.
31. Savoie FH. Arthroscopic examination of the throwing shoulder. J Orthop Sports Phys Ther. 1993;18(2):409–12.
32. Wolf EM. Anterior portals in shoulder arthroscopy. Arthroscopy. 1989;5(3):201–8. https://doi.org/10.1016/0749-8063(89)90172-2.
33. Wetzel FT, Reider B. Cervical and thoracic spine. Philadelphia, PA: W.B. Saunders Company; 1999. p. 301–42.
34. Savoie New Frontiers.
35. Neer CS, Foster CR. Inferior capsular shift for involuntary inferior and multidirectional instability of the shoulder. J Bone Joint Surg Am. 1980;62:897–908.

36. Savoie FH, Holt MS, Field LD, Ramsey JR. Arthroscopic management of posterior instability: evolution of technique and results. Arthroscopy. 2008;24(4):389–96. Copyright © 2008 Arthroscopy Association of North America
37. Fleega BA, El Shewy MT. Arthroscopic inferior capsular shift: long-term follow-up. Am J Sports Med. 2012;40(5):1126–32. https://doi.org/10.1177/0363546512438509.
38. Walch G, Boileau P. Latarjet-Bristow procedure for recurrent anterior instability. Tech Shoulder Elbow Surg. 2000;1(4):256–61.
39. Field LD, Ryu RK, Abrams JS, Provencher M. Arthroscopic management of anterior, posterior, and multidirectional shoulder instabilities. Instr Course Lect. 2016;65:411–35.
40. Frantz TL, Everhart JS, Cvetanovich GL, Neviaser A, Jones GL, et al. Are patients who undergo the latarjet procedure ready to return to play at 6 months? A Multicenter Orthopaedic Outcomes Network (MOON) Shoulder Group Cohort Study. Am J Sports Med. 2020;48(4):923–30. https://doi.org/10.1177/0363546520901538.
41. Privitera DM, Sinz NJ, Miller LR, Siegel EJ, Solberg MJ, Daniels SD, Higgins LD. Clinical outcomes following the latarjet procedure in contact and collision athletes. J Bone Joint Surg. 2018;100(6):459–65. https://doi.org/10.2106/jbjs.17.00566.
42. Treacy SH, Savoie FH, Field LD. Arthroscopic treatment of multidirectional instability. J Shoulder Elb Surg. 1999;8:345–50.
43. Latarjet M. Treatment of recurrent dislocation of the shoulder. Lyon Chir. 1954;49(8):994–7.
44. Abrams JS. Arthroscopic surgical stabilization of glenohumeral dislocations with clavicular graft and remplissage. JBJS Essential Surgical Techniques. 2019;9(1):e11. https://doi.org/10.2106/jbjs.st.17.00072.
45. Richards DP, Burkhart SS. Arthroscopic humeral avulsion of the glenohumeral ligaments (HAGL) repair. Arthroscopy. 2004;20:134–41. https://doi.org/10.1016/j.arthro.2004.04.045.
46. Ozturk BY, Maak TG, Fabricant P, et al. Return to sports after arthroscopic anterior stabilization in patients aged younger than 25 years. Arthroscopy. 2013;29(12):1922–31.
47. Uchiyama Y, Handa A, Shimpuku E, Omi H, Hashimoto H, Imai T, Watanabe M. Open Bankart repair plus inferior capsular shift versus arthroscopic Bankart repair without augmentations for traumatic anterior shoulder instability: A prospective study. J Orthop Surg. 2017;25(3):230949901772794. https://doi.org/10.1177/2309499017727947.
48. Chang ES, Greco NJ, McClincy MP, Bradley JP. Posterior Shoulder Instability in Overhead Athletes. Orthop Clin N Am. 2016;47(1):179–87. https://doi.org/10.1016/j.ocl.2015.08.026.
49. Neer CS, Satterlee CC, Dalsey RM, Flatow EL. The anatomy and potential effects of contracture of the coracohumeral ligament. Clin Orthop Relat Res. 1992;280:182–5.
50. Harryman DT, Sidles JA, Harris SL, Matsen FA. The role of the rotator interval capsule in passive motion and stability of the shoulder. J Bone Joint Surg Am. 1992;74:53–66.
51. Burkhart SS, DeBeer JF. Traumatic glenohumeral bone defects and their relationship to failure of arthroscopic Bankart repairs: significance of the inverted-pear glenoid and the humeral engaging Hill-Sachs lesion. Arthroscopy. 2000;16:677–94.
52. Burkhart SS, DeBeer JF, Tehrany AM, Parten PM. Quantifying glenoid bone loss arthroscopically in shoulder instability. Arthroscopy. 2002;18:488–91.
53. Sugaya H. Arthroscopic osseous Bankart repair for chronic recurrent traumatic anterior glenohumeral instability. J Bone Jnt Surg Am. 2005;87(8):1752.
54. Bradley JP, Arner JW, Jayakumar S, Vyas D. Risk factors and outcomes of revision arthroscopic posterior shoulder capsulolabral repair. Am J Sports Med. 2018;46(10):2457–65. https://doi.org/10.1177/0363546518785893.

Rotator Cuff Injuries in Throwing Athletes

9

Umile Giuseppe Longo, Giovanna Stelitano, Vincenzo Candela, and Vincenzo Denaro

9.1 Introduction

Rotator cuff tears represent a widespread disabling disease, which predominantly afflicts throwers. The exact incidence of cuff tears in overhead athletes remains still unclear and, probably, underappreciated, considering that several of them do not complain of symptoms [1]. However, cadaveric studies, imaging, and arthroscopic researches have attested the high prevalence of rotator cuff damages in young athletes (about 40% involving the dominant shoulders), which practice repetitive overhead activities [2]. Both partial- and full-thickness rotator cuff tears have seemed to show a significant increase in the last period, maybe thanks to the improvement in radiographic and diagnostic techniques. At the same time, the progress in the arthroscopic field has brought new operative strategies of treatment [3]. However, despite the greater capacity to detect and quantify tear extensions and the progress in surgical procedures, successful management of this pathology has not been reached. Arthroscopic repair and debridement, and surgical repair of significant partial- and full-thickness tears do not allow athletes to have predictable recovery and return to preceding levels of sport. The main problem is represented by the coexistence of concomitant pathologies, such as shoulder impingement, SLAP lesions, and subacromial conflict that worse the final outcomes and the patient's management. For each of these reasons, it is essential to understand the pathogenic mechanism about the onset of the tears, their clinical presentation, diagnostic examination, and principles of treatment.

U. G. Longo (✉) · G. Stelitano · V. Candela
V. Denaro
Department of Orthopaedic and Trauma Surgery, Campus Bio-Medico University, Rome, Italy
e-mail: g.longo@unicampus.it

9.2 Pathophysiology

For the first time, Nee described rotator cuff tears as consequence of the outlet impingement. This theory is currently outdated thanks to the advances in basic science and imaging technology, which have shown as the rotator cuff disease presupposes a multifactorial pathogenic mechanism [2]. In throwing athletes, the exact processes of the disease onset can be clearly explained: The repetitive loads of up to 108% of body weight and the humeral angular velocities upwards of 7000 degrees cause no indifferent stress on the shoulders, especially if pathologies anywhere else in the kinetic chain coexist. These strains and forces, more elevated in the acceleration and deceleration steps of the throwing cycle, provoke repeated trauma to the tendons tissue, in particular in their insertion where the vascular network is weak [4]. Exacerbation of the capsular articular stress together with the compression

G. L. Canata et al. (eds.), *Management of Track and Field Injuries*,
https://doi.org/10.1007/978-3-030-60216-1_9

due to the internal impingement cause a progressive cuff impairment, with intrinsic shear strains and undersurface fiber damage that bring to articular partial-thickness cuff tears [5]. The pathologic internal impingement can be induced by different latent factors such as recurrent microtrauma and intratendinous stress forces, in particular through the eccentric contraction of the rotator cuff in the deceleration state of throwing [6]. Slight anterior instability, weakness of the anterior band of the inferior glenohumeral ligament, contracture of the posterior capsular, reduced humeral retroversion, bad throwing mechanics, and scapular imbalance have to be considered as supplementary factors able to unmask the disease [7]. Finally, patients affected by scapular dyskinesis seem to be a higher predisposition to the development of rotator cuff tears. In essence, the protrusion of the scapula moves the posterior glenoid against the cuff, producing a mechanism of injury [8].

9.3 Type of Tears

Rotator cuff disease in throwing athletes includes a wide spectrum of injuries that goes from tendinosis to partial articular, bursal, intratendinous tears, until to full-thickness tears with presupposing the whole tendons detachment. The incidence of partial tears is almost twice higher than full-thickness ones.

9.3.1 Articular Tears

Articular tears represent the most common injuries in the rotator cuff of overhead athletes. They usually involve the posterior surface of the supraspinatus and the anterior fibers of the infraspinatus. The mechanism on the base of their onset is multifactorial. Potential risk conditions include weaker strain-to-failure ratio on the articular portion, anatomical alterations such as a lower number of collagen fibers, randomly oriented, and with reduced power compared with the bursal surface. Further, the slight vascular network in the articular cuff could predispose to the pathology. Within this category of injuries, the partial articular supraspinatus tendon avulsions, named "PASTA," have been identified by Snyder as a separated clinical entity.

9.3.2 Intratendinous Tears

Sometimes, in throwing athletes, the articular lesions can show an intratendinous expansion. These types of injuries have been identified for the first time by Yamanaka, Fukuda, and Conway. In particular, the latter founded the acronymous "PAINT" to precisely define partial-thickness tears with intratendinous extension. The leading mechanism of PAINT injuries onset has to be searched into the rotator cuff's five-layer histologic composition, which seems to influence the onset of intrinsic shear strengths.

9.3.3 Bursal-Sided Tears

Bursal-sided tears occur with higher incidence in the middle- and older-aged athletes [9]. These lesions are strictly linked to the subacromial impingement [10]. Furthermore, bursal tears can arise both as primary or secondary lesion. In this last case, the association with intra-articular or intratendinous cuff disease is often observed.

9.4 Classification

Rotator cuff tear classification is essential in the preoperative decision making. Obviously, the employing of this classification consents to evaluate the postoperative outcomes. First, Ellman classified the lesions considering their depth. He identified three different grades of tears: grade 1: <3 mm deep or 25%; grade 2: 3 to 6 mm deep or 50%; and grade 3: >6 mm deep or > 50% and tear area (in mm2). Subsequently, Snyder modified the classification system, including tear placement and severity. He divided tears into articular, bursal, or full-thickness and coined a scale from 0 to 4, ranging from normal to >3 cm severe cuff lesion.

9.5 Clinical Findings

Rotator cuff tear clinical presentation in athletes is widely changeable. Patients affected usually complain about moderate discomfort, more relevant during the throw, which implicates a reduction in throwing velocity. Not rarely, the pathology can have an abrupt onset, followed by a "pop" that indicates the potential tearing of the cuff or the labrum. This condition can develop without previous symptoms, or more frequently, as a worsening of prior painful symptoms [11]. Several other clinical findings include the reduction in upper arm strength, fatigue at the beginning of the activity, limited pitch velocity, loss of pitch location, instability, and restricted range of motion.

9.6 Physical Examination

Clinical examination of overhead athletes affected from rotator cuff tears is essential to make a differential diagnosis with other pathologies that commonly hit these patients, such as posterior capsular tightness, labral fraying or tearing, and SLAP tears. It is important to remember that in most cases, these clinical conditions coexist [12]. Physical evaluation of the rotator cuff is based on the research of Neer and Hawkins impingement signs, even if these are not exclusively indicative of rotator cuff disease [13]. Tenderness at the level of the supraspinatus insertion, the posterior glenohumeral joint capsule, the biceps tendon, and the acromioclavicular joint could be detectable through palpation. The examiner should assess every element of rotator cuff in terms of pain and force. The supraspinatus, the infraspinatus, and the subscapularis muscles should be included in the evaluation, remembering that to realize a correct assessment of the supraspinatus strength test, scapula stabilization is required. The glenohumeral internal rotation and external rotation are measured with the subject in the supine position. The examiner proceeds with scapular stabilization after that he kindly rotates the shoulder externally until the scapula starts to move, observing the range of rotation. In the same way, internal rotation is evaluated. A whole physical examination always supposes the comparison with the opposite shoulder. Usually, patients present an increased external rotation accompanied by a simultaneous limited internal rotation. Usually, patients present an increased external rotation accompanied by a simultaneous limited internal rotation. The internal rotation deficit is defined when a loss higher than 25 degrees of rotation occurs. A substantial part of the clinical examination is given by the scapular rhythm assessment. Any asymmetry needs to be accurately evaluated. Other relevant tests that should be included in throwing athlete's evaluation are the internal impingement sign modified relocation sign, and the internal rotation resistance test, useful to exclude the presence of a concomitant internal impingement condition. In any case, these tests result to be examiner dependent, and for this reason, their sensitivity and accuracy remain unclear [14]. At last, the examiner should pay attention on the AC joint, biceps tendon, labral complex (for SLAP injuries), and glenohumeral joint to exclude instability. The assessment of the cervical spine and the nervous and vascular structures surrounding completes the investigation of the upper limbs.

9.7 Radiological Evaluation

Imaging examinations are essential to confirm diagnosis. Plain radiographs are generally normal in patients affected by shoulder pain in which cuff lesions are not detectable. In case of blown rotator cuff tears, instead, different alterations have been described. Among these, the most frequent are type II or type III acromial morphology, greater tuberosity sclerosis, and cystic alterations [15]. The greater tuberosity changes usually occur in partial-thickness articular surface lesions in overhead athletes. The outlet view is essential to evaluate the acromial anatomical structure, and it becomes obligatory for the preoperative planning. However, the gold standard radiographic investigation for assessing rotator cuff tears is magnetic resonance imaging (MRI) even if the conventional MRI technology is not sufficient to distingue partial cuff tears from tendinosis and to evaluate the right extension of the injuries [16]. The introduction of the MR arthrography (MRA) has given an enormous contribution in the diagnosis and management of partial undersurface and insubstance cuff tears. The first-choice examination for patients affected by potential partial-thickness tear or labral disease is the MRA performed with the arm first abducted and then in external rotation [17]. In any case, the MRI finding interpretation results are complex: Often, as widely shown in literature, rotator cuff of the overhead athletes can present abnormal signal anomalies, despite the absence of symptoms. Current researches have demonstrated as MRI abnormalities in players after throwing return to baseline after 1 week, normalizing the MRI signals. Although it has been considered for a long time a highly sensitive and precise examination to evaluate rotator cuff, ultrasonography has an unsurpassed limitation: its operator dependence. Operator experience and abilities, in fact, can modify examination results. However, the current possibility to employ portable systems and to examine both shoulders dynamically has increased the interest in this radiological method as first choice to evaluate rotator cuff disease [18]. Ultrasound and MRI can be comparable in terms of specificity and sensitivity in the making diagnosis ability of full-thickness tears and the determination of muscle retraction and tear extensions. Anyway, MRI remains the more advantageous radiographic examination thanks to its specific ability to identify the labral tears.

9.8 Conservative Management

The choice of rotator cuff tear treatment is influenced by several different factors, which depend on both kind of patient and type of injury. Symptom severity, onset way, functional disability, response to treatment, and timing concerning season represent the athletes leading characteristics to consider in the making operative decision. On the other hand, tear size and classification, as well as the presence of concurrent shoulder pathologies, have to be widely considered. Finally, outcomes of previous investigations, procedures, and responses to prior treatment have to be involved in any management planning. However, nonoperative treatment remains the primary choice for throwing athletes with cuff rotator tears. The conservative management priority finds explanation in the high asymptomatic incidence of cuff tears in athletes' sample, the successful answer of athletes to nonoperative program, and the uncertain outcomes after surgical procedures. In several cases, in fact, athletes underwent surgery and do not manage to return to their prior level of physical functions. Conservative treatment in throwing athletes includes rest from throwing activity, use of nonsteroidal anti-inflammatory drugs, and rehabilitation program [19]. In case of posterior capsular contractures, stretching with the arm adducted and internally rotated, the so-called sleeper stretch, is required. Occasionally, subacromial corticosteroid injection could be considered. The conservative treatment duration depends on the severity of symptoms, personal player necessities, and extension of tears. Three months is usually a sufficient period for a full program. Sometimes patients need a longer rehabilitation program, especially if affected by a full-thickness tear.

9.9 Operative Management

Operative treatment is reserved for those athletes affected by partial- or full-thickness cuff tears in which conservative management has shown unsuccessful results. Nevertheless, it is essential to underline the possibility of surgical treatment failure, especially in the case of cuff repair [20]. Athletes, in fact, could not return to the previous level activity. This has been confirmed in a current study, which investigated high-level overhead athletes underwent arthroscopic SLAP repairs. Only 57% of the patients enrolled were able to restart sport at high levels. Consequently, conservative management should be exhausted before passing to the operative one. The latter presents different opportunities for partial- or full-thickness rotator cuff injuries. The arthroscopic cuff debridement and/or repair are the leading treatment options for partial cuff tears. Besides, a subacromial decompression and/or labral debridement or repair could be required. Obviously, operative procedures can be decided before surgery but usually are defined during arthroscopy. The choice suffers from the influence of several factors such as the patient age, tissue condition, tear depth, concurrent pathologies, and surgeon experience. Ordinarily, partial tears up to 75% are candidate for repairing [21].

9.10 Arthroscopic Debridement

Arthroscopic debridement is used to eliminate unstable flaps, smooth irregular borders, and allow evaluation of lesion profundity and length. Through the employing of a motorized shaver, pathologic tissue is removed from the side of articular cuff tears, recreating healthy margin [22]. The eventual presence of intratendinous tears (such as PAINT lesion) requires the elimination of unhealthy tissue to improve the healing process. After tissue debridement, cuff defect is repaired passing, with a spinal needle and a monofilament suture. At this point, the scope is retired from the glenohumeral joint. This kind of suture helps the assessment of the cuff on the corresponding bursal side. The subacromial space is evaluated to exclude the presence of subacromial impingement and bursal side damages [23]. Arthroscopic debridement of partial cuff tears has shown successful outcomes in nonthrowing athletes when lesion depth was up 50%. Current literature, however, lacks significant studies about arthroscopic debridement in throwers [24]. It was described that about 80% of high-level athletes underwent this procedure reported satisfactory outcome, while only 60% of them have returned to preinjury athletic activity.

9.11 Surgical Repair

Not always positive outcomes after arthroscopy debridement and the improvements in arthroscopic techniques have put on the foundation to consider partial cuff tears repair more frequently. Current guidelines suggest making debridement for tears <50% of the cuff's thickness and repairment for tears up 50%. This general recommendation finds its reason in the biomechanical rationale for which cuff tissue in proximity to partial tears has shown augmented pathologic loading when the damage was over than 50%. The final decision between debridement and repair should be essentially founded on the extension of partial tears, which would need an accurate system for depth evaluation. Unfortunately, a direct technique for this determination does not still exist. Furthermore, the presence of concurrent pathologies could influence the choice to make a repair. The athlete's age and position represent other critical decision factors. Throwing athletes older than 30 years affected by important partial-thickness cuff tears should be undergone to debridement alone, while surgical repair should be performed in younger pitchers or position players, considering the implications that complete functional recovery could have in their sportive careers [25]. Concern partial cuff tears up 75% and full-thickness cuff tears, repair may be performed in case of conservative treatment and/or debridement failure. The arthroscopic procedure rather than the open surgical approach for rotator cuff repair is considered the first management option for some advantages in overhead athletes' population. In essence, the lower risk of stiffness and the capacity to reproduce a more anatomically cuff footprint are widely described [26]. Recent researches have shown as pitchers undergoing miniopen cuff repair using a transosseous technique have had only a 12% chance of coming back to previous activity level, a minimal percentage if compared with the clinical outcomes of arthroscopic repair in the throwers.

9.12 Repair Techniques

The repair of full-thickness cuff tears can be performed employing both single and dual row methods, through arthroscopic or miniopen procedures. In case of partial tears, a transtendinous approach can be used. Sometimes, it could be necessary to complete the partial- to full-thickness tears repairing it subsequently. Many surgical procedures have been described for partial rotator cuff tear repair. Tear location and/or surgeon experience are the main factors, which influence the choice of the procedure [27]. Current studies have shown arthroscopic technique repair for partial-thickness bursal tears, which are generally transformed into full-thickness tears, and repaired utilizing suture anchors. The same procedure can be applied for articular-sided tears repair, even if they can be treated with a "transtendon" technique in which the articular-sided fibers are re-attached in their anatomic footprint. Intratendinous tears usually require suture plication of the delaminated layers, and subsequently a reattachment with suture anchors in their original footprint [28]. However, it is essential to consider the huge difficulty in recreating an attachment at the anatomic footprint in throwers' population. The repair could constrain athletes' shoulder, causing an obligate position of hyperabduction and external rotation, effectively altering their sportive ability. Anyway, although literature lacks a lot of studies in which high-level overhead athletes' outcomes have been assessed, arthroscopic repair results for partial and full-thickness tears in the common population are promising. On the other hand, this surgical choice is strengthened by few researches in which partial and full-thickness cuff tears have been repaired in throwers.

9.13 Treatment Algorithm

Notwithstanding enthusiasm for rotator cuff tears repair [29], the doubts about the advantage for the high-level athletes persist. The major unsolved problem remains to understand to which anatomy should be reset to normal in high-demand athletes. In fact, while the repair of the intratendinous cuff tissue could be advantageous, the progression of the articular tear toward the tuberosity could cause a joint overstress, reducing the muscle–tendon length of the cuff. On the other hand, nonanatomic repair using suture anchors risk made cuff insertion excessively medial, altering shoulder anatomy and biomechanics [30]. Throwing athletes' management represents an isolated field for rotator cuff repair because of its considerably higher relevance compared to the general population. In the decisional operative program, it is necessary to consider both the depth of the tear, and the depth and condition of the intratendinous portion. If the depth of the articular-sided tear is <75%, a debridement should be performed only. If the tear is >75%, transtendon repair should be made, considering addressing supraspinatus lesion first than infraspinatus ones. If the intratendinous segment is thin or < 1 cm, the surgeon should opt for debridement of the articular section only. If it is thick or exceeds 1 cm, a mattress intratendinous repair with or without an anchor should represent the first choice. Finally, if the depth of the intratendinous segment is 1 to 2 cm, arthroscopic repair should be the main option treatment. If it exceeds 2 cm, miniopen approach could be indicated, making repair with suture anchors.

References

1. Connor PM, Banks DM, Tyson AB, Coumas JS, D'Alessandro DF. Magnetic resonance imaging of the asymptomatic shoulder of overhead athletes: a 5-year follow-up study. Am J Sports Med. 2003;31:724–7. https://doi.org/10.1177/03635465030310051501.
2. Lesniak BP, Baraga MG, Jose J, Smith MK, Cunningham S, Kaplan LD. Glenohumeral findings on magnetic resonance imaging correlate with innings pitched in asymptomatic pitchers. Am J Sports Med. 2013;41:2022–7. https://doi.org/10.1177/0363546513491093.
3. Longo UG, Berton A, Papapietro N, Maffulli N, Denaro V. Epidemiology, genetics and biological factors of rotator cuff tears. Med Sport Sci. 2012;57:1–9. https://doi.org/10.1159/000328868.
4. Nho SJ, Yadav H, Shindle MK, Macgillivray JD. Rotator cuff degeneration: etiology and pathogenesis. Am J Sports Med. 2008;36:987–93. https://doi.org/10.1177/0363546508317344.
5. Mazzocca AD, Rincon LM, O'Connor RW, Obopilwe E, Andersen M, Geaney L, Arciero RA. Intra-articular partial-thickness rotator cuff tears: analysis of injured and repaired strain behavior. Am J Sports Med. 2008;36:110–6. https://doi.org/10.1177/0363546507307502.

6. Kvitne RS, Jobe FW, Jobe CM. Shoulder instability in the overhand or throwing athlete. Clin Sports Med. 1995;14:917–35.
7. Jobe CM. Posterior superior glenoid impingement: expanded spectrum. Arthroscopy. 1995;11:530–6. https://doi.org/10.1016/0749-8063(95)90128-0.
8. Longo UG, Huijsmans PE, Maffulli N, Denaro V, De Beer JF. Video analysis of the mechanisms of shoulder dislocation in four elite rugby players. J Orthop Sci. 2011;16:389–97. https://doi.org/10.1007/s00776-011-0087-6.
9. Clark JM, Harryman DT. Tendons, ligaments, and capsule of the rotator cuff. Gross and microscopic anatomy. J Bone Joint Surg Am. 1992;74:713–25.
10. Lohr JF, Uhthoff HK. The microvascular pattern of the supraspinatus tendon. Clin Orthop Relat Res. 1990;254:35–8.
11. Mazoué CG, Andrews JR. Repair of full-thickness rotator cuff tears in professional baseball players. Am J Sports Med. 2006;34:182–9. https://doi.org/10.1177/0363546505279916.
12. Park HB, Yokota A, Gill HS, El Rassi G, McFarland EG. Diagnostic accuracy of clinical tests for the different degrees of subacromial impingement syndrome. J Bone Joint Surg Am. 2005;87:1446–55. https://doi.org/10.2106/JBJS.D.02335.
13. MacDonald PB, Clark P, Sutherland K. An analysis of the diagnostic accuracy of the Hawkins and Neer subacromial impingement signs. J Shoulder Elb Surg. 2000;9:299–301. https://doi.org/10.1067/mse.2000.106918.
14. Zaslav KR. Internal rotation resistance strength test: a new diagnostic test to differentiate intra-articular pathology from outlet (Neer) impingement syndrome in the shoulder. J Shoulder Elb Surg. 2001;10:23–7. https://doi.org/10.1067/mse.2001.111960.
15. Pearsall AW, Bonsell S, Heitman RJ, Helms CA, Osbahr D, Speer KP. Radiographic findings associated with symptomatic rotator cuff tears. J Shoulder Elb Surg. 2003;12:122–7. https://doi.org/10.1067/mse.2003.19.
16. Nakagawa S, Yoneda M, Hayashida K, Wakitani S, Okamura K. Greater tuberosity notch: an important indicator of articular-side partial rotator cuff tears in the shoulders of throwing athletes. Am J Sports Med. 2001;29:762–70. https://doi.org/10.1177/03635465010290061501.
17. Parker BJ, Zlatkin MB, Newman JS, Rathur SK. Imaging of shoulder injuries in sports medicine: current protocols and concepts. Clin Sports Med. 2008;27:579–606. https://doi.org/10.1016/j.csm.2008.07.006.
18. Tuite MJ, Yandow DR, DeSmet AA, Orwin JF, Quintana FA. Diagnosis of partial and complete rotator cuff tears using combined gradient echo and spin echo imaging. Skelet Radiol. 1994;23:541–5. https://doi.org/10.1007/bf00223087.
19. Burkhart SS, Morgan CD, Kibler WB. The disabled throwing shoulder: spectrum of pathology part I: pathoanatomy and biomechanics. Arthroscopy. 2003;19:404–20. https://doi.org/10.1053/jars.2003.50128.
20. Tibone JE, Elrod B, Jobe FW, Kerlan RK, Carter VS, Shields CL, Lombardo SJ, Yocum L. Surgical treatment of tears of the rotator cuff in athletes. J Bone Joint Surg Am. 1986;68:887–91.
21. Neri BR, ElAttrache NS, Owsley KC, Mohr K, Yocum LA. Outcome of type II superior labral anterior posterior repairs in elite overhead athletes: effect of concomitant partial-thickness rotator cuff tears. Am J Sports Med. 2011;39:114–20. https://doi.org/10.1177/0363546510379971.
22. Esch JC, Ozerkis LR, Helgager JA, Kane N, Lilliott N. Arthroscopic subacromial decompression: results according to the degree of rotator cuff tear. Arthroscopy. 1988;4:241–9. https://doi.org/10.1016/s0749-8063(88)80038-0.
23. Andrews JR, Broussard TS, Carson WG. Arthroscopy of the shoulder in the management of partial tears of the rotator cuff: a preliminary report. Arthroscopy. 1985;1:117–22. https://doi.org/10.1016/s0749-8063(85)80041-4.
24. Reynolds SB, Dugas JR, Cain EL, McMichael CS, Andrews JR. Débridement of small partial-thickness rotator cuff tears in elite overhead throwers. Clin Orthop Relat Res. 2008;466:614–21. https://doi.org/10.1007/s11999-007-0107-1.
25. Van Kleunen JP, Tucker SA, Field LD, Savoie FH. Return to high-level throwing after combination infraspinatus repair, SLAP repair, and release of glenohumeral internal rotation deficit. Am J Sports Med. 2012;40:2536–41. https://doi.org/10.1177/0363546512459481.
26. Lo IK, Burkhart SS. Transtendon arthroscopic repair of partial-thickness, articular surface tears of the rotator cuff. Arthroscopy. 2004;20:214–20. https://doi.org/10.1016/j.arthro.2003.11.042.
27. Yoo JC, Ahn JH, Lee SH, Kim JH. Arthroscopic full-layer repair of bursal-side partial-thickness rotator cuff tears: a small-window technique. Arthroscopy. 2007;23:903.e901–4. https://doi.org/10.1016/j.arthro.2006.11.008.
28. Deutsch A. Arthroscopic repair of partial-thickness tears of the rotator cuff. J Shoulder Elb Surg. 2007;16:193–201. https://doi.org/10.1016/j.jse.2006.07.001.
29. Longo UG, Franceschi F, Spiezia F, Marinozzi A, Maffulli N, Denaro V. The low-profile Roman bridge technique for knotless double-row repair of the rotator cuff. Arch Orthop Trauma Surg. 2011;131:357–61. https://doi.org/10.1007/s00402-010-1203-3.
30. Brockmeier SF, Dodson CC, Gamradt SC, Coleman SH, Altchek DW. Arthroscopic intratendinous repair of the delaminated partial-thickness rotator cuff tear in overhead athletes. Arthroscopy. 2008;24:961–5. https://doi.org/10.1016/j.arthro.2007.08.016.

Elbow Injuries in Throwing Athletes

10

Luigi Adriano Pederzini, Matteo Bartoli, Andrea F. Cheli, and Anna Maria Alifano

10.1 Introduction: Anatomy and Biomechanics

The elbow is characterized by highly intrinsic congruity and stability. In normal conditions, elbow flexion in men ranges from 0° to 150°, whereas in women from hyperextension, 12–15° to 150°, and approximately 170° in pronation–supination. The functional range of motion consists of 30–130° in flexion–extension in order to perform activities of daily living and 20–130° for throwing patterns [1].

Elbow stability is strictly associated with static and dynamic constraints and could be compromised by repetitive exertion of the joint due to work or sport activities.

The elbow is the second most affected joint when considering the classification of major joint dislocation [2], and 15–35% of acute injuries may lead to degrees of instability [3, 4].

Static soft tissue stabilizers may involve the anterior and posterior joint capsule and the medial and LCL compounds.

Elbow stabilizing factor contribution to proper elbow kinematics and stability is strictly dependent on the degree of flexion–extension and forearm rotation.

In extension movements, the anterior capsule provides about 70% of the soft tissue restraint; in flexion, the main agent is the medial collateral ligament. In full extension, the ulnohumeral articulation, anterior joint capsule, and medial collateral ligament equally provide valgus stability. In a 74% flexion, the medial collateral ligament generates resistance. Essentially, the ulnohumeral articulation and the anterior joint capsule endure varus stress. In full extension, varus resistance is controlled equally by joint congruency (mainly the olecranon in olecranon fossa and lateral collateral ligament), which provides 55% of the stabilizing force, whereas increasing the flexion, its associated contribution increases to 75%. The radial collateral ligament provides minimal varus limitation, both in flexion (9%) and in extension (14%). In extension, the anterior capsule provides 85% of the resistance to dislocation. In flexion, the medial collateral ligament provides nearly 80% of resistance to dislocation.

Athletes involved in repetitive high-speed overhead movements and other motions entailing significant valgus stress (ex tennis players or baseball pitchers or volleyball players), experience tensile forces on their medial structures, compression forces on their lateral structures, and impingement forces in their posteromedial compartment. The valgus intensity on the ath-

L. A. Pederzini · M. Bartoli (✉) · A. F. Cheli
Nuovo Ospedale Civile di Sassuolo, Orthopedics and Arthroscopic Surgery Unit, Sassuolo (MO), Italy

A. M. Alifano
Campus Biomedico University, Orthopedics and Traumatology Unit, Rome, Italy

G. L. Canata et al. (eds.), *Management of Track and Field Injuries*,
https://doi.org/10.1007/978-3-030-60216-1_10

lete's elbow can rise up to 68 N. In pitching, the maximum pitch speed may considerably imply the risk of elbow injury. Surgery for UCL injuries has been shown to be essential for the pitchers with the highest maximum ball velocity [5]. Reduced elbow valgus torque is correlated with delayed trunk rotation, reduced shoulder external rotation, increased elbow flexion, and overhand pitching (vs sidearm delivery) [6]. The overhead gestures bring the shoulder in a maximized grade of abduction, decreasing the valgus force transmitted to the elbow compared with baseball pitching; this could explain the higher incidence of UCL injuries in baseball [7].

Baseball pitch can be divided into five main phases:

1. Windup: Initial preparation as the elbow flexes and a slight pronation of the forearm take place.
2. Early cocking: The ball is thrown with the hand covered with the glove and is complete when the forward foot comes in contact with the ground. Shoulder abduction and external rotation are initiated in this stage.
3. Late cocking: Further shoulder abduction and maximal external rotation. Moreover, the elbow flexes between 90° and 120° and the forearm pronates to 90°.
4. Rapid acceleration: It produces a large forward-directed force on the extremity, accompanied by rapid elbow extension.
5. Ball release and Follow Through: Dissipation of all excess kinetic energy as the elbow reaches full extension and completes the movement.

Multiple biomechanical studies have shown that the elbow extends over 2300°/s during the throwing cycle. This generates a medial sheer force of approximately 300 N and a lateral compressive force of nearly 900 N. Maximum valgus force along the elbow is generated during late cocking and acceleration. The elbow is flexed to 95 degrees and undergoes valgus forces up to 64 Nm. During ball release, the lateral part of the elbow undergoes greater than 500 N of force. These extreme medial and lateral forces can cause injuries that can endanger the career of the throwing athlete. Tremendous valgus stress is generated over the medial part of the elbow during the acceleration phase, the majority of which is conducted to the anterior bundle of the UCL. The remaining stress is scattered by the secondary supporting structures of the medial elbow, mainly the flexor–pronator musculature.

These extraordinary forces generated on the elbow joint by the overhead athlete result in elbow injury. The typical pattern of injury is either caused by repetitive microtrauma or chronic stress overload.

Chronic traction forces on the UCL may thicken the ligament and produce osteophytes at its proximal insertion. Repetitive valgus stress can strain the flexor–pronator muscle mass, as dynamic stabilizer to valgus stress, exposing the UCL to additional stress and potentially triggering UCL attenuation, stretching, or even rupture [8]. Conway in 1992 [9] studied a cohort of athletes describing the incidence of different UCL injury tear patterns: 87% were torn at the midsubstance, 10% were distal tears from the ulna, and 3% were torn proximally from the medial epicondyle. Stress on the medial elbow can also lead to medial epicondylitis in these athletes. Likewise, the medial dynamic stabilizers undergo further stress due to stretching or rupture of the UCL exposes, causing injury of the structures. Several other elbow pathologies are associated with UCL traumas. Valgus forces may cause traction neuritis of the ulnar nerve, which can be exacerbated by an inadequate UCL [10]. Furthermore, many baseball players have increased cubitus valgus and flexion contractures, placing greater strain on the UCL. Chronic UCL injuries also lead to thickening and calcifications of the UCL, which is the base of the cubital tunnel. Traction on the UCL may give rise to marginal osteophytes at the medial joint. Posteromedial olecranon osteophytes may develop due to valgus extension overload. All of these anatomic changes produce a restricted, irritable environment for the ulnar nerve. Valgus torque leads to lateral elbow compressive forces of approximately 500 N between the radial head and humeral capitellum exacerbated by UCL laxity, potentially triggering avascular necrosis, osteochondritis dissecans, chondral wear, or osteochondral chip fractures [11]. The posteromedial elbow is compromised because traction osteophytes on the olecranon and hypertrophy of the distal humerus (which decreases the size of the olecranon fossa) cause repetitive posteromedial impingement, possibly causing osteophytes, chondromalacia, and/or loose bodies [12].

Young athletes with open physes differ from adult athletes because their medial epicondylar apophysis is the weakest link on the medial side of the elbow. Repetitive valgus stress and tension overload of medial structures may result in "Little League elbow," a general term encompassing medial epicondylar tearing, medial epicondylar apophysitis, and increased apophyseal growth with delayed closure of the epicondylar growth plate. Furthermore, the M-UCL proximal insertion is mildly attached to the humerus through the physis. Due to valgus stress, the flexibility of the physis causes increased elbow valgus, increasing the radiocapitellar load and potentially leading to osteochondritis dissecans. UCL injuries are uncommon in the skeletally immature athlete but seem to be more frequently detected in older ones [13, 14].

10.2 Diagnosis

UCL injuries can be divided into acute, chronic, or acute on chronic. It is worth outlining the history of acute traumatic events affecting the elbow. Subjects with an acute UCL tearing typically refer to sudden onset of pain, often accompanied by a popping sensation, during a particular moment of throw. Some report inability to throw following an injury. Overuse can cause chronic valgus instability due to attenuation or complete rupture of the UCL. Athletes describe gradual onset of medial elbow pain or discomfort in throwing movements, particularly in the late cocking and acceleration phases. They may refer to decreased speed, distance, and accuracy of their pass or throw. They may complain of recurrent episodes of elbow pain treated with conservative management. Patients with chronic valgus instability may also report a sudden episode of giving way or severe elbow pain, most likely due to the rupture of a previously attenuated UCL. Athletes with chronic UCL injuries can often throw, but typically regain less than 60% to 80% of their preinjury maximal velocity.

A history of associated elbow pathologies must also be investigated. For example, loose bodies may be accompanied by mechanical symptoms. Ulnar neuritis symptoms may include medial elbow pain radiating down the ulna to the hand and tingling in the ulnar two digits. Athletes in particular may experience clumsiness or heaviness of the hand and fingers associated with, and often exacerbated by, throwing or overhead activities. Ulnar neuritis can occur in both acute and chronic UCL injuries. In acute injuries, the nerve may be irritated by hemorrhage and edema. In chronic injuries, valgus instability makes the ulnar nerve susceptible to higher tensile stress, and UCL scarring may cause a decrease in the cubital tunnel space. Symptoms could occur at first only following (physical) activity but, in time, may persist even with rest.

10.2.1 Physical Examination and Imaging

Inspection, palpation, and assessment of range of motion are the typical first passages to perform. The arm must be evaluated from the hand to the shoulder. It is essential to analyze the scapulothoracic area because a proximal alteration can alter the throwing motion. Local pain can be felt at the flexor–pronator epicondyle and subtle differences in the extent of pain may be detected between the medial epicondylitis and M-UCL lesion. The second one is 2 cm distal from the medial epicondyle. Both factors could be present. Many tests are performed to assess the valgus instability pattern; this examination produces a smaller difference than varus stress when lateral structures are injured. Even the most experienced surgeons may encounter difficulties to perfectly detect the lesion.

Conventionally, in the elbow abduction stress test, the humerus is stabilized and the elbow undergoes a valgus stress at about a 20 to 30 elbow flexion. In a positive test, there is no firm endpoint and the articular surfaces of the ulna and medial humeral condyle move apart and the forearm sways out laterally.

When testing the posterior band of the AOL, the milking maneuver involves producing a valgus force by pulling the patient's thumb with forearm supination, shoulder extension, and elbow flexion beyond 90°. Patients with a UCL injury may refer to a feeling of anxiety, instability, and medial joint pain. A modified version of this test has been described by Marc Safran [7]: The patient abducts and externally rotates the arm to be examined. The examiner places a hand on the elbow to be investigated in order to stabilize the elbow and palpate the medial joint line for medial joint gapping and for the endpoint quality. The patient flexes the examined elbow to 70°, and the examiner exerts valgus stress by pulling the ipsilateral thumb down. The examiner assesses medial joint laxity (gapping) and quality of endpoint and records pain following valgus stress. The test is repeated on the contralateral elbow for comparison.

In the moving valgus stress test as described by O'Driscoll and colleagues [15], the patient's shoulder is held at 90° of abduction and external rotation. The examiner applies and maintains a constant moderate valgus torque to the fully flexed elbow and afterward quickly extends the elbow. In case of positivity, the patient complains of maximal medial elbow pain between 120 and 70 of elbow flexion.

Magnetic resonance imaging (MRI) is 57–79% sensitive and 100% specific for UCL tears [16]. A magnetic resonance (MR) arthrogram is 97% sensitive for UCL tears although occasionally difficult to obtain in acute phases. Ultrasound imaging can be a useful instrument to dynamically detect valgus instability compared to the opposite arm. To date, standard X-rays are the first imaging step in detecting heterotopic ossifications and articular degeneration.

10.3 Treatment of M-UCL Lesions in Athletes

Rettig and collegues [17] described a conservative approach allowing up to 42% of athletes to resume previous sports levels. The study did not elicit the prognostic factors influencing the success.

Acute proximal lesions in athletes may be handled successfully by M-UCL repair using a 3.5 or 4.5 anchor or trans-osseous sutures at the medial epicondyle [9].

It is essential to perform surgery promptly and use effective techniques to accurately reach the physiological M-UCL insertion area at the humerus in order to avoid biomechanical impairment. Recent literature suggests that using a fiber tape between two swive locks as an augmentation could allow athletes to resume their sports activity quicker on the field [18].

Reconstruction procedures seem to play a pivotal role in all other conditions, particularly in chronic or "acute on chronic" patterns.

Jobe and collegues [19] first described an M-UCL reconstruction technique using an autologous palmaris longus tendon transplantation. In the original technique, called "Tommy John," the muscle mass of the flexor–pronator muscles was raised and the ulnar nerve was permanently placed underneath the muscle tissue. However, Conway and Jobe [9] have reported a high incidence of ulnar neuropathies when using this technique (21%) requiring a subsequent ulnar nerve decompression in more than half of the affected subjects.

Smith and collegues [20] described a technique ("modified Jobe") only splitting the flexor–pronator mass, using a different positioning of the bone tunnels, and tensioning the neo-ligament with a supine forearm, flexed at 60 ° and in varus stress. The transposition of the nerve was not required. Andrews in 1995 [21] described a similar technique, by simply lifting the mass of the flexor–pronators, without incising it, thus reducing procedure invasiveness.

The "docking technique," described by Althcheck and Rohrbough simplified the procedure, the tensioning and fixation of the ligament achieving outstanding results in 92% of the cases with a complication rate of 5.5% [22].

Ruland and collegues [23] compared three different surgical techniques employed in the reconstruction of the collateral ulnar ligament. By investigating the resistance to torsion strength, in groups that use the palmaris longus, the torsion strength is statistically lower compared to the native ligament. On the other hand, in the group that uses the semitendinosus tendon, the score is significantly higher compared to the other two groups.

Thompson [24] found an 82% rate of excellent results on 33 follow-up patients after reconstructive surgery with the modified Jobe technique.

Dodson [25] found 90% excellent results (out of 100 patients operated with the same technique) with a 3% complication rate (out of 100 patients operated with the same technique). Similarly, on a sample of 12 patients operated with the "docking technique" and 8 patients with the modified technique, Koh and collegues [26] found 95% excellent results with a complication rate of 5%, without notable/relevant differences in the two groups.

Hechtman [27] proposed a hybrid technique by using anchors for the reconstruction of the UCL; through this study, performed on cadavers, the authors have reported that this technique allows a reliable anatomical reconstruction of the UCL. Long-term results [28] have shown remarkable findings in 85% of the cases on 34 operated patients, with a complication rate of 3%.

Chang [29] published a bibliographic review on the various reconstruction procedures of the ulnar collateral ligament, comparing the Jobe technique (both traditional and modified), the "docking technique," and alternative techniques, indicating the docking technique as the method reporting greater solidity and muscular split with the highest sparing of the ulnar nerve as the best surgical practice.

Another suggested technique is the DANE TJ hybrid technique [30] in which a single ulnar tunnel is shaped at the sublime tubercle in which the transplant is fixed with a screw, while, at the humeral level, it is fastened with the traditional "docking technique." The advantage of this technique is the lower percentage of fractures of the ulnar tunnel using the interference screw. Through this procedure, Dines [31] reported, on a total of 22 patients, 86% excellent results and 18% complications. Lastly, Savoie [32] described the results of a retrospective study on the short-term results of 116 patients undergoing UCL reconstruction with semitendinosus allograft: The result was excellent in 80% of the patients, although the complication rate was 6%.

Most of the numerous reviews proposed in literature compare the several reconstructive techniques. Vitale and Ahmad [33] found that 83% of the patients operated with the Jobe technique or the "docking technique" reached the same level of preoperative activities. Furthermore, a muscular split approach has led to a 17% rise in excellent results, also supported by the fact that ulnar nerve transposition is not required. Additionally, the same meta-analysis shows that the results of the "docking technique" are better compared to the outcomes reported by the Jobe technique. In another quite recent review of the literature, Watson [34] has compared the clinical and biomechanical results of all techniques, as well as the Jobe technique, the "docking technique," the fixation with the interference screw and with Endobutton®. The authors observed a resuming of sports activities in 79% of the cases, while the "docking technique" presented the lowest percentage of complications. From a biomechanical point of view, they also noted that, in the docking technique and Endobutton® procedure, the main cause of failure was associated with suture failure, whereas the tunnel fracture was the main cause of failure in the Jobe technique. Lastly, in the screw fixing procedure, the cause of the failure is mainly due to the graft itself.

Complications of M-UCL reconstruction surgery are rare. A serious injury, as in the case of palmaris longus tendon samples, may affect the median nerve, which has also been reported [35, 36].

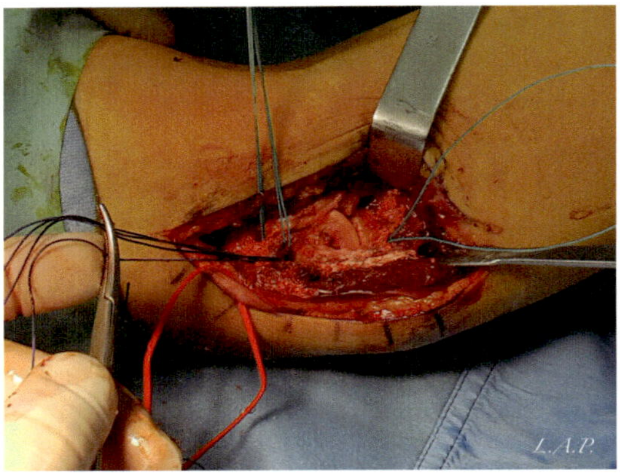

Fig. 10.1 On the left, the ulnar blind tunnel, on the right, the humeral convergent passages (one is *dark blue* and one l*ight blue*); the *red* one is the isolated ulnar nerve

Our standard treatment in active patients, particularly in athletes, is the anatomical reconstruction of the M-UCL [37].

The autologous hamstring is harvested from the ipsilateral knee and prepared with a Krackow suture at the two ends. It has to fit in a 4.5-mm caliper.

A medial, muscle splitting approach is performed. A 7-mm drill hole is made at the sublime tubercle toward the lateral and posterior cortex of the ulna. The graft is folded over onto itself and fixed with a bio-absorbable 6-mm interference screw. An additional 7-mm tunnel is prepared at the humeral side. It is a blind tunnel and positioned from anterior to superior preserving the ulnar nerve. Two supplementary tunnels, 4.5 mm in diameter, are prepared independently converging on the 7 mm tunnel (Fig. 10.1).

A soft suture passer is used to handle separately the "two bundles" of the graft through the common tunnel and dividing them into smaller ones (Fig. 10.2).

The residual part of every bundle is sutured on itself after proper tensioning: The anterior bundle is tensioned at 30° of elbow flexion and the posterior bundle at 80° of elbow flexion. A "cycling" of the elbow takes place before this phase to improve the settling of the tendons into the tunnels and provide pre-tensioning.

Isolation of the ulnar nerve is performed, but no anteposition is required.

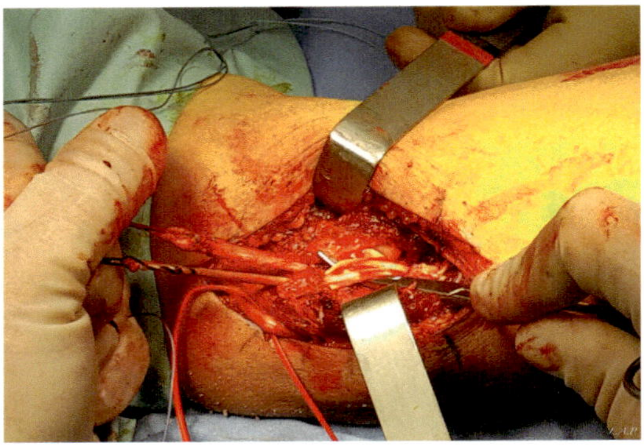

Fig. 10.2 Passage of the graft: on the right the two bundles in their independent tunnels

10.3.1 Post-op Protocol

Rehabilitation following surgical reconstruction of the UCL begins with range of motion and initial protection of the reconstruction, along with resistive exercises to strengthen the shoulder and core. This is followed by progressive exercises for resistive exercise to fully restore strength and muscular endurance in order to ensure a safe resuming of sport and overhead functional activities.

The early phases of postoperative care for UCL reconstructions involve specific time frames, limitations, and preventive measures to protect healing tissues and the surgical fixation/fastening.

The knee is maintained in full extension for 2 weeks, and the patient is allowed to bear weights as tolerated with or without two crutches (pes anserinus donor site protection).

The later stages of rehabilitation are presented in a criterion-based progression, according to which progression to subsequent levels is based on strength and control.

The resuming of competitive sports will take 6–10 months. Patients should apply ice packs on the elbow for 10–15 min after each rehab session to decrease pain and post-op swelling.

Clinical good outcomes (Fig. 10.3) [37] indicate that it is a reliable technique with a reduced incidence of complications. Resuming sports is reported as consistently successful.

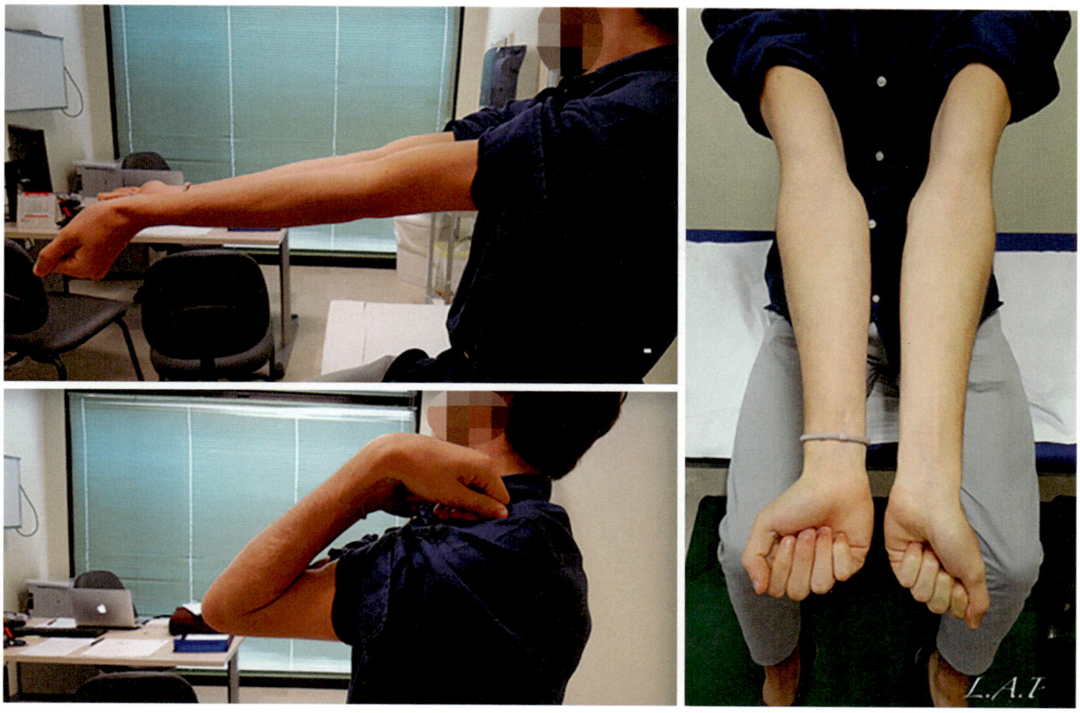

Fig. 10.3 Clinical outcome at the right elbow at mid-term FU

References

1. Morrey BF, Askew LJ, Chao EY. A biomechanical study of normal functional elbow motion. J Bone Joint Surg Am. 1981;63:872–7.
2. Safran MR, Baillargeon D. Soft-tissue stabilizers of the elbow. J Shoulder Elbow Surg. 2005;14(1 Suppl S):179S–85S. https://doi.org/10.1016/j.jse.2004.09.032.
3. Mehlhoff TL, Noble PC, Bennett JB, Tullos HS. Simple dislocation of the elbow in the adult. Results after closed treatment. J Bone Joint Surg Am. 1988;70:244–9.
4. Murthi AM, Keener JD, Armstrong AD, Getz CL. The recurrent unstable elbow: diagnosis and treatment. Instr Course Lect. 2011;60:215–26.
5. Bushnell BD, Anz AW, Noonan TJ, Torry MR, Hawkins RJ. Association of maximum pitch velocity and elbow injury in professional baseball pitchers. Am J Sports Med. 2010;38(4):728–32. https://doi.org/10.1177/0363546509350067.
6. Aguinaldo AL, Chambers H. Correlation of throwing mechanics with elbow valgus load in adult baseball pitchers. Am J Sports Med. 2009;37(10):2043–8.
7. Hariri S, Safran MR. Ulnar collateral ligament injury in the overhead athlete. Clin Sports Med. 2010;29(4):619–44.
8. Ciccotti MC, Schwartz MA, Ciccotti MG. Diagnosis and treatment of medial epicondylitis of the elbow. Clin Sports Med. 2004;23(4):693–705. https://doi.org/10.1016/j.csm.2004.04.011.
9. Conway JE, Jobe FW, Glousman RE, et al. Medial instability of the elbow in throwing athletes. Treatment by repair or reconstruction of the ulnar collateral ligament. J Bone Joint Surg Am. 1992;74(1):67–83.
10. Aoki M, Takasaki H, Muraki T, Uchiyama E, Murakami G, Yamashita T. Strain on the ulnar nerve at the elbow and wrist during throwing motion. J Bone Joint Surg Am. 2005;87(11):2508–14. https://doi.org/10.2106/JBJS.D.02989.
11. Takahara M, Shundo M, Kondo M, Suzuki K, Nambu T, Ogino T. Early detection of osteochondritis dissecans of the capitellum in young baseball players. Report of three cases. J Bone Joint Surg Am. 1998;80(6):892–7. https://doi.org/10.2106/00004623-199806000-00014.
12. Wilson FD, Andrews JR, Blackburn TA, McCluskey G. Valgus extension overload in the pitching elbow. Am J Sports Med. 1983;11(2):83–8. https://doi.org/10.1177/036354658301100206.
13. Chen FS, Diaz VA, Loebenberg M, Rosen JE. Shoulder and elbow injuries in the skeletally immature athlete. J Am Acad Orthop Surg. 2005;13(3):172–85. https://doi.org/10.5435/00124635-200505000-00004.
14. Rudzki JR, Paletta GA Jr. Juvenile and adolescent elbow injuries in sports. Clin Sports Med.

2004;23(4):581–608., ix. https://doi.org/10.1016/j.csm.2004.05.001.
15. O'Driscoll SW, Lawton RL, Smith AM. The "moving valgus stress test" for medial collateral ligament tears of the elbow. Am J Sports Med. 2005;33(2):231–9. https://doi.org/10.1177/0363546504267804.
16. Timmerman LA, Schwartz ML, Andrews JR. Preoperative evaluation of the ulnar collateral ligament by magnetic resonance imaging and computed tomography arthrography. Evaluation in 25 baseball players with surgical confirmation. Am J Sports Med. 1994;22(1):26–31. https://doi.org/10.1177/036354659402200105.
17. Rettig AC, Sherrill C, Snead DS, Mendler JC, Mieling P. Nonoperative treatment of ulnar collateral ligament injuries in throwing athletes. Am J Sports Med. 2001;29(1):15–7. https://doi.org/10.1177/03635465010290010601.
18. Wilson WT, Hopper GP, Byrne PA, MacKay GM. Repair of the ulnar collateral ligament of the elbow with internal brace augmentation: a 5-year follow-up. BMJ Case Rep. 2018;11(1):e227113. https://doi.org/10.1136/bcr-2018-227113.
19. Jobe FW, Stark H, Lombardo SJ. Reconstruction of the ulnar collateral ligament in athletes. J Bone Joint Surg Am. 1986;68:1158–63.
20. Smith GR, Altchek DW, Pagnani MJ, Keeley JR. A muscle-splitting approach to the ulnar collateral ligament of the elbow. Neuroanatomy and operative technique. Am J Sports Med. 1996;24:575–80. https://doi.org/10.1177/036354659602400503.
21. Andrews JR, Timmerman LA. Outcome of elbow surgery in professional baseball players. Am J Sports Med. 1995;23:407–13. https://doi.org/10.1177/036354659502300406.
22. Rohrbough JT, Altchek DW, Hyman J, Williams RJ III, Botts JD. Medial collateral ligament reconstruction of the elbow using the docking technique. Am J Sports Med. 2002;30:541–8. https://doi.org/10.1177/03635465020300041401.
23. Ruland RT, Hogan CJ, Randall CJ, Richards A, Belkoff SM. Biomechanical comparison of ulnar collateral ligament reconstruction techniques. Am J Sports Med. 2008;36:1565–7. https://doi.org/10.1177/0363546508319360.
24. Thompson WH, Jobe FW, Yocum LA, Pink MM. Ulnar collateral ligament reconstruction in athletes: muscle-splitting approach without transposition of the ulnar nerve. J Shoulder Elb Surg. 2001;10:152–7. https://doi.org/10.1067/mse.2001.112881.
25. Dodson CC, Thomas A, Dines JS, Nho SJ, Williams RJ III, Altchek DW. Medial ulnar collateral ligament reconstruction of the elbow in throwing athletes. Am J Sports Med. 2006;34:1926–32. https://doi.org/10.1177/0363546506290988.
26. Koh JL, Schafer MF, Keuter G, Hsu JE. Ulnar collateral ligament reconstruction in elite throwing athletes. Arthroscopy. 2006;22:1187–91. https://doi.org/10.1016/j.arthro.2006.07.024.
27. Hechtman KS, Tjin-A-Tsoi EW, Zvijac JE, Uribe JW, Latta LL. Biomechanics of a less invasive procedure for reconstruction of the ulnar collateral ligament of the elbow. Am J Sports Med. 1998;26:620–4. https://doi.org/10.1177/03635465980260050401.
28. Hechtman KS, Zvijac JE, Wells ME, Botto-van BA. Long-term results of ulnar collateral ligament reconstruction in throwing athletes based on a hybrid technique. Am J Sports Med. 2011;39:342–7. https://doi.org/10.1177/0363546510385401.
29. Chang ES, Dodson CC, Ciccotti MG. Comparison of surgical techniques for ulnar collateral ligament reconstruction in overhead athletes. J Am Acad Orthop Surg. 2016;24:135–49. https://doi.org/10.5435/JAAOS-D-14-00323.
30. Conway JE. The DANE TJ procedure for elbow medial ulnar collateral ligament insufficiency. Tech Shoulder Elbow Surg. 2006;7:36–43. https://doi.org/10.1097/00132589-200603000-00005.
31. Dines JS, ElAttrache NS, Conway JE, Smith W, Ahmad CS. Clinical outcomes of the DANE TJ technique to treat ulnar collateral ligament insufficiency of the elbow. Am J Sports Med. 2007;35:2039–44. https://doi.org/10.1177/0363546507305802.
32. Savoie FH III, Morgan C, Yaste J, Hurt J, Field L. Medial ulnar collateral ligament reconstruction using hamstring allograft in overhead throwing athletes. J Bone Joint Surg Am. 2013;95:1062–6. https://doi.org/10.2106/JBJS.L.00213.
33. Vitale MA, Ahmad CS. The outcome of elbow ulnar collateral ligament reconstruction in overhead athletes: a systematic review. Am J Sports Med. 2008;36:1193–205. https://doi.org/10.1177/0363546508319053.
34. Watson JN, McQueen P, Hutchinson MR. A systematic review of ulnar collateral ligament reconstruction techniques. Am J Sports Med. 2014;42:2510–6. https://doi.org/10.1177/0363546513509051.
35. Vastamäki M. Median nerve as free tendon graft. J Hand Surg Br. 1987;12:187–8. https://doi.org/10.1016/0266-7681(87)90010-6.
36. Weber RV, Mackinnon SE. Median nerve mistaken for palmaris longus tendon: restoration of function with sensory nerve transfers. Hand (N Y). 2007;2:1–4. https://doi.org/10.1007/s11552-006-9011-5.
37. Bartoli M, Pederzini LA, Severini G, Serafini F, Prandini M. Elbow medial ulnar collateral ligament chronic isolated insufficiency: anatomical M-UCL reconstruction technique and clinical experience in a mid-term follow-up. Musculoskelet Surg. 2018;102(Suppl 1):75–83. https://doi.org/10.1007/s12306-018-0559-3.

Wrist Injuries in Throwers

11

Margaret Woon Man Fok and Gregory I. Bain

11.1 Background

Among all injuries, wrist injuries are not common in throwers, as in shoulder and elbow. Yet for that reason, they are often overlooked as minor wrist sprains and are delayed in presentation.

The throwing motion involves a kinetic chain of events, which include the transfer of energy from the lower extremity, to the trunk, to the shoulder and elbow, and finally to the hand and wrist, before the release. Based on the size and weight of the object, e.g., a ball, a shot put, a javelin, or a discus, which the athlete is holding, the gripping mechanism, and the biomechanisms of the throwing action, different injuries may be sustained at the wrist [1]. Moreover, these injuries are not usually arisen from a single incident. Instead, they are often caused by repetitive motion. For example, a javelin is held with a circular grip with the primary load across the wrist being radial to ulnar. Athletes may complain of ulnar wrist pain after repetitive throws. A shot put is held deep in the hand leading to wrist and finger extension during cocking and acceleration phase. Athletes are susceptible to carpal tunnel syndrome due to repetitive gripping and wrist and hand sprains due to the weight of the shot and the wrist extension required to throw. A hammer and a discus create distraction loads across the wrist, but the fingers maintain a flexed position until release. As a result, athletes are prone to de Quervain's tenosynovitis and wrist sprains involving either triangular fibrocartilage complex (TFCC) or extensor carpi ulnaris tendon (ECU).

M. W. M. Fok (✉)
Department of Orthopaedics and Traumatology, Queen Mary Hospital, The University of Hong Kong, Hong Kong, Hong Kong

G. I. Bain
Department of Orthopaedic Surgery, Flinders University, and Flinders Medical Centre, Adelaide, Australia
e-mail: admin@gregbain.com.au

11.2 De Quervain's Disease

De Quervain's disease (DQV) is stenosing tenosynovitis involving the first extensor compartment, namely abductor pollicis longus (APL) and extensor pollicis brevis (EDB). It is the consequence of shear microtrauma from repetitive gliding of the two tendons beneath the sheath of the first compartment over the radial styloid. In throwers, it is caused by repeated gripping motion, e.g., in a discus throw when the wrist "snap" at the time of release or in a hammer throw.

Athletes present with radial wrist pain especially when the wrist is put in ulnar deviation (Fig. 11.1). Tenderness over the first extensor compartment is noted. Eichhoff test which is often mistaken as Finkelstein test is pathognomonic in

G. L. Canata et al. (eds.), *Management of Track and Field Injuries*,
https://doi.org/10.1007/978-3-030-60216-1_11

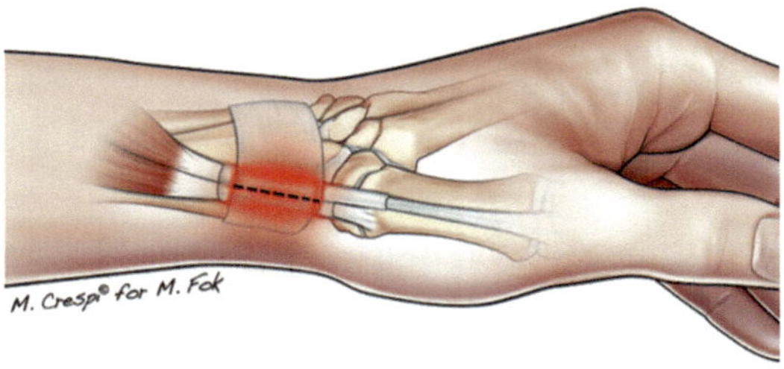

Fig. 11.1 De Quervain's disease—the stenosis tenosynovitis of the first extensor compartment—abductor pollicis longus and extensor pollicis brevis (copyright Dr. Margaret Fok)

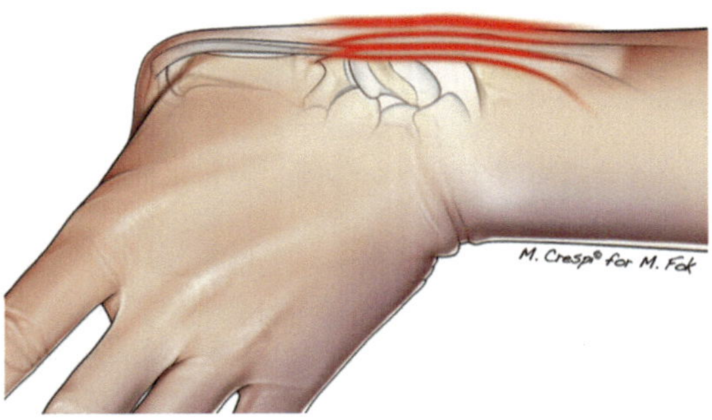

Fig. 11.2 Eichhoff test (copyright Dr. Margaret Fok)

making the diagnosis [2, 3]. It involves putting the thumb into the palm while the examiner is putting the fisted wrist in ulnar deviation (Fig. 11.2). Significant pain is elicited over the first compartment. The diagnosis is reinforced when no or minimal pain can be elicited when the thumb is not in the palm during the wrist movement. Imaging is only used to rule out other diagnoses, e.g., missed scaphoid fracture or nonunion.

The initial treatment is rest, nonsteroidal anti-inflammatory medications (NSAIDs) and immobilization with forearm-based splint [4, 5]. Physiotherapy including cryotherapy, ultrasound, and iontophoresis may be given. Kinesio taping with differential tensioning is popular in the management of DQV in recent years. In a prospective randomized control trial between physiotherapy and Kinesio taping, Kinesio taping showed a more favorable outcome than physiotherapy [6]. Yet, corticosteroid injection remains the most successful nonsurgical treatment modality with a 62–100% success rate [7–9.

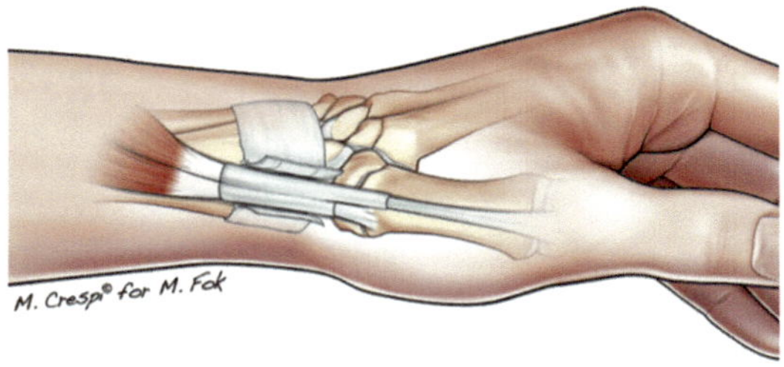

Fig. 11.3 Surgical release of first extensor compartment (copyright Dr. Margaret Fok)

Surgical release is indicated for patients who failed conservative treatment. It can be done under local anesthesia and involves the release of sheath covering the first extensor compartment, together with the sub-sheath that separates the EPB and APL tendons when present (Fig. 11.3). It is important to note that multiple slips of APL can be present. Failure to release all sub-sheaths may lead to recurrence or residual symptoms.

11.3 Extensor Carpi Ulnaris Tendinitis

Extensor carpi ulnaris tendinitis is a chronic inflammation of the ECU tendon, which is located at the most ulnar extensor compartment of the wrist (i.e., sixth extensor compartment) (Fig. 11.4) It results from chronic loading of the tendon, due to repetitive flexion and extension of the wrist, particularly in supination. Patients complain of swelling and constant dull ache on the dorso-ulnar aspect of the wrist. Sudden searing pain can also be felt along the ECU tendon on active contraction of the muscle. The ECU synergy test is a sensitive and specific test for ECU tendinitis. This test is performed with the wrist in supination [10]. Pain is felt over the ECU tendon when the examiner grasps the patient's extended thumb and middle finger while asking the patient to radially deviate his thumb against resistance (Fig. 11.5). ECU tendinitis is a clinical diagnosis. In case of uncertainty, patient may undergo ultrasound and magnetic resonance imaging (MRI) to look for thickened tendon with increased fluid in the surrounding sheath. Ultrasound can also con-

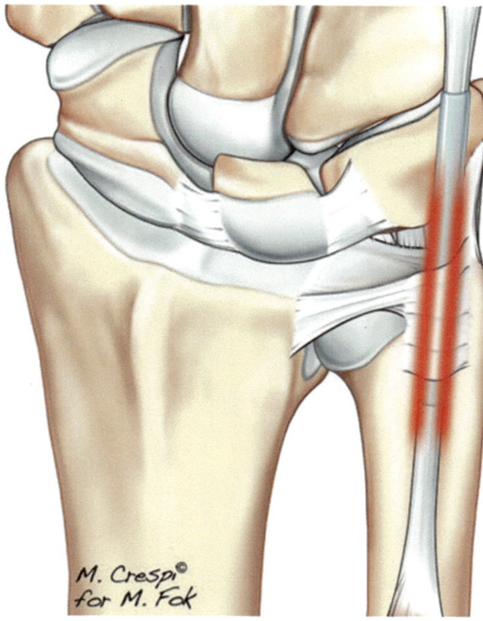

Fig. 11.4 ECU tendonitis (copyright Dr. Margaret Fok)

comitantly assess the stability of the ECU tendon (refer to ECU subluxation).

In suspected cases of ECU tendinitis, the status of TFCC needs to be evaluated. ECU tendinitis and TFCC tear may present with ulnar wrist pain in a similar fashion. Moreover, the presence of TFCC tear can lead to ECU tendonitis. If TFCC tear is not diagnosed and addressed in the management plan, patient may have persisted symptoms (refer to the TFCC section).

Management of ECU tendonitis is usually conservative with rest, NSAID, and short-arm splint to maintain the wrist at 30^0 extension and ulnar deviation for 3 weeks [4]. Ultrasound, iontophoresis, and Kinesio tape may also be used. Surgical debridement of the tendon and release of the compartment are usually not necessary and can be performed under local or regional anesthesia (Fig. 11.6).

11.4 Extensor Carpi Ulnaris Subluxation

Unlike extensor carpi ulnaris tendinitis, ECU subluxation is usually a result of a traumatic event, with forced supination, palmar flexion, and ulnar deviation. Yet, it is often missed [5]. With

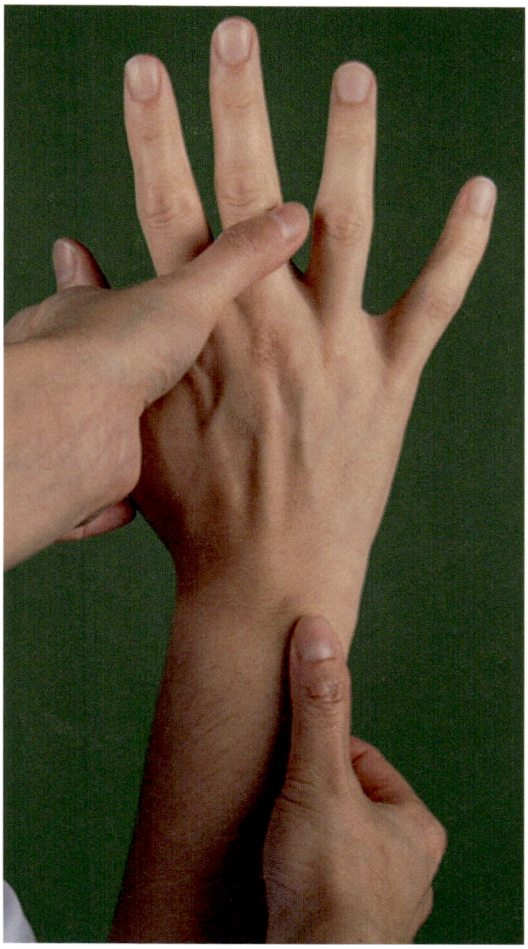

Fig. 11.5 ECU synergy test (copyright Dr. Margaret Fok)

subsequent repetitive stress on the wrist during the throwing motion, it may lead to symptomatic recurrent ECU subluxation, presented as painful snapping of the tendon during wrist rotation.

Anatomically, ECU tendon courses through the sixth extensor compartment in the wrist where it is held tightly to the ulnar groove by a sub-sheath, which is separated from the extensor retinaculum. In ECU tendon subluxation, it is the only sub-sheath that is torn. The extensor retinaculum remains intact.

Patient is presented with dorsal ulnar wrist pain and complains of clicking or snapping in pronation and supination. On physical examination, subluxation of the tendon can be elicited by active forearm supination and ulnar deviation (Fig. 11.7). The tendon is reduced with forearm

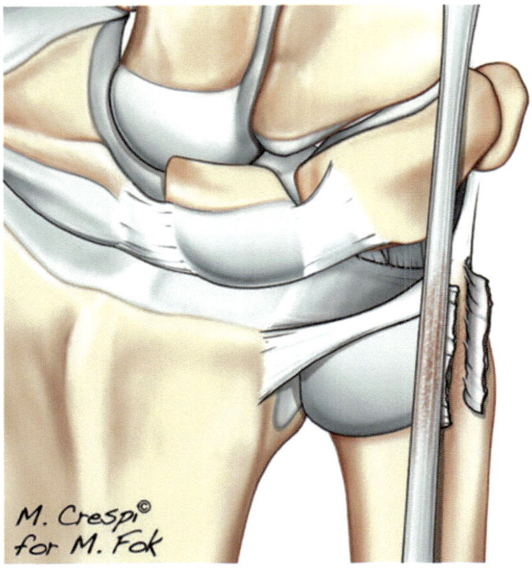

Fig. 11.6 Release of ECU compartment (copyright Dr. Margaret Fok)

in pronation. Tenderness and swelling may be presented with the ECU tendon at the ulnar head. Ultrasound can be used to evaluate the stability of the tendon during forearm rotation. MRI may show inflammation around the sheath and the malposition of the tendon. It is important to note that there are asymptomatic patients with ECU tendon displacement of up to 50% of the tendon width from the ulnar groove [11]. A comparison with the asymptomatic wrist may be beneficial.

In acute dislocation of the ECU tendon, reduction and immobilization in a long-arm splint with the forearm in pronation and the wrist in radial deviation from 6 weeks to up to 4 months can be successful [12]. Yet, most athletes are reluctant to undergo prolonged period of splinting. Surgical stabilization is indicated for patients with chronic symptomatic subluxation and for athletes who demand early mobilization. Numerous techniques (both anatomic [13] and nonanatomic [14]) have been described to reconstruct the ECU sub-sheath with satisfactory results noted. Our preferred method is using Burkhart technique, of which it uses part of the extensor retinaculum to stabilize ECU in the dorsal aspect of the wrist (Fig. 11.8). The tension of the repair must be checked by taking the wrist in full range of movement, to ensure smooth gliding of the newly stabilized tendon. Return to play depends on the healing of the sheath.

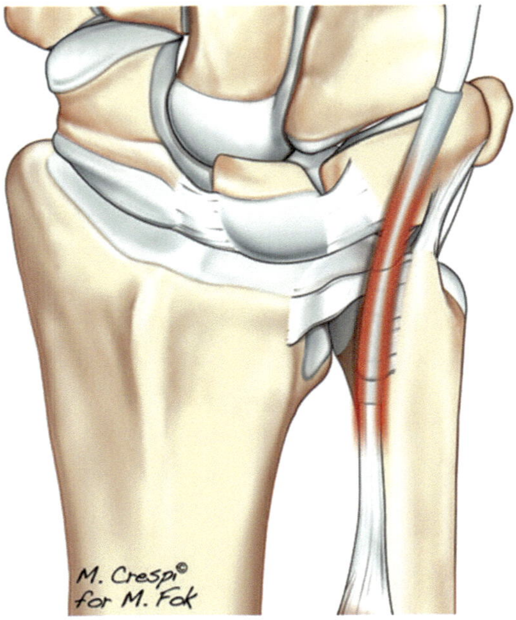

Fig. 11.7 ECU subluxation during forearm supination (copyright Dr. Margaret Fok)

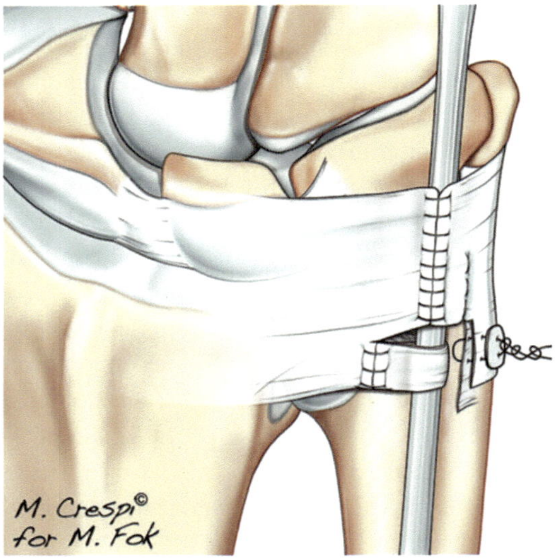

Fig. 11.8 ECU sheath reconstruction (copyright Dr. Margaret Fok)

11.5 Triangular Fibrocartilage Complex Tear

The true incidence of triangular fibrocartilage complex tear in athletes is unknown, as a significant percentage of patients are often treated as wrist sprain. It can be caused by a single traumatic incident or repetitive throwing motion, which involves gripping of the objects and moving wrist in ulnar deviation, e.g., hammer throw. It may occur as an isolated tear (usually caused by repetitive movement) or it may be associated with ECU tendonitis and DRUJ instability.

Patients present with ulnar wrist pain especially in activities that require forceful wrist flexion and rotation. Tenderness is elicited at the ulnar styloid, fovea, or distal radioulnar joint (DRUJ). DRUJ ballottement test, ulnocarpal stress test, and press test are some of the common maneuvers to elicit DRUJ joint instability [15]. These movements need to be compared with the unaffected side to differentiate normal from pathological laxity.

Radiographs are used to evaluate ulnar variance (as positive ulnar variance predisposes ulnar impaction syndrome and TFCC degenerative tear). MRI or magnetic resonance arthrogram (MRA) may be used to determine the status of TFCC, with the MRA being more superior in the diagnosis but is more uncomfortable investigation [16].

In acute traumatic event, TFCC tear with or without DRUJ dislocation can be treated conservatively with splints. A minimum of 6 weeks of immobilization with forearm in supination (a position which the ulnar is usually reduced and is most stable) is recommended, followed by a period of graded strengthening. For subacute or chronic cases, rest, activity modification, NSAID, and soft splint may be used. Surgical intervention is indicated when conservative therapy fails. In these incidences, due to the prolonged period of rehabilitation, definitive management may be deferred until the athlete is out of season.

Wrist arthroscopy is the gold standard in evaluating TFCC tear. Central tear of the TFCC is debrided, while peripheral tear may be amenable to open or arthroscopic repair. Many techniques in TFCC repair have been described with satisfactory results [17–19]. Our preferred method is the arthroscopic inside out technique using meniscal double-barrel cannula and double-arm straight needles (Fig. 11.9) [20, 21]. At least 6 weeks of immobilization are recommended postoperatively. Rehabilitation starts after this period, with the aim to return to sports after 3–6 months.

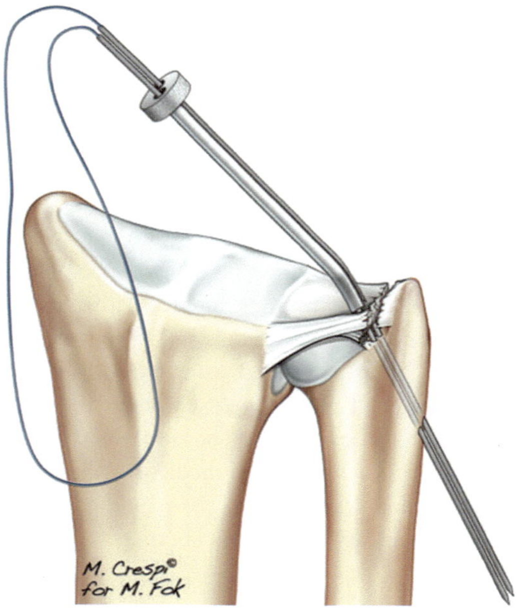

Fig. 11.9 TFCC peripheral repair (copyright Dr. Margaret Fok)

In the presence of positive ulnar variance, concomitant ulnar shortening osteotomy should be considered. In cases in which TFCC is not repairable and DRUJ is unstable, TFCC reconstruction with tendon graft should be performed.

11.6 Physeal Injury

In skeletally immature athletes, instead of sustaining injuries to their tendons around the wrist and TFCC, their physis may be more prone to injury, which may in turn result in growth arrest. De Smet et al. report a case of growth arrest of distal radial epiphysis, similar to a Madelung deformity, in a 14-year-old Javelin thrower [22]. In early cases, radiographs may not reveal the growth arrest and MRI and technetium bone scan may be needed for diagnosis. Rest is recommended, in order to halt further injury. Return to sport should not be allowed until wrist pain has resolved and motion has been regained [23]. Depending on symptoms, functional status, and extent of the physeal involvement and existing deformity, surgical intervention may be considered in order to cease further deformity and to correct the existing deformity. Prevention is the key. Restrictions on the number of hours of practice may be needed for young athletes.

11.7 Conclusion

Wrist injuries in throwers are not uncommon. Majority of these injuries involves soft tissues like tendons, tendon sheaths, and fibrocartilage. They are often dismissed as nonspecific wrist sprain. Delayed presentations are not unusual. Athletes and coaches should be alert of the potential wrist injuries. To seek medical advice in a timely manner can achieve good outcomes and can minimize downtime from sports.

References

1. Setayesh K, Mayekar E, Research BSAOSMA. Upper extremity injuries in field athletes: targeting 512 injury prevention. Ann Sports Med Res. 2017;4(1):1098.
2. Wu F, Rajpura A, Sandher D. Finkelstein's test is superior to eichhoff's test in the investigation of de Quervain's disease. J Hand Microsurg. 2018;10(02):116–8. https://doi.org/10.1055/s-0038-1626690.
3. Goubau JF, Goubau L, Van Tongel A, Van Hoonacker P, Kerckhove D, Berghs B. The wrist hyperflexion and abduction of the thumb (WHAT) test: a more specific and sensitive test to diagnose de Quervain tenosynovitis than the Eichhoff's Test. J Hand Surg Eur Vol. 2014;39(3):286–92. https://doi.org/10.1177/1753193412475043.
4. Adams JE, Habbu R. Tendinopathies of the hand and wrist. J Am Acad Orthop Surg. 2015;23(12):741–50.
5. Fok MWM, Liu B, White J, Bain G. Hand and wrist tendinopathies. In: Muscle and tendon injuries. 5th ed. Berlin, Heidelberg: Springer Berlin Heidelberg; 2017. p. 241–53.
6. Homayouni K, Zeynali L, Mianehsaz E. Comparison between kinesio taping and physiotherapy in the treatment of de quervain's disease. J Musculoskelet Res. World Scientific Publishing Company. 2014;16(04):1350019.
7. Richie CA, Briner WW. Corticosteroid injection for treatment of de Quervain's tenosynovitis: a pooled quantitative literature evaluation. J Am Board Fam Pract Am Board Fam Med. 2003;16(2):102–6.
8. Jirarattanaphochai K, Saengnipanthkul S, Vipulakorn K, Jianmongkol S, Chatuparisute P, Jung S. Treatment of de quervain disease with triamcinolone injection with or without nimesulide. J Bone Joint Surg Am. 2004;86(12):2700–6.
9. Peters-Veluthamaningal C, van der Windt DA, Winters JC, Meyboom-de JB. Corticosteroid injection for de Quervain's tenosynovitis. Cochrane Musculoskeletal Group, editor. John Wiley & Sons, Ltd. Cochrane Datab Syst Rev. 2009;27A(2):322.
10. Ruland RT, Hogan CJ. The ECU synergy test: an aid to diagnose ECU tendonitis. J Hand Surg Am. 2008;33(10):1777–82.
11. Lee KS, Ablove RH, Singh S, De Smet AA, Haaland B, Fine JP. Ultrasound imaging of normal displacement of the extensor carpi ulnaris tendon within the ulnar groove in 12 forearm–wrist positions. Am J Roentgenol. American Roentgen Ray Society. 2012;193(3):651–5.
12. Campbell D, Campbell R, O'Connor P, Hawkes R. Sports-related extensor carpi ulnaris pathology: a review of functional anatomy, sports injury and management. Br J Sports Med. 2013;47(17):1105–11.
13. MacLennan AJ, Nemechek NM, Waitayawinyu T, Trumble TE. Diagnosis and anatomic reconstruction of extensor carpi ulnaris subluxation. J Hand Surg Am. W.B. Saunders. 2008;33(1):59–64.
14. Burkhart SS, Wood MB, Linscheid RL. Posttraumatic recurrent subluxation of the extensor carpi ulnaris tendon. J Hand Surg Am. 1982;7(1):1–3.
15. Pang EQ, Yao J. Ulnar-sided wrist pain in the athlete (TFCC/DRUJ/ECU). Curr Rev Musculoskelet Med. Springer US. 2017;10(1):53–61.
16. Lee YH, Choi Y-R, Kim S, Song H-T, Suh J-S. Intrinsic ligament and triangular fibrocartilage complex (TFCC) tears of the wrist: comparison of isovolumetric 3D-THRIVE sequence MR arthrography and conventional MR image at 3 T. Magn Reson Imaging. 2013;31(2):221–6.
17. Kovachevich R, Elhassan BT. Arthroscopic and open repair of the TFCC. Hand Clinics. 2010;26(4):485–94.
18. Luchetti R, Atzei A, Cozzolino R, Fairplay T, Badur N. Comparison between open and arthroscopic-assisted foveal triangular fibrocartilage complex repair for post-traumatic distal radio-ulnar joint instability. J Hand Surg Eur Vol. SAGE Publications. 2014;39(8):845–55.
19. Andersson JK, Åhlén M, Andernord D. Open versus arthroscopic repair of the triangular fibrocartilage complex: a systematic review. J Exp Orthop. SpringerOpen. 2018;5(1):6–10.
20. Tang CYK, Fung B, Rebecca C, Lung CP. Another light in the dark: review of a new method for the arthroscopic repair of triangular fibrocartilage complex. J Hand Surg Am. 2012;37(6):1263–8.
21. Tang C, Fung B, Chan R, Fok M. The beauty of stability: distal radioulnar joint stability in arthroscopic triangular fibrocartilage complex repair. Hand Surg. 2013;18(1):21–6.
22. Smet LD, Fabry G. Growth arrest of the distal radial epiphysis in a javelin thrower: reversed madelung? J Pediatr Orthopaedics B. 1995;4(1):116.
23. Swanstrom MM, Lee SK. Open treatment of acute scapholunate instability. Hand Clinics. 2015;31(3):425–36.

Part III

Hip Injuries

12 Acute and Long-Standing Groin Injuries

Per Hölmich and Lasse Ishøi

12.1 Introduction

During the last two decades, our understanding of groin pain in athletes has evolved substantially. The Doha agreement on terminology and definitions of groin pain in athletes was published in 2015 [1], as an attempt to keep terminology more clear to support both clinical and scientific purposes. Based on the agreement, four clinical entities based on anatomical location of painful structures were defined for long-standing groin pain in athletes: adductor-related, iliopsoas-related, inguinal-related, and pubic-related groin pain [1]. The definition of these terms also meant that the expert group advised against using terms such as adductor and iliopsoas tendinitis or tendinopathy, athletic groin pain, athletic pubalgia, Gilmore's groin, osteitis pubis, sportsman's groin, and sportsman's hernia [1, 2]. For acute groin injuries, no agreement on terminology exists; however, the abovementioned clinical entities can, however, be used to describe most acute groin injuries [1] supplemented with imaging findings to describe a more specific injury location [3].

Although multidirectional field-based team sports such as soccer, football, and ice hockey are associated with the highest injury rate of sustaining a groin injury, acute and long-standing groin pain is not an uncommon problem in athletics. In a descriptive epidemiology study [4], almost 2000 hip and groin injuries across multiple collegiate sports were reported, with most of these being adductor- and or hip flexor-related injuries. Hip and groin injury rates per 100,000 athlete exposures for women's and men's outdoor/indoor track were an overall of 31–43 injuries, with an indication of higher injury rates during competition compared to practice, especially for men's outdoor track [4]. Noteworthy, the most hip and groin injuries in indoor and outdoor track athletes are associated with only limited time loss, with up to 50% of injuries lasting less than 24 h [4]. However, this is a common phenomenon from the hip and groin pain literature, as these injuries often present with a gradual onset, and thus, many athletes continue to be involved in sport despite having pain [5]. Although no data exist on this in athletics, a similar pattern is expected, and the clinician should thus be aware of this, as gradual progressive groin pain, despite not being associated with time loss in the initial phases, may likely evolve into a long-standing condition affecting performance and athletic abilities.

P. Hölmich (✉) · L. Ishøi
Sports Orthopedic Research Center — Copenhagen (SORC-C), Department of Orthopedic Surgery, Copenhagen University Hospital, Amager-Hvidovre, Denmark
e-mail: per.holmich@regionh.dk

G. L. Canata et al. (eds.), *Management of Track and Field Injuries*, https://doi.org/10.1007/978-3-030-60216-1_12

12.2 Diagnosis

Athletes presenting with groin pain should follow a standardized sequence including subjective history taking, special diagnostic tests, and assessment of self-reported and objectively measured physical function. The clinician must be aware that groin pain is the main symptom in many diagnoses surrounding the hip and groin area, including also intra-articular hip joint conditions. An extensive approach for the hip and groin examination is beyond the scope of this chapter; however, we advise that the clinician always attempts to rule out potential causes of groin pain such as intra-articular hip joint pain and/or referred lower back pain, prior to examining for specific muscle–tendon pain, especially when diagnosing long-standing groin pain. An excellent resource for a structured approach on hip joint examination can be found here [6].

12.2.1 Subjective History

The subjective history is a very important aspect of diagnosing groin pain in athletes, as it will typically narrow down the potential diagnoses into a few candidates. General questions should include type of injury being an acute or gradual onset; a direct or indirect trauma; and previous treatment and potential improvements or lack hereof. If an acute episode was the start of the injury, a precise description of the injury mechanism can be very helpful. Was the traumatic incident of a violent nature involving high forces? Or was it a more subtle incident like a stumble, a fall, or perhaps a sudden twist or wrong movement? Was any kind of contact involved? Was a snap, click, pop, or similar sensation felt or even heard? How was the function immediately after? Could the activity continue? When was the worst pain resolved and when could some activity be resumed? Furthermore, specific location of pain and pain characteristics may also provide to be useful in some cases, although this can sometimes be difficult for the athlete to describe. Sensation of deep-seated groin pain, pain mainly during excessive hip flexion, and intermittent sharp pain may point in the direction of an intra-articular problem [7], whereas superficial pain anterior on the hip and/or pain on the medial aspect around the area of the pubic symphyses may point toward a muscle–tendon problem. Furthermore, although a sensation of hip joint clicking and/or catching has traditionally been associated with a acetabular labrum tear [8], many athletes without labral tears also report similar symptoms, and thus, a recent expert group found that such symptoms are not specific for intra-articular pain [7].

Even though the history and the present symptoms in some cases may leave very little doubt on the correct diagnosis, a direct examination of the hip and groin region and its structures is always warranted.

12.3 Diagnosis of Long-Standing Adductor-Related Groin Pain

The hip adductor muscle group comprises of pectineus, gracilis, adductor brevis, adductor magnus, and adductor longus, with the most frequent cause of long-standing adductor-related pain involving the adductor longus muscle [9]. Athletes with adductor-related groin pain typically present with pain medially in the groin at the proximal tendons and/or in the area around the origin of the adductor longus just lateral to the pubic symphysis and inferior to the pubic crest [10]. The etiology of long-standing adductor-related groin pain is currently unknown; however, it has been hypothesized that excessive repetitive loading of the hip adductors may lead to microtrauma at the proximal tendon and insertion point at the pubic bone [11, 12]. This is supported by the fact that many cases of groin pain in indoor and outdoor track athletes have a gradual onset [4].

Long-standing adductor-related groin pain can be diagnosed as adductor tenderness and pain on resisted hip adduction [1]. Additionally, many athletes with current and/or previous long-standing adductor-related groin pain often have low hip adduction strength and/or limited range of motion in passive abduction and bent knee

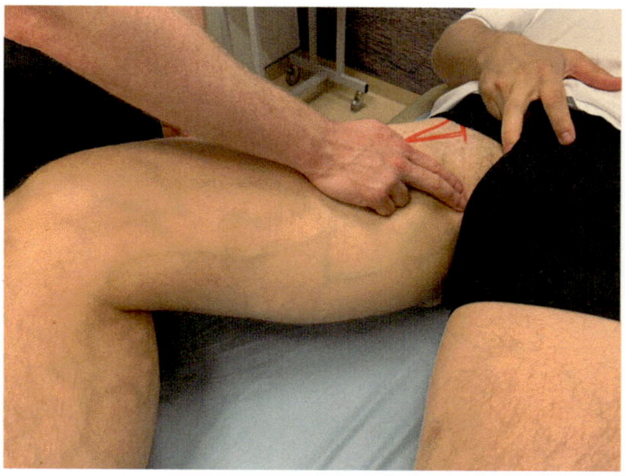

Fig. 12.1 Palpation at the origin of the adductor longus

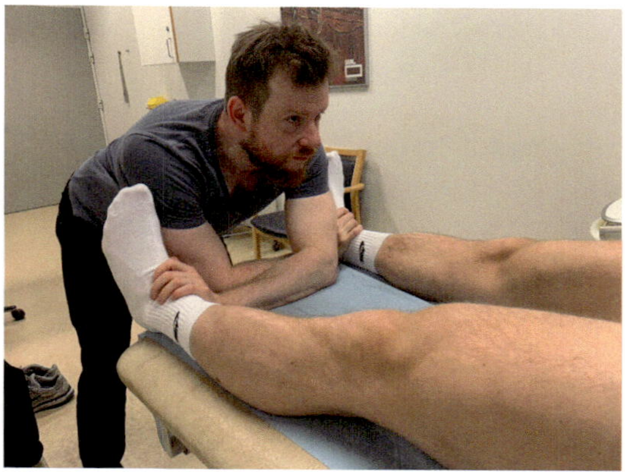

Fig. 12.2 Resisted hip adduction

fallout [13, 14]. In relation to this, deficits in hip adduction strength, measured during the long-lever squeeze test, seem to be associated with poorer sports function and higher pain [15].

The examination of adductor tenderness is performed with the patient lying supine with the hip flexed, abducted, and externally rotated, and the knee slightly flexed (Fig. 12.1). In this position, the adductor longus tendon can be easily palpated, by using the right hand on the right leg and vice versa, by following the adductor longus tendon with two fingers from the muscle belly to the insertion at the pubic bone. The insertion area, including the bone, is tested with firm pressure at a radius of about 1 cm. Pain on palpation suggests adductor-related groin pain [9, 16]. It is important to be aware that many athletes are sore on palpation in the area around the pubic bone, and thus, palpation should seek to reproduce the known pain and always be compared to the other side.

Pain on resisted hip adduction can be easily tested during the long-lever hip adduction squeeze test. The examiner stands at the end of the examination table with the lower arms between the feet placed just proximal to the medial malleolus (Fig. 12.2). By using the length of the lower arms between the legs, rather than a ball or a fist, the hips are placed in a slightly abducted position, which improves the force-generating capacity of the adductors, hence stressing the muscle–tendon unit of the adductors the most [17, 18]. The feet should be pointing straight up, and the athlete is instructed to squeeze both legs together with maximal exertion without lifting the legs or pelvis. The test is positive if it reproduces known pain from the insertion site of the adductor longus where the patient also was tender at palpation [9, 16].

12.4 Diagnosis of Acute Adductor Injuries

Similar to long-standing adductor-related groin pain, the adductor longus accounts for the majority of acute groin injuries [3]. The adductor brevis and pectineus are often injured in combination with an adductor longus injury, while obturator externus, gracilis, and adductor magnus injuries are rare causes of acute groin pain [3]. Due to the origin of the adductor magnus muscle being partly at the ischial tuberosity, acute injuries in this muscle can be mistaken for a posterior thigh injury.

A recent review [19] identified a single study, with the purpose of investigating diagnostic accuracy of clinical tests for acute adductor injuries [20]. The diagnosis of an acute adductor injury can be made with a clinical examination consisting of adductor palpation, adductor stretch, and adductor resistance tests [20]. The adductor palpation is performed as shown in Fig. 12.1; however, also the muscle belly should be palpated and the other adductor muscles even though they are more rarely injured. Palpation of the adductors (adductor longus, gracilis, and pectineus) has a high sensitivity of 96% and a low specificity of 57% when compared to MRI [20]. This generally means that adductor palpation is best suitable to rule out an acute adductor injury when no pain is present during palpation. Thus, the clinician can have great confidence that patients with a negative palpation test do not have an MRI verifiable acute adductor injury. Conversely, there is uncertainty as to whether a positive test confirms an acute adductor injury [20].

Adductor resistance tests useful for the diagnosis are squeeze test with 0° hip flexion as shown in Fig. 12.2 and outer range hip adduction, and with a sensitivity and specificity of 80% and 74% and 85% and 74%, respectively. Due to the relatively high specificity in both tests, these can be used to rule in an acute adductor injury when positive. Conversely, there is uncertainty as to whether a negative test can be used to rule out an acute adductor injury [20].

Passive stretching of the adductors is performed with the athlete lying supine and the examiner standing at the edge of the examination table facing the athletes. The examiner gently moves the affected side in a passive abduction with one hand while holding the other leg with the other hand. Reproducible pain in the adductors during stretching indicates a positive test. This test has sensitivity and specificity of 61% and 80%, respectively, and is thus best at ruling in an acute adductor injury when positive [20].

The clinician should be aware that the clinical examination does not show perfect agreement with imaging findings, and thus if a specific diagnosis and/or location is warranted MRI must be considered. A detailed MRI study of acute adductor injuries has shown that there are three characteristic locations of adductor longus injuries: (1) the proximal insertion, (2) the musculotendinous junction (MTJ) of the proximal tendon, and (3) the MTJ of the distal tendon. In the MTJ injuries at both the proximal insertion and distal insertion, there is rarely any injury to the tendon structure itself, whereas at the proximal insertion most injuries are complete avulsions [21]. The specific injury location seems to have important implications for the time to return to sport. Thus, injury at the bone–tendon junction confirmed with MRI seems to result in delayed return to sport [22], and also pain when palpating the proximal adductor longus insertion point suggests that the athlete can expect a prolonged return to sport [22, 23].

12.5 Diagnosis of Long-Standing Iliopsoas-Related Groin Pain

Iliopsoas-related groin pain is the second most common source of groin pain in athletes [9]. The pain is typically present during activities that require a large hip flexor moment such as running, sprinting, and jumping, and is located primarily in the anterior aspect of the thigh, lateral to the adductor-related pain. Iliopsoas-related pain is frequently observed in conjunction with adductor-related groin pain [24] and is also common in patients presenting with intra-articular hip joint pathology [25].

Long-standing iliopsoas-related pain is diagnosed as pain during palpation of the iliopsoas muscle belly and/or the iliopsoas tendon [1]. Some athletes may have pain during passive stretching of the iliopsoas during the Thomas test or when tested isometrically with 90° of hip flexion.

The iliopsoas palpation is done with the athlete in supine. The iliopsoas is palpable (1) proximal to the inguinal ligament at the level of the ASIS and (2) distal to the inguinal ligament, medial to the sartorius muscle, and lateral to the femoral artery (Fig. 12.3a, b). Abdominal palpation is performed with the hands positioned on each side of the prominence of the anterior iliac spine and then palpating in the area lateral to the rectus abdominis using soft gentle fingers. The fingers are gently pressed posteriorly while pushing the abdominal structures away to reach the iliopsoas muscle. The patient is then asked to elevate the leg 5 cm, and the psoas can be felt and palpated for any pain. The palpation of the distal iliopsoas tendon is most easily performed by first locating the proximal part of the sartorius muscle

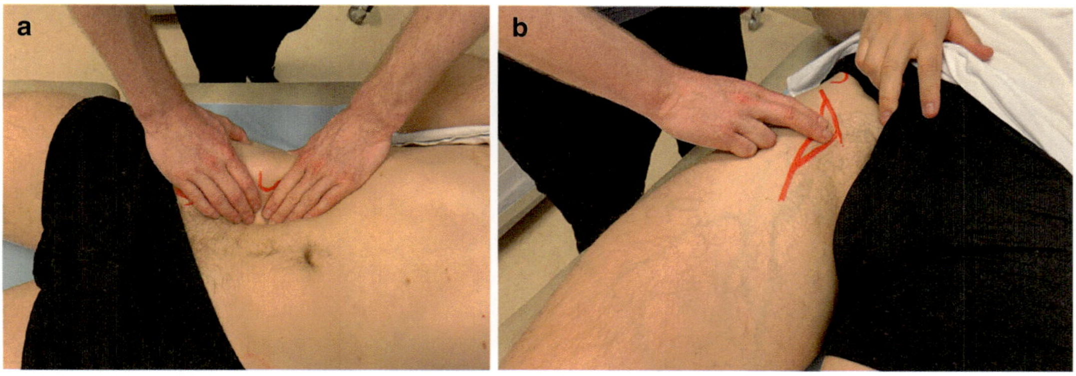

Fig. 12.3 Palpation of the iliopsoas muscle (**a**) at proximal part through the lower abdominal wall and (**b**) at the distal part just distally to the inguinal ligament and medially to the sartorius muscle

just distal to the inguinal ligament, and then moved the fingers slightly medially. The patient is then asked to elevate the examined leg 5 cm, and the finger position is adjusted until the tendon is clearly felt under the fingers. The tendon is then palpated again for any pain that reproduces the known symptoms [9, 16].

12.6 Diagnosis of Acute Iliopsoas Injuries

Around one third of acute groin injuries affect the hip flexor muscles, with the iliopsoas muscles being the primary site of injury in about half of these [3]. The clinical diagnosis of acute iliopsoas injuries can be a challenge as there is often widespread pain. As such, specific hip flexor tests (palpation, stretch, and resistance tests) are often positive without this being confirmed at MRI [3]. This results in overall poor accuracy of the clinical examination tests [20], and imaging or a delayed clinical examination may therefore be helpful. When positive, an MRI can provide detailed information on the location and extent of the injury, which may be relevant as complete tears are possible, although rare [21]. A detailed MRI study has shown that the iliacus muscle is more frequently injured than the psoas major [21]; however, it is still unclear whether this differentiation is clinically relevant for diagnosis or prognosis.

12.7 Diagnosis of Inguinal-Related Groin Pain

Inguinal-related groin injury is a rare diagnosis in the groin region [24], yet if present, the condition can be very hard to treat and may require surgery. Inguinal-related groin pain is typically characterized as pain over the inguinal canal and at the pubic tubercle that may radiate to the medial groin and the scrotum. The condition is thought to result from accumulation of shear forces leading to lesions of the fascia transversalis and the conjoined tendon, or dilatation of the inguinal ring. Inguinal-related groin pain is diagnosed as tenderness at the insertion of the conjoined tendon at the pubic tubercle and pain when palpating the inguinal canal through the scrotum with the patient standing [1].

12.8 Diagnosis of Acute Rectus Femoris Injuries

Quadriceps muscle strains, and in particular the rectus femoris portion, are a common source of complaints in athletics. Injuries in the rectus femoris muscle can be located in the distal and proximal muscle–tendon junction, the deep muscle–tendon junction at the central and/or indirect part of the tendon, and at the proximal muscle origin of the direct and indirect tendon. There are wide variations in the time to return to sport indicating that the injury location may have an impact on the rehabilitation period [26]. Athletes with rectus femoris injuries located in the deep muscle–tendon junction appear to experience a significant time of absence from sport compared with injuries located either in the proximal or distal muscle–tendon junction of the rectus femoris or in the vastus lateralis [26].

The diagnosis of an acute rectus femoris injury can be made through a clinical examination consisting of rectus femoris palpation and resistance testing in an elongated position [20]. There is evidence to suggest that a negative palpation test for rectus femoris and a negative knee extension test for resistance in a modified Thomas Test position have high diagnostic ability to rule out a positive MRI finding, and these tests can thus be used to exclude an acute rectus femoris injury. Furthermore, a positive palpation test for rectus femoris has high diagnostic ability to confirm a positive MRI finding, and can therefore be used to confirm an acute rectus femoris injury [3].

12.9 Treatment

Treatment of both long-standing and acute groin injuries, and rectus femoris injuries is centered around exercise-based treatment to improve load-tissue capacity of the involved structure/structures while also targeting other muscles in the hip and groin area to improve overall muscular function and stability of the pelvic. As such, a basic understanding of muscular function in relation to different activities is an important aspect of the treatment. For example, while many track and field activities, such as sprint running and various jumping, mainly occur in the sagittal plane, the clinician must be aware of the hip adductors' role in these movements. Although the primary function of the hip adductors is to generate hip adduction torque, the moment arm of the adductor muscles (which changes with hip angle) makes them an important synergist to hip flexor and extensor muscles. In an extended hip position, such as during to-off in sprinting, the adductor muscles are important hip flexor synergists. Conversely, in a flexed femur position, such as during the upward movement when jumping, the adductor muscles will have a line of force posterior to the rotational center and thus contribute to hip extension torque with the adductor magnus muscle considered as a substantial contributor to an effective hip extension movement [27]. Likewise in an adducted hip position, the iliopsoas muscle is considered not only as a hip flexor muscle, but also as an adductor muscle [28].

12.9.1 Long-Standing Adductor-Related Groin Pain

The treatment of athletes with long-standing adductor-related groin pain is centered around an active exercise approach with the aim to restore optimal hip adductor muscle function and increase load capacity [29]. There is consistent evidence that athletes with long-standing adductor-related groin pain typically have reduced hip adduction strength [14, 30]. While this can easily be measured isometrically using a handheld dynamometer, some athletes only demonstrate muscle deficits when measured eccentrically. This is possibly due to the more stressful nature of maximal eccentric contractions, and thus, a systematic examination of hip adduction strength is warranted to get a clear picture of injury severity and muscular deficits [30]. Passive treatment modalities or wait and see as the sole treatment approach does not seem to resolve pain effectively [5, 29].

Only few high-quality studies on the treatment of long-standing adductor-related pain exist [2], with a randomized controlled trial showing exercise therapy to be more effective in comparison with passive treatment modalities, such as massage or laser therapy [29]. Bony morphologies such as cam and pincer related to femoroacetabular impingement syndrome do not seem to prevent a successful treatment outcome at long-term follow-up [31]; however, if the athlete does not respond adequately on treatment, potential bony morphologies should be considered as a contributing factor to pain and may as well be a sign of intra-articular hip injuries [32] that may need surgical treatment.

The treatment program for adductor-related groin pain is structured in two modules, with the first module lasting approximately 2 weeks; here, the goal is to gradually activate the adductor muscles using isometric and low-load exercises. The second module includes more demanding exercises targeting both the adductor muscles specifically and the stability of the lumbo-pelvic region (Table 12.1).

The athlete and clinician should be aware that at least 8–12 weeks of focused exercise therapy may usually be needed to resolve all symptoms and allow full return to previous sporting activities [29]. During the treatment period, it is important to modify and/or restrict some aspect of athletic activity, such as high-speed running and forceful jumping and landing activities as these may expose the hip and groin structure to excessive load. Such activities should be introduced gradually considering both intensity and volume, and the clinician should be cautious not to re-integrate maximal sprint running and jumping to soon to avoid recurrence. After return to athletic activities, maintenance and/or further improvement of eccentric hip adductor strength should be included in the general strength and conditioning program. This can be easily done using the Copenhagen adduction exercise [33] or hip adduction with an elastic band [34]. Both of these exercises target the adductor longus muscle [35] and results in substantial strength gains following an 8-week period of progressive training [33, 34].

Table 12.1 The Hölmich treatment program for long-standing adductor-related groin pain

Module 1 (first 2 weeks)	Adductor squeeze (ball between feet), 10 × 30 s Adductor squeeze (ball between knees), 10 × 30 s Abdominal sit-ups (straight and oblique), 5 × 10 reps Folding knife (ball between knees), 5 × 10 reps Balance (wobble board), 5 min One-foot sliding board, 5 × 1 min
Module 2 (from third week)	Side-lying hip adduction/abduction, 5 × 10 reps Hip extension, 5 × 10 reps Standing hip adduction/abduction (elastic band), 5 × 10 reps Abdominal sit-ups (straight and oblique), 5 × 10 reps Cross-country skiing, 5 × 10 reps Sideward motion on "fitter," 5 min Balance (wobble board), 5 min Skating (sliding board), 5 × 1 min

12.9.2 Treatment of Acute Adductor Injury

Management of acute groin injuries should follow a progressive exercise approach starting with active flexibility, such as dynamic hip adduction and abduction leg swings, progressing into low-load resistance training and hip-load high-speed exercises [23]. Exercises may include hip standing adduction and flexion with elastic band or cable, combined hip flexion and contralateral arm flexion to create a tension arc across the core and pelvic area, and the Copenhagen adduction exercise [23, 33]. Load in exercises should be closely monitored and adjusted based on pain; that is, if pain is less than 3 out of 10 during a specific exercise, the clinician should be considered increasing the load and/or number of repetitions [23]. Besides specific groin exercises, training of other muscles groups relevant for the athletes, as well as progressive running with increasing intensity, should be scheduled on alternate days to prepare the athlete to return to sport.

12.9.3 Long-Standing and Acute Iliopsoas-Related Groin Pain

There is currently no evidence-based treatment of long-standing iliopsoas-related groin pain. As such, the clinician is recommended to adopt an active exercise program focusing on strengthening the iliopsoas muscle [36]. This can be done using a systematic and gradual strengthening program with a simple hip flexion exercise using an elastic band as external resistance [36]. Preferable, exercises with the sole aim of targeting the iliopsoas muscle should be performed through full range of motion and above 90 degrees of hip flexion to limit the contribution from the iliacus muscle and the rectus femoris [28, 37]. Running exercises should be planned cautiously in the initial treatment phase due to the large forces acting across the iliopsoas muscle, and it is recommended that the athlete gradually build up running volume and speed over a period of 8–12 weeks [38].

12.9.4 Long-Standing Inguinal-Related Groin Injury

The management of inguinal-related groin pain follows similar principles as for long-standing adductor-related groin pain, with an exercise-based approach superior to passive modalities such as massage and laser therapy [39]. The aim of the treatment is to strengthen the muscles of the inguinal canal, using exercises for the oblique abdominals and the rectus abdominis both in the outer and inner ranges (Table 12.2). If the exer-

Table 12.2 Treatment program for inguinal-related groin pain [39]

Module 1 (first 2 weeks)	• Static adduction against soccer ball placed between feet, 30 s × 10 reps. • Static adduction against soccer ball placed between knees, 30 s × 10 reps. • Bridging on the floor, 5 × 10 reps. • Sitting on ball, positioning knee and hips at 90° with hands on thighs while trying to maintain pelvic and trunk stability. • Abdominal sit-ups, both in straightforward direction and in oblique direction, 5 × 10 reps. • Combined abdominal sit-up and hip flexion, starting from supine position and with soccer ball placed between knees (folding knife exercise), 5 × 10 reps. • Balance training on wobble board, 5 min.
Module 2 (weeks 2–6)	• Cardiovascular warm-up: Bike or elliptical. • Leg abduction and adduction exercises lying on side, 5 × 10 reps. • One-leg weight-pulling abduction/adduction standing, 5 × 10 reps. • Abdominal sit-ups, both in straightforward direction and in oblique direction, 5 × 10 reps. • Bridging on ball: Place a physioball under legs and apply downward pressure to the ball as the legs straighten allowing the pelvis to rise from the surface. • Hip conditioning and core stabilization exercises: Sitting on the ball with the opposite upper extremity placing opposing pressure on raised knee while the other upper extremity is raised in the air for additional stabilizing challenge. • Quadriped hip extension with neutral spine, 2 × 15 reps. • Quadriped alternating opposite arm and leg extension with neutral spine, 2 × 15 reps. • Forward/backward walking lunges with medicine ball lift. 2–3 × 10–15 reps. • Single leg balance on 360° balance board with knees and hips flexed.
Module 2 (weeks 6–8)	• Cardiovascular warm-up on bike or elliptical with higher speed and resistance. • Clam exercise: The patient in side-lying position with the target hip on top in 30° flexion, externally rotated and abducted. A resistance band is used to perform isometric contraction, 5 × 10 reps. • Standing adduction with leg pulley: Attach cable to ankle, perform adduction movement standing next to machine, 5 × 10 reps. • Bridging coupled with lower extremity lift: The patient is on ball, lifts one leg into the air while keeping knee extended and trunk stabilized. • Front plank: Align shoulders with elbows and lift into forearm plank keeping pelvis in alignment and then progress to placing hands aligned with shoulders and fingers pressing into surface keeping pelvis aligned with plank position. • Side plank: Lying on side, align shoulder, elbow, hips, and ankles and raise up into plank position, maintaining alignment. • Pelvic stability on unstable surface: The patient sits on an air-filled balance disk, maintains balance while lifting one knee toward chest, and then lifts both knees. The same exercise was repeated with a ball toss. • Forward/backward walking lunges with medicine ball lift, 2–3 × 10–15 reps. • Single leg balance on 360° balance board with knees and hips flexed with ball toss.

cise therapy is not sufficient, surgical treatment with various techniques often quite similar to those used for regular hernia treatment can be used.

References

1. Weir A, Brukner P, Delahunt E, Ekstrand J, Griffin D, Khan KM, Lovell G, Meyers WC, Muschaweck U, Orchard J, Paajanen H, Philippon M, Reboul G, Robinson P, Schache AG, Schilders E, Serner A, Silvers H, Thorborg K, Tyler T, Verrall G, de Vos RJ, Vuckovic Z, Holmich P. Doha agreement meeting on terminology and definitions in groin pain in athletes. Br J Sports Med. 2015;49:768–74. https://doi.org/10.1136/bjsports-2015-094869.
2. Serner A, van Eijck CH, Beumer BR, Holmich P, Weir A, de Vos RJ. Study quality on groin injury management remains low: a systematic review on treatment of groin pain in athletes. Br J Sports Med. 2015b;49:813. https://doi.org/10.1136/bjsports-2014-094256.
3. Serner A, Tol JL, Jomaah N, Weir A, Whiteley R, Thorborg K, Robinson M, Holmich P. Diagnosis of acute groin injuries: a prospective study of 110 athletes. Am J Sports Med. 2015a;43:1857–64. https://doi.org/10.1177/0363546515585123.
4. Kerbel YE, Smith CM, Prodromo JP, Nzeogu MI, Mulcahey MK. Epidemiology of hip and groin injuries in collegiate athletes in the United States. Orthop J Sports Med. 2018;6:2325967118771676. https://doi.org/10.1177/2325967118771676.
5. Thorborg K, Rathleff MS, Petersen P, Branci S, Holmich P. Prevalence and severity of hip and groin pain in sub-elite male football: a cross-sectional cohort study of 695 players. Scand J Med Sci Sports. 2017;27:107–14. https://doi.org/10.1111/sms.12623.
6. Reiman MP, Thorborg K. Clinical examination and physical assessment of hip joint-related pain in athletes. Int J Sports Phys Ther. 2014;9:737–55.
7. Reiman MP, Thorborg K, Covington K, Cook CE, Holmich P. Important clinical descriptors to include in the examination and assessment of patients with femoroacetabular impingement syndrome: an international and multi-disciplinary Delphi survey. Knee Surg Sports Traumatol Arthrosc. 2017; https://doi.org/10.1007/s00167-017-4484-z.
8. Narvani AA, Tsiridis E, Kendall S, Chaudhuri R, Thomas P. A preliminary report on prevalence of acetabular labrum tears in sports patients with groin pain. Knee Surg Sports Traumatol Arthrosc. 2003;11:403–8. https://doi.org/10.1007/s00167-003-0390-7.
9. Holmich P. Long-standing groin pain in sportspeople falls into three primary patterns, a "clinical entity" approach: a prospective study of 207 patients. Br J Sports Med. 2007;41:247–52.; discussion 252. https://doi.org/10.1136/bjsm.2006.033373.
10. Schilders E, Bharam S, Golan E, Dimitrakopoulou A, Mitchell A, Spaepen M, Beggs C, Cooke C, Holmich P. The pyramidalis–anterior pubic ligament–adductor longus complex (PLAC) and its role with adductor injuries: a new anatomical concept. Knee Surg Sports Traumatol Arthrosc. 2017;25:3969–77. https://doi.org/10.1007/s00167-017-4688-2.
11. Branci S, Thorborg K, Bech BH, Boesen M, Nielsen MB, Holmich P. MRI findings in soccer players with long-standing adductor-related groin pain and asymptomatic controls. Br J Sports Med. 2015;49:681–91. https://doi.org/10.1136/bjsports-2014-093710.
12. Dimitrakopoulou A, Schilders E. Current concepts of inguinal-related and adductor-related groin pain. Hip Int. 2016;26(Suppl 1):2–7. https://doi.org/10.5301/hipint.5000403.
13. Esteve E, Rathleff MS, Vicens-Bordas J, Clausen MB, Holmich P, Sala L, Thorborg K. Preseason adductor squeeze strength in 303 Spanish male soccer athletes: a cross-sectional study. Orthop J Sports Med. 2018;6:2325967117747275. https://doi.org/10.1177/2325967117747275.
14. Mosler AB, Agricola R, Weir A, Holmich P, Crossley KM. Which factors differentiate athletes with hip/groin pain from those without? A systematic review with meta-analysis. Br J Sports Med. 2015;49:810. https://doi.org/10.1136/bjsports-2015-094602.
15. Wörner T, Thorborg K, Eek F. Five-second squeeze testing in 333 professional and semiprofessional male ice hockey players: how are hip and groin symptoms, strength, and sporting function related? Orthop J Sports Med. 2019;7:2325967119825858. https://doi.org/10.1177/2325967119825858.
16. Holmich P, Holmich LR, Bjerg AM. Clinical examination of athletes with groin pain: an intraobserver and interobserver reliability study. Br J Sports Med. 2004;38:446–51. https://doi.org/10.1136/bjsm.2003.004754.
17. Delp S, Malony W. Effects of hip center location on the moment-generating capacity of the muscles - ScienceDirect. J Biomech. 1993;26:485–99.
18. Light N, Thorborg K. The precision and torque production of common hip adductor squeeze tests used in elite football. J Sci Med Sport. 2016;19:888–92. https://doi.org/10.1016/j.jsams.2015.12.009.
19. Ishøi L, Krommes K, Husted RS, Juhl CB, Thorborg K. Diagnosis, prevention and treatment of common lower extremity muscle injuries in sport – grading the evidence: a statement paper commissioned by the Danish Society of Sports Physical Therapy (DSSF). Br J Sports Med. 2020; https://doi.org/10.1136/bjsports-2019-101228.
20. Serner A, Weir A, Tol JL, Thorborg K, Roemer F, Guermazi A, Holmich P. Can standardised clinical examination of athletes with acute groin injuries predict the presence and location of MRI findings? Br J Sports Med. 2016;50:1541–7. https://doi.org/10.1136/bjsports-2016-096290.
21. Serner A, Weir A, Tol JL, Thorborg K, Roemer F, Guermazi A, Yamashiro E, Holmich P. Characteristics

of acute groin injuries in the adductor muscles: a detailed MRI study in athletes. Scand J Med Sci Sports. 2017;28(2):667–76. https://doi.org/10.1111/sms.12936.
22. Serner A, Weir A, Tol JL, Thorborg K, Yamashiro E, Guermazi A, Roemer FW, Hölmich P. Associations between initial clinical examination and imaging findings and return-to-sport in male athletes with acute adductor injuries: a prospective cohort study. Am J Sports Med. 2020b;48:1151–9. https://doi.org/10.1177/0363546520908610.
23. Serner A, Weir A, Tol JL, Thorborg K, Lanzinger S, Otten R, Hölmich P. Return to sport after criteria-based rehabilitation of acute adductor injuries in male athletes: a prospective cohort study. Orthop J Sports Med. 2020a;8:2325967119897247. https://doi.org/10.1177/2325967119897247.
24. Taylor R, Vuckovic Z, Mosler A, Agricola R, Otten R, Jacobsen P, Holmich P, Weir A. Multidisciplinary assessment of 100 athletes with groin pain using the Doha agreement: high prevalence of adductor-related groin pain in conjunction with multiple causes. Clin J Sport Med. 2017;28(4):364–9. https://doi.org/10.1097/JSM.0000000000000469.
25. Jacobsen JS, Hölmich P, Thorborg K, Bolvig L, Jakobsen SS, Søballe K, Mechlenburg I. Muscle-tendon-related pain in 100 patients with hip dysplasia: prevalence and associations with self-reported hip disability and muscle strength. J Hip Preserv Surg. 2017;5(1):39–46. https://doi.org/10.1093/jhps/hnx041.
26. Cross TM, Gibbs N, Houang MT, Cameron M. Acute quadriceps muscle strains: magnetic resonance imaging features and prognosis. Am J Sports Med. 2004;32:710–9. https://doi.org/10.1177/0363546503261734.
27. Neumann DA. Kinesiology of the hip: a focus on muscular actions. J Orthop Sports Phys Ther. 2010;40:82–94. https://doi.org/10.2519/jospt.2010.3025.
28. Blemker SS, Delp SL. Three-dimensional representation of complex muscle architectures and geometries. Ann Biomed Eng. 2005;33:661–73. https://doi.org/10.1007/s10439-005-1433-7.
29. Holmich P, Uhrskou P, Ulnits L, Kanstrup IL, Nielsen MB, Bjerg AM, Krogsgaard K. Effectiveness of active physical training as treatment for long-standing adductor-related groin pain in athletes: randomised trial. Lancet. 1999;353:439–43. https://doi.org/10.1016/S0140-6736(98)03340-6.
30. Thorborg K, Branci S, Nielsen MP, Tang L, Nielsen MB, Holmich P. Eccentric and isometric hip adduction strength in male soccer players with and without adductor-related groin pain: an assessor-blinded comparison. Orthop J Sports Med. 2014;2:2325967114521778. https://doi.org/10.1177/2325967114521778.
31. Holmich P, Thorborg K, Nyvold P, Klit J, Nielsen MB, Troelsen A. Does bony hip morphology affect the outcome of treatment for patients with adductor-related groin pain? Outcome 10 years after baseline assessment. Br J Sports Med. 2014;48:1240–4. https://doi.org/10.1136/bjsports-2013-092478.
32. Ishøi L, Thorborg K, Kraemer O, Lund B, Mygind-Klavsen B, Hölmich P. Demographic and radiographic factors associated with intra-articular hip cartilage injury: a Cross-sectional study of 1511 hip arthroscopy procedures. Am J Sports Med. 2019;EP15.6:363546519861088. https://doi.org/10.1177/0363546519861088.
33. Ishoi L, Sorensen CN, Kaae NM, Jorgensen LB, Holmich P, Serner A. Large eccentric strength increase using the Copenhagen adduction exercise in football: a randomized controlled trial. Scand J Med Sci Sports. 2016;26:1334–42. https://doi.org/10.1111/sms.12585.
34. Jensen J, Holmich P, Bandholm T, Zebis MK, Andersen LL, Thorborg K. Eccentric strengthening effect of hip-adductor training with elastic bands in soccer players: a randomised controlled trial. Br J Sports Med. 2014;48:332–8. https://doi.org/10.1136/bjsports-2012-091095.
35. Serner A, Jakobsen MD, Andersen LL, Holmich P, Sundstrup E, Thorborg K. EMG evaluation of hip adduction exercises for soccer players: implications for exercise selection in prevention and treatment of groin injuries. Br J Sports Med. 2014;48:1108–14. https://doi.org/10.1136/bjsports-2012-091746.
36. Thorborg K, Bandholm T, Zebis M, Andersen LL, Jensen J, Holmich P. Large strengthening effect of a hip-flexor training programme: a randomized controlled trial. Knee Surg Sports Traumatol Arthrosc. 2016;24:2346–52. https://doi.org/10.1007/s00167-015-3583-y.
37. Visser JJ, Hoogkamer JE, Bobbert MF, Huijing PA. Length and moment arm of human leg muscles as a function of knee and hip-joint angles. Eur J Appl Physiol. 1990;61:453–60. https://doi.org/10.1007/BF00236067.
38. Schache AG, Blanch PD, Dorn TW, Brown NA, Rosemond D, Pandy MG. Effect of running speed on lower limb joint kinetics. Med Sci Sports Exerc. 2011;43:1260–71. https://doi.org/10.1249/MSS.0b013e3182084929.
39. Abouelnaga WA, Aboelnour NH. Effectiveness of active rehabilitation program on sports hernia: randomized control trial. Ann Rehabil Med. 2019;43:305–13. https://doi.org/10.5535/arm.2019.43.3.305.

13 Femoral Neck Stress Fractures and Avascular Necrosis of the Femoral Head

W. Michael Pullen and Marc Safran

13.1 Introduction

Hip pain is a common complaint among training athletes. These symptoms may spawn from many different causes, ranging from muscular strains and tears, anatomic abnormalities such as femoroacetabular impingement, and intrinsic issues with the bone such as stress fractures or avascular necrosis. Though uncommon in the track and field athlete, early recognition, diagnosis, and treatment of femoral neck stress fractures or avascular necrosis (AVN) are critical to reducing significant complications that may be the result of the untreated natural history of these problems, maximizing patient outcomes and increasing the chance that the athlete can return to sport. This chapter focuses on the pathology, risk factors, evaluation, and management of femoral neck stress fractures and avascular necrosis.

13.2 Femoral Neck Stress Fractures

Femoral neck stress fractures represent an uncommon, but potentially devastating injury affecting track and field athletes. Left unrecognized and untreated, there is risk of propagation to a completed or displaced fracture and decreased ability to return to full activity [1]. Early recognition and treatment are paramount in returning athletes to prior sporting activities, with high rates of return to sport associated with early treatment [2]. In contrast, delayed diagnosis can lead fracture completion, displacement, and subsequent increase in the risk of avascular necrosis, resulting in a decreased chance of return to sport (even without AVN) [1–4]. While femoral neck stress fractures are commonly studied in the military personnel, an increasing body of evidence demonstrates that knowledge of this pathologic spectrum is critical to providers managing athletes [5, 6]. More importantly, through increased awareness and identification of prodromal symptoms, the incidence of dis-

The views expressed in this chapter are those of the author(s) and do not necessarily reflect the official policy or position of the Department of the Navy, Department of Defense, or the US Government. W Michael Pullen is a military service member. This work was prepared as part of his official duties. Title 17 U.S.C. 105 provides that "Copyright protection under this title is not available for any work of the USA Government." Title 17 U.S.C. 101 defines a US Government work as a work prepared by a military service member or employee of the US Government as part of that person's official duties.

W. M. Pullen
Department of Orthopaedic Surgery,
Medical University of South Carolina, Charleston,
South Carolina, USA

M. Safran (✉)
Department of Orthopaedic Surgery, Stanford
University, Redwood City, VA, USA
e-mail: msafran@stanford.edu

G. L. Canata et al. (eds.), *Management of Track and Field Injuries*,
https://doi.org/10.1007/978-3-030-60216-1_13

placed femoral neck stress fractures can be demonstrably decreased [7].

13.2.1 Epidemiology and Risk Factors

Femoral neck stress fractures account for 3% of stress fractures within athletes and account for 50% of stress fractures within the femur [2]. While sports training in general can place an individual at risk for a femoral neck stress fracture, endurance training, specifically cross country and track athletes, has been shown to be at high risk for developing stress fractures [8, 9]. An additional correlative risk factor for femoral neck stress fracture is poor baseline physical fitness level [3, 10]. While this may not seem to apply directly to a conditioned athlete, an alteration in intensity or duration of training, such as increasing mileage or transition from off-season to in-season, can place an athlete at increased risk [6, 8]. Additionally, change in training surface, from trail running or composite track to concrete surfaces, may also increase the risk of stress fractures.

There are anatomical considerations that may place an athlete at increased risk for development of femoral neck stress fractures. Acetabular abnormalities have been associated with femoral neck stress fractures, with studies finding higher rates of acetabular retroversion and coxa profunda in femoral neck stress fracture patients when compared with normal controls [11, 12]. It has been proposed these acetabular variations create a levering or fulcrum effect, which, especially when combined with hip abductor weakness or fatigue, can increase the mechanical stresses seen at the femoral neck [12]. The impact of femoral head–neck junction abnormalities, or CAM deformities, is unclear, with mixed results reported within the literature regarding risk and association with femoral neck stress fractures [11–13].

Female athletes are disproportionally at risk for femoral neck stress fractures, with studies demonstrating a 4–10 times increased risk in female athlete when contrasted to their same sport male counterparts [8, 14–16]. Specifically, female athletes are subject to increased risk due to higher association with relative energy deficiency in sports (RED-S), formally referred to as the female athlete triad [14, 17]. Commonly seen in endurance athletes and runners, this low-energy availability (LEA) state results in impaired physiologic function, leading to impaired metabolic rate, menstrual function, bone health, immunity, protein synthesis, and cardiovascular health [17]. Specifically, studies of female endurance athletes with clinical evidence of LEA have demonstrated abnormal bone remodeling potential, decreased bone mineral density, and increased risk for stress fractures compared with controls [14, 17, 18]. Men are also susceptible to RED-S, and those that have it are at increased risk for developing femoral neck stress injuries as well.

13.2.2 Pathogenesis

Femoral neck stress fractures occur as a result of high-frequency repetitive sub-maximal loads applied to the femoral neck [4, 7, 16, 19]. The femoral neck region is particularly susceptible as it is a region of high stress during activity. In fact, stresses of up to 3–5 times body weight can be seen across the femoral neck with activities such as jogging [2, 20]. These increased stresses are often the result of an increase in training regimen in terms of duration or intensity outside of the patient's normal activity. Most commonly, these are seen as compressive forces across the neck on the inferior, or compressive portion of the neck [2, 16, 20]. Less commonly, tension or distraction forces can be seen on the superior portion of the neck. These less common tension sided fractures have been proposed to occur secondary to gluteus medius and minimus weakness or fatigue resulting in an inability to counterbalance the superiorly directed forces [2, 16, 20]. When these mechanical forces, either compression or tension, occur in the absence of rest, it can exceed the bone inherent metabolic repair rate, thus resulting in a stress reaction within the bone as it tries to repair [14]. If the stresses are allowed to continue, the rebuilding process is unable to outpace osteoclast activity, ultimately resulting in a fracture line within the femoral neck. This line can then propagate to the full length of the neck ultimately resulting in a completed, and potentially displaced, femoral neck fracture [2, 7, 16]. As such, the majority of these fractures can be classified as fatigue fracture, as the normal bone is experiencing an abnormal stress which it is unable to overcome. The exception to this is athletes with RED-S, which can be considered a combination of a fatigue fracture and, to varying degrees, an insufficiency fracture [14].

13.2.3 Classification

Multiple classification systems have been developed and modified since the 1960s (Table 13.1). Devas published some of the earliest work, utilizing plain radiographs to classifying fractures as either compression type, which he deemed required no treatment, and transverse type (distraction type), which required treatment to prevent the increased risk of fracture completion [21]. Blickenstaff and Morris utilized radiographs to classify patients as those with endosteal or periosteal callus without fracture line (type 1), a non-displaced fracture line (type 2), or a displaced fracture (type 3) [19]. Modifying the aforementioned work, Fullerton and Snowdy incorporated bone scintigraphy for earlier detection and created a three-type classification system—compression, tension, and displaced [22]. Shin and Gillingham used MRI for the detection and classification of femoral neck fractures and subdivided compression injuries to those with a fracture line less than 50% of the neck width and those with fracture line greater than 50% neck width to the above classifications [16].

MRI classifications have also been developed to further subclassify fracture patterns in an attempt to predict for return to sport and help guide treatment recommendations [3, 23, 24]. Grade 1 injuries consist of MRI signal changes present only on short tau inversion recovery (STIR) imaging. Grade 2 injuries include STIR findings and T2 changes in MRI. Grade 3 injuries demonstrate signal changes in STIR, T2, and T1, but without a definitive fracture line. Grade 4 injuries demonstrate changes in T1 and T2 with the presence of a fracture line. Rohena-Quinquilla and colleagues looked at femoral neck stress fractures suggesting a division into low grade, consisting of marrow edema only, and high grade, consisting of a macroscopic fracture line on imaging [3]. They subdivided high grade into those with a fracture line, which is less than 50% or greater than 50%, recommending surgical stabilization for those greater than 50%.

13.2.4 Clinical Evaluation

Athletes with femoral neck stress fractures will often present with vague, insidious hip pain, which is worse with activities or training, and will decrease with rest [2, 3, 6, 16, 22]. The location of pain may vary to include the anterior hip, proximal thigh, or groin. Additionally, patients often report increased pain with extremes of range of motion [2, 6, 22]. Regional tenderness can be a nonspecific and misleading finding when coupled with a low level of suspicion, as this can be misconstrued as a muscular or soft tissue

Table 13.1 Classification systems for femoral neck stress fractures

	Devas [21]	Blickenstaff and Morris [19]	Fullerton and Snowdy [22]	Shin and Gillingham [16]
Imaging modality	Plan radiographs	Plain radiographs	Plain radiographs Bone scintigraphy	Plain radiographs MRI
Classification	Compression	Type 1—Endosteal and/or periosteal callus, no fracture	Compression side	Compression <50%
	Transverse	Type 2—Fracture without displacement	Tension side	Compression >50%
		Type 3—Fracture with displacement	Displaced fracture	Tension
				Displaced fracture

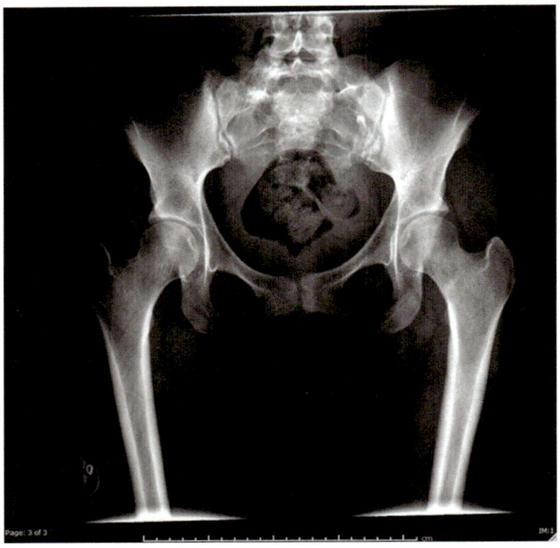

Fig. 13.1 Anteroposterior pelvis radiograph demonstrating left-sided femoral neck fracture. There is sclerosis along the compressive portion of the left femoral neck which is not the full width of the femoral neck

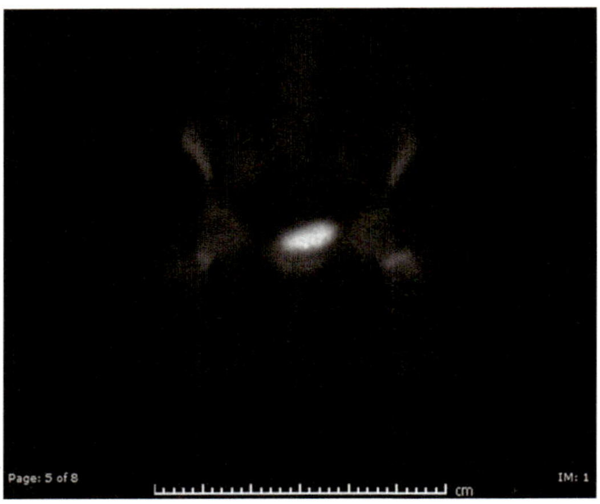

Fig. 13.2 Bone scintigraphy of a patient with bilateral hip pain which identifies bilaterally increased uptake at the femoral neck, greater on the left than the right, consistent with bilateral femoral neck stress fractures

injury [25]. With a high level of suspicion, athletes at risk for femoral neck stress fractures should undergo diagnostic work-up.

Plain radiographs are often utilized as initial screening tool for patients with insidious groin pain. These initial imaging studies should include both an anteroposterior view of the pelvis and a lateral view of the proximal femur [16, 19, 20, 22]. If present, plain radiographs can demonstrate periosteal or endosteal callus formation, linear sclerotic changes along the femoral neck (Fig. 13.1), and/or a fracture line [3, 26]. Plain radiographs, however, may remain normal for several weeks after start of symptoms and have been found to be negative up to two-thirds of the time [2]. As track athletes are often at higher risk for femoral neck stress fractures, further diagnostic work-up is required to ensure the diagnosis is not missed.

Bone scintigraphy has been used to aid in the identification of femoral neck stress fractures in the setting of benign radiographs [22]. While its use has decreased due to the ease, specificity, and accuracy of MRI, it can be utilized in institutions where MRI is not available or in cases where there are multiple body locations with suspicion for stress injuries [22, 26]. Findings on bone scintigraphy include increased uptake at the femoral neck (Fig. 13.2). While it boasts a sensitivity of 92% or greater, it is less specific and has a false-positive rate as high as 32% [26, 27]. Coupling this with the required radiation exposure, bone scintigraphy has been supplanted by MRI in most situations.

MRI has become the gold standard imaging modality in the diagnostic work-up of femoral neck stress fractures [3, 16, 23, 24, 26, 28]. MRI has been shown to have up to a 100% specificity, sensitivity, and accuracy for the diagnosis of femoral neck stress fracture [26, 27]. Moreover, rapid MRI sequences have been developed to include a coronal fast spin-echo T1 sequence and coronal STIR sequence to decrease imaging time without compromise of fracture detection [27]. Typical findings include diffuse hypo-intense signal on T1-weighted imaging and correlative hyperintense signal on T2-weighted or STIR sequencing (Fig. 13.3).

13.2.5 Nonoperative Management

Nonoperative management is relegated for cases, which are deemed low risk for progression or completion. These typically consist of patients with compression sided (medial neck) lesions with MRI stress reaction only (Fig. 13.4) or MRI demonstrated fracture line of less than 50% width of the femoral neck. This typically involves treatment with limited weight-bearing until dissipation of symptoms followed by activity restriction [2, 6, 16]. Though a typical time frame is 6–8 weeks, some studies have demonstrated up to 14 weeks of treatment may be necessary [2, 22, 29]. Moreover, it is imperative that these patients be followed by clinically and with possible repeat imaging as those with fracture lines are at risk for progression [28]. Should the patient have progression, then consideration would be given to a repeat weight-bearing restrictions or surgical stabilization [2, 28].

Nutritional and biochemical evaluation should also be considered to aid in further understanding the etiology and to potentially prevent recurrent stress fractures [17, 30]. Nutritional evaluation should be performed to ensure adequate energy and micronutrient availability [30]. Importantly, this assessment needs to account for both short- and long-term dietary needs and should account not only for an athlete's lean body mass, but also for a relative high exercise energy expenditure [30, 31]. Laboratory evaluation may be beneficial to identify markers of bone turnover, endocrine abnormalities, or micronutrient deficiencies, which can predispose continued risk. Of particular importance is evaluation for vitamin D deficiency, as low circulating levels of vitamin D have been associated with increased risk of stress fractures in runners [30–33]. Moreover, micronutrient supplementation with calcium and vitamin D has demonstrated benefit in athletes, with studies demonstrating decreased stress fracture incidence and reduced bone turnover markers [32, 33]. These supplements serve to aid in bone health through directly supporting bone mineralization and indirectly through suppression of parathyroid hormone axis activity [32, 33]. Finally, consultation with a bone endocrinologist should be performed in patients with recurrence or multiple stress fractures.

Biomechanical assessment can also be considered in the track and field athlete. This includes assessment for leg length inequality and other bony predisposition for stress fractures [5, 12, 13]. Gait and/or running analysis can be performed to identify asymmetry, muscular imbalance, and/or muscle weakness, which can potentially be addressed with physical therapy, selective strengthening, or gait retraining [5, 34]. Specific to the femoral neck, this may include a

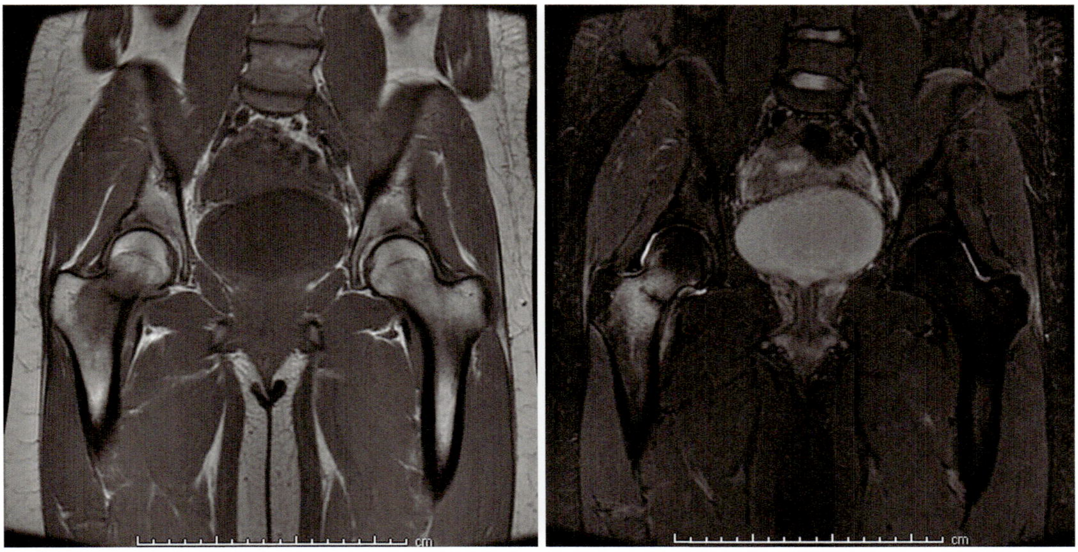

Fig. 13.3 Coronal T1 and STIR sequences which demonstrate characteristic findings of a femoral neck stress fracture. Hypo-intense signal is seen on T1 with the presence of a compression sided fracture line. Correlative hyperintense signal is seen on the STIR sequence

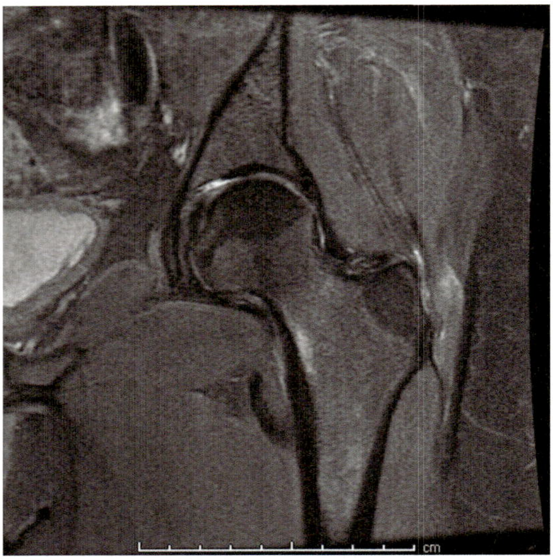

Fig. 13.4 Left hip STIR sequence which demonstrates stress reaction without fracture line on the compression portion of the femoral neck. This was successfully treated with conservative treatment with an ultimate return to sport

core and peritrochanteric hip strengthening regiment with a focus on gluteal strength [6].

13.2.6 Operative Management

Operative intervention is indicated for patients with an incomplete fracture line of >50% of the femoral neck width, complete compression sided fractures, all tension sided injuries, and displaced femoral neck fractures [2, 3, 6, 10, 16, 20, 22, 28]. For non-displaced compression sided fractures, most support the use of multiple cannulated compression screws, most often placed in an inverted triangle configuration as seen in Fig. 13.5 [1, 2, 4, 16]. There is debate within the literature as to the proper treatment for tension sided fractures, as the more vertical nature of the fracture may require more robust fixation. As such, consideration should be given to dynamic hip screw fixation for patients with tension sided fractures [2, 7, 35, 36].

Displaced fractures should be managed with anatomic reduction and fixation. Both multiple cannulated screws and dynamic hip screws have been described in the management of displaced fractures [2, 3, 7, 22, 35, 36]. Controversy exists as to whether an open reduction is necessary; however, fractures treated with closed reductions have been associated with higher rates of avascular necrosis [37].

Following operative fixation, patients are typically treated with limited weight-bearing for 6 to 12 weeks after which the patients progressed as tolerated. Full return to sport can often be achieved by 6 months, though there are reports of more prolonged postoperative courses [1, 2, 10, 23, 37].

13.2.7 Outcomes and Complications

Functional outcomes following femoral neck stress fractures in athletes are primarily relegated to case reports and small case series. Overall, return to previous level of function has been shown to be greater in those patients who have non-displaced fractures when compared with displaced fractures, further demonstrating the importance of early recognition and treatment [1, 10, 37, 38]. Ramey and colleagues reported outcomes from nonoperatively treated fractures in 27 patients. They showed an average return to running of 14.1 weeks and found increased MRI grading was associated with prolonged return to running time [23]. In a cohort of 23 athletes, Johansson et al. showed a 40% return to prior level of sport in displaced fractures and a 62% return with non-displaced fractures [39]. In the military population, studies have demonstrated approximately 50% return to previously level of duty, independent of self-reported post-injury pain scores [1, 10]. Recently, there have been multiple case reports, which, when aggregated, show a combined return rate of 9/11 for non-displaced nonsurgically managed fractures, 3/3 for surgically treated non-displaced fractures, and 11/11 for surgically treated displaced fractures [2].

Complications following femoral neck stress fractures are often dependent on nature and treatment of the fracture. For non-displaced, nonoperatively managed fractures, complications typically consist of progression or re-fracture

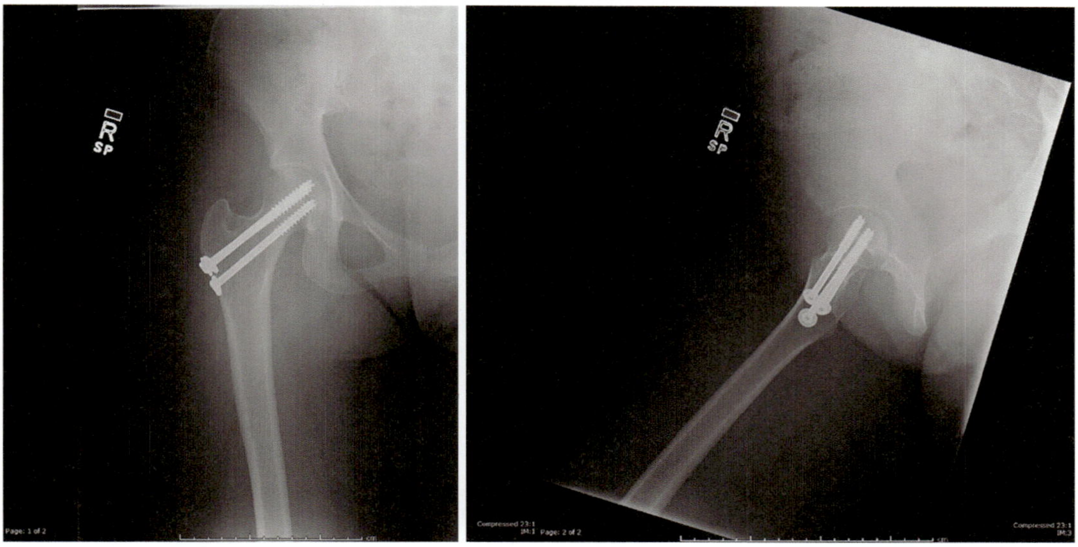

Fig. 13.5 Anteroposterior and lateral view of a right hip stress fracture treated with three partially threaded cannulated screws in an inverted triangle formation

[10, 16, 28, 40]. Complications following surgical management include avascular necrosis, nonunion, delayed union, malunion, fixation failure, and osteoarthritis [1, 2, 6, 7, 16, 19–22, 36–40].

13.3 Avascular Necrosis of the Femoral Head

Avascular necrosis, or osteonecrosis, is another cause of insidious hip pain, which can present in the athlete. It most commonly presents within the third to fifth decades of life and can remain asymptomatic in early stages [41]. There has not been a clearly defined etiology, however, proposed risk factors include trauma (to include femoral neck fractures and hip dislocations), corticosteroid use, alcohol consumption, blood disorders (including sickle cell), autoimmune disorders, and lysosomal storage disorders [41, 42]. These mechanisms can contribute to femoral head necrosis by way of ischemia, vascular disruption, occlusion, or constriction [41, 42]. Athletes would be at particular risk from trauma-related causes, to include displaced femoral neck fracture, with avascular necrosis rates up to 50%, and hip dislocation, with rates up to 25% [1, 2, 7, 29, 36, 37, 41]. While non-displaced traumatic hip fractures are at risk for avascular necrosis, non-displaced femoral neck stress fractures are at very low risk for avascular necrosis [2, 16].

13.3.1 Clinical Evaluation and Classification

Avascular necrosis may be asymptomatic in its early stages. Once symptoms develop, they typically consist of deep groin pain, but may also include back, buttock, or knee pain [41]. A high index of suspicion is key, as symptoms are often vague and early diagnosis is beneficial for long-term outcomes. Initial evaluation is performed with plain radiographs; however, much like femoral neck stress fractures, these may be negative in early cases [41, 43]. If there is concern due to persistent symptoms, advanced imaging is recommended with MRI. MRI is the primary diagnostic tool utilized both in early diseases where radiographs are normal, but also serves utility in later stage disease to identify the extent of the involvement [41, 43]. Importantly, in atraumatic cases the contralateral hip should be evaluated as bilateral disease can occur in up to 75% of cases [44]. MRI typically demonstrates subcortical changes to include hypo-intense T1 signal and hyperintense T2 signal (Fig. 13.6). There may be associated bone marrow edema and/or joint effusions [45].

Multiple staging systems have been developed in an attempt to create a common language and to guide treatment. The most commonly cited classification system is that from Ficat and Arlet (Table 13.2). It has since been modified to include MRI, and other classification systems have emerged to address size, location, and articular involvement [46]. All of the systems have limitations with no one classification system being used alone to guide treatment [43]. As such, four factors are often considered when guiding treatment—pre-collapse vs post-collapse, lesion size, amount of depression, and acetabular involvement or osteoarthritis [43, 47].

13.3.2 Nonoperative Treatment

There is a limited role for nonsurgical treatment for symptomatic osteonecrosis. For asymptomatic lesions, most commonly identified as incidental findings on imaging, approximately 60% of cases will progress to become symptomatic [41, 43, 46, 48]. Size of the lesion is a predictor for progression to a symptomatic lesion, with lesions <30% of the femoral head progressing in 5% of cases, whereas large lesions of >50% of the head progress in 83% of cases [48]. As such, initial observation and monitoring may be appropriate for small asymptomatic lesions, however, should be accompanied by appropriate counseling based on the extent of the lesion.

Pharmacologic treatments and biophysical modalities have been trialed, to include anticoagulants, lipid-lowering medications, vasodilators, bisphosphonates, extracorporeal shock waves, and electromagnetic fields [41, 43, 45]. These treatments, however, are mainly experimental with short-term or inconclusive results [45]. Moreover, it is established that those patients with early-stage disease fair better with surgical management and head preservation than those with advanced-stage disease [41, 44–46, 48, 49]. At this time, there is poor evidence to support that nonoperative treatment will prevent disease progression once symptomatic [49]. As such, the current recommendations would support early surgical intervention for patients with early, symptomatic disease.

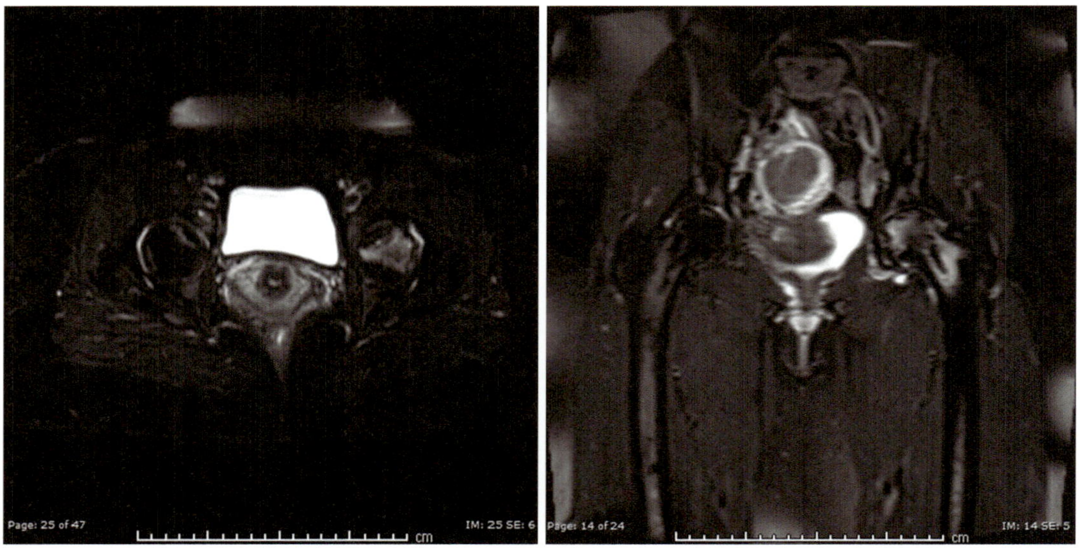

Fig. 13.6 Axial T2 and coronal fat saturated images which demonstrate bilateral avascular necrosis, with the left side being more advanced and more symptomatic than the right side

Table 13.2 Ficat and Arlet radiographic classification for avascular necrosis

Stage 1	Normal radiographs (abnormal findings only on MRI or bone scintigraphy)
Stage 2	Radiographs show sclerosis or cystic changes
Stage 3	Subchondral fracture, "crescent sign," with or without head collapse
Stage 4	Femoral head collapse, acetabular involvement, osteoarthritis

13.3.3 Operative Treatment

Surgical management can be broadly classified into femoral head preserving procedures and arthroplasty. Femoral head preserving procedures typically include core decompression, nonvascularized or vascularized bone grafting, and rotational osteotomies [41, 43, 50]. These techniques are typically implored in lesions, which are pre-collapse lesions [41, 44, 51]. Patients who fail these techniques, or more advanced cases with collapse or acetabular changes, are managed with hip arthroplasty [43].

Core decompression has been widely cited as a treatment for pre-collapse lesions [44]. Multiple techniques have been utilized to include both small-caliber drilling and large-caliber drilling [41]. Utilizing modern techniques, 70% of patients do not require additional procedures, with radiographic success occurring in 63% of patients. [52]. Not surprisingly, core decompression is more successful in small, early-stage lesions [52]. Core decompression has more recently been augmented with the use of nonvascularized graft, stem cells, and other biologic adjuncts, which have shown promise in increasing the effectiveness of these techniques [45, 50, 52].

Vascularized bone grafts and proximal osteotomies are less commonly applied techniques, which have demonstrated success. Vascularized bone grafts serve to provide both structural support to the subchondral bone and aid in revascularization of the necrotic segment [45, 50, 53]. Success rates have been reported as high as 88% in pre-collapse lesions (stages 1 and 2) and 78% in post-collapse lesions [54]. Moreover, return to athletic activity has been reported as high as 75% after postoperative recovery [55]. It is a technically complex operation, with higher volume centers reporting a 16.9% overall complication rate, with 4.3% being major complications [44, 56]. An underreported complication is donor site morbidity, and return to high-level sporting activities is not reported.

Rotational osteotomies have been utilized with success in Japan; however, these results with this technically difficult procedure have not been replicated in Europe and the USA [44]. Angular osteotomies have reported success rates as high as 72–87%, but complications are common and most results are relegated to small, single-surgeon case series [44, 57]. Given these results, these procedures are reasonable when performed in pre-collapse lesions by a surgeon experienced with the procedure [41, 44, 53, 54, 56, 57].

In lesions that have failed joint preserving techniques or have progressed to later stages to include collapse, treatment is typically relegated to total hip arthroplasty or resurfacing, depending on the quality of the residual bone [41]. Certainly, most surgeons do not recommend return to impact sports activities, like track, after total joint arthroplasty, or resurfacing.

13.4 Summary

Femoral neck stress fractures and avascular necrosis of the femoral head are uncommon causes of hip pain in the athlete. A high index of suspicion is needed in the diagnosis as symptoms are often vague and initial radiographs may fail to demonstrate the pathology. Early recognition and treatment are critical to improve functional outcomes and decrease short- and long-term complications.

References

1. Kusnezov NA, Eisenstein ED, Dunn JC, Waterman BR. Functional outcomes following surgical management of femoral neck stress fractures. Orthopedics. 2017;40(3):e395–e9.

2. Robertson GA, Wood AM. Femoral neck stress fractures in sport: a current concepts review. Sports Med Int Open. 2017;1(2):E58–68.
3. Rohena-Quinquilla IR, Rohena-Quinquilla FJ, Scully WF, Evanson JRL. Femoral neck stress injuries: analysis of 156 cases in a U.S. military population and proposal of a New MRI classification system. AJR Am J Roentgenol. 2018;210(3):601–7.
4. Waterman BR, Gun B, Bader JO, Orr JD, Belmont PJ Jr. Epidemiology of lower extremity stress fractures in the United States military. Mil Med. 2016;181(10):1308–13.
5. Behrens SB, Deren ME, Matson A, Fadale PD, Monchik KO. Stress fractures of the pelvis and legs in athletes: a review. Sports Health. 2013;5(2):165–74.
6. McInnis KC, Ramey LN. High-risk stress fractures: diagnosis and management. PM R. 2016;8(3 Suppl):S113–24.
7. Pihlajamaki HK, Ruohola JP, Kiuru MJ, Visuri TI. Displaced femoral neck fatigue fractures in military recruits. J Bone Joint Surg Am. 2006;88(9):1989–97.
8. Rizzone KH, Ackerman KE, Roos KG, Dompier TP, Kerr ZY. The epidemiology of stress fractures in collegiate student-athletes, 2004–2005 through 2013–2014 academic years. J Athl Train. 2017;52(10):966–75.
9. Hame SL, LaFemina JM, McAllister DR, Schaadt GW, Dorey FJ. Fractures in the collegiate athlete. Am J Sports Med. 2004;32(2):446–51.
10. Kupferer KR, Bush DM, Cornell JE, Lawrence VA, Alexander JL, Ramos RG, et al. Femoral neck stress fracture in air force basic trainees. Mil Med. 2014;179(1):56–61.
11. Kuhn KM, Riccio AI, Saldua NS, Cassidy J. Acetabular retroversion in military recruits with femoral neck stress fractures. Clin Orthop Relat Res. 2010;468(3):846–51.
12. Goldin M, Anderson CN, Fredericson M, Safran MR, Stevens KJ. Femoral neck stress fractures and imaging features of Femoroacetabular impingement. PM R. 2015;7(6):584–92.
13. Carey T, Key C, Oliver D, Biega T, Bojescul J. Prevalence of radiographic findings consistent with femoroacetabular impingement in military personnel with femoral neck stress fractures. J Surg Orthop Adv. 2013;22(1):54–8.
14. Feingold D, Hame SL. Female athlete triad and stress fractures. Orthop Clin North Am. 2006;37(4):575–83.
15. Schwartz O, Malka I, Olsen CH, Dudkiewicz I, Bader T. Overuse injuries among female combat warriors in the Israeli defense forces: a cross-sectional study. Mil Med. 2018;183(11-12):e610–e6.
16. Shin AY, Gillingham BL. Fatigue fractures of the femoral neck in athletes. J Am Acad Orthop Surg. 1997;5(6):293–302.
17. Mountjoy M, Sundgot-Borgen JK, Burke LM, Ackerman KE, Blauwet C, Constantini N, et al. IOC consensus statement on relative energy deficiency in sport (RED-S): 2018 update. Br J Sports Med. 2018;52(11):687–97.
18. Ackerman KE, Nazem T, Chapko D, Russell M, Mendes N, Taylor AP, et al. Bone microarchitecture is impaired in adolescent amenorrheic athletes compared with eumenorrheic athletes and nonathletic controls. J Clin Endocrinol Metab. 2011;96(10):3123–33.
19. Blickenstaff LD, Morris JM. Fatigue fracture of the femoral neck. J Bone Joint Surg Am. 1966;48(6):1031–47.
20. Egol KA, Koval KJ, Kummer F, Frankel VH. Stress fractures of the femoral neck. Clin Orthop Relat Res. 1998;348:72–8.
21. Devas MB. Stress fractures of the femoral neck. J Bone Joint Surg Br. 1965;47(4):728–38.
22. Fullerton LR Jr, Snowdy HA. Femoral neck stress fractures. Am J Sports Med. 1988;16(4):365–77.
23. Ramey LN, McInnis KC, Palmer WE. Femoral neck stress fracture: can MRI grade help predict return-to-running time? Am J Sports Med. 2016;44(8):2122–9.
24. Arendt EA, Griffiths HJ. The use of MR imaging in the assessment and clinical management of stress reactions of bone in high-performance athletes. Clin Sports Med. 1997;16(2):291–306.
25. Clough TM. Femoral neck stress fracture: the importance of clinical suspicion and early review. Br J Sports Med. 2002;36(4):308–9.
26. Shin AY, Morin WD, Gorman JD, Jones SB, Lapinsky AS. The superiority of magnetic resonance imaging in differentiating the cause of hip pain in endurance athletes. Am J Sports Med. 1996;24(2):168–76.
27. May LA, Chen DC, Bui-Mansfield LT, O'Brien SD. Rapid magnetic resonance imaging evaluation of femoral neck stress fractures in a U.S. active duty military population. Mil Med. 2017;182(1):e1619–e25.
28. Steele CE, Cochran G, Renninger C, Deafenbaugh B, Kuhn KM. Femoral neck stress fractures: MRI risk factors for progression. J Bone Joint Surg Am. 2018;100(17):1496–502.
29. Pihlajamaki HK, Ruohola JP, Weckstrom M, Kiuru MJ, Visuri TI. Long-term outcome of undisplaced fatigue fractures of the femoral neck in young male adults. J Bone Joint Surg Br. 2006;88(12):1574–9.
30. Giffin KL, Knight KB, Bass MA, Valliant MW. Predisposing risk factors and stress fractures in division I cross country runners. J Strength Cond Res. 2021;35(1):227–32.
31. Close GL, Sale C, Baar K, Bermon S. Nutrition for the prevention and treatment of injuries in track and field athletes. Int J Sport Nutr Exerc Metab. 2019;29(2):189–97.
32. Armstrong RA, Davey T, Allsopp AJ, Lanham-New SA, Oduoza U, Cooper JA, et al. Low serum 25-hydroxyvitamin D status in the pathogenesis of stress fractures in military personnel: an evidenced link to support injury risk management. PLoS One. 2020;15(3):e0229638.
33. Gaffney-Stomberg E, Nakayama AT, Guerriere KI, Lutz LJ, Walker LA, Staab JS, et al. Calcium and vitamin D supplementation and bone health in marine recruits: effect of season. Bone. 2019;123:224–33.

34. Crowell HP, Davis IS. Gait retraining to reduce lower extremity loading in runners. Clin Biomech (Bristol, Avon). 2011;26(1):78–83.
35. Liporace F, Gaines R, Collinge C, Haidukewych GJ. Results of internal fixation of Pauwels type-3 vertical femoral neck fractures. J Bone Joint Surg Am. 2008;90(8):1654–9.
36. Ly TV, Swiontkowski MF. Treatment of femoral neck fractures in young adults. Instr Course Lect. 2009;58:69–81.
37. Lee CH, Huang GS, Chao KH, Jean JL, Wu SS. Surgical treatment of displaced stress fractures of the femoral neck in military recruits: a report of 42 cases. Arch Orthop Trauma Surg. 2003;123(10):527–33.
38. Neubauer T, Brand J, Lidder S, Krawany M. Stress fractures of the femoral neck in runners: a review. Res Sports Med. 2016;24(3):185–99.
39. Johansson C, Ekenman I, Tornkvist H, Eriksson E. Stress fractures of the femoral neck in athletes. The consequence of a delay in diagnosis. Am J Sports Med. 1990;18(5):524–8.
40. Patel KM, Handal BA, Payne WK. Early diagnosis of femoral neck stress fractures may decrease incidence of bilateral progression and surgical interventions: a case report and literature review. Int J Surg Case Rep. 2018;53:189–92.
41. Zalavras CG, Lieberman JR. Osteonecrosis of the femoral head: evaluation and treatment. J Am Acad Orthop Surg. 2014;22(7):455–64.
42. Zalavras C, Dailiana Z, Elisaf M, Bairaktari E, Vlachogiannopoulos P, Katsaraki A, et al. Potential aetiological factors concerning the development of osteonecrosis of the femoral head. Eur J Clin Investig. 2000;30(3):215–21.
43. Mont MA, Jones LC, Hungerford DS. Nontraumatic osteonecrosis of the femoral head: ten years later. J Bone Joint Surg Am. 2006;88(5):1117–32.
44. Atilla B, Bakircioglu S, Shope AJ, Parvizi J. Joint-preserving procedures for osteonecrosis of the femoral head. EFORT Open Rev. 2019;4(12):647–58.
45. Cohen-Rosenblum A, Cui Q. Osteonecrosis of the femoral head. Orthop Clin North Am. 2019;50(2):139–49.
46. Steinberg ME, Steinberg DR. Classification systems for osteonecrosis: an overview. Orthop Clin North Am. 2004;35(3):273–83. vii–viii
47. Mont MA, Marulanda GA, Jones LC, Saleh KJ, Gordon N, Hungerford DS, et al. Systematic analysis of classification systems for osteonecrosis of the femoral head. J Bone Joint Surg Am. 2006;88(Suppl 3):16–26.
48. Nam KW, Kim YL, Yoo JJ, Koo KH, Yoon KS, Kim HJ. Fate of untreated asymptomatic osteonecrosis of the femoral head. J Bone Joint Surg Am. 2008;90(3):477–84.
49. Klumpp R, Trevisan C. Aseptic osteonecrosis of the hip in the adult: current evidence on conservative treatment. Clin Cases Miner Bone Metab. 2015;12(Suppl 1):39–42.
50. Lieberman JR, Berry DJ, Mont MA, Aaron RK, Callaghan JJ, Rajadhyaksha AD, et al. Osteonecrosis of the hip: management in the 21st century. Instr Course Lect. 2003;52:337–55.
51. Eward WC, Rineer CA, Urbaniak JR, Richard MJ, Ruch DS. The vascularized fibular graft in precollapse osteonecrosis: is long-term hip preservation possible? Clin Orthop Relat Res. 2012;470(10):2819–26.
52. Marker DR, Seyler TM, Ulrich SD, Srivastava S, Mont MA. Do modern techniques improve core decompression outcomes for hip osteonecrosis? Clin Orthop Relat Res. 2008;466(5):1093–103.
53. Garberina MJ, Berend KR, Gunneson EE, Urbaniak JR. Results of free vascularized fibular grafting for femoral head osteonecrosis in patients with systemic lupus erythematosus. Orthop Clin North Am. 2004;35(3):353–7.
54. Aldridge JM 3rd, Urbaniak JR. Avascular necrosis of the femoral head: role of vascularized bone grafts. Orthop Clin North Am. 2007;38(1):13–22. v
55. Sabesan VJ, Pedrotty DM, Urbaniak JR, Ghareeb GM, Aldridge JM. Free vascularized fibular grafting preserves athletic activity level in patients with osteonecrosis. J Surg Orthop Adv. 2012;21(4):242–5.
56. Gaskill TR, Urbaniak JR, Aldridge JM 3rd. Free vascularized fibular transfer for femoral head osteonecrosis: donor and graft site morbidity. J Bone Joint Surg Am. 2009;91(8):1861–7.
57. Marker DR, Seyler TM, McGrath MS, Delanois RE, Ulrich SD, Mont MA. Treatment of early stage osteonecrosis of the femoral head. J Bone Joint Surg Am. 2008;90(Suppl 4):175–87.

Femoroacetabular Impingement Syndrome and Labral Injuries

14

Rintje Agricola, Michiel van Buuren, and Pim van Klij

14.1 Introduction

Over the past two decades, the hip joint has been increasingly recognised as a cause of pain in athletes. Although the concept of femoroacetabular impingement (FAI) has been described in older literature, it was first popularised by Ganz et al. in 2003 [1]. Based on their clinical observations with open dislocations of the hip, they described two types of FAI, matching the acetabular chondrolabral damage pattern and osseous morphology [2]. A typical pattern of chondral and labral damage at the anterosuperior acetabular portion was often observed together with an aspherical femoral head; this was referred to as cam-type impingement. A more circumferential acetabular damage pattern was observed together with a deep acetabular socket, acetabular retroversion or an overcoverage of the acetabulum; this was referred to as pincer-type impingement. Ganz et al. proposed a motion-dependent mechanism in which the osseous morphology creates intra-articular soft tissue injury by an abnormal contact between the proximal femur and acetabulum during certain movements of the hip. This is different from the pathomechanism of acetabular dysplasia—another cause of chondrolabral damage—in which a more static axial loading of the hip is thought to create the damage.

The motion-dependent aspect of FAI is an important reason that FAI was suddenly increasingly recognised as a cause of hip-related pain in athletes. Especially, athletes practising high-impact sports and sports where a large range of hip motion is required are probably at risk of developing FAI. By the advances of hip arthroscopy and hip joint imaging such as magnetic resonance imaging (MRI), a better understanding and definition of FAI have now been established. However, the exact mechanism of FAI is still not fully unravelled.

In 2016, an international multidisciplinary group published a consensus statement on FAI syndrome [3]. This consensus statement proposed to use uniform terminology when referring to the bony characteristics underlying FAI,

R. Agricola (✉)
Department of Orthopaedics and Sports Medicine, Erasmus MC, University Medical Center, Rotterdam, The Netherlands

Department of Orthopaedic Surgery, Sports and Orthopaedics Eindhoven, St. Anna Hospital, Eindhoven, The Netherlands
e-mail: r.agricola@erasmusmc.nl

M. van Buuren
Department of Orthopaedics and Sports Medicine, Erasmus MC, University Medical Center, Rotterdam, The Netherlands

P. van Klij
Department of Orthopaedics and Sports Medicine, Erasmus MC, University Medical Center, Rotterdam, The Netherlands

Department of Sports Medicine, Isala Hospital, Zwolle, The Netherlands

G. L. Canata et al. (eds.), *Management of Track and Field Injuries*,
https://doi.org/10.1007/978-3-030-60216-1_14

namely 'cam morphology' in case of an aspherical femoral head and 'pincer morphology' in case of acetabular overcoverage. These definitions were agreed upon, rather than previously used terminology such as 'cam deformity', 'cam abnormality' or 'asymptomatic FAI'. It is important to realise that these types of bony morphology are frequently found—especially in athletes—and do not necessarily lead to FAI syndrome and/or pathology. That is also the reason that the term 'FAI syndrome' was agreed upon rather than 'FAI', as FAI syndrome reflects a triad of symptoms, positive clinical signs and imaging findings of cam and/or pincer morphology.

In the past decade, the clinical and scientific interest in FAI syndrome and labral injuries has significantly increased. The aetiology, clinical presentation, treatment and prognosis have been studied in more detail, and high-quality studies, including large prospective cohorts and randomised controlled trials, have been published or are underway. A large portion of recent literature has been focusing on athletes, including track and field athletes, as FAI syndrome and labral injuries are often seen in this population.

14.2 Aetiology

14.2.1 Femoroacetabular Impingement Syndrome

The aetiology of FAI syndrome is complex, as it is a motion-related disorder and therefore difficult to quantify on static imaging. The theoretical concept of FAI syndrome is that either cam or pincer morphology creates a premature abutment during hip motion and thereby damage to the labrum and cartilage (Fig. 14.1). This mechanism can be reproduced during surgery and the location of the cam morphology corresponds with the site of acetabular cartilage damage [2, 4]. However, cam and pincer morphology are prevalent in up to 80% of the athletic population [5, 6],

Fig. 14.1 The mechanism of FAI syndrome. A hurdler with cam morphology of the left hip experiencing impingement during hip motion is shown

but the majority will not develop FAI syndrome [7, 8]. It is important to realise that other anatomical factors such as femoral and acetabular version and orientation are also important to consider in the mechanism of FAI syndrome. However, the reason why some athletes will develop FAI syndrome while others with similar bony morphology and similar exposure to athletic activities will not develop FAI syndrome remains unclear. Although the mechanism of FAI syndrome is still a topic of research, there is evidence on the aetiology of the underlying bony morphology, particularly on that of cam morphology.

14.2.2 Cam Morphology

The most important cause of cam morphology development is loading of the hip joint during growth [6, 9–11]. This can result in shear stresses on the anterolateral side of the head–neck junction of the femur, which can cause extra bone formation in that area [8]. Finite element analysis suggests that specific repetitive movements, such as deep hip flexion and external rotation, might be the trigger for extra bone formation on the anterolateral head–neck junction [12]. Also, several other aspects of bony hip morphology are associated with cam morphology, such as a varus position and an extended proximal femoral growth plate orientation towards the femoral neck. Changes in hip morphology mostly occur during adolescence as the first femoral chondral changes in athletes can be observed from the age of 10 years [9] and the first bony changes from an age of 12 to 14 years [10, 11]. A prospective study with 5-year follow-up showed that cam morphology gradually arises during growth and did not change after proximal femoral growth plate closure [6]. The prevalence of cam morphology varies widely over several populations. It is more frequently observed in males [13–16] and in professional athletes [17]. Specific for track and field athletes, a cam morphology prevalence of 27–34% is observed [18, 19]. Current literature is highly supportive of the fact that physical activity during adolescence is the main risk factor for developing cam morphology. However, it is likely that other factors might also play a role, such as metabolic factors (growth hormones) and genetic background. To date, only indirect evidence for genetic involvement in cam morphology development has been found [20, 21].

14.2.3 Pincer Morphology

Less is known about the aetiology of pincer morphology and parameters related to its development. Pincer morphology can theoretically result in impingement between the femoral head–neck junction and the acetabular rim during flexion of the hip. The prevalence of pincer morphology has a very wide spread, which is partly due to the current heterogeneous definition of pincer morphology. Most probably, there is no difference in pincer morphology prevalence between gender [22, 23] and ethnicity [20, 24]. A prevalence in athletes of around 50% is presented in two systematic reviews, which might be even higher in the general population [25, 26]. Participation in track and field is not associated with an increased pincer morphology prevalence [18, 19], while in other sports the reported prevalence is highly variable [14, 20, 27–29].

14.2.4 Labral Tears

FAI syndrome can cause increased shear forces on several soft tissues, such as the labrum. As the labrum is a fibrocartilaginous rim, which can increase the depth of the acetabulum and stabilise the hip, labral damage can have consequences for hip joint function. Labral tears are most often observed on the anterior side, as this is the usual location where the abnormal contact occurs [30, 31]. In high-level running, labral damage is probably caused by traumatic twisting or by overuse/repetitive impingement. The prevalence of labral tears in athletes is high. It might be equally prevalent for males and females and independent of symptomatology. A labral tear prevalence of up to around 70% is reported [32–35]. Specific for track and field athletes, only one study, with a limited amount of participants, reported a per hip prevalence of labral tears of 4.5% in asymptomatic athletes [19].

14.3 Diagnosis

When an athlete presents with pain in the groin area, the differential diagnosis can be broad. The groin area contains not only the hip joint, but also many muscles and connective tissues. This makes a diagnosis often challenging. The 'Doha agreement meeting on terminology and definitions in groin pain in athletes' has shed some light on this complex problem in 2015 [36]. The consensus group has distinguished three categories of groin pain: defined clinical entities for groin pain, hip-related groin pain, and other conditions causing groin pain. Clinical entities for groin pain comprise adductor-related, iliopsoas-related, inguinal-related and pubic-related groin pain. There are numerous examples of other causes in the third category. In this chapter, we focus on hip-related groin pain, specifically FAI syndrome and labral tears, while some other causes of groin pain are described in Chaps. 12, 13 and 15. The Doha agreement has acknowledged that hip-related groin pain may be hard to distinguish from the other causes of groin pain, because symptoms may overlap, and because most clinical tests and signs are more sensitive than specific. This makes them more useful for ruling out certain hip-related pathologies than to diagnose them [36]. The Doha agreement did not aim to further classify the possible causes of hip-related groin pain. On the other hand, the Warwick agreement on FAI syndrome [3] has further elaborated on the terminology, diagnosis and treatment options for FAI syndrome; the Zurich agreement [37] has elaborated on the definition and diagnostic criteria of hip-related pain.

14.3.1 Medical History

The primary complaint of an athlete with FAI syndrome is usually hip-related pain, aggravated with hip motion [3]. The presentation of this pain may vastly differ between athletes though. Most patients refer to the groin area, but pain may also be felt at the greater trochanter, in the lower back, buttock or posterior thigh, or in the anterior thigh, all the way to the knee. The pain in FAI syndrome is typically motion-related or position-related [38]. Various track and field activities may therefore trigger this pain, from vigorous activity within normal range of motion (ROM), to movements with supraphysiological ROM. Examples of both could be the starting position of a sprint, the hurdling motion and various jumps including long, high and triple jumps (Fig. 14.1). Note that most of these movements require extreme flexion in the hip, sometimes combined with internal or external rotation. These are particularly 'at risk' types of motions as cam morphology is mostly located in the anterolateral head–neck junction. The characteristics of the pain are often described as sharp or aching, with an insidious onset in two thirds of patients. Mechanical features can also be present in two thirds of patients, varying from popping, snapping, catching and locking, to giving way [38]. Symptoms in patients with a labral tear can be exactly the same as those in FAI syndrome, with an insidious onset in two thirds of patients, a sharp or dull pain in most cases, and activity-related pain in almost all patients. Mechanical features may only be present in half of the patients though, slightly less than in FAI syndrome patients [39]. This makes differentiation between the two almost impossible from history alone. In FAI syndrome patients, there may also be gender-specific differences in symptomatology, with females having more symptoms with milder morphological features [23].

Besides the present complaint, the past medical history can also be helpful. Sports practised during childhood and adolescence are worth noting, especially sports that mechanically load the hip joint, like some track and field sports, running, football and basketball. High-impact loading of the hip has been connected to the development of cam morphology in adolescents who have practised these sports [6, 9–11]. History of trauma, childhood hip disease or previous surgery should also be noted, as well as risk factors for other causes of groin pain such as osteonecrosis, osteopenia, osteoporosis or stress fractures [40].

14.3.2 Clinical Signs

Physical examination is usually an important first step in the diagnosis of orthopaedic pathology. However, there used to be little consensus about its value in diagnosing hip-related groin pain. The International Hip-related Pain Research Network (IHiPRN) has recently made consensus recommendations on the classification, definition and diagnostic criteria of hip-related pain [37]. Physical examination alone is of limited value; a comprehensive examination of symptoms, signs and imaging is recommended.

In addition to a general physical examination, it is recommended to examine gait, single leg balance, muscle tenderness, hip strength and ROM, and to do specific impingement tests [41]. Muscles around the hip may be weaker in FAI syndrome [42]. Hip ROM may also be decreased, especially in flexion and internal rotation, while adduction and extension are usually not impaired [42, 43]. Patients with FAI syndrome may develop an abnormal movement pattern in the sagittal and frontal plane, due to the impaired ROM [44, 45]. Impingement tests such as the flexion–adduction–internal rotation (FADIR) and flexion–abduction–external rotation (FABER) are considered positive if they reproduce the patient's typical pain [3]. The FADIR test has good sensitivity, but poor specificity, making it useful for excluding FAI syndrome if the test is negative [37, 46]. There is very limited evidence for the clinical utility of other clinical tests, such as the Thomas test, prone instability test, ligamentum teres tear test and max squat test [37].

14.3.3 Imaging Findings

As with physical examination, the use of imaging alone is not recommended and imaging findings have to be evaluated in the light of the patient's symptoms and clinical signs [37]. It is recommended to start with a pelvic radiograph in anteroposterior (AP) direction, in conjunction with a lateral femoral head–neck view [3, 37]. The lateral view is needed because most cams are located anterolateral and can be missed on the AP view. The primary goal of imaging is morphological assessment of the hip, and identification of a cam or pincer morphology, which is a requirement for the diagnosis of FAI syndrome. Additionally, a plain radiograph is useful for excluding other causes of pain. Computed tomography (CT) is recommended for better evaluation of 3D bony morphology, such as cam and pincer morphology, especially when surgery is being considered. When a labral tear or other soft tissue pathology is suspected, magnetic resonance imaging (MRI) with intra-articular contrast (MR arthrography) is recommended [3, 37].

Common measurements used to diagnose cam or pincer morphology are the α angle (cam) and the centre-edge angle (CEA) or cross-over sign (pincer). For the α angle, a cut-off of 60 degrees has been proposed as the preferred threshold [47]. Such a threshold may especially be valuable for research purposes, where comparing findings is important. In a clinical setting, a clear threshold may be undesirable though, and focus should lie on the triad of symptoms, clinical signs and imaging findings [3]. A CEA of 40 degrees or higher is considered to be representative of a pincer morphology [48].

14.4 Treatment

In the Warwick agreement on FAI syndrome, consensus was reached on three treatment options: conservative care, rehabilitation or surgery [3]. Each of these may have a place depending on the type of patient. Conservative care includes education, watchful waiting and lifestyle or activity modification. However, for young adult, active patients, rehabilitation or surgery will be a more likely treatment option.

14.4.1 Rehabilitation

As physical activity is non-invasive and important to maintain physical and mental health, a physiotherapist-led rehabilitation programme is advised for young adult patients with hip-related pain as their first treatment option. Patients with hip-related pain at first need to undergo optimal conservative therapy to strengthen hip, trunk and functional components. Advised is to perform resistance and strengthening exercises under physiotherapist guidance. Specific muscle target training can focus on the deep hip stabilisers and gluteus maximus muscle, but can also consist of more general exercises to improve balance and proprioception, and optimise gait biomechanics and functional task performance. This can finally result in improvement of pain, weight-bearing function and quality of life. Exercises could also include the careful, manual release of soft tissue, needling or stretching to try to increase the hip ROM [49]. It must be acknowledged that the optimal effective type, dose, loading and exercise progression are yet unknown. To evaluate the results of this conservative therapy, exercises must be adequately fulfilled during at least 3 months [50]. The response to any type of treatment must be evaluated by the use of patient-reported outcome measures (PROMs), such as the Hip and Groin Outcome Score (HAGOS) or the International Hip Outcome Tool (iHOT) questionnaires. These measures can be used to guide the clinician and patient in the process to return to psychical activity and eventually to sports and performance, where patient expectations must be quantified and guided properly. Sport-specific activities should be assessed to guide this return to sport. In a recent consensus meeting on hip-related pain, most of the aforementioned recommendations are described in detail for the clinician [41, 51, 52].

14.4.2 Surgical Treatment

When a sufficient rehabilitation programme is unable to relieve the patient's symptoms, hip surgery is a good option. In two recent randomised controlled trials, it was shown that both a rehabilitation programme and surgical treatment could improve symptoms in patients with FAI syndrome. Hip surgery had significantly better, and clinically meaningful, outcomes than a rehabilitation programme [53, 54]. Surgery aims to correct the bony morphology in order to create an impingement-free ROM. Also, the labrum can be restored. Cam morphology can be addressed by removing the extra bone formation, thereby creating a spherical femoral head. In case of pincer morphology, the acetabular rim can be trimmed. In both types of FAI syndrome, the orientation of the proximal femur and acetabulum should be taken in mind and corrected if necessary. These procedures can be done by either open or arthroscopic surgery. Although there is limited evidence on the long-term outcomes of hip surgery for FAI syndrome and/or labral tears, it is generally believed that patients without (severe) chondropathy and/or first signs of hip osteoarthritis (OA) have favourable outcomes [55]. Other predictors for a favourable outcome may include younger age, male sex, lower BMI and pain relief from preoperative intra-articular hip injections [55].

14.5 Prognosis

The short-term prognosis of FAI syndrome can be relatively good as long as it is treated. Most patients have improvement of symptoms after treatment and can return to previous activities including sports. However, after arthroscopic treatment only a little over half of the athletes returned to preinjury sports at a preinjury level, and only one third of those athletes reported optimal sports performance [56]. There are many more reports on the long-term outcomes of surgical treatment [57–61] than on the outcomes of rehabilitation [62, 63]. The probability of returning to preinjury sports level after rehabilitation is therefore still unknown, but studies are underway [50, 52]. When left untreated, patients may experience deteriorating symptoms on the short term [3].

The long-term prognosis of FAI syndrome is not entirely clear. Cam morphology has been associated with hip OA on the long term in numerous prospective cohort studies, whether patients had symptoms or not. This may be attributed to repetitive impingement motions, causing shear stress and impaction of labral tissue and articular cartilage [64–68]. It is still unclear if treatment of FAI syndrome will actually prevent the development of hip OA, as comparative trials with long-term follow-up are lacking. For pincer morphology, epidemiological studies have not proved an association with the development of hip OA [5, 69].

The symptom-related prognosis of labral tears is unclear. The prevalence of labral tears is highly variable in both symptomatic and asymptomatic persons, which indicates a discordant relationship between labral tears and hip-related pain [32, 33]. The relationship between labral tears and other intra-articular damage such as cartilage damage is also not fully unravelled. In high-level runners, one study found that 6 out of 8 young athletes (75%) with a labral tear had underlying acetabular cartilage damage. However, this evidence is limited to case series with few participants and therefore difficult to generalise [70].

References

1. Ganz R, et al. Femoroacetabular impingement: a cause for osteoarthritis of the hip. Clin Orthop Relat Res. 2003;417:112–20.
2. Ito K, et al. Femoroacetabular impingement and the cam-effect. A MRI-based quantitative anatomical study of the femoral head-neck offset. J Bone Joint Surg Br. 2001;83(2):171–6.
3. Griffin DR, et al. The Warwick agreement on femoroacetabular impingement syndrome (FAI syndrome): an international consensus statement. Br J Sports Med. 2016;50(19):1169–76.
4. Beck M, et al. Hip morphology influences the pattern of damage to the acetabular cartilage: femoroacetabular impingement as a cause of early osteoarthritis of the hip. J Bone Joint Surg Br. 2005;87(7):1012–8.
5. van Klij P, et al. The prevalence of cam and pincer morphology and its association with development of hip osteoarthritis. J Orthop Sports Phys Ther. 2018;48(4):230–8.
6. van Klij P, et al. Cam morphology in young male football players mostly develops before proximal femoral growth plate closure: a prospective study with 5-year-follow-up. Br J Sports Med. 2019;53(9):532–8.
7. Agricola R, et al. Cam impingement of the hip: a risk factor for hip osteoarthritis. Nat Rev Rheumatol. 2013;9(10):630–4.
8. Agricola R, Weinans H. What causes cam deformity and femoroacetabular impingement: still too many questions to provide clear answers. Br J Sports Med. 2016;50(5):263–4.
9. Palmer A, et al. Physical activity during adolescence and the development of cam morphology: a cross-sectional cohort study of 210 individuals. Br J Sports Med. 2018;52(9):601–10.
10. Agricola R, et al. The development of cam-type deformity in adolescent and young male soccer players. Am J Sports Med. 2012;40(5):1099–106.
11. Agricola R, et al. A cam deformity is gradually acquired during skeletal maturation in adolescent and young male soccer players: a prospective study with minimum 2-year follow-up. Am J Sports Med. 2014;42(4):798–806.
12. Roels P, et al. Mechanical factors explain development of cam-type deformity. Osteoarthr Cartil. 2014;22(12):2074–82.
13. Jung KA, et al. The prevalence of cam-type femoroacetabular deformity in asymptomatic adults. J Bone Jt Surg Ser B. 2011;93B(10):1303–7.
14. Kapron AL, et al. Radiographic prevalence of femoroacetabular impingement in collegiate football players: AAOS Exhibit Selection. J Bone Joint Surg Am. 2011;93(19):e111(1–10).
15. Leunig M, et al. Prevalence of cam and pincer-type deformities on hip MRI in an asymptomatic young Swiss female population: a cross-sectional study. Osteoarthr Cartil. 2013;21(4):544–50.
16. Pollard TC, et al. Femoroacetabular impingement and classification of the cam deformity: the reference interval in normal hips. Acta Orthop. 2010;81(1):134–41.
17. Nepple JJ, Vigdorchik JM, Clohisy JC. What is the association between sports participation and the development of proximal femoral cam deformity? A systematic review and meta-analysis. Am J Sports Med. 2015;43(11):2833–40.
18. Kapron AL, et al. The prevalence of radiographic findings of structural hip deformities in female collegiate athletes. Am J Sports Med. 2015;43(6):1324–30.
19. Lahner M, et al. Prevalence of femoro-acetabular impingement in international competitive track and field athletes. Int Orthop. 2014;38(12):2571–6.
20. Mosler AB, et al. Ethnic differences in bony hip morphology in a cohort of 445 professional male soccer players. Am J Sports Med. 2016;44(11):2967–74.
21. Pollard TC, et al. Genetic influences in the aetiology of femoroacetabular impingement: a sibling study. J Bone Joint Surg Br. 2010;92(2):209–16.
22. Li Y, et al. Prevalence of Femoroacetabular impingement morphology in asymptomatic adolescents. J Pediatr Orthop. 2017;37(2):121–6.
23. Nepple JJ, et al. Clinical presentation and disease characteristics of femoroacetabular impingement are sex-dependent. J Bone Joint Surg Am. 2014;96(20):1683–9.
24. Tannenbaum E, et al. Gender and racial differences in focal and global acetabular version. J Arthroplast. 2014;29(2):373–6.
25. Frank JM, et al. Prevalence of Femoroacetabular impingement imaging findings in asymptomatic volunteers: a systematic review. Arthroscopy. 2015;31(6):1199–204.
26. Mascarenhas VV, et al. Imaging prevalence of femoroacetabular impingement in symptomatic patients, athletes, and asymptomatic individuals: a systematic review. Eur J Radiol. 2016;85(1):73–95.
27. Gerhardt MB, et al. The prevalence of radiographic hip abnormalities in elite soccer players. Am J Sports Med. 2012;40(3):584–8.
28. Harris JD, et al. Radiographic prevalence of dysplasia, cam, and pincer deformities in elite ballet. Am J Sports Med. 2016;44(1):20–7.
29. Lerebours F, et al. Prevalence of cam-type morphology in elite ice hockey players. Am J Sports Med. 2016;44(4):1024–30.
30. Lewis CL, Sahrmann SA. Acetabular labral tears. Phys Ther. 2006;86(1):110–21.

31. Narvani AA, et al. Acetabular labrum and its tears. Br J Sports Med. 2003;37(3):207–11.
32. Heerey JJ, et al. What is the prevalence of hip intra-articular pathologies and osteoarthritis in active athletes with hip and groin pain compared with those without? A systematic review and meta-analysis. Sports Med. 2019;49(6):951–72.
33. Heerey JJ, et al. What is the prevalence of imaging-defined intra-articular hip pathologies in people with and without pain? A systematic review and meta-analysis. Br J Sports Med. 2018;52(9):581–93.
34. Jayakar R, et al. Magnetic resonance arthrography and the prevalence of acetabular labral tears in patients 50 years of age and older. Skelet Radiol. 2016;45(8):1061–7.
35. Mayes S, et al. Similar prevalence of acetabular labral tear in professional ballet dancers and sporting participants. Clin J Sport Med. 2016; 26(4):307–13.
36. Weir A, et al. Doha agreement meeting on terminology and definitions in groin pain in athletes. Br J Sports Med. 2015;49(12):768–74.
37. Reiman MP, et al. Consensus recommendations on the classification, definition and diagnostic criteria of hip-related pain in young and middle-aged active adults from the International Hip-related Pain Research Network, Zurich 2018. Br J Sports Med. 2020.
38. Clohisy JC, et al. Clinical presentation of patients with symptomatic anterior hip impingement. Clin Orthop Relat Res. 2009;467(3):638–44.
39. Burnett RS, et al. Clinical presentation of patients with tears of the acetabular labrum. J Bone Joint Surg Am. 2006;88(7):1448–57.
40. Nepple JJ, et al. Clinical diagnosis of femoroacetabular impingement. J Am Acad Orthop Surg. 2013;21(Suppl 1):S16–9.
41. Mosler AB, et al. Standardised measurement of physical capacity in young and middle-aged active adults with hip-related pain: recommendations from the first International Hip-related Pain Research Network (IHiPRN) meeting, Zurich, 2018. Br J Sports Med. 2019.
42. Freke MD, et al. Physical impairments in symptomatic femoroacetabular impingement: a systematic review of the evidence. Br J Sports Med. 2016;50(19):1180.
43. van Klij P, et al. The relationship between cam morphology and hip and groin symptoms and signs in young male football players. Scand J Med Sci Sports. 2020.
44. Botha N, et al. Movement patterns during a small knee bend test in academy footballers with femoroacetabular impingement (FAI). Health Sciences Working Papers. 2014;1:10.
45. Diamond LE, et al. Physical impairments and activity limitations in people with femoroacetabular impingement: a systematic review. Br J Sports Med. 2015;49(4):230–42.
46. Reiman MP, et al. Diagnostic accuracy of clinical tests for the diagnosis of hip femoroacetabular impingement/labral tear: a systematic review with meta-analysis. Br J Sports Med. 2015;49(12):811.
47. Sutter R, et al. How useful is the alpha angle for discriminating between symptomatic patients with cam-type femoroacetabular impingement and asymptomatic volunteers? Radiology. 2012;264(2):514–21.
48. Tannast M, Siebenrock KA, Anderson SE. Femoroacetabular impingement: radiographic diagnosis--what the radiologist should know. AJR Am J Roentgenol. 2007;188(6):1540–52.
49. Harris-Hayes M, et al. Movement-pattern training to improve function in people with chronic hip joint pain: a feasibility randomized clinical trial. J Orthop Sports Phys Ther. 2016;46(6):452–61.
50. Kemp JL, et al. The physiotherapy for Femoroacetabular impingement Rehabilitation STudy (physioFIRST): A Pilot Randomized Controlled Trial. J Orthop Sports Phys Ther. 2018;48(4): 307–15.
51. Impellizzeri FM, et al. Patient-reported outcome measures for hip-related pain: a review of the available evidence and a consensus statement from the International Hip-related Pain Research Network, Zurich 2018. Br J Sports Med. 2020;54(14):848–57.
52. Kemp JL, et al. Physiotherapist-led treatment for young to middle-aged active adults with hip-related pain: consensus recommendations from the International Hip-related Pain Research Network, Zurich 2018. Br J Sports Med. 2019;54(9):504–11.
53. Griffin DR, et al. Hip arthroscopy versus best conservative care for the treatment of femoroacetabular impingement syndrome (UK FASHIoN): a multicentre randomised controlled trial. Lancet. 2018;391(10136):2225–35.
54. Palmer AJR, et al. Arthroscopic hip surgery compared with physiotherapy and activity modification for the treatment of symptomatic femoroacetabular impingement: multicentre randomised controlled trial. BMJ. 2019;364:l185.
55. Sogbein OA, et al. Predictors of outcomes after hip arthroscopic surgery for Femoroacetabular impingement: a systematic review. Orthop J Sports Med. 2019;7(6):2325967119848982.
56. Ishoi L, et al. Return to sport and performance after hip arthroscopy for Femoroacetabular impingement in 18- to 30-year-old athletes: a cross-sectional cohort study of 189 athletes. Am J Sports Med. 2018;46(11):2578–87.
57. Levy DM, et al. High rate of return to running for athletes after hip arthroscopy for the treatment of Femoroacetabular impingement and capsular plication. Am J Sports Med. 2017;45(1):127–34.
58. Murata Y, et al. A comparison of clinical outcome between athletes and nonathletes undergoing hip arthroscopy for Femoroacetabular impingement. Clin J Sport Med. 2017;27(4):349–56.

59. Nho SJ, et al. Outcomes after the arthroscopic treatment of Femoroacetabular impingement in a mixed Group of High-Level Athletes. Am J Sports Med. 2011;39:14S–9S.
60. Philippon M, et al. Femoroacetabular impingement in 45 professional athletes: associated pathologies and return to sport following arthroscopic decompression. Knee Surg Sports Traumatol Arthroscopy. 2007;15(7):908–14.
61. Tranovich MJ, et al. A review of femoroacetabular impingement and hip arthroscopy in the athlete. Phys Sportsmed. 2014;42(1):75–87.
62. Emara K, et al. Conservative treatment for mild femoroacetabular impingement. J Orthop Surg (Hong Kong). 2011;19(1):41–5.
63. Wall PD, et al. Nonoperative treatment for femoroacetabular impingement: a systematic review of the literature. PM R. 2013;5(5):418–26.
64. Agricola R, et al. Cam impingement causes osteoarthritis of the hip: a nationwide prospective cohort study (CHECK). Ann Rheum Dis. 2013;72(6):918–23.
65. Nelson AE, et al. Measures of hip morphology are related to development of worsening radiographic hip osteoarthritis over 6 to 13 year follow-up: the Johnston County osteoarthritis project. Osteoarthr Cartil. 2016;24(3):443–50.
66. Nicholls AS, et al. The association between hip morphology parameters and nineteen-year risk of end-stage osteoarthritis of the hip: a nested case-control study. Arthritis Rheum. 2011;63(11):3392–400.
67. Saberi Hosnijeh F, et al. Cam deformity and acetabular dysplasia as risk factors for hip osteoarthritis. Arthritis Rheumatol. 2017;69(1):86–93.
68. Thomas GE, et al. Subclinical deformities of the hip are significant predictors of radiographic osteoarthritis and joint replacement in women. A 20 year longitudinal cohort study. Osteoarthr Cartil. 2014;22(10):1504–10.
69. Agricola R, et al. Pincer deformity does not lead to osteoarthritis of the hip whereas acetabular dysplasia does: acetabular coverage and development of osteoarthritis in a nationwide prospective cohort study (CHECK). Osteoarthr Cartil. 2013;21(10):1514–21.
70. Guanche CA, Sikka RS. Acetabular labral tears with underlying chondromalacia: a possible association with high-level running. Arthroscopy. 2005;21(5):580–5.

Peritrochanteric Disorders in Athletes

15

Yosef Sourugeon, Baris Kocaoglu, Yaron Berkovich, Yaniv Yonai, and Lior Laver

15.1 Introduction

Hip injuries in the athletic population have gained more focus and attention in recent years. Track and field athletes—particularly, those involved in running, have been suggested to have an increased risk of a hip injury [1]. Therefore, it is important for clinicians to recognize, diagnose, and appropriately manage these injuries. Lateral hip pain and peritrochanteric pain have presented a significant challenge to clinicians over the years. Lateral hip pain in athletes is more commonly caused by overuse and can often cause frustration to the athlete due to the longevity and persistence of symptoms. The term "trochanteric bursitis," used for many years for the diagnosis of any focal tenderness over the greater trochanter or lateral hip, has been challenged in recent years by inconsistent or even lack of sufficient supportive findings on imaging, histological, and surgical findings [2–4]. Therefore, the term "greater trochanteric pain syndrome" (GTPS) has been coined to cover the spectrum of disorders causing lateral hip pain, including tendinopathies, strains, and tears of the hip abductor complex—mainly gluteus medius (GMed) and gluteus minimus (GMin)—as well as the tensor fascia latae, trochanteric bursitis, external snapping hip syndrome, and proximal iliotibial band syndrome. Diagnosis of these conditions may be challenging due to variability and sometimes overlap in their clinical presentations [5–7]. Insertional tendinopathy of the GMed and/or GMin is considered the main underlying pathology in GTPS and the main reason for lateral hip pain [8, 9]. Co-existence between more than one of these pathologies is not uncommon. Although GTPS is more common in sedentary individuals, it can be quite common in athletes as well, particularly in runners [10, 11].

This chapter aims to review the current evidence for the underlying causes for peritrochanteric disorders in athletes, their pathomechanics, assessment, and management.

Y. Sourugeon
The Ruth and Bruce Rappaport Faculty of Medicine, Technion Medical School, Technion, Haifa, Israel

B. Kocaoglu
Acibadem MAA University Faculty of Medicine, Department of Orthopedic Surgery, Istanbul, Turkey

Y. Berkovich
Department of Orthopaedics, Hillel Yaffe Medical Center (HYMC), A Technion University Hospital, Hadera, Israel
e-mail: yaronb@hy.health.gov.il

Y. Yonai
HYMC Sports Medicine Unit, Department of Orthopaedics, Hillel Yaffe Medical Center (HYMC), A Technion University Hospital, Hadera, Israel

L. Laver (✉)
HYMC Sports Medicine Unit, Department of Orthopaedics, Hillel Yaffe Medical Center (HYMC), A Technion University Hospital, Hadera, Israel

ArthroSport Sports Medicine, ArthroSport Clinic, Tel-Aviv, Israel

G. L. Canata et al. (eds.), *Management of Track and Field Injuries*,
https://doi.org/10.1007/978-3-030-60216-1_15

15.2 Gluteal Tendinopathy

Gluteal tendinopathy, a condition that is a part of the GTPS spectrum, is now considered the most common cause for lateral hip pain. Over the years, this condition was mistaken with other conditions from the GTPS family, trochanteric bursitis in particular. However, as mentioned earlier, noninflammatory insertional tendinopathy of the GMed and/or GMin is the underlying pathology in most cases of lateral hip pain [2, 8, 9, 12–14]. Gluteal tendinopathy has been reported to be the most prevalent of all lower limb tendinopathies [15]. This condition often occurs in middle-aged sedentary people, but it can also affect athletes. Middle-aged females are more susceptible with reports of up to 23.5% being diagnosed with it compared to 8.5% of males [16]. GTPS interferes with common weight-bearing tasks and sleep, having negative impact on one's health, employment, and well-being, thus making it an extremely debilitating condition [17–19].

15.2.1 Pathomechanics

Tendon injuries often result from disrupted homeostasis due to excessive mechanical loading or lack thereof, which in turn impairs the normal function of the local tendon cell population [20, 21]. It is thought that the combination of high tensile and compression loads is the most damaging [22]. It has been long understood that mechanical loading is the main factor influencing the biological processes occurring within tendons and is responsible for changes in the balance between the catabolic and anabolic processes within the tendon. Changes in loading type, frequency, and intensity may disturb this balance.

Several bony and muscle factors and their interactions are relevant for understanding the pathomechanics of this disorder and provide better tools for optimal management of GTPS.

The GMin and GMed tendons are inserted in the anterior and posterolateral aspects of the greater trochanter (GT), respectively, overlaid by the iliotibial band (ITB), thus making them and the associated bursae vulnerable to mechanical compression (Fig. 15.1) [23]. During hip adduction, the GMin and the GMed tendons are placed under increased tensile load as they move further away from their respective origins in the ilium [24]. Additionally, the ITB applies higher compression loads at GT as the hip adducts, resulting in increased compressive loads at the GMin and the GMed tendons [25]. The accumulation of these forces may be the result of excessive hip adduction during static tasks ("hanging hip" when standing leaning on one hip during standing) and/or dynamic and sports tasks (such as running or landing from repetitive jumping with midline or cross-midline foot–ground contact pattern) [22, 26, 27]. Other aggravating dynamic patterns may include over-striding when running or fast walking, as it thought to increase impact forces on the hip; running with a narrow support base (feet close together), as this pattern exacerbates hip adduction angles, leading to increased compressive and tensile loads on the abductors; running on a camber may also increase hip adduction angles, especially on the higher side; and uphill running is another potential mechanism, as it combines additional hip flexion to the adduction pattern, therefore requiring more pelvic control, which may be compromised in subjects with gluteal tendinopathy and lateral hip pain.

Another suggested mechanism originates from muscle weakness of the hip abductor muscles (GMin, GMed, and tensor fascia lata) [28]. The hip abductor mechanism includes the trochanteric abductors (GMin and GMed) and the ITB-tensing muscles (vastus lateralis, TFL, and the upper portion of the gluteus maximus—GMax) [25, 29–31]. Recent studies have shown that the most common pathology seen on MRI in individuals with lateral hip pain is fatty atrophy in the GMed and/or GMin muscles and defects in their corresponding tendons [13], resulting in hip abductor muscles weakness in symptomatic patients. Moreover, the TFL has been shown to hypertrophy in individuals with tendon pathologies [32]. It is unclear whether these changes in balance between the trochanteric abductors and the ITB tensors precede or result from tendinopathy; however, unbalanced activity of the ITB ten-

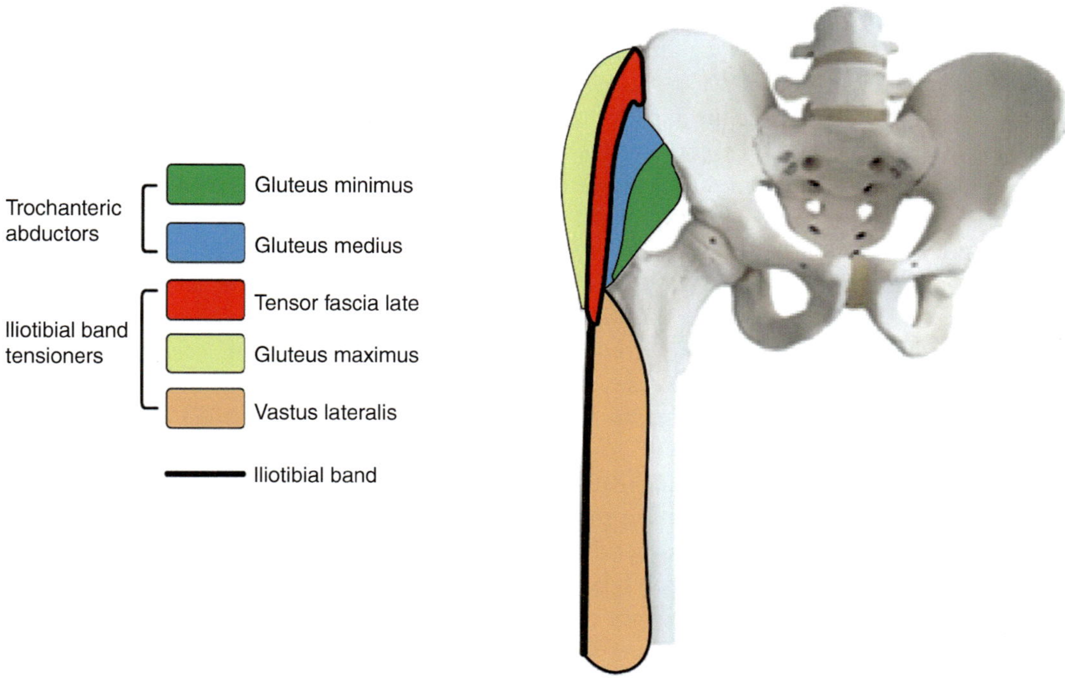

Fig. 15.1 Anatomic relationship between the hip abductors and iliotibial band tensioners

Table 15.1 Description of commonly used diagnostic clinical test for peri-trochanteric pain

Clinical test	Method	Positive test result
Single-leg stance/ Trendelenburg rest (Fig. 15.3)	This test should be performed bilaterally as part of the standing evaluation, with the non-affected side examined first to establish a baseline. The patient needs to elevate the nonbearing side for 30 s	Pelvic drop toward the nonbearing side and/or shift of more than 2 cm toward the bearing side—All within 30 s
FABER (Fig. 15.4)	While the patient is in the supine position, the hip is flexed and externally rotated. The lateral malleolus of the evaluated leg is placed over the patella of the opposite leg in a "figure of 4" position.	Reproduction of pain—Anterior pain is suggestive of hip joint pathology, while posterior pain is suggestive of sacroiliac pathology. Lateral pain over the greater trochanter may suggest for GTPS.
Hip lag sign (Fig. 15.5a, b)	The patient lies in the lateral position with the affected leg on top. With one hand, the examiner stabilizes the hip while he passively extends (10°) and abducts (20°) the hip with the other. The patient's knee must always be kept flexed at 45°.	Failure to keep the leg in the aforementioned position and/or the foot drops for more than 10 cm
Ober's rest (Fig. 15.6)	The patient lies in the lateral position with the affected leg on top. The examiner stabilizes the pelvis from behind with one hand, while the other hand slightly abducts the hip and extends it to the end of the range. The examiner then slowly releases the support from the upper leg.	This test is considered positive if the upper leg remains in abduction/does not go into adduction past midline after the examiner stops supporting the weight of the leg. The patient may also report lateral knee pain in a positive test. It is important to note there is no compensatory internal rotation at the hip, which may result in a false-negative result. A positive test suggests ITB and TFL tightness

sioners could potentially exert higher compressive forces over the GT during hip abduction.

Bony morphology may also affect the compressive forces transmitted at the hip by the ITB. At the typical physiological femoral neck angle of 128°, the ITB has been shown to exert a compressive force of 656 N at the GT, while at 115° (coxa vara), the compressive force exerted was 997 N (Fig. 15.2) [25, 33]. The neck–shaft angle might also affect the offset between the iliac wings and GTs as an increased offset, may further increase the compression subjected by the ITB against the gluteals [34].

Lastly, there is growing evidence that hip dysplasia may cause tendon-related pain and their prevalence in the population may be higher than perceived, especially for borderline hip dysplasia [35, 36]. Resulting biomechanical changes can elicit higher loads on the lateral hip muscles and tendons, which may lead to overuse-related injuries [35].

15.2.1.1 Diagnosis

Patients usually present with pain and tenderness localized to the greater trochanter (GT). Radiation of pain down the lateral thigh toward the knee is also a common complaint. The onset of pain is gradual and usually worsens over time and is triggered by exercise, overuse of sport activities, falls, and prolonged weight-bearing activities. Pain usually worsens at night, and those affected have trouble sleeping on the affected side, which could negatively impact their sleep quality.

Other common complaints include lateral hip pain during single-leg loading tasks such as climbing stairs and walking and running uphill, as well as lateral hip pain and stiffness on extending the hip when rising to stand.

Clinical hip tests generally possess weak diagnostic properties but may provide useful information in the differential diagnosis of lateral hip pain. Direct palpation is considered as the main clinical sign and the most important sign, reported by Grimaldi et al. to have a 83% PPV for positive

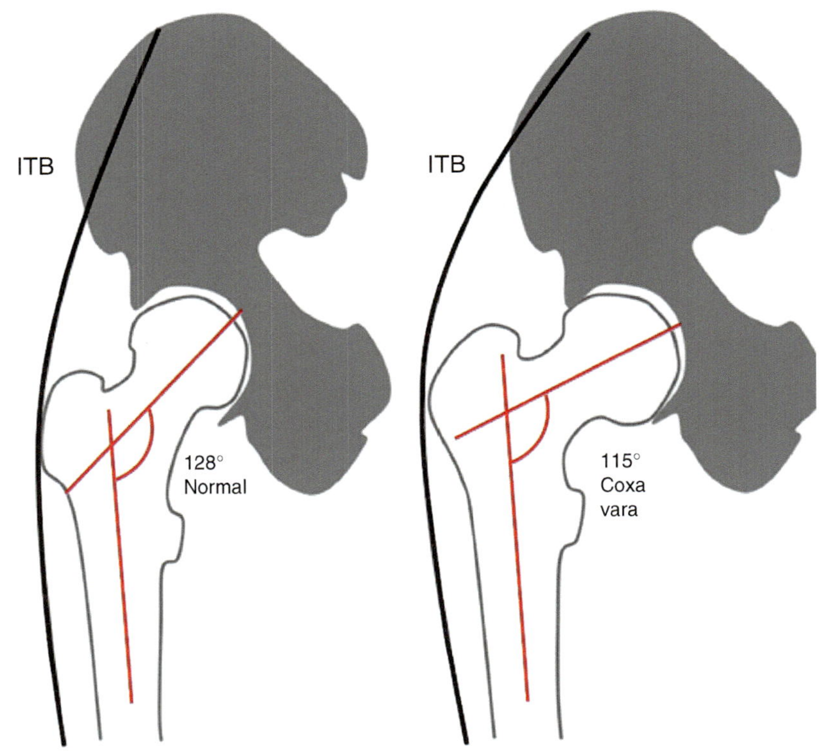

Fig. 15.2 Effect of femoral neck angles on compressive loads at the greater trochanter. Lower neck angles (coxa vara) result in higher compressive forces than normal neck angles [25]

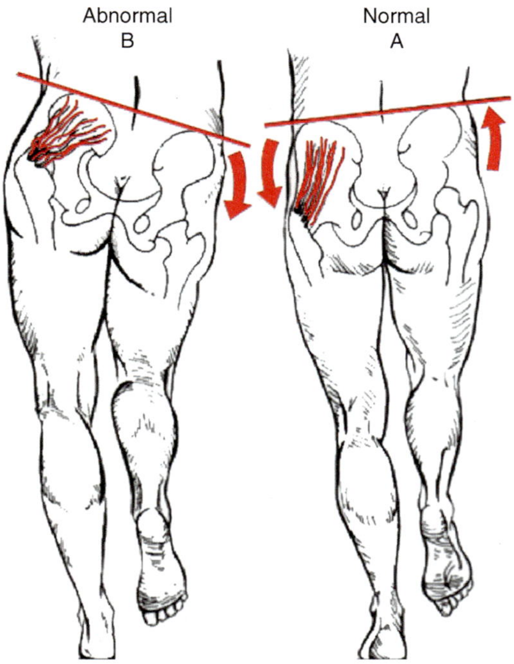

Fig. 15.3 Single-leg stance/Trendelenburg test

MRI findings [12]. If there is no pain on palpation, it is unlikely that the patient has GTPS. Another test is the single-leg stance (100% PPV for positive MRI findings) [12]; however, a negative test does not exclude the possibility of GTPS. The FABER (flexion, abduction, and external rotation) test is considered positive if it reproduces pain in the lateral hip without limited range of motion (ROM). Patients with limited ROM with or without pain during the FABER test should be suspected of intra-articular hip joint-based pathologies rather than GTPS. Finally, a positive hip lag sign has been reported to yield sensitivity and specificity for abductor tendon damage as high as 89% and 96%, respectively [37].

In GMed and GMin tears, in addition to the above mentioned, the patient may present with visible symptoms of hip abductor weakness, including Trendelenburg gait and sign [14].

GTPS is generally acknowledged as being a clinical diagnosis [38]; however, imaging can be used if the differential diagnosis is unclear. Hip X-ray is a useful tool to exclude common differentials such as osteoarthritis of the hip and fractures [39].

Ultrasound (US) and MRI scans are the dominant imaging tools for radiological investigation in GTPS. Generally, MRI is more sensitive in detecting gluteal tendon pathologies, and it can also exclude other etiologies for lateral hip pain and is considered the golden standard for gluteal tendon assessment [2, 9]. While US scans are easily available are substantially more affordable, and allow for a dynamic evaluation, they are very much "operator dependent." The sensitivity of standard US to surgically confirmed GMed tendon tears has been reported to be as low as 61%, with lacking evidence regarding GMin tears [4]. Newer sonographic techniques such as ultrasound tissue characterization (UTC) [40] and elastography [41] have been suggested to offer improved tendon structural characterization and assessment; however, further research is necessary to establish their clinical utility. MRI can detect direct signs of gluteal tendinopathy such as focal tendon discontinuity, soft tissue edema, tendon thickening, and intrasubstance abnormality—as well as associated indirect signs such as gluteal fatty atrophy and bursal inflammation [9].

15.2.1.2 Other Related Conditions

Primary trochanteric bursitis is rare, and it may coexist as a secondary process [9, 42]. Diagnostic features are not specific and are similar to those discussed above. MRI and US may reveal evidence of bursal inflammation. Treatment is usually conservative and consists of physical therapy that is focused on core strengthening and stretching [43, 44] and injections (CSI, PRP). Surgical intervention with bursal debridement should be considered in refractory cases who failed conservative treatment.

Proximal ITB syndrome is a proximal IT band strain that is commonly mistaken for hip-related pathology and may be confused with other GTPS conditions. This overuse injury is predominant in women and is relatively common in triathletes [45–47]. Pathophysiology of this condition is based on thickening of the proximal portion of the ITB [48]. Patients will present with symptoms of

Table 15.2 Description of common clinical tests used in the diagnosis of Piroformis Syndrome

Clinical test	Method	Positive test result
Seated piriformis stretch test (Fig. 15.8)	The patient is in the seated position, with 90 degrees of hip flexion. The examiner extends the knee (engaging the sciatic nerve) and passively adducts the flexed hip with internal rotation while palpating about 1 cm lateral to the ischium and proximally toward the sciatic notch	Reproduction of buttock or sciatic pain
Pace abduction test (Fig. 15.9)	The patient is seated with the knees adducted in a normal position, flexed at 90°. The patient is asked to abduct the hips against the examiner's resistance placed on the lateral aspect of each knee	Pain in the piriformis area and weakness with resisted abduction in the seated position
Freiberg test (Fig. 15.10)	While the patient is in the prone position and the knee of the affected side is flexed to 90°, the examiner moves the affected leg into internal rotation and so stretches the piriformis muscle	Reproduction of buttock or sciatic pain, as well as internal rotation tightness on the affected side when comparing sides
Beatty test (Fig. 15.11a)	The patient is in the lateral position with the affected leg on top and the affected side knee flexed and resting on the examination bed. The patient is then asked to lift and hold the knee approximately 10 cm above the table	Reproduction of buttock or sciatic pain
Modified Beatty test (Fig. 15.11b)	Adding resistance to the Beatty test and asking the patient to abduct against resistance.	Reproduction of buttock or sciatic pain
Active piriformis test (Fig. 15.12)	The patient lies in the lateral position on the unaffected hip. The patient is asked to flex the affected side knee and to push the heel down into the table while actively abducting the hip with external rotation against resistance	Reproduction of buttock or sciatic pain
FAIR test (flexion, adduction, and internal rotation) test (Fig. 15.13)	The patient lies in lateral position with the affected leg on top. With one hand the examiner stabilizes the hip while he passively internally rotates and adducts the hip. The patient's knee must always be kept flexed at 90°.	Reproduction of buttock or sciatic pain

pain and tenderness around lateral hip with or without associated knee stiffness, and sometimes will be diagnosed with GTPS refractory to hip injections and physical therapy. This condition may also be confused with intra-articular pathologies. The primary imaging modality is the MRI scan, which can detect strain injury at the iliac tubercle enthesis of the ITB and its thickening [48]. However, one might consider using US to detect structural changes and signs of inflammation in this area [49]. This condition is unique and thus far has little to no literature about it. Treatment is limited and includes rest and physical therapy for stretching and strengthening the muscles of the ITB [50]. In patients that fail conservative treatment, surgical ITB release should be considered.

TFL strain is an injury that usually affects runners and athletes, and it may or may not involve the ITB. The TFL rises from the ASIS, extends posteriorly along the iliac crest, and inserts to the ITB and the lateral condyle of the tibia. The TFL consists of two heads—anteromedial (AM) and posterolateral (PL) [51]. Strain of the PL head of the TFL can mimic the symptomology of GTPS and should be considered when discussing lateral hip pain. Strain of the AL head will present with pain in the groin area. Patient history may include running recently on a banked surface. Physical examination can detect focal tenderness over the TFL and a positive Ober's test [52]. Treatment should consist of rest and physical therapy.

TFL tears pose a unique challenge for the clinician since reports on this condition are scarce. Patients that were diagnosed with this condition were usually asymptomatic but sometimes suffered from pain in the buttock and groin area that was exacerbated by standing or walking. Patient history is unspecific and may or may not include an account of trauma or overuse. Physical examination is unspecific as well and may include pain during hip extension and a palpable soft tissue mass on the anterolateral region of the hip. MRI scan is the imaging modality of choice as it can detect TFL tears with ease [53]. TFL tears are almost always located close to the proximal insertion to the anterior aspect of the iliac crest. Treatment is conservative and consists of physical therapy. US-guided PRP injections may enhance healing in acute tears and persistent TFL injuries, which do not resolve by physical therapy alone.

15.2.1.3 Management

Clear evidence-based guidelines and protocols for the management of GTPS are yet to be established [10]. So far, conservative treatment is considered the gold standard with over 90% success rate [54]. Treatments include anti-inflammation medication, exercise, and strategies to manage tendon load, shockwave therapy, and surgical interventions [22, 55].

.In the acute phase, pain can be managed with ice, taping, and anti-inflammatory medication. Topical and oral nonsteroidal anti-inflammatory drugs (NSAIDs) have equal benefits [56].

ITB and piriformis stretch exercises proved to be largely unhelpful [57]. For athletes, controlling load management is key in the process of rehabilitation. Complete rest may prove to be catabolic for tendons, but reducing high-intensity activities such as higher speed and longer distance running may be helpful [23]. Alteration of the running technique, specifically reducing peak hip adduction, might be required and can be achieved using biofeedback [58]. Stretching exercises for lower limb tendinopathies are not recommended due to high compressive and tensile loads on gluteal tendon insertions [22, 23].

Radial shockwave therapy (SWT) emits shock waves that can penetrate soft tissue to a depth of 40 mm [59], inducing a mechanobiological effect that has been suggested to promote healing and an analgesic reaction on painful tendons [57, 60, 61]. However, no high-quality randomized trials have proved superior therapeutic capabilities of SWT over other methods of treatment [62], nor have they taken into consideration the effect that difference in adiposity between patients can have on SWT [22]. Thus, the net outcomes of this treatment remain unclear.

Corticosteroid injection (CSI) has been shown to be very useful in short-term pain relief, with up to 75% response in the first 4 weeks [63, 64]. Alas, CSI does not always alleviate the pain completely, and in the medium and longer term, the positive response drops to 41–55% [57, 65, 66]. Pain recurrence following CSI suggests that this treatment does not target and treat the underlying pathology associated with longer-term tendon pain as seen in other tendinopathies [67]. Furthermore, studies suggest that CSI might limit the tendon ability to respond to loads, especially around the enthesis, by downregulating the production of collagen by fibroblasts [68]. CSI should be used to reduce pain in a manner that would allow a return to a moderate physical activity and physical therapy [60, 69].

PRP (platelet-rich plasma) injections contain various endogenous growth factors that have been shown to possess the potential to accelerate the natural process of healing and alleviate pain. Recent high-level studies suggest that patients who failed conservative treatment might benefit from PRP, and in the long term proved to be more effective than CSI [70–72]. Fitzpatrick et al. reported that a single intra-tendinous leukocyte-rich PRP (LR-PRP) injection performed under ultrasound guidance results in greater improvement in pain and function compared to a single CSI. The improvement after LR-PRP injection was sustained at 2 years, whereas the improvement from a CSI was maximal at 6 weeks and was not maintained beyond 24 weeks [70]. However, one should keep in mind that different PRP preparation protocols might have different ingredients and may thus have an impact on efficacy [73].

Surgical intervention should be considered for individuals who failed to rehabilitate following appropriate conservative treatment [10]. Surgical solutions include bursectomy, ITB release, and gluteal tendon repair. Often, a combination of interventions is incorporated during surgery.

When evaluating the effectiveness of surgical management for gluteal tendinopathy, one of the main limitations lies in the relatively low methodological quality of the relevant studies. Therefore, a level of caution is required when assessing the various surgical techniques. Additionally, studies that were made on athletes were scarce and the vast majority of results were reported on surgical interventions in the general population.

Gluteal Tendon Repair

Tears can be partial, full thickness, or intrasubstance, most commonly involving the lateral portion of the GMed tendon [74, 75]. Both endoscopic and open techniques show good results [74–79]. Not many comparisons between the two techniques regarding outcome superiority have been done so far, and no technique has been shown superior over the other [79]. Endoscopic surgery benefits over open surgery include small incisions, quicker healing time, less postoperative pain, and shorter theater and hospitalization time [80]. The main drawbacks of the endoscopic technique are the greater surgical skill required and the limited use in cases of larger tears or tendon detachments, where better visualization and exposure are required [81]. Postoperative complications have been reported to be relatively high with rates reaching up to 19%, most commonly including deep vein thrombosis (DVT) and tendon re-tears [76, 77].

ITB Release/Lengthening

As previously discussed, the ITB exerts anatomical pressure onto the soft tissue structures it envelops and therefore ITB surgical lengthening could potentially reduce this pressure. However, no study has shown that this population suffered from primary ITB tightness [22]. It is possible that the excessive ITB pressure applied on the gluteal tendons is secondary to a combination of weakened hip abductor muscles and excessive adduction. In this case, lengthening of the ITB will provide pain relief through immediate reduction in pressure, but it will not treat the underlying pathology and may cause additional complications: (1) This procedure does not solve the underlying pathomechanics of weak hip abductors and poor hip control; (2) herniation of underlying soft tissue and painful external snapping may occur due to excessive ITB resection; and (3) further worsening the abductor muscles function due to reduced ITB control potential affecting the muscles attached to it (TFL and GMax) [82].

With that in mind, studies have shown that this technique provides good long-term outcomes, with low postoperative complications [55, 80].

Trochanteric Bursectomy

Endoscopic trochanteric bursectomy has been reported to provide good long-term outcomes [83, 84] and is often performed in conjunction with ITB lengthening. Pain relief is explained by the removal of the inflamed bursa, but this procedure does not address the underlying pathomechanics discussed earlier.

15.3 "Hip Pointer" Injuries

Hip pointer injuries refer to a contusion of the iliac crest and/or the GT area following a direct impact or collision [85]. This term has also been used to describe fractures or favulsions around the lateral hip area. A hematoma often develops following the direct impact to the iliac crest and GT, and consequently, varying degrees of bleeding may occur into the hip abductor musculature (Fig. 15.7). While these injuries are more common in contact and collision sports, they are probably more common than reported in track and field athletes as falls during various hurdle running is not uncommon, In addition, track and field athletes are usually very lean and the iliac crest and GT area in this population are less padded by natural tissue and are less protected during falls. The iliac crest is the origin of several muscles that can be affected: the sartorius, TFL, GMed, and even the adjacent abdominal musculature—specifically the transverse or oblique muscles. Radiographs are often necessary to rule out fractures. A resultant subperiosteal hematoma can develop in these cases and may lead to myositis ossificans (MO), leading to significant disability and pain. Additional radiographs and ultrasound (can detect MO formation in its early stages) should be performed if MO formation is suspected. A high degree of attention is required in high school and college athletes since the anterior superior iliac spine (ASIS) may fuse as late as the third decade of life. This population should be evaluated to rule out potential avulsion injuries in the area, including ASIS avulsion injuries (sartorius and/or TFL avulsion), Iliac crest avulsion injuries secondary to abdominal musculature avulsions, and less common in this mechanism—anterior inferior iliac spine (AIIS) avulsion injuries (rectus femoris avulsion). A CT scan or an MRI should be considered if the patient has continued pain or pain, which exceeds that expected from examination findings. Treatment is often symptomatic, including ice, compression, rest, and potentially protected ambulation. Avoiding vigorous activity for 48 h may reduce recurrent bleeding. Surgical intervention is uncommon for these injuries and is only considered in avulsion cases with large retraction (>2 cm). In very rare cases, persistent significant swelling may develop in the peritrochanteric area, and in these occasions, a Morel-Lavallée lesion should be suspected. This is a closed degloving soft tissue injury, as a result of abrupt separation of skin and subcutaneous tissue from the underlying fascia, which disrupts perforating vessels and lymphatics, thus creating a potential space filled with serosanguinous fluid, blood, and necrotic fat. If persistent bone edema is present on MRI, hyperbaric oxygen therapy has been suggested as a potential therapy to enhance healing [86]. Ultrasound-guided PRP injections and in specific persistent cases a local anesthetic injection combined with CSI may facilitate return to activity (however require a gradual return to activity, especially following a CSI) [85, 87].

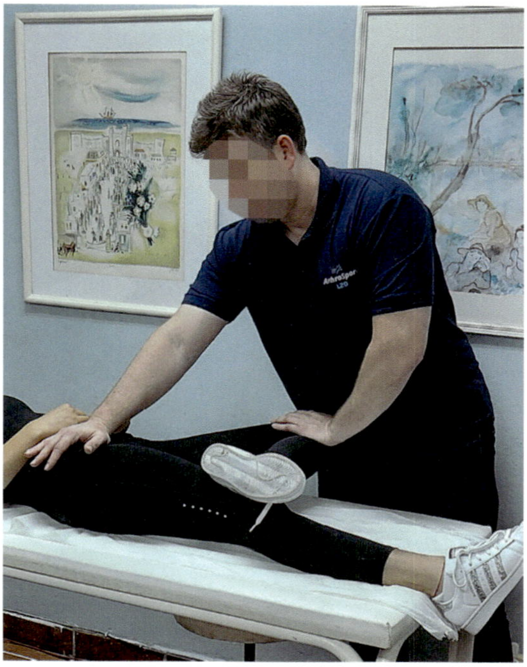

Fig. 15.4 FABER test

15.4 Painful Snapping Hip Syndrome

Snapping hip syndrome (SHS), also known as *coxa saltans* (or dancer's hip), is a clinical condition characterized by an audible or palpable snap of the hip. The prevalence among the general population is up to 10% [88]. However, in selected populations, especially those who require higher hip ROM like ballet dancers, prevalence can reach up to 90% [89]. SHS has multiple etiologies that can be classified into two main subcategories based on the anatomic origins of the snapping sensation—extra-articular (ESHS) and intra-articular (ISHS). ISHS is usually caused by loose bodies and labrum tears and will not be discussed in this chapter.

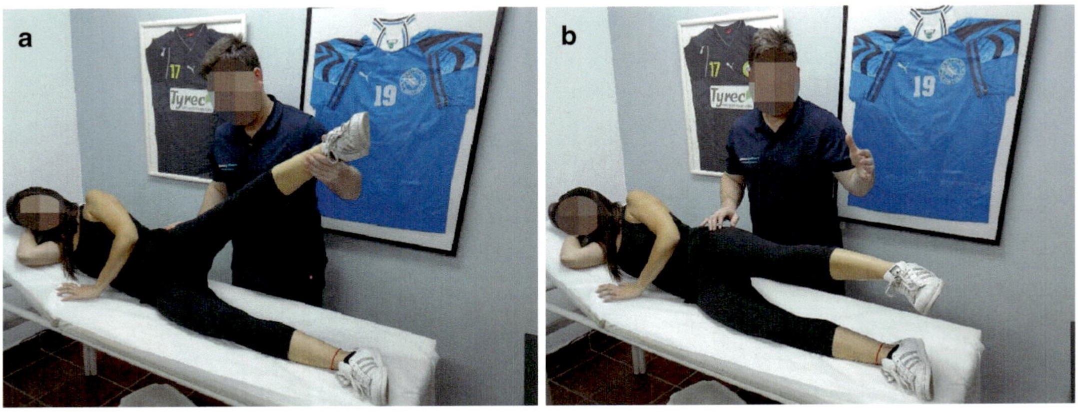

Fig. 15.5 (**a**) Hip lag sign, (**b**) Hip lag sign

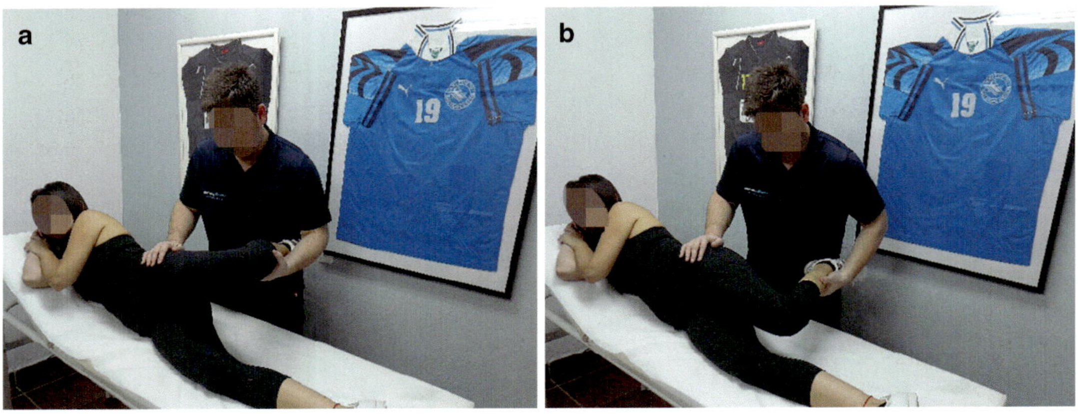

Fig. 15.6 (**a**, **b**) Ober's test

Fig. 15.7 Hip pointer injuries with hematomas around the iliac crest (**a**) and greater trochanter (**b**)

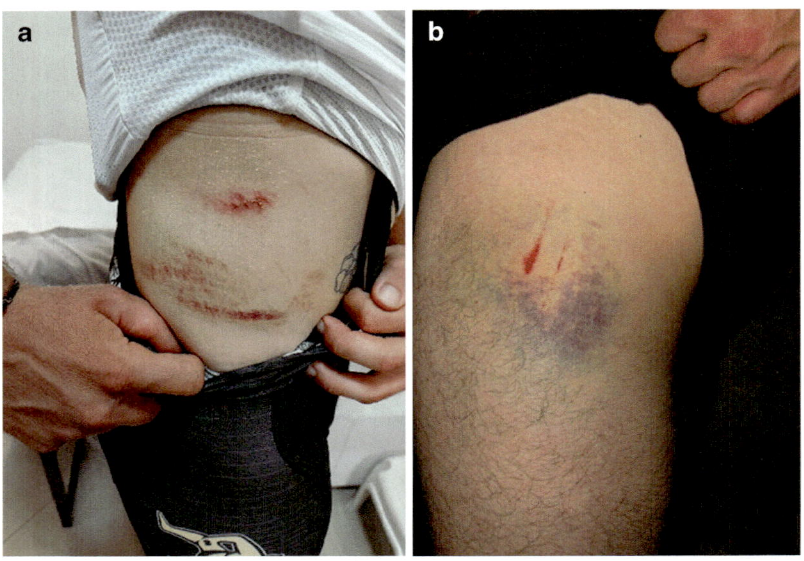

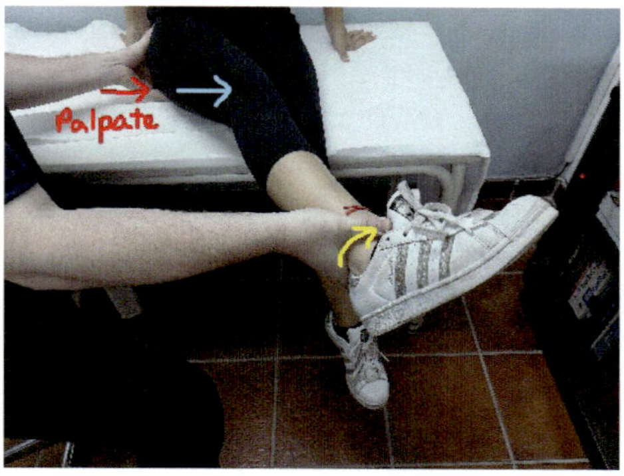

Fig. 15.8 Seated piriformis stretch test

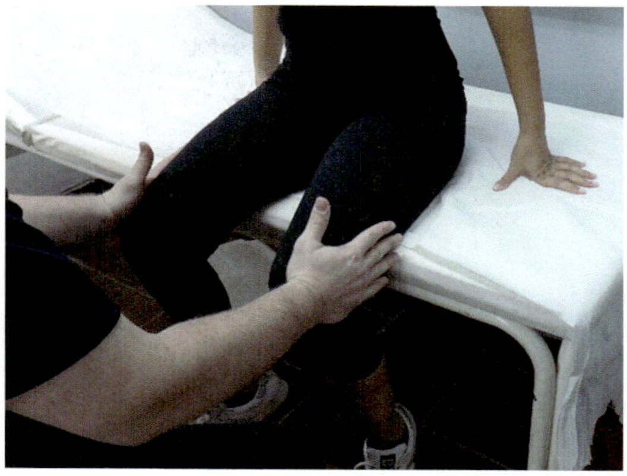

Fig. 15.9 Pace abduction test

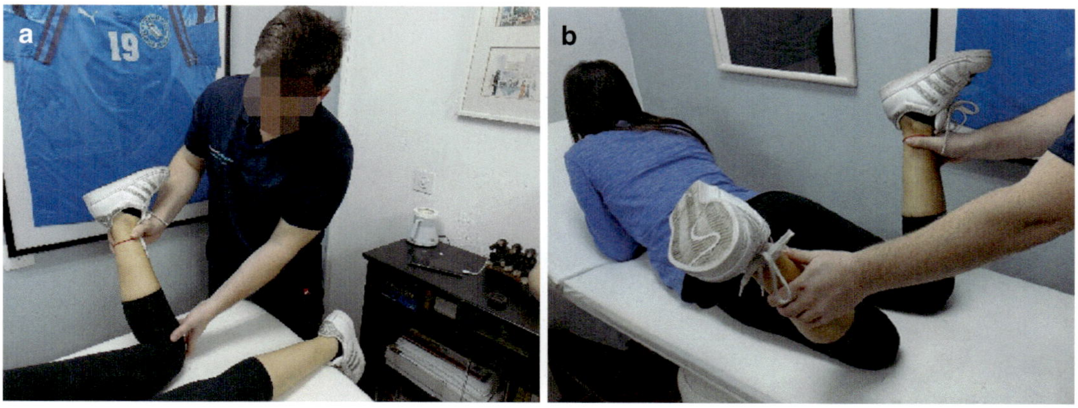

Fig. 15.10 (**a**) Freiberg test. (**b**) Freiberg test

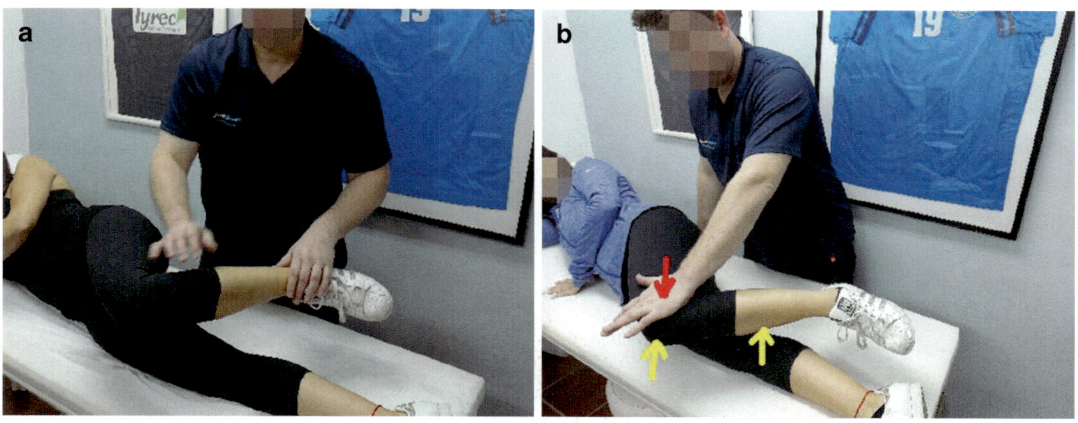

Fig. 15.11 (**a**) Beatty test. (**b**) Modified Beatty test

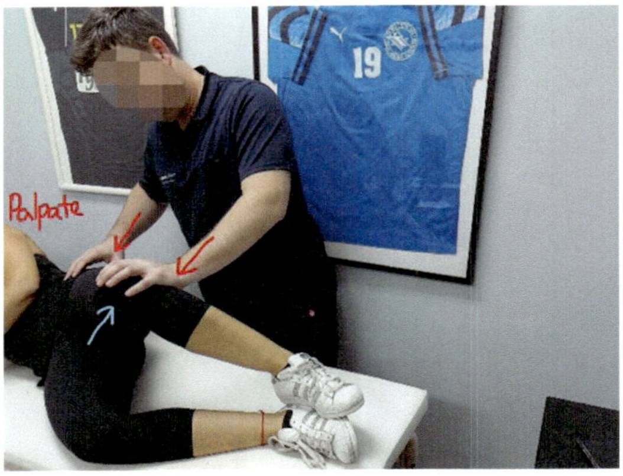

Fig. 15.12 Active piriformis test

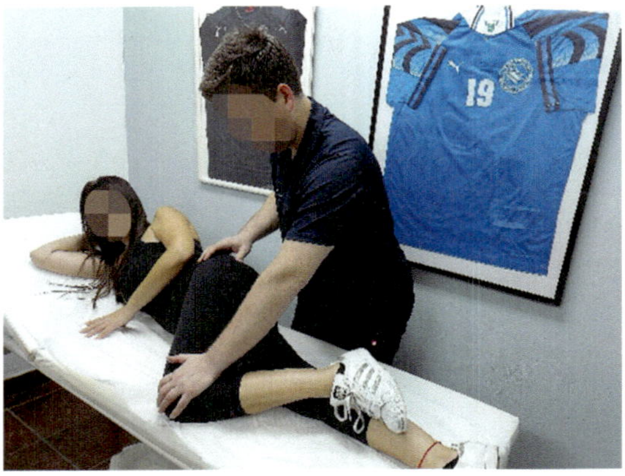

Fig. 15.13 FAIR test

15.4.1 Pathomechanics

There are two forms of ESHS that are generally accepted—internal and external, with the external form being more prevalent [90].

Most commonly, the external form of ESHS is caused by the sliding of the ITB over the greater trochanter during hip movements such as flexion and extension [90, 91], but it can also be caused by the snapping of the gluteus maximus itself [92]. The GMax and the TFL attach to the ITB posteriorly and anteriorly, respectively. Thickening of the posterior insertion of the GMax or its anterior aspect of the GMax may further accentuate the sound and the clicking sensation [93]. Pain is usually absent in ESHS, but it may be provoked due to the compression of the trochanteric bursae between the ITB and the GT, thus making it one of the causes for GTPS.

In the internal form of ESHS, the snap is generally attributed to the movement of the iliopsoas tendon passing anterior to the hip joint [88]. It is believed that the most common mechanism involves the iliopsoas tendon snapping over the iliopectineal eminence and the femoral head [88, 94]. It can also be caused by the iliacus muscle itself, snapping between the iliopsoas tendon and the pubic bone [89, 95].

15.4.1.1 Diagnosis

Careful history should be taken. Patients will describe a sensation as snapping or cracking and will often direct the physician to the region of interest and the underlying pathomechanism. The external form of ESHS is often described as a feeling that the hip dislocates. The internal form of ESHS is described as a deep snapping or "getting stuck" and is usually localized to the anterior aspect of the hip [90, 93]. It is generally accepted the external form of ESHS produces an audible snap, while the internal form only produces a snapping sensation.

Tests to determine ESHS are provocative tests, aiming to reproduce the characteristic audible snap [96]. Such tests typically include femoral rotation and/or flexion, flexion, and/or extension of the hip [91, 92, 97, 98]. For the internal form of ESHS, the test generally requires iliopsoas contraction [99, 100]. To reproduce the internal snap, the patient's leg is moved from the FABER (flexion, abduction, and external rotation) position to EAdIR (extension, adduction, internal rotation) [90]. Patients should be evaluated for ROM and hip stability tests. Assessment should include the FADIR test (flexion, adduction, and internal rotation or anterior impingement test) to examine the possibility of intra-articular pathologies as internal ESHS can often be secondary to an intra-articular pathology leading to secondary iliopsoas tightness and snapping. Visible muscle weakness of the GMed is often common with SHS [101].

ESHS diagnosis is clinical, and while plain radiographs will usually have no meaningful findings, they may sometimes reveal anatomical conditions that can perpetuate this condition, i.e., coxa vara [101]. MRI may show ITB and GMax thickening alongside unspecific signs of inflammation [102]. Iliopsoas bursography can be used to diagnose internal ESHS by filling the bursa with contrast, and under fluoroscopy, the tendon can be visualized flipping back and forth. Ultrasound can be used to visualize the dynamic motion of the iliopsoas tendon, the ITB, or the GMax as the hip moves and can also detect signs of inflammation [91, 95, 103]. Finally, if the patient is in pain, an US-guided anesthetic injection into the iliopsoas bursa can be diagnostic of internal ESHS when this procedure provides temporary pain relief [104, 105].

15.4.1.2 Management

Most patients are not symptomatic and do not require treatment. If the snapping becomes symptomatic, conservative management is attempted first. Conservative management includes rest, icing, avoidance of aggressive activities, anti-inflammation medication, and physical therapy [88].

A combined injection of CSI and anesthetic to bursal tissue or around the tendon sheath can give symptomatic relief [104–106].

Physical therapy can help regain normal function within 6–12 months, focusing on the underlying cause for SHS. If the muscles are too short, stretching exercises should be done combined with correction of habitual movement patterns. If excessive muscular activation exists, intervention should be directed at the neuromuscular control over movement [90].

Surgery is done on patients with painful snapping refractory to conservative treatment. The main goal, both in the internal and external forms, is to relax the involved tendon to eliminate snapping. This is accomplished by various types of lengthening procedures and can be done both open and arthroscopically. These procedures are rather rare, and the research on the subject is even rarer.

In the external form of ESHS, the goal is to relax the ITB, and the predominant procedures are Z-shape release, formal Z-lengthening, a cross-shaped release, and release of the gluteus maximus tendon femoral insertion. The only study conducted on physically active patients evaluated the Z-lengthening procedure, with 8 patients being active-duty soldiers [97, 107–109]. Complications include mild-to-moderate Trendelenburg gait, which could be disastrous to athlete [97].

In the internal form of ESHS, the goal is to relax the iliopsoas tendon through various techniques of fractional lengthening or by complete release-based. Current literature favors arthroscopic procedures, with some studies conducted on athletes [100, 106, 110].

15.5 Piriformis Syndrome

The concept behind this somewhat controversial syndrome focuses on the piriformis muscle as a potential reason for sciatica and unilateral gluteal pain [111]. The pathomechanics of this syndrome are still poorly understood, with theories spanning from piriformis anatomical variations to repetitive trauma to the buttocks area [112–115]. At the core of this condition, there is an irritation of the sciatic nerve by the piriformis muscle. It has been reported that this syndrome is responsible for at least 6% of all cases of lower back pain with/without sciatica [116, 117].

Pain emergence is usually insidious and gradually worsens. Common symptoms include lower back pain, tenderness around the buttocks area, difficulty with activities that strain the gluteal region like prolonged sitting, and sciatica-like symptoms. It is important to note that this condition might mimic GTPS symptomology, thus carefully examining the patient's history is paramount.

Piriformis syndrome is often a diagnosis of exclusion. Physical examination is quite inconclusive as many tests elicit local buttock pain and shooting leg pain, but none can rebut other sources of pain from the lumbosacral region for example. Commonly used tests are described in Table 2. Imaging and EMG studies are used to exclude other conditions.

15.5.1 Treatment

Conservative therapy is the most effective treatment and includes administration of muscle relaxants, NSAIDs, rest, and physical therapy.

Physical therapy, based on stretching exercises, can help alleviate symptoms [118–120]. For sciatica symptoms, a local CSI with or without anesthetics [114, 121], or a botulinum type A injection might help [122]. The idea behind the injections is also diagnostic, as well as to provide pain relief and relaxation (in the case of botulinum injections) in order to better engage in stretching protocols for the structures surrounding the sciatic nerve.

Surgical treatment should be reserved for patients who have failed conservative treatment. Techniques include nerve decompression if nerve impingement is present, removal of adhesions, and scars from the nerve and piriformis release. Results of these procedures are unpredictable, and some patients continue to feel pain [123].

15.6 Summary

Various reasons exist for lateral hip and peritrochanteric pain in athletes. It is important to understand the differential diagnosis and potential conditions causing lateral hip and peritrochanteric pain in this population, as well as the underlying mechanisms for each condition to optimize management strategies. While most conditions and cases respond to conservative treatments, refractory cases may sometimes require surgical intervention. The underlying mechanisms for the various conditions described in this chapter are not fully understood and a better understanding of the pathomechanics in each condition could aid in devising more concise and efficient treatment strategies, with a faster and uncomplicated return to sport.

References

1. Scopp JM, Moorman CT. The assessment of athletic hip injury. Clin Sports Med. 2001;20(4):647–59. https://doi.org/10.1016/S0278-5919(05)70277-5.
2. Bird PA, Oakley SP, Shnier R, Kirkham BW. Prospective evaluation of magnetic resonance imaging and physical examination findings in patients with greater trochanteric pain syndrome. Arthritis Rheum. 2001;44(9):2138–45. https://doi.org/10.1002/1529-0131(200109)44:9<2138::AID-ART367>3.0.CO;2-M.
3. Connell DA, Bass C, Sykes CJ, Young D, Edwards E. Sonographic evaluation of gluteus medius and minimus tendinopathy. Eur Radiol. 2003;13(6):1339–47. https://doi.org/10.1007/s00330-002-1740-4.
4. Fearon AM, Scarvell JM, Cook JL, Smith PNF. Does ultrasound correlate with surgical or histologic findings in greater trochanteric pain syndrome? A pilot study. Clin Orthop Relat Res. 2010;468(7):1838–44. https://doi.org/10.1007/s11999-009-1174-2.
5. Williams BS, Cohen SP. Greater trochanteric pain syndrome: a review of anatomy, diagnosis and treatment. Anesth Analg. 2009;108(5):1662–70. https://doi.org/10.1213/ane.0b013e31819d6562.
6. Hugo D, de Jongh HR. Greater trochanteric pain syndrome. South African Orthop J. 2012;11:1. http://www.scielo.org.za/scielo.php?script=sci_arttext&pid=S1681-150X2012000100005. Accessed 10 Aug 2020
7. Klauser A, Martinoli C, Tagliafico A, Bellmann-Weiler R, Feuchtner G, Wick M, Jaschke W. Greater trochanteric pain syndrome. Semin Musculoskelet Radiol. 2013;17(1):43–8. https://doi.org/10.1055/s-0033-1333913.
8. Kingzett-Taylor A, Tirman PFJ, Feller J, McGann W, Prieto V, Wischer T, Cameron JA, Cvitanic O, Genant HK. Tendinosis and tears of gluteus medius and minimus muscles as a cause of hip pain: MR imaging findings. Am J Roentgenol. 1999;173(4):1123–6. https://doi.org/10.2214/ajr.173.4.10511191.
9. Kong A, Vliet A, Zadow S. MRI and US of gluteal tendinopathy in greater trochanteric pain syndrome. Eur Radiol. 2007;17(7):1772–83. https://doi.org/10.1007/s00330-006-0485-x.
10. Del Buono A, Papalia R, Khanduja V, Denaro V, Maffulli N. Management of the greater trochanteric pain syndrome: a systematic review. Br Med Bull. 2012;102:115–31. https://doi.org/10.1093/bmb/ldr038.
11. Fields KB. Running injuries V changing trends and demographics. Curr Sports Med Rep. 2011;10(5):299–303. https://doi.org/10.1249/JSR.0b013e31822d403f.
12. Grimaldi A, Mellor R, Nicolson P, Hodges P, Bennell K, Vicenzino B. Utility of clinical tests to diagnose MRI-confirmed gluteal tendinopathy in patients presenting with lateral hip pain. Br J Sports Med. 2017;51(6):519–24. https://doi.org/10.1136/bjsports-2016-096175.
13. Pfirrmann CWA, Notzli HP, Dora C, Hodler J, Zanetti M. Abductor tendons and muscles assessed at MR imaging after total hip arthroplasty in asymptomatic and symptomatic patients. Radiology. 2005;235(3):969–76. https://doi.org/10.1148/radiol.2353040403.
14. Brukner P, Khan K. Brukner and Khan's clinical sports medicine. 4th ed. North Ryde: N.S.W. McGraw-Hill; 2012.
15. Albers IS, Zwerver J, Diercks RL, Dekker JH, Van Den Akker-Scheek I. Incidence and prevalence of lower extremity tendinopathy in a Dutch general practice population: a cross sectional study. BMC

Musculoskelet Disord. 2016;17:6. https://doi.org/10.1186/s12891-016-0885-2.
16. Segal NA, Felson DT, Torner JC, Zhu Y, Curtis JR, Niu J, Nevitt MC. Greater trochanteric pain syndrome: epidemiology and associated factors. Arch Phys Med Rehabil. 2007;88(8):988–92. https://doi.org/10.1016/j.apmr.2007.04.014.
17. Mellor R, Grimaldi A, Wajswelner H, Hodges P, Abbott JH, Bennell K, Vicenzino B. Exercise and load modification versus corticosteroid injection versus “wait and see” for persistent gluteus medius/minimus tendinopathy (the LEAP trial): a protocol for a randomised clinical trial. BMC Musculoskelet Disord. 2016;17:1. https://doi.org/10.1186/s12891-016-1043-6.
18. Fearon AM, Stephens S, Cook JL, Smith PN, Neeman T, Cormick W, Scarvell JM. The relationship of femoral neck shaft angle and adiposity to greater trochanteric pain syndrome in women. A case control morphology and anthropometric study. Br J Sports Med. 2012;46(12):888–92. https://doi.org/10.1136/bjsports-2011-090744.
19. Lequesne M. From “periarthritis” to hip “rotator cuff” tears. Trochanteric tendinobursitis. Jt Bone Spine. 2006;73(4):344–8. https://doi.org/10.1016/j.jbspin.2006.04.002.
20. Killian ML, Cavinatto L, Galatz LM, Thomopoulos S. The role of mechanobiology in tendon healing. J Shoulder Elb Surg. 2012;21(2):228–37. https://doi.org/10.1016/j.jse.2011.11.002.
21. Magnusson SP, Langberg H, Kjaer M. The pathogenesis of tendinopathy: balancing the response to loading. Nat Rev Rheumatol. 2010;6(5):262–8. https://doi.org/10.1038/nrrheum.2010.43.
22. Grimaldi A, Mellor R, Hodges P, Bennell K, Wajswelner H, Vicenzino B. Gluteal tendinopathy: a review of mechanisms, assessment and management. Sport Med. 2015;45(8):1107–19. https://doi.org/10.1007/s40279-015-0336-5.
23. Cook JL, Purdam C. Is compressive load a factor in the development of tendinopathy? Br J Sports Med. 2012;46(3):163–8. https://doi.org/10.1136/bjsports-2011-090414.
24. Dobson F, Allison K, Diamond L, Hall M. Contemporary non-surgical considerations in the management of People with extra- and intra-articular hip pathologies. IntechOpen; 2019. https://doi.org/10.5772/intechopen.81821.
25. Birnbaum K, Siebert CH, Pandorf T, Schopphoff E, Prescher A, Niethard FU. Anatomical and biomechanical investigations of the iliotibial tract. Surg Radiol Anat. 2004;26(6):433–46. https://doi.org/10.1007/s00276-004-0265-8.
26. Anderson K, Strickland SM, Warren R. Hip and groin injuries in athletes. Am J Sports Med. 2001;29(4):521–33. https://doi.org/10.1177/03635465010290042501.
27. Clancy WG. Runners& injuries part two. Evaluation and treatment of specific injuries. Am J Sports Med. 1980;8(4):287–9. https://doi.org/10.1177/036354658000800415.
28. Allison K, Vicenzino B, Wrigley TV, Grimaldi A, Hodges PW, Bennell KL. Hip abductor muscle weakness in individuals with gluteal tendinopathy. Med Sci Sports Exerc. 2016;48(3):346–52. https://doi.org/10.1249/MSS.0000000000000781.
29. Grimaldi A. Assessing lateral stability of the hip and pelvis. Man Ther. 2011;16(1):26–32. https://doi.org/10.1016/j.math.2010.08.005.
30. Grimaldi A, Richardson C, Stanton W, Durbridge G, Donnelly W, Hides J. The association between degenerative hip joint pathology and size of the gluteus medius, gluteus minimus and piriformis muscles. Man Ther. 2009;14(6):605–10. https://doi.org/10.1016/j.math.2009.07.004.
31. Johnsson H. Movement, stability and low back pain. The essential role of the pelvis. Curr Orthop. 1998;45(1):67–8. https://doi.org/10.1016/s0268-0890(98)90009-3.
32. Sutter R, Kalberer F, Binkert CA, Graf N, Pfirrmann CWA, Gutzeit A. Abductor tendon tears are associated with hypertrophy of the tensor fasciae latae muscle. Skelet Radiol. 2013;42(5):627–33. https://doi.org/10.1007/s00256-012-1514-2.
33. Birnbaum K, Prescher A, Niethard FU. Hip centralizing forces of the iliotibial tract within various femoral neck angles. J Pediatr Orthop Part B. 2010;19(2):140–9. https://doi.org/10.1097/BPB.0b013e328332355e.
34. Viradia NK, Berger AA, Dahners LE. Relationship between width of greater trochanters and width of iliac wings in tronchanteric bursitis. Am J Orthop (Belle Mead NJ). 2011;40(9):E159–62.
35. Jacobsen JS, Hölmich P, Thorborg K, Bolvig L, Jakobsen SS, Søballe K, Mechlenburg I. Muscle-tendon-related pain in 100 patients with hip dysplasia: prevalence and associations with self-reported hip disability and muscle strength. J Hip Preserv Surg. 2018;5(1):39–46. https://doi.org/10.1093/jhps/hnx041.
36. Henak CR, Ellis BJ, Harris MD, Anderson AE, Peters CL, Weiss JA. Role of the acetabular labrum in load support across the hip joint. J Biomech. 2011;44(12):2201–6. https://doi.org/10.1016/j.jbiomech.2011.06.011.
37. Kaltenborn A, Bourg CM, Gutzeit A, Kalberer F. The hip lag sign - prospective blinded trial of a new clinical sign to predict hip abductor damage. PLoS One. 2014;9(3):e91560. https://doi.org/10.1371/journal.pone.0091560.
38. Barratt PA, Brookes N, Newson A. Conservative treatments for greater trochanteric pain syndrome: a systematic review. Br J Sports Med. 2017;51(2):97–104. https://doi.org/10.1136/bjsports-2015-095858.
39. Chowdhury R, Naaseri S, Lee J, Rajeswaran G. Imaging and management of greater trochanteric pain syndrome. Postgrad Med J. 2014;90(1068):576–81. https://doi.org/10.1136/postgradmedj-2013-131828.
40. Docking SI, Daffy J, van Schie HTM, Cook JL. Tendon structure changes after maximal exercise in the thoroughbred horse: use of ultrasound

tissue characterisation to detect in vivo tendon response. Vet J. 2012;194(3):338–42. https://doi.org/10.1016/j.tvjl.2012.04.024.
41. Ooi CC, Malliaras P, Schneider ME, Connell DA. "Soft, hard, or just right?" Applications and limitations of axial-strain sonoelastography and shear-wave elastography in the assessment of tendon injuries. Skeletal Radiol. 2014;43(1):1–12. https://doi.org/10.1007/s00256-013-1695-3.
42. Silva F, Adams T, Feinstein J, Arroyo RA. Trochanteric bursitis: refuting the myth of inflammation. J Clin Rheumatol. 2008;14(2):82–6. https://doi.org/10.1097/RHU.0b013e31816b4471.
43. Krout RM, Anderson TP. Trochanteric bursitis: management. Arch Phys Med Rehabil. 1959;40(1):8–14.
44. Gordon EJ. Trochanteric bursitis and tendinitis. Clin Orthop. 1961;20:193–202.
45. Noble CA. Iliotibial band friction syndrome in runners. Am J Sports Med. 1980;8(4):232–4. https://doi.org/10.1177/036354658000800403.
46. Fredericson M, Cookingham CL, Chaudhari AM, Dowdell BC, Oestreicher N, Sahrmann SA. Hip abductor weakness in distance runners with iliotibial band syndrome. Clin J Sport Med. 2000;10(3):169–75. https://doi.org/10.1097/00042752-200007000-00004.
47. Fredericson M, Weir A. Practical management of iliotibial band friction syndrome in runners. Clin J Sport Med. 2006;16(3):261–8. https://doi.org/10.1097/00042752-200605000-00013.
48. Khoury AN, Brooke K, Helal A, Bishop B, Erickson L, Palmer IJ, Martin HD. Proximal iliotibial band thickness as a cause for recalcitrant greater trochanteric pain syndrome. J Hip Preserv Surg. 2018;5(3):296–300. https://doi.org/10.1093/jhps/hny025.
49. Decker G, Hunt D. Proximal iliotibial band syndrome in a runner: a case report. PM R. 2019;11(2):206–9. https://doi.org/10.1016/j.pmrj.2018.06.017.
50. Sher I, Umans H, Downie SA, Tobin K, Arora R, Olson TR. Proximal iliotibial band syndrome: what is it and where is it? Skelet Radiol. 2011;40(12):1553–6. https://doi.org/10.1007/s00256-011-1168-5.
51. Pare EB, Stern JT, Schwartz JM. Functional differentiation within the tensor fasciae latae. A telemetered electromyographic analysis of its locomotor roles. J Bone Jt Surg - Ser A. 1981;63(9):1457–71. https://doi.org/10.2106/00004623-198163090-00013.
52. Cooperman JM. Case studies: isolated strain of the tensor fasciae latae. J Orthop Sports Phys Ther. 1984;5(4):201–3. https://doi.org/10.2519/jospt.1984.5.4.201.
53. Ilaslan H, Wenger DE, Shives TC, Unni KK. Unilateral hypertrophy of tensor fascia lata: a soft tissue tumor simulator. Skelet Radiol. 2003;32(11):628–32. https://doi.org/10.1007/s00256-003-0687-0.
54. Pretell J, Ortega J, García-Rayo R, Resines C. Distal fascia lata lengthening: an alternative surgical technique for recalcitrant trochanteric bursitis. Int Orthop. 2009;33(5):1223–7. https://doi.org/10.1007/s00264-009-0727-z.
55. Reid D. The management of greater trochanteric pain syndrome: a systematic literature review. J Orthop. 2016;13(1):15–28. https://doi.org/10.1016/j.jor.2015.12.006.
56. Sarno D, Sein M, Singh J. (364) The effectiveness of topical diclofenac for greater trochanteric pain syndrome: a retrospective study. J Pain. 2015;16(4):S67. https://doi.org/10.1016/j.jpain.2015.01.283.
57. Rompe JD, Segal NA, Cacchio A, Furia JP, Morral A, Maffulli N. Home training, local corticosteroid injection, or radial shock wave therapy for greater trochanter pain syndrome. Am J Sports Med. 2009;37(10):1981–90. https://doi.org/10.1177/0363546509334374.
58. Noehren B, Scholz J, Davis I. The effect of real-time gait retraining on hip kinematics, pain and function in subjects with patellofemoral pain syndrome. Br J Sports Med. 2011;45(9):691–6. https://doi.org/10.1136/bjsm.2009.069112.
59. van der Worp H, Zwerver J, Hamstra M, van den Akker-Scheek I, Diercks RL. No difference in effectiveness between focused and radial shockwave therapy for treating patellar tendinopathy: a randomized controlled trial. Knee Surg Sports Traumatol Arthrosc. 2014;22(9):2026–32. https://doi.org/10.1007/s00167-013-2522-z.
60. Kaux JF, Forthomme B, le Goff C, Crielaard JM, Croisier JL. Current opinions on tendinopathy. J Sport Sci Med. 2011;10(2):238–53.
61. Seo KH, Lee JY, Yoon K, Do JG, Park HJ, Lee SY, Park YS, Lee YT. Long-term outcome of low-energy extracorporeal shockwave therapy on gluteal tendinopathy documented by magnetic resonance imaging. PLoS One. 2018;13(7):e0197460. https://doi.org/10.1371/journal.pone.0197460.
62. Carlisi E, Cecini M, Di Natali G, Manzoni F, Tinelli C, Lisi C. Focused extracorporeal shock wave therapy for greater trochanteric pain syndrome with gluteal tendinopathy: a randomized controlled trial. Clin Rehabil. 2019;33(4):670–80. https://doi.org/10.1177/0269215518819255.
63. Labrosse JM, Cardinal É, Leduc BE, Duranceau J, Rémillard J, Bureau NJ, Belblidia A, Brassard P. Effectiveness of ultrasound-guided corticosteroid injection for the treatment of gluteus medius tendinopathy. Am J Roentgenol. 2010;194(1):202–6. https://doi.org/10.2214/AJR.08.1215.
64. Sayegh F, Potoupnis M, Kapetanos G. Greater trochanter bursitis pain syndrome in females with chronic low back pain and sciatica. Acta Orthop Belg. 2004;70(5):423–8.
65. Brinks A, van Rijn RM, Willemsen SP, Bohnen AM, Verhaar JAN, Koes BW, Bierma-Zeinstra SMA. Corticosteroid injections for greater trochanteric pain syndrome: a randomized controlled trial in primary care. Ann Fam Med. 2011;9(3):226–34. https://doi.org/10.1370/afm.1232.

66. Cohen SP, Strassels SA, Foster L, Marvel J, Williams K, Crooks M, Gross A, Kurihara C, Nguyen C, Williams N. Comparison of fluoroscopically guided and blind corticosteroid injections for greater trochanteric pain syndrome: multicentre randomised controlled trial. BMJ. 2009;338:b1088. https://doi.org/10.1136/bmj.b1088.
67. Coombes BK, Bisset L, Brooks P, Khan A, Vicenzino B. Effect of corticosteroid injection, physiotherapy, or both on clinical outcomes in patients with unilateral lateral epicondylalgia: a randomized controlled trial. JAMA. 2013;309(5):461–9. https://doi.org/10.1001/jama.2013.129.
68. Rees JD, Stride M, Scott A. Tendons - time to revisit inflammation. Br J Sports Med. 2014;48(21):1553–7. https://doi.org/10.1136/bjsports-2012-091957.
69. Andres BM, Murrell GAC. Treatment of tendinopathy: what works, what does not, and what is on the horizon. Clin Orthop Relat Res. 2008;466(7):1539–54. https://doi.org/10.1007/s11999-008-0260-1.
70. Fitzpatrick J, Bulsara MK, O'Donnell J, Zheng MH. Leucocyte-rich platelet-rich plasma treatment of gluteus Medius and Minimus tendinopathy: a double-blind randomized controlled trial with 2-year follow-up. Am J Sports Med. 2019;1:8. https://doi.org/10.1177/0363546519826969.
71. Fitzpatrick J, Bulsara MK, O'Donnell J, McCrory PR, Zheng MH. The effectiveness of platelet-rich plasma injections in gluteal tendinopathy: a randomized, double-blind controlled trial comparing a single platelet-rich plasma injection with a single corticosteroid injection. Am J Sports Med. 2018;46(4):933–9. https://doi.org/10.1177/0363546517745525.
72. Walker-Santiago R, Wojnowski NM, Lall AC, Maldonado DR, Rabe SM, Domb BG. Platelet-rich plasma versus surgery for the Management of Recalcitrant Greater Trochanteric Pain Syndrome: a systematic review. Arthrosc - J Arthrosc Relat Surg. 2020;36(3):875–88. https://doi.org/10.1016/j.arthro.2019.09.044.
73. Dhurat R, Sukesh M. Principles and methods of preparation of platelet-rich plasma: a review and author's perspective. J Cutan Aesthet Surg. 2014;7(4):189–97. https://doi.org/10.4103/0974-2077.150734.
74. Kagan A. Rotator cuff tears of the hip. Clin Orthop Relat Res. 1999;368:135–40. https://doi.org/10.1302/0301-620X.79B4.0790618.
75. Lequesne M, Djian P, Vuillemin V, Mathieu P. Prospective study of refractory greater trochanter pain syndrome. MRI findings of gluteal tendon tears seen at surgery Clinical and MRI results of tendon repair. Jt Bone Spine. 2008;75(4):458–64. https://doi.org/10.1016/j.jbspin.2007.12.004.
76. Walsh MJ, Walton JR, Walsh NA. Surgical repair of the gluteal tendons. A report of 72 cases. J Arthroplasty. 2011;26(8):1514–9. https://doi.org/10.1016/j.arth.2011.03.004.
77. Davies JF, Stiehl JB, Davies JA, Geiger PB. Surgical treatment of hip abductor tendon tears. J Bone Jt Surg - Ser A. 2013;95(15):1420–5. https://doi.org/10.2106/JBJS.L.00709.
78. Byrd JWT, Jones KS. Endoscopic repair of hip abductor tears: outcomes with two-year follow-up. J Hip Preserv Surg. 2017;4(1):80–4. https://doi.org/10.1093/jhps/hnw047.
79. Chandrasekaran S, Lodhia P, Gui C, Vemula SP, Martin TJ, Domb BG. Outcomes of open versus endoscopic repair of abductor muscle tears of the hip: a systematic review. Arthrosc - J Arthrosc Relat Surg. 2015;31(10):2057–67. https://doi.org/10.1016/j.arthro.2015.03.042.
80. Govaert LHM, van Dijk CN, Zeegers AVCM, Albers GHR. Endoscopic Bursectomy and iliotibial tract release as a treatment for refractory greater trochanteric pain syndrome: a new endoscopic approach with early results. Arthrosc Tech. 2012;1(2):e161–4. https://doi.org/10.1016/j.eats.2012.06.001.
81. Voos JE, Shindle MK, Pruett A, Asnis PD, Kelly BT. Endoscopic repair of gluteus medius tendon tears of the hip. Am J Sports Med. 2009;37(4):743–7. https://doi.org/10.1177/0363546508328412.
82. Domb BG, Carreira DS. Endoscopic repair of full-thickness gluteus Medius tears. Arthrosc Tech. 2013;36(8):2160–9. https://doi.org/10.1016/j.eats.2012.11.005.
83. Wiese M, Rubenthaler F, Willburger RE, Fennes S, Haaker R. Early results of endoscopic trochanter bursectomy. Int Orthop. 2004;28(4):218–21. https://doi.org/10.1007/s00264-004-0569-7.
84. Baker CL, Massie RV, Hurt WG, Savory CG. Arthroscopic Bursectomy for recalcitrant trochanteric bursitis. Arthrosc - J Arthrosc Relat Surg. 2007;23(8):827–32. https://doi.org/10.1016/j.arthro.2007.02.015.
85. Varacallo M, Bordoni B. Hip pointer injuries (Iliac Crest Contusions). Bethesda, MD: National Center for Biotechnology Information, U.S. National Library of Medicine; 2019.
86. Barata P, Cervaens M, Resende R, Camacho O, Marques F. Hyperbaric oxygen effects on sports injuries. Ther Adv Musculoskelet Dis. 2011;3(2):111–21. https://doi.org/10.1177/1759720X11399172.
87. Nuccion S, Hunter DMFG. Hip and pelvis. 2nd ed. Philadelphia, PA: Lippincott Williams & Wilkins; 2003.
88. Byrd JWT. Snapping hip. Oper Tech Sports Med. 2005;13(1):46–54. https://doi.org/10.1053/j.otsm.2004.09.003.
89. Winston P, Awan R, Cassidy JD, Bleakney RK. Clinical examination and ultrasound of self-reported snapping hip syndrome in elite ballet dancers. Am J Sports Med. 2007;35(1):118–26. https://doi.org/10.1177/0363546506293703.
90. Yen YM, Lewis CL, Kim YJ. Understanding and treating the snapping hip. Sports Med Arthrosc Rev. 2015;23(4):194–9. https://doi.org/10.1097/JSA.0000000000000095.
91. Pelsser V, Cardinal E, Hobden R, Aubin B, Lafortune M. Extraarticular snapping hip: sonographic findings. Am J Roentgenol. 2001;176(1):67–73. https://doi.org/10.2214/ajr.176.1.1760067.

92. Binnie JF. Snapping hip (hanche a ressort: schnellende hufte). Ann Surg. 1913;58(1):59–66. https://doi.org/10.1097/00000658-191307000-00006.
93. Lewis CL. Extra-articular snapping hip: a literature review. Sports Health. 2010;2(3):186–90. https://doi.org/10.1177/1941738109357298.
94. Allen WC, Cope R. Coxa Saltans: the snapping hip revisited. J Am Acad Orthop Surg. 1995;3(5):303–8. https://doi.org/10.5435/00124635-199509000-00006.
95. Deslandes M, Guillin R, Cardinal É, Hobden R, Bureau NJ. The snapping iliopsoas tendon: new mechanisms using dynamic sonography. Am J Roentgenol. 2008;190(3):576–81. https://doi.org/10.2214/AJR.07.2375.
96. Giai Via A, Fioruzzi A, Randelli F. Diagnosis and management of snapping hip syndrome: a comprehensive review of literature. Rheumatol Curr Res. 2017;7:4. https://doi.org/10.4172/2161-1149.1000228.
97. Provencher MT, Hofmeister EP, Muldoon MP. The surgical treatment of external coxa Saltans (the snapping hip) by Z-plasty of the iliotibial band. Am J Sports Med. 2004;32(2):470–6. https://doi.org/10.1177/0363546503261713.
98. Ilizaliturri VM, Martinez-Escalante FA, Chaidez PA, Camacho-Galindo J. Endoscopic iliotibial band release for external snapping hip syndrome. Arthrosc - J Arthrosc Relat Surg. 2006;22(5):505–10. https://doi.org/10.1016/j.arthro.2005.12.030.
99. Wahl CJ, Warren RF, Adler RS, Hannafin JA, Hansen B. Internal coxa saltans (snapping hip) as a result of overtraining: a report of 3 cases in professional athletes with a review of causes and the role of ultrasound in early diagnosis and management. Am J Sports Med. 2004;32(5):1302–9. https://doi.org/10.1177/03363546503258777.
100. Ilizaliturri VM, Villalobos FE, Chaidez PA, Valero FS, Aguilera JM. Internal snapping hip syndrome: treatment by endoscopic release of the iliopsoas tendon. Arthrosc - J Arthrosc Relat Surg. 2005;21(11):1375–80. https://doi.org/10.1016/j.arthro.2005.08.021.
101. Larsen E, Johansen J. Snapping hip. Acta Orthop. 1986;57(2):168–70. https://doi.org/10.3109/17453678609000894.
102. Krishnamurthy G, Connolly BL, Narayanan U, Babyn PS. Imaging findings in external snapping hip syndrome. Pediatr Radiol. 2007;37(12):1272–4. https://doi.org/10.1007/s00247-007-0602-2.
103. Choi YS, Lee SM, Song BY, Paik SH, Yoon YK. Dynamic sonography of external snapping hip syndrome. J Ultrasound Med. 2002;21(7):753–8. https://doi.org/10.7863/jum.2002.21.7.753.
104. Adler RS, Buly R, Ambrose R, Sculco T. Diagnostic and therapeutic use of sonography-guided iliopsoas peritendinous injections. Am J Roentgenol. 2005;185(4):940–3. https://doi.org/10.2214/AJR.04.1207.
105. Blankenbaker DG, De Smet AA, Keene JS. Sonography of the iliopsoas tendon and injection of the iliopsoas Bursa for diagnosis and management of the painful snapping hip. Skelet Radiol. 2006;35(8):565–71. https://doi.org/10.1007/s00256-006-0084-6.
106. Wettstein M, Jung J, Dienst M. Arthroscopic psoas Tenotomy. Arthrosc - J Arthrosc Relat Surg. 2006;22(8):907.e1–4. https://doi.org/10.1016/j.arthro.2005.12.064.
107. Ilizaliturri VM, Camacho-Galindo J. Endoscopic treatment of snapping hips, iliotibial band, and iliopsoas tendon. Sports Med Arthrosc Rev. 2010;18(2):120–7. https://doi.org/10.1097/JSA.0b013e3181dc57a5.
108. White RA, Hughes MS, Burd T, Hamann J, Allen WC. A new operative approach in the correction of external coxa saltans: the snapping hip. Am J Sports Med. 2004;32(6):1504–8. https://doi.org/10.1177/0363546503262189.
109. Polesello GC, Queiroz MC, Domb BG, Ono NK, Honda EK. Surgical technique: endoscopic gluteus maximus tendon release for external snapping hip syndrome hip. Clin Orthop Relat Res. 2013;471(8):2471–6. https://doi.org/10.1007/s11999-012-2636-5.
110. Anderson SA, Keene JS. Results of arthroscopic iliopsoas tendon release in competitive and recreational athletes. Am J Sports Med. 2008;36(12):2363–71. https://doi.org/10.1177/0363546508322130.
111. Yeoman W. The relation of ARTHRITIS of the SACRO-iliac joint to sciatica, with an analysis of 100 cases. Lancet. 1928;212(5492):1119–23. https://doi.org/10.1016/S0140-6736(00)84887-4.
112. Wun-Schen C. Bipartite piriformis muscle: an unusual cause of sciatic nerve entrapment. Pain. 1994;58(2):269–72. https://doi.org/10.1016/0304-3959(94)90208-9.
113. Sayson SC, Ducey JP, Maybrey JB, Wesley RL, Vermilion D. Sciatic entrapment neuropathy associated with an anomalous piriformis muscle. Pain. 1994;59(1):149–52. https://doi.org/10.1016/0304-3959(94)90060-4.
114. Benzon HT, Katz JA, Benzon HA, Iqbal MS. Piriformis syndrome: anatomic considerations, a new injection technique, and a review of the literature. Anesthesiology. 2003;98(6):1442–8. https://doi.org/10.1097/00000542-200306000-00022.
115. Benson ER, Schutzer SF. Posttraumatic piriformis syndrome: diagnosis and results of operative treatment. J Bone Jt Surg - Ser A. 1999;81(7):941–9. https://doi.org/10.2106/00004623-199907000-00006.
116. Boyajian-O'Neill LA, McClain RL, Coleman MK, Thomas PP. Diagnosis and management of piriformis syndrome: an osteopathic approach. J Am Osteopath Assoc. 2008;108(11):657–64. https://doi.org/10.7556/jaoa.2008.108.11.657.
117. Pace JB, Nagle D. Piriformis syndrome. West J Med. 1976;124(6):435–9.

118. Keskula DR, Tamburello M. Conservative management of piriformis syndrome. J Athl Train. 1992;27(2):102–10.
119. Mayrand N, Fortin J, Descarreaux M, Normand MC. Diagnosis and management of posttraumatic piriformis syndrome: a case study. J Manip Physiol Ther. 2006;29(6):486–91. https://doi.org/10.1016/j.jmpt.2006.06.006.
120. Cox JM, Bakkum BW. Possible generators of retrotrochanteric gluteal and thigh pain: the gemelli-obturator internus complex. J Manip Physiol Ther. 2005;28(7):534–8. https://doi.org/10.1016/j.jmpt.2005.07.012.
121. Mullin V, De Rosayro M. Caudal steroid injection for treatment of piriformis syndrome. Anesth Analg. 1990;71(6):705–7. https://doi.org/10.1213/00000539-199012000-00023.
122. Fannucci E, Masala S, Sodani G, Varrucciu V, Romagnoli A, Squillaci E, Simonetti G. CT-guided injection of botulinic toxin for percutaneous therapy of piriformis muscle syndrome with preliminary MRI results about denervative process. Eur Radiol. 2001;11(12):2543–8. https://doi.org/10.1007/s003300100872.
123. Nazlikul H, Ural FG, Oztürk GT, Oztürk ADT. Evaluation of neural therapy effect in patients with piriformis syndrome. J Back Musculoskelet Rehabil. 2018;31(6):1105–10. https://doi.org/10.3233/BMR-170980.

Part IV

Muscle Injuries

Acute and Chronic Hamstring Injuries

16

Robin Vermeulen, Anne D. van der Made, Johannes L. Tol, and Gino M. M. J. Kerkhoffs

16.1 Introduction

It was supposed to be Usain Bolt's fabled farewell in 2017. The world's fastest sprinter participated in the last event of his last ever World Championships: the 4 × 100 m relay. Instead of a legendary gold medal, the world witnessed an all too common occurrence in track and field athletes. He sustained an acute hamstring injury.

Acute hamstring muscle injury is the most common injury seen at outdoor and indoor athletics championships [1, 2]. These injuries account for 17.1% of all injuries with an overall incidence of 22.4 and 11.5 injuries per 1000 athletes for men and women, respectively [2]. They predominantly affect athletes in the sprinting, jumping and hurdle disciplines [2]. These injuries impose a high burden on the athlete and their medical team because of a high prevalence and a high re-injury rate (up to 63%) [3].

Far less common is the proximal hamstring tendinopathy. Proximal hamstring tendinopathy is a chronic type injury that can occur in the same athletics disciplines as acute hamstring injuries. Although less common, this injury typically has a prolonged convalescence period and poor response to treatment.

This chapter focuses on the diagnosis and treatment of acute and chronic hamstring injuries. To better understand these injuries, it is important to first understand the anatomy and the injury distribution. Diagnosis and treatment

R. Vermeulen (✉)
Aspetar, Orthopaedic and Sports Medicine Hospital, Doha, Qatar

Department of Orthopaedic Surgery, Amsterdam UMC, University of Amsterdam, Amsterdam Movement Sciences, Amsterdam, The Netherlands

Academic Center for Evidence-based Sports medicine (ACES), Amsterdam Movement Sciences Amsterdam UMC, Amsterdam, The Netherlands

Amsterdam Collaboration for Health and Safety in Sports (ACHSS), AMC/VUmc IOC Research Center, Amsterdam, The Netherlands

A. D. van der Made · G. M. M. J. Kerkhoffs
Department of Orthopaedic Surgery, Amsterdam UMC, University of Amsterdam, Amsterdam Movement Sciences, Amsterdam, The Netherlands

Academic Center for Evidence-based Sports medicine (ACES), Amsterdam Movement Sciences Amsterdam UMC, Amsterdam, The Netherlands

Amsterdam Collaboration for Health and Safety in Sports (ACHSS), AMC/VUmc IOC Research Center, Amsterdam, The Netherlands
e-mail: a.d.vandermade@amsterdamumc.nl; g.m.kerkhoffs@amsterdamumc.nl

J. L. Tol
Academic Center for Evidence-based Sports medicine (ACES), Amsterdam Movement Sciences Amsterdam UMC, Amsterdam, The Netherlands

Amsterdam Collaboration for Health and Safety in Sports (ACHSS), AMC/VUmc IOC Research Center, Amsterdam, The Netherlands
e-mail: j.l.tol@amsterdamumc.nl

G. L. Canata et al. (eds.), *Management of Track and Field Injuries*,
https://doi.org/10.1007/978-3-030-60216-1_16

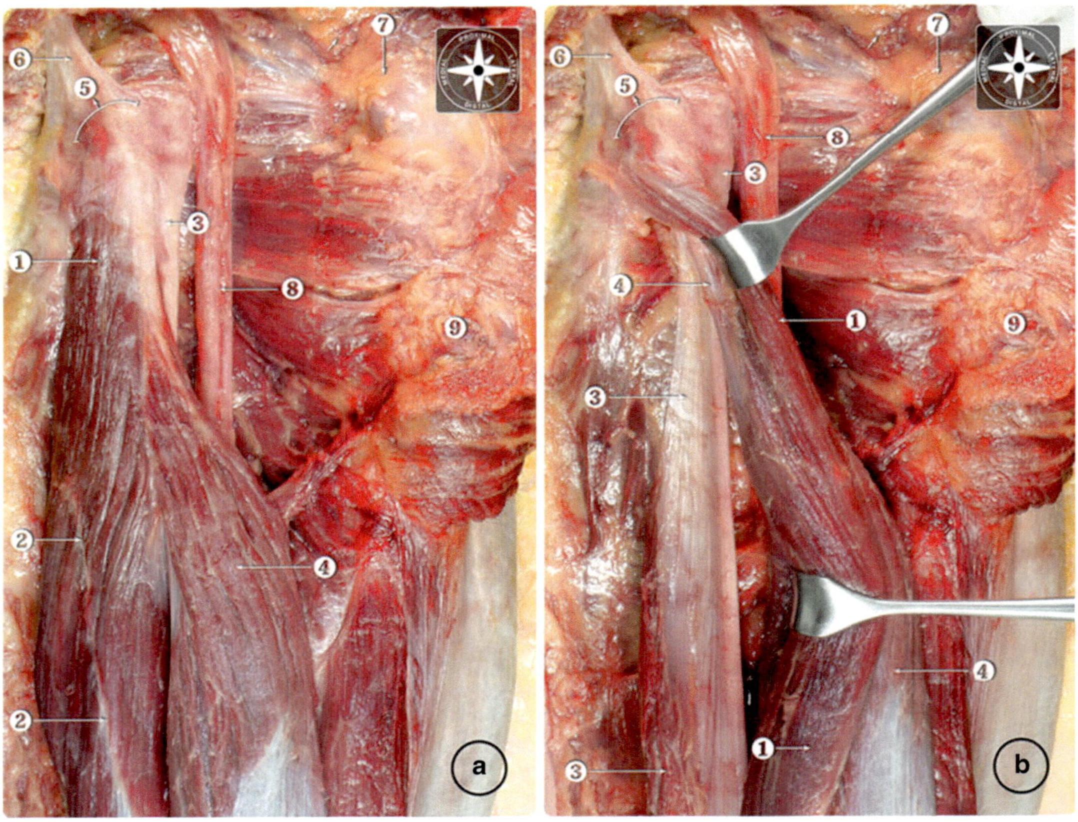

Fig. 16.1 Proximal anatomy of the hamstring muscles. (**a**) Normal proximal anatomy. (**b**) The semitendinosus and biceps femoris long head muscle have been reflected laterally to expose the proximal semimembranosus muscle. 1 Semitendinosus muscle. 2 Raphe of semitendinosus. 3 Semimembranosus muscle. 4 Biceps femoris long head muscle. 5 Ischial tuberosity. 6 Sacrotuberous ligament. 7 Greater trochanter. 8 Sciatic nerve. 9 Gluteus maximus (cut and reflected). (Reproduced from van der Made et al. [10] with permission of copyright owner)

are more easily understood once an understanding of the anatomy and injury distributions is established.

16.1.1 Anatomy and Injury Distribution

The hamstrings group consists of four muscles and is divided into a lateral and a medial complex. The lateral complex consists of the biceps femoris (long head and short head). The medial complex consists of the semitendinosus and semimembranosus muscles (see Fig. 16.1). Acute hamstring injuries occur mostly at the level of the proximal musculotendinous junction (MTJ) [4]. The chronic injuries mostly concern the proximal free tendons [5] (Table 16.1).

16.1.1.1 Biceps Femoris Long Head and Short Head

For the acute injuries, the biceps femoris long head is the most commonly injured hamstring muscle. It is involved in up to 80% of acute injuries [6]. The most common mechanism of injury is high-speed running [4]. The biceps femoris originates from the medial facet of the upper region of the ischial tuberosity as the conjoint tendon (that it shares with the ST, hence 'conjoint'). It courses laterally until it terminates in the muscle belly as the intramuscular tendon. The common distal tendon of the biceps femoris overlaps with the proximal tendon. This distal tendon has a bifurcated insertion. A direct arm inserts into the posterolateral aspect of the head of the fibula and an anterior arm inserts into the lateral edge of the head of the fibula [7]. The short head

Table 16.1 Injury locations, types and incidence in acute and chronic hamstring injuries

Onset	Location	Type	Incidence[a]
Acute	Free tendon	Partial- or full-thickness avulsion/ rupture	Rare (3–11%)
	Intramuscular tendon	Partial- or full-thickness rupture	Common (15–24%)
	Musculotendinous junction	Generally partial thickness injury	Very common (up to 80%)
Insidious/ chronic	Free tendon	Tendinopathy	Unknown

[a]All hamstring injuries, collated from current best available evidence

of the biceps femoris originates from the linea aspera midway on the femur. It has no proximal tendon as the muscle fibres directly originate from the bone. As mentioned, it shares its distal tendon with the long head. The short head is a uni-articular and flat muscle that fans out. It is the least injured (in isolation) of the hamstring muscle group and accounts for only 7% of all acute hamstring injuries [8].

16.1.1.2 Semitendinosus

For acute injuries, the semitendinosus is equally injured in isolation or in conjunction with the biceps in sprinting type injuries. It is involved in 14–15% of these injuries [9]. Its tendinous origin is shared with the biceps femoris long head, but a portion of its muscle fibres originate directly from the ischial tuberosity. The proximal part of the conjoint tendon is occupied mostly by the semitendinosus. Muscle fibres from the biceps femoris long head start attaching onto it around five centimetres from the ischial tuberosity [10]. Further distally, there is a tendinous inscription called the 'raphe'. The function of the raphe is currently unclear, but it might function as a strut that divides the semitendinosus into two regions. This is reflected by the fact that these separate regions are innervated by two different motor branches of the sciatic nerve [11]. The distal tendon of the semitendinosus inserts on the anteromedial side of the proximal tibia as part of the pes anserinus. This distal tendon is the longest of the hamstring group [10]. Due to its length, it is commonly harvested for use in ACL autografts.

16.1.1.3 Semimembranosus

The semimembranosus muscle originates from the lateral facet of the upper region of the ischial tuberosity. It starts out as an asymmetrical and long free tendon. This free tendon courses anteromedially to the conjoint tendon, and muscle fibres start attaching around 11 cm from the ischial tuberosity [12]. A varied and complex distal insertion pattern has been described. The main insertions of the distal tendon are onto the posterior aspect of the medial tibial condyle and the fascia of the popliteus muscle [10, 11, 13]. The proximal and distal MTJ of the semimembranosus overlap, and most acute injuries occur in the proximal MTJ. Proximal free tendon injuries are also common and occur during slow-speed stretching situations [14].

16.1.1.4 The Proximal Hamstring Tendon Complex

The anatomy of the proximal hamstring free tendons is of special interest for the acute full-thickness tendon injuries and chronic proximal hamstring tendinopathy. In acute injuries, the proximal hamstring tendon complex can sustain a partial-thickness or full-thickness free tendon injury. The most common injury mechanism is a combination of forced hip flexion and knee extension [15]. This is a relatively rare (3–11% of acute hamstring injuries [15]), but potentially career threatening injury due to residual functional impairment if not treated properly [15].

The development of proximal hamstring tendinopathy is not fully understood. One of the current theories is based on compressive forces [16]. It supposes that the proximal tendons are compressed during hip flexion due to their position on the ischial tuberosity. This compressive loading might be a key factor in the pathogenesis of chronic injury. Its progression could be due to its self-reinforcing nature; as the tendon thickens, compression increases [16]. Direct clinical evidence for this theory is currently lacking.

16.1.1.5 The Intramuscular Tendon

The clinical relevance of the intramuscular tendon is a recent addition to the hamstring injury literature [3, 17–20]. The intramuscular tendon (sometimes referred to as the central tendon) is defined as '*the part of the tendon to which the muscle fibres attach*' [18]. An injury involving the intramuscular tendon delayed return to sport times and increased re-injury rates drastically in elite track and field athletes, up to an average of 84 (±49.4) days and 63% recurrence [3]. It must be noted that this is a retrospective study based on 15 athletes. Other recent studies in other sports demonstrated only a moderate increase of 1 week in return to sport times. There was an average return of 31.6 ± 10.9 days with a full-thickness tendon discontinuity and no difference in re-injury rates in soccer players [18, 19]. A possible explanation for the difference in return to sports period is the biomechanical demand placed on the intramuscular tendon in different types of sports. Increasing speed seems to be the most contributing factor to a higher biomechanical load for the (lateral) knee flexors [21, 22]. Track and field sprint events typically require a short but maximal output for the athlete, whereas soccer athletes can pick and choose their efforts more tactically during a game.

It has been suggested that the intramuscular tendon is not susceptible to overuse injuries in the same way as the proximal free tendons. This might be due to its higher vascular perfusion as compared to a free tendon. However, direct evidence to confirm this hypothesis is still lacking [23].

16.1.2 Diagnosis

A quick and accurate diagnosis of an acute or chronic hamstring injury is paramount for professional athletes to ensure appropriate prognosis and treatment. The cornerstone of the diagnosis lies in the combination of a comprehensive history and clinical examination, often supplemented by imaging.

16.1.2.1 History: Acute Versus Chronic

Most acute hamstring injuries have a typical history and injury situation. The athlete is engaged in their sport and during a sprint or (forced) stretch they suddenly feel a sharp pain in their posterior thigh [4, 14]. This is occasionally accompanied by an audible pop or popping/tearing sensation. Usually, the athlete cannot continue their sporting activity and there is loss of function. The partial- or full-thickness proximal free tendon injuries usually present with even more dramatic pain and loss of function [24]. They often report a mechanism with forced hip flexion and knee extension such as a slip or fall. Walking is often difficult, sitting is painful and athletes report extensive bruising that appears within days after injury. This is in contrast with a chronic injury such as a proximal hamstring tendinopathy, which resembles the general history of a tendinopathy. Patients mainly report pain in the region of the ischial tuberosity with or without radiating pain towards (but not below) the knee. Stiffness in the morning or after a prolonged period of resting can be present. Symptom onset is gradual and provoked by/or worsened when commencing exercise. The symptoms usually reduce or even resolve after warming up. The symptoms are usually worse again after cessation of exercise and can last for several days.

Nerve-related symptoms such as a burning sensation, numbness and tingling with or without radiation to the leg or foot can occur in both conditions. It is more frequent and well understood in acute full-thickness free tendon injuries [25]. This is due to the proximity of the proximal tendons to the sciatic nerve. This proximity can make the distinction between proximal hamstring tendinopathy and other causes of nerve-related symptoms challenging.

16.1.2.2 Physical Examination

Physical examination of the hamstrings is relatively straightforward and overlaps for both the acute and chronic injuries. Diagnostic effectiveness of the physical examination of acute hamstring injuries is low with imaging as a reference standard [26–28]. However, commonly used tests are reliable [29, 30]. Scientific evidence for the value of physical examination for proximal hamstring tendinopathy is lacking compared to the evidence for acute injuries [31, 32]. The aim of the examination is to reproduce the injury pain through either compressive or tensile loads and to assess the degree of functional limitation(s).

The order of examination is not set in stone. It typically starts with assessment of the gait pattern and functional examination of the lumbar spine, hips and knees. Inspection of the injured hamstring focuses on identifying (subtle or more extensive) bruising. In the case of a proximal full-thickness free tendon injury, a loss of muscle contour compared to the uninjured leg can be seen. The hamstring is palpated to determine the location of the injury (proximal versus distal, medial vs. lateral). Special care is taken to palpate and assess the proximal bone–tendon continuity during resisted knee flexion to avoid missing a full-thickness free tendon injury [24]. Palpation of the ischial tuberosity and the proximal hamstring tendons can provoke the pain of proximal hamstring tendinopathy. Assessment of the range of motion of the hamstrings is done through active and passive flexibility tests of the hip and knee joint. Common range of motion tests for acute hamstring injuries is the active/passive straight leg raise and the active/passive knee extension test [27, 29, 33–35]. Basic strength testing includes isometric knee flexion and hip extension against resistance. Both ranges of motion and strength testing are compared to the uninjured side. Pain during testing can be assessed with a simple numeric rating scale (NRS) question on the scale of 1–10 [33]. Tools such as a goniometer or a handheld dynamometer can be used to assess side-to-side differences with more accuracy [33].

For proximal hamstring tendinopathy, only three tests have been identified as useful (moderate to high validity). These are the bent knee stretch test, the modified bent knee stretch test and the Puranen-Orava test [31]. Pain during range of motion or strength testing in positions with hip flexion is suggestive for proximal tendinopathy, but also fit in the wider differential diagnosis. The differential diagnosis of buttock pain includes sciatic nerve irritation, ischiofemoral impingement, partial- or full-thickness injury of the proximal hamstring tendon(s) and other diagnoses [36].

16.1.2.3 Imaging

The clinical diagnosis of a hamstring injury is relatively straightforward. Imaging should be seen as an adjunct to the diagnostic arsenal. It is only necessary if it is expected to change clinical management, for example in case of a suspected full-thickness free tendon injury.

In elite athlete settings, other factors such as external pressures on the medical team can play a role in the decision for the use of imaging. When imaging is considered, magnetic resonance imaging (MRI) is the gold standard for acute hamstring injuries [37].

The role of imaging as a diagnostic aid for proximal hamstring tendinopathy is more complex. MRI and ultrasound are both capable of visualizing tendinopathic changes in the proximal tendons. Ultrasound is a cheaper alternative but heavily dependent on operator skill. If alternative diagnoses are considered, ultrasound is also less sensitive in detecting other changes such as bone marrow oedema, partial-thickness free tendon injuries and peritendinous fluid [38]. The problem with the use of MRI for proximal hamstring tendinopathy lies in the fact that the common MRI findings are non-specific and/or false positive. These findings are commonly seen in asymptomatic patients and are also increasingly common with advancing age [39–41]. It is our opinion that imaging in proximal hamstring tendinopathy is best used to exclude other conditions in the differential diagnosis [36]. The use of MRI as a prognostic tool to predict return to sport is of limited value [28, 42]. For acute injuries, there is moderate evidence for MRI-negative injuries (faster return to sport) and the presence of free tendon injury (longer return to sport) [42]. For proximal hamstring tendinopathy, there is currently no evidence for MRI findings and association with return to sport.

16.1.3 Treatment

The treatment for these two types of hamstring injuries is different on many levels. For the partial-thickness MTJ acute hamstring injuries, there are 14 RCTs to guide our evidence-based rehabilitation. For both proximal full-thickness tendon injury and chronic proximal tendinopathy, we are left in the dark with no RCTs. In-depth protocols for treatment of these injuries are beyond the scope of this chapter, but general principles are described.

16.1.3.1 Physiotherapy

Acute hamstring injuries—physiotherapy has received a lot of research attention and is the mainstay treatment for acute hamstring injuries. A delay in starting physiotherapy can significantly lengthen the time to return to sports [43]. Programmes with multifactorial, criteria-based progression [44] and an eccentric overload component (e.g. Askling lengthening exercises [9] or Nordic hamstring exercise) are effective for treating partial-thickness MTJ injuries with or without intramuscular tendon involvement [45].

Proximal hamstring tendinopathy—due to its stubborn and drawn out nature, managing expectations of the athlete (and other stakeholders) is key. It is important to emphasize that the road to recovery may take a long time and will have (un) expected setbacks. The cornerstone of treatment is activity modification and progressive loading as tolerated [36]. The aim is to reduce pain and increase energy storage capacity [36]. Activity modification can be done by modifying or avoiding activities/positions with increased hip flexion [36]. General tendinopathy and specific proximal hamstring tendinopathy treatments have been described in the literature [36, 46].

16.1.3.2 Surgery

Acute hamstring injuries—surgery is only advocated for full-thickness injuries of the hamstring tendons. Surgery seems to lead to good functional (strength) and good satisfaction scores [15]. Evidence comparing outcomes of operative versus non-operative treatments is lacking. In the scientific literature, for every 27 hamstrings that are operated, one is treated non-operatively (possibly indicating publication bias) [47].

Proximal hamstring tendinopathy—surgery has been advocated for patients that 'fail' non-operative treatment. Only one retrospective case series on the outcomes of surgery exists [48]. Surgery is not recommended as a first-line treatment with the current knowledge of the convalescence period of this injury and the paucity of evidence.

16.1.3.3 Extracorporeal Shockwave Therapy

It is unsure what the role of extracorporeal shockwave therapy (ESWT) is in acute hamstring injuries and proximal hamstring tendinopathy. It is unknown if it can be considered as an adjunct or even replacement to exercise therapy. For proximal hamstring tendinopathy, ESWT has been shown to decrease pain significantly more than conventional exercise therapy in an RCT [49]. However, this study was at a substantial risk of bias due to the lack of blinding. There is currently no evidence for the use of ESWT in acute hamstring injuries.

16.1.3.4 Platelet-Rich Plasma

Platelet-rich plasma (PRP) injections are a popular type of medical treatment since the results of animal studies showed increased muscle regeneration and a lack of adverse effects. Unfortunately, this effect is not seen in human subjects, with evidence against its effectiveness for partial-thickness MTJ acute hamstring injuries [50]. There is little to no evidence for its effectiveness in proximal hamstring tendinopathy [51] or acute full-thickness free tendon injury.

16.1.3.5 Non-steroidal Anti-Inflammatory Drugs and Corticosteroids

Non-steroidal anti-inflammatory drugs (NSAIDs) are commonly used in the inflammatory stages of the muscle healing response. There is no additive effect of NSAIDs on the healing of acute partial-thickness MTJ hamstring injuries [52]. Detrimental effects were seen in animal models (oral use; increased fibrosis [53]) and human subjects (injections; myotoxicity [54]). This combination of data makes it difficult to recommend it as a treatment modality.

Corticosteroid injections are used as an anti-inflammatory therapy in the inflammatory stages of muscle healing and generally in various tendinopathies. Side effects of corticosteroids include increased necrotic tissue, decreased regeneration and (local) atrophy. There is only one retrospective study looking at intramuscular corticosteroid injections for partial-thickness MTJ acute hamstring injuries in elite (American football) athletes. There were no side effects or re-injuries reported for the 58 athletes in this study [55]. Interpretation of this study result is limited due to its design. For tendinopathies, there is increasing evidence for the short-term benefits but mid-to-long-term detrimental effects [48, 56]. This short-term gain for long-term detriment is best avoided.

16.1.4 Conclusion

Acute and chronic hamstring injuries are common, but heterogeneous conditions in athletics. Between them, they are vastly different in their incidences, approaches to management and outcomes. Both are primarily clinical diagnoses with imaging as an adjunct if necessary. History and patient presentation are key differentiators: acute injuries are a sudden occurrence that happens during high-speed running or stretching situations. The onset proximal hamstring tendinopathy is more gradual, possibly due to excessive compressive loads.

For the partial-thickness MTJ acute hamstring injuries, there are 14 RCTs to guide our therapy. Prospective case series and consensus reports suggest that surgery might be indicated for full-thickness free tendon acute hamstring injuries in elite athletes.

For the chronic hamstring tendinopathy, there is little scientific evidence to guide treatment. The cornerstones of treatment are physiotherapy-based interventions with progressive (eccentric) loading and activity modification, combined with expectation management. Extra-corporeal shockwave therapy could be considered as an adjunct to exercise treatment. Other treatments such as injections (PRP, corticosteroids) and non-steroidal anti-inflammatory medication have no strong evidence for use in both acute and chronic hamstring injuries and are not recommended.

References

1. Feddermann-Demont N, Junge A, Edouard P, Branco P, Alonso J-M. Injuries in 13 International athletics championships between 2007–2012. Br J Sports Med. 2014;48:513–22.
2. Edouard P, Branco P, Alonso J-M. Muscle injury is the principal injury type and hamstring muscle injury is the first injury diagnosis during top-level international athletics championships between 2007 and 2015. Br J Sports Med. 2016;50:619–30.
3. Pollock N, Patel A, Chakraverty J, Suokas A, James SLJ, Chakraverty R. Time to return to full training is delayed and recurrence rate is higher in intratendinous ('c') acute hamstring injury in elite track and field athletes: clinical application of the British athletics muscle injury classification. Br J Sports Med. 2016;50:305–10.
4. Askling CM, Tengvar M, Saartok T, Thorstensson A. Acute first-time hamstring strains during high-speed running: a longitudinal study including clinical and magnetic resonance imaging findings. Am J Sports Med. 2006;35:197–206.
5. Fredericson M, Moore W, Guillet M, Beaulieu C. High hamstring tendinopathy in runners meeting the challenges of diagnosis, treatment, and rehabilitation. Physician Sports Med. 2005;33(5):32–43.
6. Hallén A, Ekstrand J. Return to play following muscle injuries in professional footballers. J Sports Sci. 2014;32:1229–36.
7. Terry GC, LaPrade RF. The biceps Femoris muscle complex at the knee: its anatomy and injury patterns associated with acute anterolateral-anteromedial rotatory instability. Am J Sports Med. 1996;24:2–8.
8. Entwisle T, Ling Y, Splatt A, Brukner P, Connell D. Distal musculotendinous T junction injuries of the biceps Femoris: an MRI case review. Orthop J Sports Med. 2017;5:232596711771499.
9. Askling CM, Tengvar M, Tarassova O, Thorstensson A. Acute hamstring injuries in Swedish elite sprinters and jumpers: a prospective randomised controlled clinical trial comparing two rehabilitation protocols. Br J Sports Med. 2014;48:532–9.
10. van der Made AD, Wieldraaijer T, Kerkhoffs GM, Kleipool RP, Engebretsen L, van Dijk CN, et al. The hamstring muscle complex. Knee Surg Sports Traumatol Arthrosc. 2015;23:2115–22.
11. Stępień K, Śmigielski R, Mouton C, Ciszek B, Engelhardt M, Seil R. Anatomy of proximal

attachment, course, and innervation of hamstring muscles: a pictorial essay. Knee Surg Sports Traumatol Arthrosc. 2019;27:673–84.
12. Woodley SJ, Mercer SR. Hamstring muscles: architecture and innervation. Cells Tissues Organs. 2005;179:125–41.
13. Benninger B, Delamarter T. Distal semimembranosus muscle-tendon-unit review: morphology, accurate terminology, and clinical relevance. Folia Morphol (Warsz). 2013;72:1–9.
14. Askling CM, Tengvar M, Saartok T, Thorstensson A. Acute first-time hamstring strains during slow-speed stretching: clinical, magnetic resonance imaging, and recovery characteristics. Am J Sports Med. 2007;35:1716–24.
15. van der Made AD, Reurink G, Gouttebarge V, Tol JL, Kerkhoffs GM. Outcome after surgical repair of proximal hamstring avulsions: a systematic review. Am J Sports Med. 2015;43:2841–51.
16. Cook J, Purdam C. Is compressive load a factor in the development of tendinopathy? Br J Sports Med. 2012;46:163–8.
17. Comin J, Malliaras P, Baquie P, Barbour T, Connell D. Return to competitive play after hamstring injuries involving disruption of the central tendon. Am J Sports Med. 2013;41:111–5.
18. van der Made AD, Almusa E, Whiteley R, Hamilton B, Eirale C, van Hellemondt F, et al. Intramuscular tendon involvement on MRI has limited value for predicting time to return to play following acute hamstring injury. Br J Sports Med. 2018;52:83–8.
19. van der Made AD, Almusa E, Reurink G, Whiteley R, Weir A, Hamilton B, et al. Intramuscular tendon injury is not associated with an increased hamstring reinjury rate within 12 months after return to play. Br J Sports Med. 2018;52:1261–6.
20. Brukner P, Connell D. 'Serious thigh muscle strains': beware the intramuscular tendon which plays an important role in difficult hamstring and quadriceps muscle strains. Br J Sports Med. 2016;50:205–8.
21. Schache AG, Blanch PD, Dorn TW, Brown NAT, Rosemond D, Pandy MG. Effect of running speed on lower limb joint kinetics: medicine & science in sports & exercise. Med Sci Sports Exerc. 2011;43:1260–71.
22. Hansen C, Einarson E, Thomson A, Whiteley R. Peak medial (but not lateral) hamstring activity is significantly lower during stance phase of running. An EMG investigation using a reduced gravity treadmill. Gait Posture. 2017;57:7–10.
23. Brukner P, Cook JL, Purdam CR. Does the intramuscular tendon act like a free tendon? Br J Sports Med. 2018;52:1227–8.
24. van der Made AD, Tol JL, Reurink G, Peters RW, Kerkhoffs GM. Potential hamstring injury blind spot: we need to raise awareness of proximal hamstring tendon avulsion injuries. Br J Sports Med. 2019;53(7):390–2.
25. Wilson TJ, Spinner RJ, Mohan R, Gibbs CM, Krych AJ. Sciatic nerve injury after proximal hamstring avulsion and repair. Orthop J Sports Med. 2017;5(7):2325967117713685.
26. Reiman MP, Loudon JK, Goode AP. Diagnostic accuracy of clinical tests for assessment of hamstring injury: a systematic review. J Orthop Sports Phys Ther. 2013;43:222–31.
27. Schneider-Kolsky ME. A comparison between clinical assessment and magnetic resonance imaging of acute hamstring injuries. Am J Sports Med. 2006;34:1008–15.
28. Wangensteen A, Almusa E, Boukarroum S, Farooq A, Hamilton B, Whiteley R, et al. MRI does not add value over and above patient history and clinical examination in predicting time to return to sport after acute hamstring injuries: a prospective cohort of 180 male athletes. Br J Sports Med. 2015;49:1579–87.
29. Reurink G, Goudswaard GJ, Oomen HG, Moen MH, Tol JL, Verhaar JAN, et al. Reliability of the active and passive knee extension test in acute hamstring injuries. Am J Sports Med. 2013;41:1757–61.
30. Reurink G, Goudswaard GJ, Moen MH, Tol JL, Verhaar JAN, Weir A. Strength measurements in acute hamstring injuries: Intertester reliability and prognostic value of handheld dynamometry. J Orthop Sports Phys Ther. 2016;46:689–96.
31. Cacchio A, Borra F, Severini G, Foglia A, Musarra F, Taddio N, et al. Reliability and validity of three pain provocation tests used for the diagnosis of chronic proximal hamstring tendinopathy. Br J Sports Med. 2012;46:883–7.
32. Kerkhoffs GMMJ, van Es N, Wieldraaijer T, Sierevelt IN, Ekstrand J, van Dijk CN. Diagnosis and prognosis of acute hamstring injuries in athletes. Knee Surg Sports Traumatol Arthrosc. 2013;21:500–9.
33. Askling C. Type of acute hamstring strain affects flexibility, strength, and time to return to pre-injury level. Br J Sports Med. 2006;40:40–4.
34. Warren P, Gabbe BJ, Schneider-Kolsky M, Bennell KL. Clinical predictors of time to return to competition and of recurrence following hamstring strain in elite Australian footballers. Br J Sports Med. 2010;44:415–9.
35. Maniar N, Shield AJ, Williams MD, Timmins RG, Opar DA. Hamstring strength and flexibility after hamstring strain injury: a systematic review and meta-analysis. Br J Sports Med. 2016;50:909–20.
36. Goom TSH, Malliaras P, Reiman MP, Purdam CR. Proximal hamstring tendinopathy: clinical aspects of assessment and management. J Orthop Sports Phys Ther. 2016;46:483–93.
37. Crema MD, Yamada AF, Guermazi A, Roemer FW, Skaf AY. Imaging techniques for muscle injury in sports medicine and clinical relevance. Curr Rev Musculoskelet Med. 2015;8:154–61.
38. Zissen MH, Wallace G, Stevens KJ, Fredericson M, Beaulieu CF. High hamstring tendinopathy: MRI and ultrasound imaging and therapeutic efficacy of percutaneous corticosteroid injection. Am J Roentgenol. 2010;195:993–8.

39. De Smet AA, Blankenbaker DG, Alsheik NH, Lindstrom MJ. MRI appearance of the proximal hamstring tendons in patients with and without symptomatic proximal hamstring tendinopathy. Am J Roentgenol. 2012;198:418–22.
40. Thompson SM, Fung S, Wood DG. The prevalence of proximal hamstring pathology on MRI in the asymptomatic population. Knee Surg Sports Traumatol Arthrosc. 2017;25:108–11.
41. Bowden DJ, Byrne CA, Alkhayat A, Kavanagh EC, Eustace SJ. Differing MRI appearances of symptomatic proximal hamstring tendinopathy with ageing: a comparison of appearances in patients below and above 45 years. Clin Radiol. 2018;73:922–7.
42. Reurink G, Brilman EG, de Vos R-J, Maas M, Moen MH, Weir A, et al. Magnetic resonance imaging in acute hamstring injury: can we provide a return to play prognosis? Sports Med. 2015;45:133–46.
43. Bayer ML, Magnusson SP, Kjaer M. Early versus delayed rehabilitation after acute muscle injury. N Engl J Med. 2017;377:1300–1.
44. Mendiguchia J, Martinez-Ruiz E, Edouard P, Morin J, Martinez-Martinez F, Idoate F, et al. A multifactorial, criteria-based progressive algorithm for hamstring injury treatment. Med Sci Sports Exerc. 2017;49(7):1482–92.
45. Ishøi L, Krommes K, Husted RS, Juhl CB, Thorborg K. Diagnosis, prevention and treatment of common lower extremity muscle injuries in sport – grading the evidence: a statement paper commissioned by the Danish Society of Sports Physical Therapy (DSSF). Br J Sports Med. 2020;54(9):528–37.
46. Cook JL, Rio E, Purdam CR, Docking SI. Revisiting the continuum model of tendon pathology: what is its merit in clinical practice and research? Br J Sports Med. 2016;50:1187–91.
47. Bodendorfer BM, Curley AJ, Kotler JA, Ryan JM, Jejurikar NS, Kumar A, et al. Outcomes after operative and nonoperative treatment of proximal hamstring avulsions: a systematic review and meta-analysis. Am J Sports Med. 2018;46:2798–808.
48. Lempainen L, Sarimo J, Mattila K, Vaittinen S, Orava S. Proximal hamstring tendinopathy: results of surgical management and histopathologic findings. Am J Sports Med. 2009;37:727–34.
49. Cacchio A, Rompe JD, Furia JP, Susi P, Santilli V, De Paulis F. Shockwave therapy for the treatment of chronic proximal hamstring tendinopathy in professional athletes. Am J Sports Med. 2011;39:146–53.
50. Grassi A, Napoli F, Romandini I, Samuelsson K, Zaffagnini S, Candrian C, et al. Is platelet-rich plasma (PRP) effective in the treatment of acute muscle injuries? A systematic review and meta-analysis. Sports Med. 2018;48:971–89.
51. Davenport KL, Campos JS, Nguyen J, Saboeiro G, Adler RS, Moley PJ. Ultrasound-guided Intratendinous injections with platelet-rich plasma or autologous whole blood for treatment of proximal hamstring tendinopathy: a double-blind randomized controlled trial. J Ultrasound Med. 2015;34:1455–63.
52. Reynolds JF, Noakes TD, Schwellnus MP, Windt A, Bowerbank P. Non-steroidal anti-inflammatory drugs fail to enhance healing of acute hamstring injuries treated with physiotherapy. S Afr Med J. 1995;85:517–22.
53. Duchesne E, Dufresne SS, Dumont NA. Impact of inflammation and anti-inflammatory modalities on skeletal muscle healing: from fundamental research to the clinic. Phys Ther. 2017;97:807–17.
54. Reurink G, Goudswaard GJ, Moen MH, Weir A, Verhaar JAN, Tol JL. Myotoxicity of injections for acute muscle injuries: a systematic review. Sports Med. 2014;44:943–56.
55. Levine WN, Bergfeld JA, Tessendorf W, Moorman CT. Intramuscular corticosteroid injection for hamstring injuries a 13-year experience in the National Football League. Am J Sports Med. 2000;28:297–300.
56. Coombes BK, Bisset L, Vicenzino B. Efficacy and safety of corticosteroid injections and other injections for management of tendinopathy: a systematic review of randomised controlled trials. Lancet. 2010;376:1751–67.

Regenerative Medicine (Biological) Therapies for Acute Muscle Injury

17

Kenny Lauf, Anne D. van der Made,
Gustaaf Reurink, Johannes L. Tol,
and Gino M. M. J. Kerkhoffs

17.1 Introduction

Muscle injuries are the most common injuries in professional athletes forced to high-intensity sprinting efforts [1, 2]. In international track and field competitions between 2007 and 2015, muscle injuries accounted for 41% of all injuries. The hamstrings were the most commonly affected muscle group [3–5]. Muscle injuries lead to absence from training and competition and to loss of performance, with financial and potentially lasting athletic consequences. Due to a high rate of recurrence of muscle injuries, it is one of the most challenging tasks for a sports medicine team to prepare a professional athlete for a return to competition and ultimately performance [4]. A recurrent injury leads to 30% longer absence, before athletes can return to competitive matches [6].

In the literature, a variety of treatments for muscle injuries is described and yet the search for new treatments to improve and stimulate muscle healing is an ongoing process. In this chapter, we describe the basics of muscle healing and we discuss biological therapies and the scientific evidence on their efficacy.

17.2 Muscle Structure

Skeletal muscle is composed of two main components, muscle fibers, and the connective tissue. Muscle contraction is induced by the muscle fibers and the innervating nerves of these muscle fibers. The connective tissue is responsible for interconnecting all muscle cells and to shield the capillaries and nerves during a muscle contraction [7].

Muscle fibers originate from numerous myoblasts or (mononucleated) myogenic progenitor cells that are fused to build multinucleated myotubes. These myotubes will mature into the muscle fibers [8, 9]. For muscle contractions, contractile units (sarcomeres) contract by interaction ("sliding mechanism") of the filamentary proteins (actin and myosin). These sarcomeres are the fundamentals of a myofibril, and myofibrils are the main elements of a muscle fiber [9, 10].

Now that the composition of muscle fibers is delineated, we can describe the organization of

K. Lauf (✉) · A. D. van der Made · G. Reurink
J. L. Tol · G. M. M. J. Kerkhoffs
Department of Orthopaedic Surgery, Amsterdam UMC, University of Amsterdam, Amsterdam Movement Sciences, Amsterdam, The Netherlands

Academic Center for Evidence-based Sports medicine (ACES), Amsterdam UMC, Amsterdam, The Netherlands

Amsterdam Collaboration for Health and Safety in Sports (ACHSS), AMC/VUmc IOC Research Center, Amsterdam, The Netherlands
e-mail: k.lauf@amsterdamumc.nl

G. L. Canata et al. (eds.), *Management of Track and Field Injuries*,
https://doi.org/10.1007/978-3-030-60216-1_17

the connective tissue. The connective tissue organizes the muscle fibers on three levels: the endomysium, the perimysium, and the epimysium. The endomysium (basement membrane) envelops an individual muscle fiber and includes arteries and veins. The perimysium is a sheath of connective tissue that surrounds a group of muscle fibers (fascicles), and the epimysium is the outer layer of connective tissue that envelops the entire muscle. The connective tissue is not only a supportive skeleton for the muscle fibers. It unites the contractions of all muscle fibers into a joint effort and thus converts all individual contractions into efficient locomotion [7, 11]. Musculotendinous junctions (MTJs) are responsible for the transmission of forces generated by contracting the muscle fibers to the tendon and eventually to the bone. The MTJs are located at both ends of the muscle fibers [12].

Motor neurons are responsible for initiation of muscle contraction. The motor point is the location where the motor neuron enters the muscle. Neuromuscular junctions connect muscle fibers with axon terminals. The muscle fibers innervated by a nerve axon and the axon itself are referred to as a "motor unit." The amount of motor units per muscle and the amount of muscle fibers per motor unit differ between skeletal muscles [9, 12] (Fig. 17.1).

17.3 Muscle Healing

Skeletal muscle injury will heal with scar tissue, which is different from normal skeletal muscle tissue. Different causes of muscle injuries are described in the literature. For a contusion type of muscle injury, the rupture of muscle fibers occurs at or adjacent to the location of impact. In the muscle strain type of injury, the rupture of muscle fibers is located close to the MTJ [7]. The healing process is similar for muscle injuries resulting from different mechanisms of injury. The healing process is divided into the following phases: degeneration, inflammation, regeneration, and remodeling [7, 14].

17.3.1 Degeneration and Inflammation

Following injury, the resulting gap between the ruptured muscle fibers is filled with hematoma, due to hemorrhage from the torn blood vessels surrounding the muscle fibers [15].

Necrosis of the muscle fibers is initiated due to disruption of the plasma membrane. Cell permeability is increased and will result in a higher influx of calcium and an increase in activation of calcium-dependent proteases [16–18].

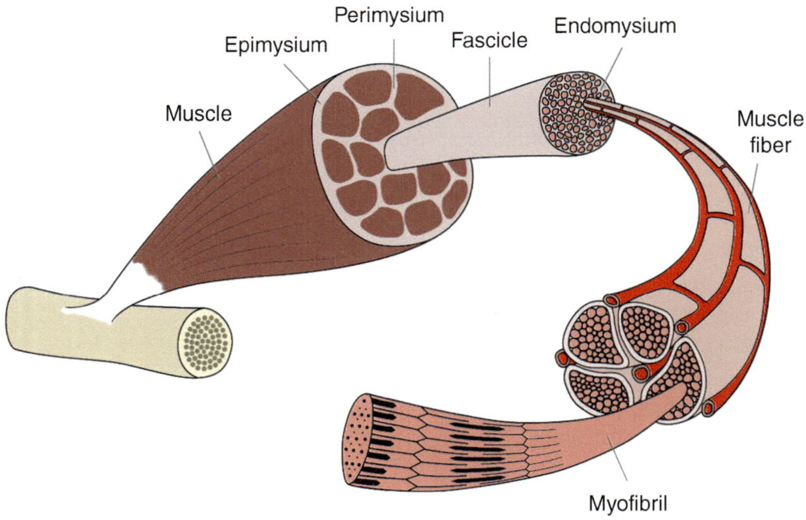

Fig. 17.1 Schematic overview of skeletal muscle structure. Reprinted from "Emerging Biological Approaches to Muscle Injuries" in Bio-orthopaedics (p 228), by van der Made A.D., Reurink G., Tol J.L., Marotta M., Rodas G., Kerkhoffs G.M., 2017, Berlin: Springer. Copyright ISAKOS 2017. Reprinted with permission [13]

The inflammatory cells in blood from the torn blood vessels have direct access to the injured site. This, in combination with the released substances of the necrotized parts of the muscle fibers that serve as chemoattractants, results in an extravasation of inflammatory cells [7, 15]. In the early acute phase after a muscle injury, polymorphonuclear leukocytes are the most abundant cells at the injury site. These leukocytes are replaced by monocytes within a day. The monocytes differentiate into macrophages that actively engage in the proteolysis and phagocytosis of the necrotic material by release of lysosomal enzymes [7, 19]. Because of the ability to adapt to the microenvironment and the multiple states of activation, macrophages have been associated with different (in vitro) phenotypes and functions [19, 20]. After several days in the healing process of muscle injuries, the macrophages switch to an anti-inflammatory profile and will contribute further in the cascade of muscle healing [17, 19–21].

17.3.2 Regeneration and Remodeling

After the destructive phases (degeneration and inflammation), the repair of the muscle injury starts with new processes: the healing process of the disrupted muscle fibers and the formation of the connective scar tissue [7].

Satellite cells are a divergent group of cells adjacent to the muscle fibers and consist of tissue-resident myogenic precursor cells. The satellite cells are located between the basal lamina and the plasma membrane (sarcolemma) and are essential cells in the cascade of the healing process of the muscles [7–10, 16, 22–24].

During the healing process of the muscles, satellite cells become activated through multiple stimuli and will migrate to the location of injury. Normally, satellite cells are in a quiescent state, which means that there is no cell cycling. At the site of injury, the satellite cells will re-enter the cell cycle to form myogenic precursor cells (myoblasts) that will differentiate into multinucleated myotubes that will adhere to the existing damaged muscle fibers [7, 10, 25]. Revascularization of the injured site is also an essential process of muscle healing. The formation of new capillaries from surrounding blood vessels is one of the first signs of muscle healing [7].

Simultaneously with the regeneration phase, the remodeling phase will start. Due to the inflammatory process, the hematoma at the injured site will form a blood clot. The blood-derived fibrin and fibronectin will form early granulation tissue, which functions as an anchorage site for fibroblasts to invade [7, 24]. Fibroblasts are activated by the release of pro-fibrotic factors. One of these pro-fibrotic factors is transforming growth factor-β (TGF-β). These pro-fibrotic factors can be released by anti-inflammatory macrophages [19, 26]. Activated fibroblasts produce remodeling factors and extracellular matrix components (EMCs) such as collagen [26]. This gives the scar tissue its initial strength to cope with the forces that will be applied during the muscle healing [7, 24].

The new muscle fibers will form mini-MTJs between the regenerated muscle fibers and the scar tissue. Gradually, the scar tissue decreases in size and will bring the ends of the damaged muscle fibers at the injury site closer to each other [23, 24]. The muscle fibers will mature, and the rise of newly formed axons will stimulate the formation of new neuromuscular junctions (NMJs). The formation of new NMJs and thus re-innervation plays a key role in muscle healing and the recovery of muscle function [22, 24, 27].

17.4 Biological Treatments

In this paragraph, we will discuss the most important biological treatments used for acute muscle injuries. We will provide a summary of the composition, the working mechanism, and the results based on the evidence available for each biological treatment.

17.4.1 Platelet-Rich Plasma (PRP)

In the media, products with autologous blood concentrates have received increasing attention over the years. Platelet-derived products like platelet-rich plasma (PRP) have gained popularity among professional and recreational athletes [28, 29]. PRP is defined as a suspension of platelets in plasma with a higher concentration in comparison with the physiological concentration in blood. When platelets are activated, they release growth factors (GFs) that play a role in regenerative processes [30].

PRP is obtained from autologous peripheral blood out of patients. A centrifuge is used to separate the platelet-rich plasma from other blood components, which result in a higher concentration of platelets in a smaller volume of plasma [29]. The platelet levels in autologous concentrated plasma could increase up to eightfold [31]. Multiple PRP products are used in different studies. Various autologous platelet-rich products are available. These products differ in preparation methods, biomolecular characteristics, and composition of cellular components, such as platelets, growth factors, cytokines, red blood cells, and leukocytes. Due to the sample variability, the interpretation of the effect of PRP is difficult [30, 32].

The rationale for the use of PRP for muscle injuries is that growth factors such as transforming growth factor-β (TGF-β), platelet-derived growth factor (PDGF), insulin-like growth factor (IGF-I, IGF-II), fibroblast growth factor (FGF), epidermal growth factor, vascular endothelial growth factor (VEGF), and endothelial cell growth factor may improve tissue recovery. These growth factors may enhance the healing of tissue and improve angiogenesis, which could stimulate the healing process [29].

Multiple randomized controlled trials (RCTs) have been conducted to examine the effect of PRP on muscle injuries. The hamstrings are the most frequently studied muscle group for the effect of PRP. One RCT studied the effect of PRP for gastrocnemius and rectus femoris injuries [33]. Most studies showed no superiority of PRP in treating muscle injuries on the time to return to pre-injury activities [34]. One RCT found a shortened time (4 days) to return to play for patients treated with PRP in hamstring muscle injuries in comparison with the control group with patients that did not receive an injection [35]. This study is at risk of bias due to the lack of presence of a placebo group, and no effect was found on the re-injury rate. In the placebo-controlled studies, no significant effect was found. A meta-analysis showed no superiority of PRP over placebo injections in hamstring injuries [34]. In one study with rats, the muscle force and the size of regenerating muscle fibers were adversely affected by the use of PRP injections as an addition to active rehabilitation [36].

In conclusion, given the lack of high-level evidence to support the efficacy of the use of PRP injections and the potential negative effect in an animal study, we do not recommend the use of PRP injections as a treatment for acute muscle injuries.

17.4.2 Actovegin

Actovegin is a drug that is used as an injection therapy for muscle injuries. Actovegin is a deproteinized hemodialysate of ultrafiltered calf serum from animals under 8 months of age. A recent in vitro study suggested that Actovegin could improve the intrinsic mitochondrial respiratory capacity in injured human skeletal muscle fibers [37]. Still, the exact working mechanism of Actovegin is unknown.

One pilot study with 11 football players diagnosed with hamstring injuries described a reduction of 8 days in return to playtime after intramuscular injections with Actovegin. These injections were an addition to a specific rehabilitation protocol for hamstring injuries. The control group consisted of patients following the specific rehabilitation protocol [38]. However, there is a high risk of bias, as there was no randomization, no blinding, and no placebo control group. Currently, there is insufficient evidence regarding its efficacy and safety profile to support the use of Actovegin for (acute) muscle injuries.

17.4.3 Traumeel

Traumeel is a fixed combination of diluted plant and mineral extracts that are currently used to treat acute muscle injuries. Traumeel has an anti-inflammatory effect because of the activity of various components that seize on different phases of the inflammatory response [39]. In vitro studies found that the systemic interleukin-6 production decreases and edema reduces, off-setting an unregulated inflammatory response. Furthermore, Traumeel inhibited the secretion of the pro-inflammatory mediators interleukin-1β (IL-1β), tumor necrosis factor-α (TNF-α), and interleukin-8 (IL-8). This suggests that Traumeel may have the potential to stabilize immune cells [39, 40].

Until now, no clinical trials are performed to examine the efficacy of the use of Traumeel in treating acute muscle injuries. Therefore, the level of scientific evidence is considered as low [39]. In conclusion, there is no scientific evidence that supports the use of Traumeel as treatment for acute muscle injuries.

17.4.4 Stem Cell Therapy

Stem cells are undifferentiated cells that can divide, under activation of specific stimuli, into an identical stem cell and a cell that can contribute to growth or regeneration. This ability of stem cells is an interesting characteristic regarding the use of stem cells as treatment for muscle injuries [41].

Research has shown the presence of several stem cell populations in skeletal muscles. Muscle-derived stem cells (MDSCs), which possibly represent satellite cell predecessors, have the ability to differentiate into cells of the myogenic lineage. The MDSCs are relatively easy to harvest and can express growth factors or anti-fibrotic molecules, like decorin, by genetic modification [42–44]. As mentioned before, these cells can theoretically contribute in the regeneration phase in muscle healing.

The therapeutic use of stem cells for muscle injury could be an interesting approach, but for now the literature to support use of stem cells is mainly focused on degenerative muscle disorders. The effect of MDSC transplantation on acute muscle injuries is studied in two studies utilizing murine contusion injury models [45, 46]. The use of intramuscular transplantation of MDSCs in mice yielded better angiogenesis and a significantly higher number of regenerative muscle fibers with a larger diameter at the fourth day post-injury in comparison with the control group or transplantations at other points in time. The MDSCs also significantly decreased fibrosis compared to the control group. When the MDSCs were transplanted during the inflammatory phase in muscle healing, a stimulation of fibrosis development occurs due to the differentiation of MDSCs in fibroblasts by the high expression of TGF-β1 [45]. These results from animal studies cannot directly be translated to humans. Thus, research in humans should be conducted.

Due to the potential tumorigenicity, there are concerns on the application of stem cell transplantation. Therefore, it is necessary to evaluate the safety of the use of stem cell transplantation as treatment for acute muscle injuries in humans.

Tissue engineering is a concept with potential for treating muscle injuries in the future. The goal of tissue engineering is to design a matrix where stem cells, such as MDSCs, will differentiate into the required tissue through the activation of signaling molecules [47].

In conclusion, in murine studies the use of stem cells provided interesting findings, but the evidence advocating the use of stem cells as treatment for muscle injuries in humans is not available. Further development and evaluation of the potential concepts are needed to provide a deliberate advice on the (intramuscular) use of stem cells in humans. Accordingly, we do not advocate the use of stem cells in muscle injuries, because of the unidentified (long-term) efficacy and safety of its use in humans.

17.4.5 Anti-Fibrotic Therapy

As mentioned before, the formation of scar tissue in muscle injuries leads to fibrosis in the affected muscle and is part of healing process in muscles.

An overstimulation of scar tissue development may lead to disproportionate accumulation of fibrosis. Fibrosis can restrict the formation and re-innervation of new muscle fibers at the injured site because it may function as a mechanical barrier [7]. This could inhibit the recovery of the injured muscle tissue and muscle function [26, 48, 49].

TGF-β1 plays a key role in formation of scar tissue by the activation of the fibrotic cascades [26, 49, 50]. With this in mind, anti-fibrotic therapies are mainly focused on the pathway of TGF-β1 to enhance muscle healing [49].

The most pro-fibrotic growth factor identified in the literature is TGF-β1. In the pathway of TGF-β1, ligand binding activates the phosphorylation of receptor-regulated SMADs (R-SMADs), such as SMAD2 and SMAD3. Subsequently, the R-SMADs bind to the common mediator SMAD (SMAD 4). This activates the transcription of collagen by the translocation of the nucleus. SMAD7 suppresses the collagen transcription [51]. To inhibit to working mechanism of TGF-β1, the anti-fibrotic therapies will aim on one of the upper mentioned steps in its pathway.

The various anti-fibrotic therapies described in the literature will be discussed.

17.4.5.1 Decorin

Decorin is a human proteoglycan serving as an anti-fibrotic agent and prevents TGF-β1 action by binding on its receptor [48, 52]. In one murine study, which used direct injections of decorin into skeletal muscle, a significant decrease in fibrosis and a significant increase in the amount of regenerating muscle fibers were described. The comparison was made with skeletal muscle of mice treated with a direct injection with saline [48]. Although a significant improvement in muscle healing was observed, a large amount of decorin was required to enhance healing process in a very small mouse muscle. This, in combination with the unknown safety of the use of decorin agents on human beings, may limit the use of direct injections with decorin as treatment for muscle injuries in the future.

17.4.5.2 Suramin

Suramin was originally designed as an anti-parasitic drug, but suramin also has an anti-fibrotic function by competitively binding the receptor of TGF-β1. Therefore, it inhibits the TGF-β1 pathway [50]. The anti-proliferative effect on fibroblasts is described in in vitro studies, and in murine models, it is shown that suramin enhances muscle healing and reduces the formation of connective scar tissue [50, 53]. Comparable to the use of decorin, the effects and the safety of the use of suramin in human beings are unknown. Therefore, more research should be done to provide a clear recommendation for the use of suramin.

17.4.5.3 Losartan

Losartan is an antihypertensive medication and has a well-tolerated profile of side effects. It works as an angiotensin-II receptor blocker. Angiotensin-II induces the formation of collagen type I via the TGF-β pathway that is mediated by the angiotensin-II type 1 (AT1) receptor. Losartan reduces fibrosis through upregulation of SMAD7, which inhibits the activation of the earlier mentioned R-SMADs [46]. Another effect of the use of losartan is the increase in follistatin at the site of injury. Follistatin is a secreted protein and is able to neutralize the actions of the TGF-β superfamily proteins and stimulates the satellite cell proliferation [46]. These effects of losartan are shown in murine models, where oral use of losartan reduced the amount of fibrosis and enhanced muscle healing [54, 55]. The dosage of losartan used in mice was an equivalent of the dosage used for hypertension in human beings and was proven to be effective [55]. These results were also found in studies in which losartan was used as an additional therapy to PRP [56] and the use of stem cells [46, 57].

Losartan tablets are generally used as antihypertensive therapy in human beings. With the positive effects on muscle healing in mice, the use of losartan could be a promising therapy in muscle healing in human beings. However, the use of losartan should be examined in human skeletal muscle before incorporating losartan as a treatment for muscle injury.

17.4.5.4 Interferon-γ

The working mechanism of interferon-γ on muscle healing is supposedly through inducing the expression of SMAD7. This inhibits the TGF-β1 pathway and thus the formation of fibrosis. A murine study found a decrease in the amount of fibrosis, an increase in muscle fibers, and an improved muscle strength [58].

Despite the proven effect of the use of interferon-γ as treatment for acute muscle healing by blocking the TGF-β1 pathway in murine models, the effects on human beings are unknown. Therefore, the efficacy and safety of interferon-γ should be evaluated in human beings before it can be integrated as treatment for acute muscle injury.

17.4.6 Safety of Intramuscular Injections

Intramuscular injection may have side effects that should be considered before it is applied in clinical practice. The myotoxic effects are evaluated in a systematic review that was performed in 2014 [59]. Evidence was found for myotoxicity of corticosteroids, local anesthetics, and nonsteroidal anti-inflammatory drugs (NSAIDs). For PRP, the evidence found for myotoxicity was ambiguous. One study found necrosis, edema, increase in inflammatory cells, and fibrosis after intramuscular injections of PRP, which were not reported in the control group. Other studies reported increased formation of muscle fibers, decrease in necrosis, and granulomatous tissue in muscle injected with PRP when compared to the control group.

For the intramuscular injections of Actovegin or Traumeel as treatment for acute muscle injuries, there is no evidence available on the myotoxicity. Due to the lack of high-level evidence on the efficacy of the use of these potential treatments in muscle injuries, more evidence is required to consider these therapies as a useful therapy in human beings.

17.4.7 Conclusion

In conclusion, multiple biological treatments for acute muscle injury are discussed. The knowledge on mechanisms of accelerating muscle tissue healing is described in the present chapter. To improve the standard of treating athletes with muscle injuries to achieve their full potential, high-quality evidence on the efficacy and the safety of these treatments should be assembled before incorporating these options into the standard of care for acute muscle injury.

As various treatments are promising, additional studies should be performed to provide this evidence. For now, the use of PRP, Actovegin, Traumeel, stem cell therapy, or anti-fibrotic agents are not advised as treatment for acute muscle injury.

References

1. Orchard J, Best TM, Verrall GM. Return to play following muscle strains. Clin J Sport Med. 2005;15(6):436–41.
2. Pollock N, James SL, Lee JC, Chakraverty R. British athletics muscle injury classification: a new grading system. Br J Sports Med. 2014;48(18):1347–51.
3. Macdonald B, McAleer S, Kelly S, Chakraverty R, Johnston M, Pollock N. Hamstring rehabilitation in elite track and field athletes: applying the British athletics muscle injury classification in clinical practice. Br J Sports Med. 2019;53(23):1464–73.
4. Hall MM. Return to play after thigh muscle injury: utility of serial ultrasound in guiding clinical progression. Curr Sports Med Rep. 2018;17(9):296–301.
5. Chu SK, Rho ME. Hamstring injuries in the athlete: diagnosis, treatment, and return to play. Curr Sports Med Rep. 2016;15(3):184–90.
6. Ekstrand J, Hagglund M, Walden M. Epidemiology of muscle injuries in professional football (soccer). Am J Sports Med. 2011;39(6):1226–32.
7. Jarvinen TA, Jarvinen TL, Kaariainen M, Kalimo H, Jarvinen M. Muscle injuries: biology and treatment. Am J Sports Med. 2005;33(5):745–64.
8. Gharaibeh B, Chun-Lansinger Y, Hagen T, Ingham SJ, Wright V, Fu F, et al. Biological approaches to improve skeletal muscle healing after injury and disease. Birth Defects Res C Embryo Today. 2012;96(1):82–94.
9. Huard J, Li Y, Fu FH. Muscle injuries and repair: current trends in research. J Bone Joint Surg Am. 2002;84(5):822–32.

10. Dumont NA, Bentzinger CF, Sincennes MC, Rudnicki MA. Satellite cells and skeletal muscle regeneration. Compr Physiol. 2015;5(3):1027–59.
11. Takala TE, Virtanen P. Biochemical composition of muscle extracellular matrix: the effect of loading. Scand J Med Sci Sports. 2000;10(6):321–5.
12. Tidball JG. Myotendinous junction injury in relation to junction structure and molecular composition. Exerc Sport Sci Rev. 1991;19:419–45.
13. van der Made AD, Reurink G, Tol JL, Marotta M, Rodas G, Kerkhoffs GM. Emerging biological approaches to muscle injuries. In: Gobbi A, Espregueira-Mendes J, Lane JG, Karahan M, editors. Bio-orthopaedics. Berlin, Heidelberg: Springer; 2017.
14. Hurme T, Kalimo H, Lehto M, Jarvinen M. Healing of skeletal muscle injury: an ultrastructural and immunohistochemical study. Med Sci Sports Exerc. 1991;23(7):801–10.
15. Jarvinen TA, Kaariainen M, Jarvinen M, Kalimo H. Muscle strain injuries. Curr Opin Rheumatol. 2000;12(2):155–61.
16. Charge SB, Rudnicki MA. Cellular and molecular regulation of muscle regeneration. Physiol Rev. 2004;84(1):209–38.
17. Ciciliot S, Schiaffino S. Regeneration of mammalian skeletal muscle. Basic mechanisms and clinical implications. Curr Pharm Des. 2010;16(8):906–14.
18. Tidball JG. Mechanisms of muscle injury, repair, and regeneration. Compr Physiol. 2011;1(4):2029–62.
19. Chazaud B, Brigitte M, Yacoub-Youssef H, Arnold L, Gherardi R, Sonnet C, et al. Dual and beneficial roles of macrophages during skeletal muscle regeneration. Exerc Sport Sci Rev. 2009;37(1):18–22.
20. Arnold L, Henry A, Poron F, Baba-Amer Y, van Rooijen N, Plonquet A, et al. Inflammatory monocytes recruited after skeletal muscle injury switch into antiinflammatory macrophages to support myogenesis. J Exp Med. 2007;204(5):1057–69.
21. Saclier M, Yacoub-Youssef H, Mackey AL, Arnold L, Ardjoune H, Magnan M, et al. Differentially activated macrophages orchestrate myogenic precursor cell fate during human skeletal muscle regeneration. Stem Cells. 2013;31(2):384–96.
22. Jarvinen TA, Jarvinen M, Kalimo H. Regeneration of injured skeletal muscle after the injury. Muscles Ligaments Tendons J. 2013;3(4):337–45.
23. Jarvinen TA, Jarvinen TL, Kaariainen M, Aarimaa V, Vaittinen S, Kalimo H, et al. Muscle injuries: optimising recovery. Best Pract Res Clin Rheumatol. 2007;21(2):317–31.
24. Kaariainen M, Jarvinen T, Jarvinen M, Rantanen J, Kalimo H. Relation between myofibers and connective tissue during muscle injury repair. Scand J Med Sci Sports. 2000;10(6):332–7.
25. Tedesco FS, Dellavalle A, Diaz-Manera J, Messina G, Cossu G. Repairing skeletal muscle: regenerative potential of skeletal muscle stem cells. J Clin Invest. 2010;120(1):11–9.
26. Mann CJ, Perdiguero E, Kharraz Y, Aguilar S, Pessina P, Serrano AL, et al. Aberrant repair and fibrosis development in skeletal muscle. Skelet Muscle. 2011;1(1):21.
27. Rantanen J, Ranne J, Hurme T, Kalimo H. Denervated segments of injured skeletal muscle fibers are reinnervated by newly formed neuromuscular junctions. J Neuropathol Exp Neurol. 1995;54(2):188–94.
28. Sheth U, Simunovic N, Klein G, Fu F, Einhorn TA, Schemitsch E, et al. Efficacy of autologous platelet-rich plasma use for orthopaedic indications: a meta-analysis. J Bone Joint Surg Am. 2012;94(4):298–307.
29. Moraes VY, Lenza M, Tamaoki MJ, Faloppa F, Belloti JC. Platelet-rich therapies for musculoskeletal soft tissue injuries. Cochrane Database Syst Rev. 2014;4:CD010071.
30. Magalon J, Bausset O, Serratrice N, Giraudo L, Aboudou H, Veran J, et al. Characterization and comparison of 5 platelet-rich plasma preparations in a single-donor model. Arthroscopy. 2014;30(5):629–38.
31. Hamilton BH, Best TM. Platelet-enriched plasma and muscle strain injuries: challenges imposed by the burden of proof. Clin J Sport Med. 2011;21(1):31–6.
32. Oh JH, Kim W, Park KU, Roh YH. Comparison of the cellular composition and cytokine-release kinetics of various platelet-rich plasma preparations. Am J Sports Med. 2015;43(12):3062–70.
33. Martinez-Zapata MJ, Orozco L, Balius R, Soler R, Bosch A, Rodas G, et al. Efficacy of autologous platelet-rich plasma for the treatment of muscle rupture with haematoma: a multicentre, randomised, double-blind, placebo-controlled clinical trial. Blood Transfus. 2016;14(2):245–54.
34. Pas HI, Reurink G, Tol JL, Weir A, Winters M, Moen MH. Efficacy of rehabilitation (lengthening) exercises, platelet-rich plasma injections, and other conservative interventions in acute hamstring injuries: an updated systematic review and meta-analysis. Br J Sports Med. 2015;49(18):1197–205.
35. Rossi LA, Molina Romoli AR, Bertona Altieri BA, Burgos Flor JA, Scordo WE, Elizondo CM. Does platelet-rich plasma decrease time to return to sports in acute muscle tear? A randomized controlled trial. Knee Surg Sports Traumatol Arthrosc. 2017;25(10):3319–25.
36. Contreras-Munoz P, Torrella JR, Serres X, Rizo-Roca D, De la Varga M, Viscor G, et al. Postinjury exercise and platelet-rich plasma therapies improve skeletal muscle healing in rats but are not synergistic when combined. Am J Sports Med. 2017;45(9):2131–41.
37. Brock J, Golding D, Smith PM, Nokes L, Kwan A, Lee PYF. Update on the role of Actovegin in musculoskeletal medicine: a review of the past 10 years. Clin J Sport Med. 2018;30(1):83–90.
38. Lee P, Rattenberry A, Connelly S, Nokes L. Our experience on Actovegin, is it cutting edge? Int J Sports Med. 2011;32(4):237–41.
39. Schneider C. Traumeel - an emerging option to nonsteroidal anti-inflammatory drugs in the management of acute musculoskeletal injuries. Int J Gen Med. 2011;4:225–34.

40. Lussignoli S, Bertani S, Metelmann H, Bellavite P, Conforti A. Effect of Traumeel S, a homeopathic formulation, on blood-induced inflammation in rats. Complement Ther Med. 1999;7(4):225–30.
41. Gates CB, Karthikeyan T, Fu F, Huard J. Regenerative medicine for the musculoskeletal system based on muscle-derived stem cells. J Am Acad Orthop Surg. 2008;16(2):68–76.
42. Deasy BM, Jankowski RJ, Huard J. Muscle-derived stem cells: characterization and potential for cell-mediated therapy. Blood Cells Mol Dis. 2001;27(5):924–33.
43. Usas A, Huard J. Muscle-derived stem cells for tissue engineering and regenerative therapy. Biomaterials. 2007;28(36):5401–6.
44. Qu-Petersen Z, Deasy B, Jankowski R, Ikezawa M, Cummins J, Pruchnic R, et al. Identification of a novel population of muscle stem cells in mice: potential for muscle regeneration. J Cell Biol. 2002;157(5):851–64.
45. Ota S, Uehara K, Nozaki M, Kobayashi T, Terada S, Tobita K, et al. Intramuscular transplantation of muscle-derived stem cells accelerates skeletal muscle healing after contusion injury via enhancement of angiogenesis. Am J Sports Med. 2011;39(9):1912–22.
46. Kobayashi M, Ota S, Terada S, Kawakami Y, Otsuka T, Fu FH, et al. The combined use of losartan and muscle-derived stem cells significantly improves the functional recovery of muscle in a young mouse model of contusion injuries. Am J Sports Med. 2016;44(12):3252–61.
47. McCullagh KJ, Perlingeiro RC. Coaxing stem cells for skeletal muscle repair. Adv Drug Deliv Rev. 2015;84:198–207.
48. Fukushima K, Badlani N, Usas A, Riano F, Fu F, Huard J. The use of an antifibrosis agent to improve muscle recovery after laceration. Am J Sports Med. 2001;29(4):394–402.
49. Lieber RL, Ward SR. Cellular mechanisms of tissue fibrosis. 4. Structural and functional consequences of skeletal muscle fibrosis. Am J Physiol Cell Physiol. 2013;305(3):C241–52.
50. Chan YS, Li Y, Foster W, Horaguchi T, Somogyi G, Fu FH, et al. Antifibrotic effects of suramin in injured skeletal muscle after laceration. J Appl Physiol. 1985;95(2):771–80.
51. Garg K, Corona BT, Walters TJ. Therapeutic strategies for preventing skeletal muscle fibrosis after injury. Front Pharmacol. 2015;6:87.
52. Li Y, Foster W, Deasy BM, Chan Y, Prisk V, Tang Y, et al. Transforming growth factor-beta1 induces the differentiation of myogenic cells into fibrotic cells in injured skeletal muscle: a key event in muscle fibrogenesis. Am J Pathol. 2004;164(3):1007–19.
53. Nozaki M, Ota S, Terada S, Li Y, Uehara K, Gharaibeh B, et al. Timing of the administration of suramin treatment after muscle injury. Muscle Nerve. 2012;46(1):70–9.
54. Bedair HS, Karthikeyan T, Quintero A, Li Y, Huard J. Angiotensin II receptor blockade administered after injury improves muscle regeneration and decreases fibrosis in normal skeletal muscle. Am J Sports Med. 2008;36(8):1548–54.
55. Kobayashi T, Uehara K, Ota S, Tobita K, Ambrosio F, Cummins JH, et al. The timing of administration of a clinically relevant dose of losartan influences the healing process after contusion induced muscle injury. J Appl Physiol (1985). 2013;114(2):262–73.
56. Terada S, Ota S, Kobayashi M, Kobayashi T, Mifune Y, Takayama K, et al. Use of an antifibrotic agent improves the effect of platelet-rich plasma on muscle healing after injury. J Bone Joint Surg Am. 2013;95(11):980–8.
57. Lee EM, Kim AY, Lee EJ, Park JK, Lee MM, Hwang M, et al. Therapeutic effects of mouse adipose-derived stem cells and losartan in the skeletal muscle of injured mdx mice. Cell Transplant. 2015;24(5):939–53.
58. Foster W, Li Y, Usas A, Somogyi G, Huard J. Gamma interferon as an antifibrosis agent in skeletal muscle. J Orthop Res. 2003;21(5):798–804.
59. Reurink G, Goudswaard GJ, Moen MH, Weir A, Verhaar JA, Tol JL. Myotoxicity of injections for acute muscle injuries: a systematic review. Sports Med. 2014;44(7):943–56.

18 Compartment Syndrome and Shin Splints

Matteo Maria Tei, Giacomo Placella, Marta Sbaraglia, Pierluigi Antinolfi, and Giuliano Cerulli

18.1 Introduction

Compartment syndrome (CS) represents an emergency involving both muscles and tendons, and their clinical diagnosis is not always easy. CS occurs when interstitial pressure increases in a fascial space, resulting in the impairment of microcirculation, thereby causing tissue ischemia. If it is not recognized and treated early, it can lead to muscle necrosis, rhabdomyolysis, and systemic disease in severe cases. The most common causes of compartment syndrome are as follows:

- Fractures caused by high-energy trauma;
- Crush injuries;
- Severe bruising,
- Snakebites;
- Dressings that are too tight;
- Plaster casts.

Pain is the earliest and most sensitive symptom, and it appears out of proportion compared to the severity of the injury. Circulatory stasis around the nerves may cause paresthesia, which may lead to progressive muscle paralysis and death. When intracompartmental pressure exceeds the blood pressure, the limb becomes pale and it is impossible to feel peripheral pulses. For its diagnosis, there is the "5P rule" as reported in the English literature: [1–7].

- Pain;
- Paresthesias;
- Pallor;
- Paralysis;
- Pulselessness.

The patient complains of severe and increasing pain and requires frequent doses of analgesic drugs. The pain increases during passive stretching of the limb; moreover, patients report tingling

M. M. Tei · G. Cerulli (✉)
Istituto di Ricerca Traslazionale per l'Apparato Locomotore Nicola Cerulli – LPMRI, Perugia/Arezzo, Italy

G. Placella
Istituto di Ricerca Traslazionale per l'Apparato Locomotore Nicola Cerulli – LPMRI, Perugia/Arezzo, Italy

Department of Orthopaedic Surgery, Ospedale San Raffaele, Milan, Italy

M. Sbaraglia
Istituto di Ricerca Traslazionale per l'Apparato Locomotore Nicola Cerulli – LPMRI, Perugia/Arezzo, Italy

Department of Pathology, Azienda Ospedaliera di Padova, Padua, Italy

P. Antinolfi
Istituto di Ricerca Traslazionale per l'Apparato Locomotore Nicola Cerulli – LPMRI, Perugia/Arezzo, Italy

Department of Orthopaedic Surgery, Ospedale Santa Maria della Misericordia, Perugia, Italy

G. L. Canata et al. (eds.), *Management of Track and Field Injuries*,
https://doi.org/10.1007/978-3-030-60216-1_18

along the nerve distribution passing through the affected compartment. A key point is that CS is a progressively developing condition. Maximum swelling occurs at about 30–36 h after the traumatic event; therefore, it is essential to pay careful attention to high-risk limbs during the early period post-trauma. In patients with altered sensitivity, clinical signs and symptoms are less useful. These patients must be closely monitored; if there is a suspicion of compartment syndrome, it is necessary to measure intracompartmental pressure. The measurements must be taken in all compartments using specific instruments and it should be measured as closely as possible to the fracture, as in this anatomical area the pressure is highest.

Normal intracompartmental pressure is about 5–8 mmHg. When intracompartmental pressure reaches 20 mmHg, tissue perfusion can decrease. Tissue perfusion is based on the local perfusion pressure (diastolic pressure—compartmental pressure), and if the difference between these pressures (delta P) is less than 30 mmHg, a fasciotomy is indicated. Early treatment of CS should include the removal of circumferential dressings, loosening tight bandages, and raising the limb above chest level, which decreases the perfusion pressure on the muscle. If these procedures reduce symptoms, the patient should then be carefully monitored and re-evaluated frequently. If these precautions are not beneficial, the patient should be taken to the operating room for fasciotomy.

18.2 Compartment Syndrome of the Upper Limbs

Upper limb compartment syndrome may develop as a result of:

- Fractures of the distal radius;
- Forearm shaft fractures;
- Crush injuries of soft tissues.
- Several less common causes include the following:
  - Snakebite;
  - Gunshot wounds;
  - Toxic shock syndrome;
  - Leukemic infiltration;
  - Viral myositis;
  - Arthroscopic infusion fluid;
  - Nephrotic syndrome.

Patients classically experience constant and oppressive pain. For low-energy injuries, the pain may seem to be out of proportion. Nerve dysfunction in the compartment involved can lead to paresthesia: burning, numbness, and tingling. In patients with fractures, the pain persists and worsens despite reduction and immobilization. In CS of the forearm, patients experience excruciating pain during flexion and extension of the fingers. The patient suffers from a state of discomfort secondary to muscle compartment tension. In addition, there is a reduction in sensitivity in the distribution of the peripheral nerves and widespread edema. Later, there is numbness, loss of peripheral pulses, and pallor of the limb. Even if the pain is the best clinical indicator of CS, some patients are unable to report it. If the patient is a child or the patient has received large amount of analgesics, is unconscious, inhibited, or sedated, he or she may not be able to refer clearly about the pain.

In these situations, it is recommended to measure the intracompartmental pressure. Elliott et al. [8] reported that 23% of the cases with forearm compartmental syndrome are caused by soft tissue injuries without fractures and 18% are caused by fractures. In our experience, there is limited amount of available evidence regarding causes, treatment, suture wound methods, functional result, and complications of forearm CS. It has been associated with various etiologies; however, fractures of the distal radius are reported as the most common cause of forearm CS. This is contrary to what has been reported in the past by Grottkau et al. [1], and the authors suggested that supracondylar fractures were the predominant cause of forearm CS in children. In a study by the National Pediatric Trauma Registry evaluating 131 cases of pediatric CS, it was found that 74% of the cases of upper limb CS were caused by forearm fractures and only 15% were secondary to supracondylar fractures [9]. Bae et al. [10], studying 33 consecutive pediatric patients with 36 cases of acute compartmental syndrome, suggested that a possible reason for this decrease in CS after supracondylar fracture could be due to the changes in fracture management, such as percutaneous pin osteosynthesis.

Patients under 35 years of age involved in a high-energy trauma and polytrauma have an increased risk of developing forearm CS. Hwang et al. [11] noted that patients with distal radius fractures and ipsilateral elbow fractures developed CS in 15% of the cases, much higher than the risk (0.25%) to develop CS after an isolated fracture of the distal radius. Upper limb CS is generally diagnosed with a careful clinical examination. The removal of any tight dressings is a critical step to enable an accurate assessment of the limb. Regarding intracompartmental pressure measuring, there is almost an equal distribution between the number of patients diagnosed by clinical examination as those diagnosed by intracompartmental pressure measure [10].

Many authors consider the measurement of intracompartmental pressure unnecessary for diagnostic purposes [5, 6, 12–17]. Others recommend its use only in patients with impaired communication capabilities or in patients whose clinical findings have an ambiguous interpretation [17–21]. With regard to the treatment of forearm CS, different skin incisions have been proposed. The typical ventral incision begins 1 cm proximal and 2 cm laterally at the forearm, and then obliquely across the antecubital fossa on the volar forearm. Incision starts just radial to flexor carpi ulnaris (FCU) at wrist and extends proximally to medial epicondyle extended distally to release carpal tunnel; in the medial direction, the incision reaches the middle line at the average distal third of the forearm. Here, the incision is continued only to the ulnar side of the long handheld tendon to avoid the palmar skin cord of the median nerve. The incision then passes through the wrist and extends into the medial portion of the palm for the concurrent release of the carpal tunnel (Fig. 18.1). The overall rate of complications of forearm CS is about 42%. Many studies report neurological deficits as the most common complication [5, 12, 22, 23]. Without treatment, CS results in contractures, neurological deficits, and severe cases of complete loss of function in the forearm and hand. Therefore, emergency treatment is necessary to prevent serious consequences.

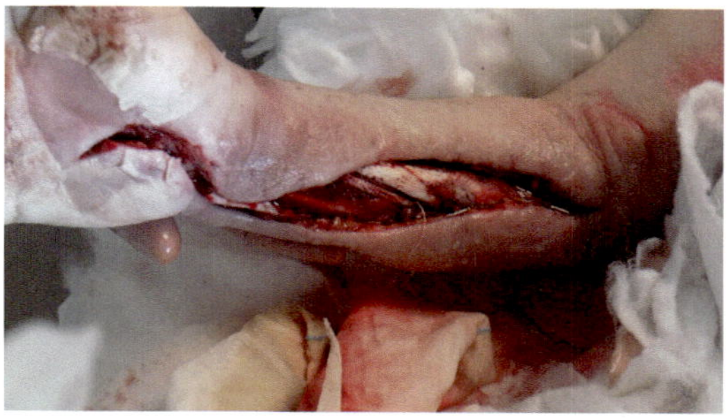

Fig. 18.1 The typical ventral incision at the forearm that starts just radial to flexor carpi ulnaris (FCU) at wrist and extends proximally to medial epicondyle extended distally to release carpal tunnel

18.3 Compartment Syndrome of the Lower Limbs

Acute compartment syndrome of the lower limb is a complication of fractures, soft tissue trauma, and reperfusion after acute arterial occlusion. It can be caused by bleeding or swelling in a muscle compartment. The long-term consequences of CS have already been described by Richard von Volkmann [9] in the late nineteenth century as a result of a too tight plaster cast, but only after a few years was a connection made with high intracompartmental pressure. The incidence of foot CS is about 6% in patients with foot injuries caused by motorcycle accidents. However, the incidence of leg CS seems lower (1.2% after closed diaphyseal fractures of the tibia) [8]. The lower limb compartment syndrome (excluding the foot) and its treatment were already described in 1958 [9], whereas, until a few years ago, compartment syndrome of the foot was largely unknown and was described only in some case reports. Myerson first described this clinical entity in 1988 and presented surgical decompression as a therapeutic intervention [24]. The leg is composed of four compartments: anterior, lateral, surface, and deep posterior. However, there is no consensus with regard to the number of anatomical compartments of the foot. At the end of 1920, three compartments were described and these were later confirmed by Kamel and Sakla in 1961 [25]. Myerson et al. later identified four compartments [14]. However, more recently, nine compartments were identified in a cadaveric study [26]. In a cadaveric study performed in 2008, the authors could not identify any distinct forefoot myofascial compartments, and therefore, it was concluded that a fasciotomy of the hindfoot compartments through a modified medial incision would be sufficient to decompress the whole foot [5]. However, studies on cadavers cannot simulate physiological conditions. Therefore, the conclusions of these studies should be interpreted with caution. The typical clinical presentation of leg and foot CS is not different from any other regions of the body. In a systematic review of the literature, the pain has been identified as the earliest and most sensitive clinical sign of CS [27]. In a retrospective study, moreover, foot pain was present in all patients with foot CS [28].

Anamnesis: When acute compartment syndrome is suspected, a careful examination is needed.

Physical examination: Medical recommendations based on evidence-based medicine (EBM) cannot be made. Serial laboratory tests should be performed as soon as possible as it is widely recognized that muscle necrosis usually occurs within the first 3 h [17]. However, contrary to what was thought in the past, muscle strength is not a good parameter to be assessed as it is difficult to determine whether the loss of strength is due to the pain or muscle necrosis. Even the examination of peripheral pulses is not reliable for the diagnosis of lower limb CS, because there may be false negatives whenever the intracompartmental pressure reaches the systolic blood pressure.

Diagnostic tests: Invasive measurements of intracompartmental pressure are a rapid and safe procedure to reach a definite diagnosis. It should be emphasized that in a cohort study with more than 200 patients with diaphyseal fractures of the tibia, the continuous monitoring of intracompartmental pressure showed no differences in outcomes or possible delays in performing fasciotomy compared to the simple clinical examination of the patient [29]. Another study showed that the rate of late complications was similar in patients having undergone continuous monitoring of the intracompartmental pressure [18]. Since nine compartments in the foot have been identified, it is not feasible to monitor the pressure for patients at high risk of developing CS in this anatomical area. It is also important to remember that intracompartmental pressure must be correlated with the diastolic pressure.

Treatment: Fasciotomy threshold is still under debate. While some authors suggest that for intracompartmental pressure the threshold for fasciotomy should be an absolute value of 30 mmHg [21], others indicate 20 mmHg less than the diastolic pressure as a threshold [30]. However, currently the indication for fasciotomy should be based on clinical findings (neurological deficits) or on a difference between intracompartmental pressure and diastolic pressure lower than 30 mmHg [30]. Although most of these recommendations derive from studies of other anatomic regions, there is no reason to assume a different pathophysiological background for foot CS.

Clinical Results: It is important to remember that clinical results should be compared over time. In short, a history of trauma and the presence of serious injuries should make the physician consider the possibility of CS. Although the management of CS consists of immediate surgical treatment, bandages and casts should be completely open in patients with severe postoperative pain. In the case of impending CS, the limb should not be raised because it reduces the blood supply that is already compromised. McQueen demonstrated that in patients with tibial fractures, the time between the onset of compartment syndrome and fasciotomy influences the outcome, rather than the time between trauma and osteosynthesis [31]. Generally, the existing literature is lacking in regard to the optimal management of tibial fractures in the presence of CS. On the other hand, multiple approaches have been used to decompress the compartments of the foot [4].

Although the etiology, pathophysiology, and treatment of CS are well described, little has been published about the long-term results. CS of the leg and foot has a low incidence rate (1.2% after closed tibial fractures, 6% after open tibial fractures); studies on a greater number of patients are, however, not available. One study has examined the quality of life after CS using the "EQ-5D score" [23].

In a study of 30 cases, patients with leg compartment syndrome had lower EQ-5D scores than the control group with isolated fracture without compartment syndrome at 12 months after treatment, although their health status was not statistically different [23]. In addition, the authors reported that patients with faster wound closure times were healthier than those with longer wound closure times [23]. In another study on the results of follow-up in 26 patients with traumatic leg CS, 15.4% complained of pain at rest and 26.9% reported pain under stress at 1–7 years after the trauma [22]. In this population, more than 50% of the patients had reduced joint ROM and reported a reduction in sensitivity. Infections due to fasciotomy were described in up to 38% of the patients. Patients who had undergone a surgical flap with skin grafting for wound closure presented a lower incidence of infections. In another study, the presence of associated lesions seemed not to affect the long-term outcome after traumatic CS of the leg with regard to the joint ROM, sensory dysfunction, and loss of muscle strength [26].

In a series of 14 patients, Myerson [28] described the return to the previous working activity after trauma in 4 patients, 6 patients had only occasional symptoms that had developed during some daily activities, whereas 3 patients developed contractures with clawed fingers. No patients, however, needed amputation (25). Paresthesia and numbness of scars distal to the compartments involved were common long-term sequelae in 8 patients.

Complications: Our experience shows that the literature available is quite limited in this specific field of orthopedics and traumatology [23]. Therefore, we believe that further studies are needed to describe long-term results. Although the pathophysiology of CS is well described, it is not yet clear when there is irreversible damage. Recent studies in animal models reported muscle necrosis after less than 3 h [32]. Moreover, the information available in the literature is inconsistent and we believe further studies are necessary. Although clinical signs are well described [32], we believe that the most important factor in the CS diagnosis is the key figure of the doctor, who must put the patient at the center of the attention and base treatment on a "holistic-like approach."

Moreover, the physician should be aware that the pain, defined as a clinical sign of CS, could be masked in patients with a reduced state of consciousness or if previously treated with analgesics. Although the literature lacks recommendations about the intervals at which serial examinations should be performed in patients at risk, we believe they should be performed at least every hour, as irreversible damage has been reported to occur within the first 3 h [32]. Recommendations for surgical treatment of foot CS are controversial as the literature lacks comparative studies. In conclusion, lower limb CS is a rare, but serious complication of which the surgeon must be aware. Although immediate fasciotomy is the undisputed treatment for patients with CS, the literature lacks evidence-based clinical guidelines.

18.4 Chronic Exertional Compartment Syndrome

A separate paragraph should be dedicated to the treatment of chronic exertional compartment syndrome (CECS). There is uncertainty about the development of the syndrome in the majority of affected patients. CECS is not commonly considered as a cause of muscle pain. Typically, there is a delay of 22 months in the diagnosis of the disease. Studies on the etiology of chronic pain in the anterior leg indicate that CECS is the causal factor in 27% of the cases [33].

Anamnesis: The delay in diagnosis, combined with the relative frequency, underlines the attention that physicians, not only specialists in orthopedics and traumatology, should pay toward

CECS as a possible diagnosis. The diagnosis affects the patient's performance of sport and work activities. The pathophysiology of CECS is connected to an increase in compartmental pressure occurring during exercise due to an increased muscle volume. The prevailing theory is that during activities, the muscle suffers a gradual increase in intracompartmental pressure with the consequent impairment of muscle tissue perfusion [34].

Incidence: In the general population, the exact incidence rate is unclear because of the difficulty to diagnose it and the delay in seeking medical care. CECS should be suspected in any athlete who presents with chronic anterior leg pain that worsens with physical activity, but it is resolved upon cessation of activity. 95% of the cases of CECS occur in the anterior and lateral compartments of the leg [35]. CECS is more frequent in young adult amateur runners and military recruits, but it is not uncommon in athletes participating in contact sports. There are no demonstrated differences in incidence between men and women [36]. The average age of onset is 20 years [13]. The risk factors for the development of CECS include use of anabolic steroids and the use of creatinine increasing muscle volume. Aberrant biomechanical factors in a runner, such as wrong foot support or overpronation, can lead to an increased risk of compartment syndrome secondary to differences between weight/load and to high pressure on individual muscle groups in the lower leg.

Physical examination: Acquiring a thorough history for compartment syndrome is important because the physical examination may be irrelevant. Classically, there is the development of pain described as a burning or pressuring sensation, in a compartment of the leg at the same time, at the same distance, or at the same intensity [37]. The pain increases in intensity as the patient continues to exercise. Symptoms occur bilaterally in 70% of 80 cases [38]. Other symptoms include numbness and tingling in the dermatomal distribution of the nerve conduction through the involved compartment. Weakness of the affected muscle is also a symptom that is reported by patients. A classic presentation of CECS is a runner that experiences burning in the leg and numbness on the back foot after about 15 min of continuous running, with absolutely no symptoms within 30 min of stopping.

The physical examination can be used to differentiate CECS from other causes of chronic pain in the lower legs. The athlete should be examined after he or she has completed the exercise provoking the pain. An important diagnostic procedure could be biomechanical functional assessments, thereby allowing stabilometric, electromyographic, and isokinetic parameters to be studied. Functional imaging studies can also give precise information about the joint kinematics and the ability to perform simple or complex gestures. Biomechanical evaluations offer a possibility for orthopedic specialists to express a precise opinion on the functional state of the musculoskeletal system and its various components through simple and more sophisticated and expensive instruments such as force plates, 16-channel EMG telemetry, instruments for isokinetic evaluation, and 3D systems. Biomechanical laboratories for the musculoskeletal system (Fig. 18.2) offer accurate and reproducible data regarding some locomotor parameters, such as reaction to the ground, proprioception, the peak of flexor and extensors of the knee muscle strength, electrical activity of various muscles of the thigh and leg being assessed in dynamic conditions, and, finally, the functional capacity during the most simple or more complex movements (Fig. 18.3). Athletes and other patients presenting with movement disorders should be assessed in dynamic conditions rather than in static conditions.

Diagnosis: Golden standard for the diagnosis of CECS is the measurement of intracompartmental pressure.

Treatment: The only certain treatment of CECS is fasciotomy [13]. Nonetheless, conservative treatment has also been described, such as avoiding activities that can generate symptoms or decreasing the workout intensity. Athletes may be advised to rest and then slowly increase their athletic training. Specifically designed orthopedic insoles might be prescribed, which give plantar arch support and correct pronation while

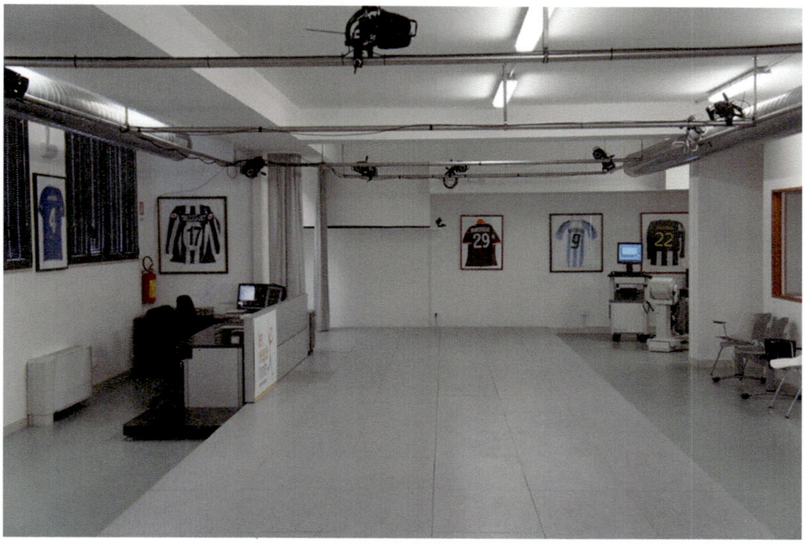

Fig. 18.2 Biomechanical laboratories for the musculoskeletal system, fully equipped. Athletes and other patients presenting with movement disorders should be assessed in dynamic conditions

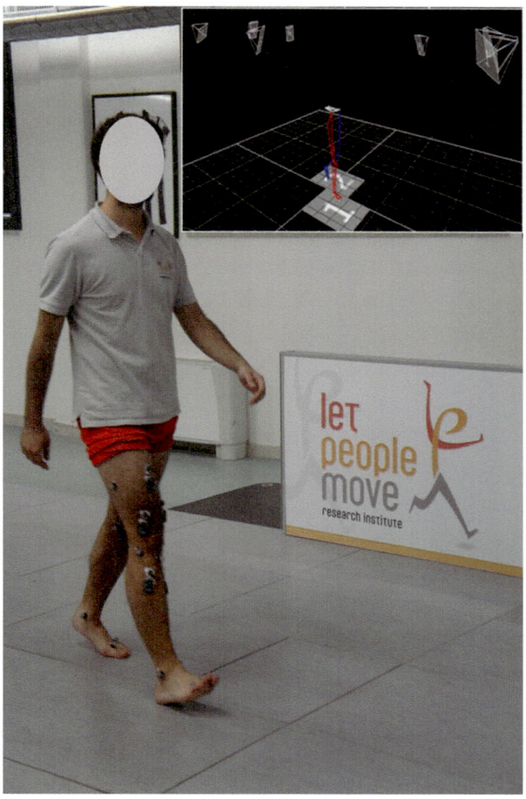

Fig. 18.3 Gait analysis with 16-channel EMG telemetry performed at the biomechanical laboratory

running. Other conservative treatment methods include avoiding running on hard surfaces, wearing appropriate footwear, and aiming at changing specific sport movements based on an objective biomechanical assessment. Massage therapy of the involved muscle tissue, ultrasound, and stretching before exercise are all treatment strategies that may prolong the time before symptoms appear. If athletes do not get any relief from conservative measures and they do wish to continue practicing sport at the same level and intensity, fasciotomy is the treatment of choice [13].

This lower percentage is attributed to the posterior compartment's complex anatomy. Several types of fasciotomies have been described: Open and subcutaneous fasciotomies are the most commonly performed surgeries. The advantage of fasciotomy in open is the full view of the compartment. Some types of open fasciotomy include the removal of band flaps to reduce the formation of aberrant scars and relapses [13].

On the other hand, subcutaneous fasciotomy involves 1–2 small incisions. Several case reports of endoscopically assisted fasciotomies have been described, but an increase in frequency of complications and relapses was reported [19]. A compressive dressing is applied postoperatively for 2–3 days. Patients are requested to perform different types of rehabilitation exercises after surgery in order to prevent the formation of tissue adhesions. Patients can swim as soon as surgical wounds are completely healed, whereas physical therapy usually beings 1–2 weeks after surgery.

The athlete can return to full sport activity within 6–8 weeks if he/she is asymptomatic and has recovered fully concerning muscle strength and elasticity as assessed according to a postoperative biomechanical evaluation [39].

18.5 Medial Tibial Stress Syndrome (MTSS, Shin Splints)

The medial tibia shin splint (MTSS) is pain occurring along the inner edge of the tibia (Fig. 18.4). The lower two-thirds of the anterior and medial part of the tibia is the most common site of pain. MTSS is a common injury in runners of long distances and in athletes with repetitive and prolonged efforts such as gymnasts, dancer, or military recruits [40]. MTSS is responsible for 35% of runner injuries [41]. Shin splints are an overuse injury affecting over three million athletes.

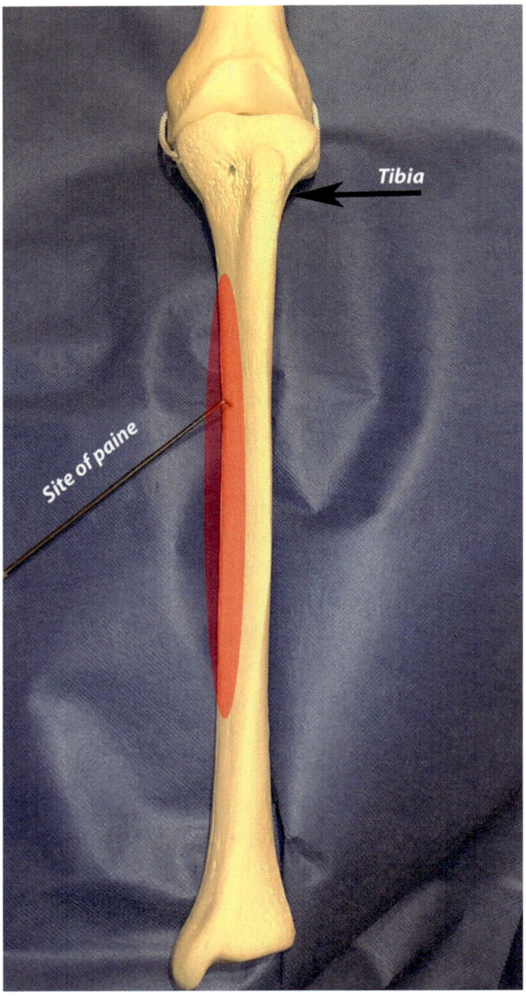

Fig. 18.4 Example of most common site of pain in medial tibia shin splint

Diagnosis: There are several serious causes for MTSS: Compartmental syndrome and tibia stress fracture are the most severe, but the most common causes in professional athletes are probably the irritation and degeneration of the soft tissue around the bone (periosteum). Anterior leg pain can be caused by other problems such as sural or peroneal nerve entrapment, tendinopathy, and popliteal artery entrapment.

The most severe problem is acute compartmental syndrome: This diagnosis often causes unrecognized MTSS by athletes' health team. The second condition often overlooked is biological tissue fatigue; often, biomechanical stress is only taken into consideration, leaving aside that chronic pain can be triggered by tissue suffering. Rehabilitation in these pains must be slow because it is guided by biological principles that cannot be asked for discounts. Often, the only problem is the haste to be able to return to sport. In case of acute tibial pain, the first thing to do is to exclude acute compartment syndrome (CS).

Risk Factors: There is only one accepted risk factor for MTSS: excessive physical stress. Overload, overuse, or misuse is always present in this disease [40]. In addition to this common and fundamental factor, there are other individual predispositions: modification of the type of training, the type of devices used, and sport frequency: Running for longer distances, on climbs, for more frequent periods with different shoes is often important factor to consider. Other factors that contribute to shin splints include flat feet or abnormally rigid arches, exercising with improper or worn-out footwear, individuals with inflexibility, and tightness of lower leg muscles.

Treatment: Treatment of MTTS is complicated because there are several overlapping causes [42]. Today, this disease is considered to be poorly treated and often left unresolved because the lack of knowledge and old treatments that are no longer reliable are often used [43, 44]. Shin splints treatment includes several weeks of rest from activity that can be substituted with lower impact types of aerobic activity, which supports an intense circulation without structural tissue overload. Ice is a common therapy as it decreases inflammation and pain, elevation of the leg can decrease the swelling to the area. Additional swelling can be treated with compression bandage or antithrombus stockings.

Once the pain has decreased, strengthening exercises should be performed focusing on the lower leg and hip muscles. Shin splints usually resolve with rest and the treatments described above. Before returning to exercise, the patient should be pain-free for at least 2 weeks. Return to exercise must be at lower level of intensity [42].

References

1. Grottkau BE, Epps HR, Di Scala C. Compartment syndrome in children and adolescents. J Pediatr Surg. 2005;40(4):678–82. https://doi.org/10.1016/j.jpedsurg.2004.12.007.
2. Hasnain M, WV RV. Acute effects of newer antipsychotic drugs on glucose metabolism. Am J Med. 2008;121(10):e17. https://doi.org/10.1016/j.amjmed.2008.04.017. author reply e9
3. El-Mir MY, Detaille D, Gloria RV, Delgado-Esteban M, Guigas B, Attia S, et al. Neuroprotective role of antidiabetic drug metformin against apoptotic cell death in primary cortical neurons. J Mol Neurosci. 2008;34(1):77–87. https://doi.org/10.1007/s12031-007-9002-1.
4. Fulkerson E, Razi A, Tejwani N. Review: acute compartment syndrome of the foot. Foot Ankle Int. 2003;24(2):180–7.
5. Ling ZX, Kumar VP. The myofascial compartments of the foot: a cadaver study. J Bone Joint Surg. 2008;90(8):1114–8. https://doi.org/10.1302/0301-620X.90B8.20836.
6. McDonald S, Bearcroft P. Compartment syndromes. Semin Musculoskelet Radiol. 2010;14(2):236–44. https://doi.org/10.1055/s-0030-1253164.
7. Tei MM, Placella G, Sbaraglia M, Tiribuzi R, Georgoulis A, Cerulli G. Does Manual Drilling Improve the Healing of Bone-Hamstring Tendon Grafts in Anterior Cruciate Ligament Reconstruction? A Histological and Biomechanical Study in a Rabbit Model. Orthop J Sports Med. 2020;8(4):2325967120911600. https://doi.org/10.1177/2325967120911600.
8. Elliott KG, Johnstone AJ. Diagnosing acute compartment syndrome. J Bone Joint Surg. 2003;85(5):625–32.
9. Volkmann G, Volkmann V, Liu XQ. Site-specific protein cleavage in vivo by an intein-derived protease. FEBS Lett. 2012;586(1):79–84. https://doi.org/10.1016/j.febslet.2011.11.028.
10. Bae DS, Kadiyala RK, Waters PM. Acute compartment syndrome in children: contemporary diagnosis, treatment, and outcome. J Pediatr Orthop. 2001;21(5):680–8.
11. Hwang RW, de Witte PB, Ring D. Compartment syndrome associated with distal radial fracture and ipsilateral elbow injury. J Bone Joint Surg Am. 2009;91(3):642–5. https://doi.org/10.2106/JBJS.H.00377.
12. Iankov ID, Petrov DP, Mladenov IV, Haralambieva IH, Ivanova R, Valeri RV, et al. Production and characterization of monoclonal immunoglobulin A antibodies directed against Salmonella H:g,m flagellar antigen. FEMS Immunol Med Microbiol. 2002;33(2):107–13.
13. Kutz JE, Singer R, Lindsay M. Chronic exertional compartment syndrome of the forearm: a case report. J Hand Surg Am. 1985;10(2):302–4.
14. Myerson M, Manoli A. Compartment syndromes of the foot after calcaneal fractures. Clin Orthop Relat Res. 1993;290:142–50.
15. Ronel DN, Mtui E, Nolan WB 3rd. Forearm compartment syndrome: anatomical analysis of surgical approaches to the deep space. Plast Reconstr Surg. 2004;114(3):697–705.
16. Salgado DM, Panqueba CA, Castro D, Martha RV, Rodriguez JA. [Myocarditis in children affected by dengue hemorrhagic fever in a teaching hospital in Colombia]. Revista Salud Publica. 2009;11(4):591–600.
17. Vaillancourt C, Shrier I, Vandal A, Falk M, Rossignol M, Vernec A, et al. Acute compartment syndrome: how long before muscle necrosis occurs? CJEM. 2004;6(3):147–54.
18. Harris IA, Kadir A, Donald G. Continuous compartment pressure monitoring for tibia fractures: does it influence outcome? J Trauma. 2006;60(6):1330–5.; discussion 5. https://doi.org/10.1097/01.ta.0000196001.03681.c3.
19. Jini R, Swapna HC, Rai AK, Vrinda R, Halami PM, Sachindra NM, et al. Isolation and characterization of potential lactic acid bacteria (LAB) from freshwater fish processing wastes for application in fermentative utilisation of fish processing waste. Braz J Microbiol. 2011;42(4):1516–25. https://doi.org/10.1590/S1517-83822011000400039.
20. Saha S, Sudershan RV, Mendu VV, Gavaravarapu SM. Knowledge and practices of using food label information among adolescents attending schools in Kolkata, India. J Nutr Educ Behav. 2013;45(6):773–9. https://doi.org/10.1016/j.jneb.2013.07.011.
21. Willy C, Sterk J, Volker HU, Sommer C, Weber F, Trentz O, et al. [Acute compartment syndrome. Results of a clinico-experimental study of pres-

sure and time limits for emergency fasciotomy]. Unfallchirurg. 2001;104(5):381–391.
22. Frink M, Klaus AK, Kuther G, Probst C, Gosling T, Kobbe P, et al. Long term results of compartment syndrome of the lower limb in polytraumatised patients. Injury. 2007;38(5):607–13. https://doi.org/10.1016/j.injury.2006.12.021.
23. Giannoudis PV, Nicolopoulos C, Dinopoulos H, Ng A, Adedapo S, Kind P. The impact of lower leg compartment syndrome on health related quality of life. Injury. 2002;33(2):117–21.
24. Myerson MS. Management of compartment syndromes of the foot. Clin Orthop Relat Res. 1991;271:239–48.
25. Kamel R, Sakla FB. Anatomical compartments of the sole of the human foot. Anat Rec. 1961;140:57–60. https://doi.org/10.1002/ar.1091400109.
26. Prasarn ML, Ouellette EA. Acute compartment syndrome of the upper extremity. J Am Acad Orthop Surg. 2011;19(1):49–58.
27. Ojike NI, Roberts CS, Giannoudis PV. Foot compartment syndrome: a systematic review of the literature. Acta Orthop Belg. 2009;75(5):573–80.
28. Myerson M. Diagnosis and treatment of compartment syndrome of the foot. Orthopedics. 1990;13(7):711–7.
29. Al-Dadah OQ, Darrah C, Cooper A, Donell ST, Patel AD. Continuous compartment pressure monitoring vs. clinical monitoring in tibial diaphyseal fractures. Injury. 2008;39(10):1204–9. https://doi.org/10.1016/j.injury.2008.03.029.
30. Olson SA, Glasgow RR. Acute compartment syndrome in lower extremity musculoskeletal trauma. J Am Acad Orthop Surg. 2005;13(7):436–44.
31. McQueen MM, Christie J, Court-Brown CM. Compartment pressures after intramedullary nailing of the tibia. J Bone Joint Surg. 1990;72(3):395–7.
32. Verleisdonk EJ, Schmitz RF, van der Werken C. Long-term results of fasciotomy of the anterior compartment in patients with exercise-induced pain in the lower leg. Int J Sports Med. 2004;25(3):224–9. https://doi.org/10.1055/s-2003-45255.
33. Blackman PG. A review of chronic exertional compartment syndrome in the lower leg. Med Sci Sports Exerc. 2000;32(3 Suppl):S4–10.
34. Goldfarb SJ, Kaeding CC. Bilateral acute-on-chronic exertional lateral compartment syndrome of the leg: a case report and review of the literature. Clinical J Sport Med. 1997;7(1):59–61. discussion 2
35. Martens MA, Backaert M, Vermaut G, Mulier JC. Chronic leg pain in athletes due to a recurrent compartment syndrome. Am J Sports Med. 1984;12(2):148–51.
36. Detmer DE, Sharpe K, Sufit RL, Girdley FM. Chronic compartment syndrome: diagnosis, management, and outcomes. Am J Sports Med. 1985;13(3):162–70.
37. Shah SN, Miller BS, Kuhn JE. Chronic exertional compartment syndrome. Am J Orthop. 2004;33(7):335–41.
38. Mouhsine E, Garofalo R, Moretti B, Gremion G, Akiki A. Two minimal incision fasciotomy for chronic exertional compartment syndrome of the lower leg. Knee Surg Sports Traumatol Arthrosc. 2006;14(2):193–7. https://doi.org/10.1007/s00167-004-0613-6.
39. McDermott AG, Marble AE, Yabsley RH, Phillips MB. Monitoring dynamic anterior compartment pressures during exercise. A new technique using the STIC catheter. Am J Sports Med. 1982;10(2):83–9.
40. Soligard T, Schwellnus M, Alonso JM, Bahr R, Clarsen B, Dijkstra HP, et al. How much is too much? (Part 1) International Olympic Committee consensus statement on load in sport and risk of injury. Br J Sports Med. 2016;50(17):1030–41. https://doi.org/10.1136/bjsports-2016-096581.
41. Ferber R, Hreljac A, Kendall KD. Suspected mechanisms in the cause of overuse running injuries: a clinical review. Sports Health. 2009;1(3):242–6. https://doi.org/10.1177/1941738109334272.
42. Edwards PH Jr, Wright ML, Hartman JF. A practical approach for the differential diagnosis of chronic leg pain in the athlete. Am J Sports Med. 2005;33(8):1241–9. https://doi.org/10.1177/0363546505278305.
43. Grant HM, Tjoumakaris FP, Maltenfort MG, Freedman KB. Levels of evidence in the clinical sports medicine literature: are we getting better over time? Am J Sports Med. 2014;42(7):1738–42. https://doi.org/10.1177/0363546514530863.
44. Starman JS, Gettys FK, Capo JA, Fleischli JE, Norton HJ, Karunakar MA. Quality and content of Internet-based information for ten common orthopaedic sports medicine diagnoses. J Bone Joint Surg Am. 2010;92(7):1612–8. https://doi.org/10.2106/JBJS.I.00821.

Part V

Common Knee Injuries

19 Management of Track and Field: Knee Meniscal and Chondral Injuries

Giacomo Zanon, Enrico Ferranti Calderoni, and Alberto Vascellari

19.1 Epidemiology

In a large study of more than 21 million athlete exposures, the incidence of meniscal injuries in high school track and field was lower when compared to football, soccer, basketball, and wrestling. Mitchell et al. demonstrated that for females participating in high school track and field, meniscus injury risk was twice that of their male counterparts (2.0 and 1.0 injury rate per 100,000 athlete exposures, respectively) [1]. A case series analyzed 378 isolated meniscal lesions in athletes and found that medial meniscal tears predominated in track and field athletes (71.4%) [2]. Horizontal and complex tears are most common and typically exist on the osteoarthritis spectrum. Radial and vertical tears are common in acute injuries, whereas root tears and ramp lesions are typically higher energy injuries associated with ACL tears. Traumatic injury to the articular cartilage of the knee is increasingly recognized among athletes where the physical demands of sport result in significant stresses on joints. The overall prevalence of focal chondral defects in the knee is 36% among all athletes compared with 16% of the general population. Chondral defects occur in association with 9–60% of acute anterior cruciate ligament (ACL) ruptures and 95% of patellar dislocations. Knee chondral lesions carry a high morbidity: Athletes are up to 12 times more likely to develop osteoarthritis than the general population [3].

G. Zanon
Orthopedic and Sport Traumatology Department - Habilita Group, Bergamo, Italy
e-mail: zanon.g@libero.it

E. Ferranti Calderoni (✉)
Ospedale Carlo Alberto Pizzardi Maggiore, Bologna, Italy
e-mail: e_ferranti@libero.it

A. Vascellari
Kinè Physiotherapic & Orthopedic Center, Treviso, Italy
e-mail: info@albertovascellari.it

19.2 Pathogenesis of Chondral Injuries

Most of the running injuries are classified as "overuse" injuries, as chondral lesion, defined as an injury of the musculoskeletal system that results from the combined fatigue effect over a period of time beyond the capabilities of the specific structure that has been stressed. Several risk factors are associated with overuse injuries, but they could be classified into three main categories: training, anatomic, and biomechanical factors [4]. Excessive running distance and intensity are identified as the main training errors, correlated with greater stresses on bones, joints, muscles, and tendons. A recent study found that runners with a body mass index (BMI) of ≥26 kg/m^2 had a reduced risk of sustaining a running-related injury when compared to run-

G. L. Canata et al. (eds.), *Management of Track and Field Injuries*,
https://doi.org/10.1007/978-3-030-60216-1_19

ners with a lower BMI [5]. About sex difference, several studies showed evidence that young men had a higher risk of running-related injuries [6]. Anatomic or anthropometric variables as high longitudinal arches (pes cavus), ankle range of motion, leg length discrepancies, and lower extremity alignment abnormalities are identified as risk factors for an overuse running, increasing amounts of internal stresses applied to various musculoskeletal structures. However, there is a huge debate among researchers regarding the effect of each of these variables, particularly the ankle range of motion. Finally, biomechanical factor is the last main cause of overuse running injuries and could be classified as kinetic or rear-foot kinematic variables [7].

19.3 Pathogenesis of Meniscal Tears

According to Snoeker et al. in a recent review, minimal evidence was found for running as a risk factor for meniscal tear, despite the need for a greater load absorption by the menisci. This could be explained by the absence of pivoting motion on a semi-flexed knee during running and the lack of contact with other players [8]. BMI has been identified as a modifiable factor associated with meniscus injury in general population and athletes. Conversely, nonmodifiable risk factors for meniscal tear include age, gender, and anatomic factors. The prevalence of meniscus tears increases with age with a prevalence of meniscal abnormalities and degenerative tears, whereas traumatic tears decrease with age, due to a decreased activity level in older population. In regard to sex, male athletes may be at greater risk of meniscus injuries than female athletes. Anatomic factors may increase the risk of medial meniscus injury. These factors include posterior tibial slope (PTS), medial meniscal slope (MMS), a biconcave medial tibial plateau, and knee malalignment. PTS >13° may increase risk of posterior horn medial meniscus tears in ACL-deficient knees, whereas MMS >3.5° may increase risk of ramp lesion in patients with ACL tear [1]. In patients with acute ACL tears, meniscus tears have been reported in 40–82% of cases. Several factors have been correlated with meniscal tears during jumping. The most commonly reported action causing injury was rotation around a planted foot (Fig. 19.1). In the case report of long jumping, it is hypothesized that these injuries result from abnormal forces on the knee caused by fixation of the distal limb by cleats, thus preventing normal tibial medial rotation during flexion from the "lock-extended" position. Tear forces on the menisci could be associated with feet anchored impacting into the sand of a long jump pit. Another example could be that during a high-impact landing, the femur is also restricted in rotation by bracing muscle tension over the hip joint [9].

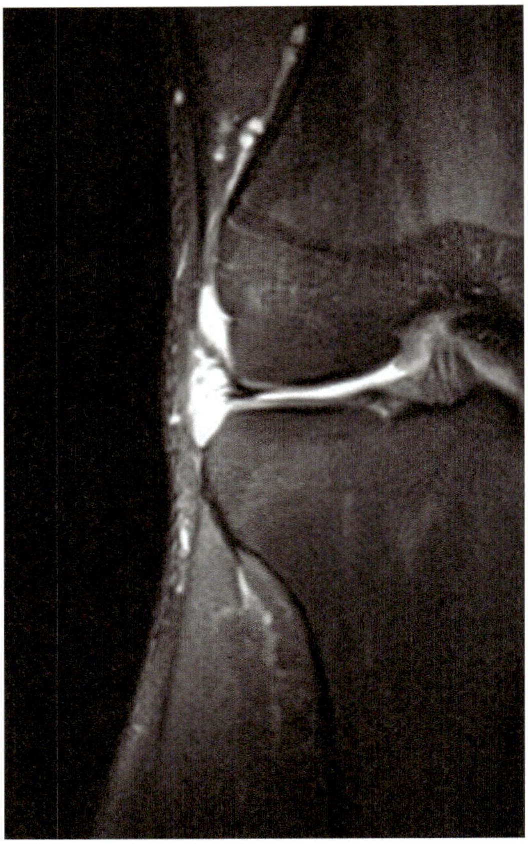

Fig. 19.1 Post-traumatic meniscus tear in track and field athlete

19.4 Management Chondral Lesions

In partial-thickness defects, there is no involvement of the vasculature. Chondroprogenitor cells in blood and marrow cannot enter the damaged region, and local articular chondrocytes do not migrate to the lesion. As such, the defect is not repaired and will progress [10]. However, when the thickness of the defect is not complete, any type of surgical treatment would seem exaggerated and is not supported by scientific evidence. In these cases, it is more appropriate to adopt existing cartilage protection strategies such as minimization of high-impact joint loading and injury prevention protocols with the possible addition of injection treatments. Viscosupplementation with hyaluronic acid is a recommended treatment for osteoarthritis of the knee in both national and international guidelines [11]. In athletes, intra-articular injection of hyaluronic acid for symptomatic treatment of osteochondral lesions has been shown to improve function and reduce pain. Platelet-rich plasma (PRP) has been proposed for the treatment of chondral lesions or osteoarthritis due to proposed healing properties attributed to the increased concentrations of autologous growth factors and secretory proteins that may enhance tissue regeneration. Nonetheless, few studies evaluated platelet aggregates in the treatment of chondral lesions, although they reported more and longer efficacy than hyaluronic acid injections in reducing pain and symptoms as well as recovering articular function with better results in younger and more active patients who had a low degree of cartilage degeneration [12].

Full-thickness chondral defects that penetrate subchondral bone have the potential for intrinsic repair due to communication with chondroprogenitors in bone marrow. These differentiating cells produce a "repair cartilage" with a high content in type I collagen, resulting in fibrocartilage rather than hyaline cartilage. Fibrocartilage is less robust and has poor wear characteristics, and is associated with reduced durability of cartilage tissue and tendency for outcomes to worsen with time. Intrinsic repair of chondral lesions in athletes can be facilitated with different techniques of stimulation or restoration of the hyaline articular cartilage; repaired and regenerated cartilage should resemble as close as possible and function like normal hyaline cartilage, and this ability may be the most significant factor for the return to sport.

Microfracture is a surgical method aimed to facilitate migration of mesenchymal stem cells into the injury site through perforation of the subchondral bone, generating conduits to the vascularized bone marrow. Microfracture has been suggested as a first-line treatment option for lesions <2 cm^2 in the absence of underlying osseous defect [13] and has gained popularity during the past 2 decades because of its minimally invasive approach, technical simplicity, low surgical morbidity, and relatively low cost [14]. Excellent short-term (first 24 months) clinical outcomes and functional improvement have been demonstrated after microfracture, particularly in younger patients with smaller lesions [15]. However, the clinical durability of bone marrow stimulated repair tissue has shown an objective and functional decline over time in young athletes [16]. A study by Steadman et al. [17] on 25 athletes who underwent microfracture for knee chondral lesions revealed a 76% rate of return to play, with improvement in pain and function. Conversely, Mithoefer et al. [18] analyzing the outcomes in 32 professional athletes with focal full-thickness lesions of the femoral condyle showed that only 25% returned to regular sport participation at pre-injury level: The outcome scores subsequently deteriorated in almost 40% of the athletes. Effectiveness of microfracture improvement has been the objective of recent technique implementation, such as the utilization of polysaccharide polymers, biodegradable hydrogels, or 3D scaffolds to improve clot stability, or concomitant use of PRP or bone marrow or adipose tissue aspirate concentrate.

The autologous matrix-induced chondrogenesis (AMIC) foresees the additional introduction of a solid acellular type I/III collagen membrane in cartilage defects after treatment with microfractures [19]. The advertised advantages are a possible stabilization of the so-called super-clot within the cartilage defect following microfracture and an improved cartilage repair. Although there is a paucity of high-quality studies testing the AMIC technique versus established procedures such as microfracture or ACI for knee chondral defects, in the majority of the available studies patients experienced decreased pain and improved knee functional scores within the first 2 years following AMIC [20]. For knee cartilage defects with a mean defect size of 3.6 cm^2, a randomized controlled bicenter trial compared AMIC with microfracture: No significant differences were found at 1 and 2 years postoperatively regarding improvements in the modified Cincinnati and International Cartilage Repair Society (ICRS) scores [21]. The same authors reported 5-year outcomes of 39 patients similarly randomized in a prospective bicenter clinical trial and found that the modified Cincinnati score was stable in AMIC groups, whereas it significantly decreased in the microfracture group [22]. Chondral lesions larger than 2 cm^2 or lesion with underlying osseous defect should be addressed with reconstructive procedures such as mosaicplasty (osteochondral autograft transplantation) and allograft transplantation. These are implantations of well-formed osteochondral tissue (unit of osteochondral plugs or constructs), and no regeneration of cartilage is necessary.

Mosaicplasty, or osteochondral autograft transplantation, is a surgical technique that has been developed to treat small- or medium-sized symptomatic focal chondral or osteochondral defects in the knee. Osteochondral implantation provides replacement of mature hyaline cartilage together with underlying subchondral bone. Mosaicplasty involves the harvesting of cylindrical osteochondral plugs from a minimally weight-bearing zone of the knee (e.g., the intercondylar notch or the femoral periphery of the patellofemoral joint) and transplanted to areas of symptomatic full-thickness cartilage or osteochondral injury. Harvesting and grafting may be conducted through a mini-arthrotomy or arthroscopically [23]. Due to tissue availability and donor-site morbidity, autologous osteochondral mosaicplasty is indicated for limited-size defects. Based on promising clinical results, mosaicplasty has been used to treat the athlete population and has been demonstrated to be a useful alternative in the treatment of focal full-thickness cartilage damages of professional athletes. Hangody et al. reported good to excellent results in 91% of femoral, 86% of tibial, and 74% of patellofemoral mosaicplasty in athletic patients, after an average follow-up time of 9.6 years [23]. 63% of the patients returned to the same level of sports activity, and 28% of the patients were able to return to a lower level of sports activity, whereas 9% of the operated patients had to give up any kind of sports activity. In a prospective randomized study of osteochondral autologous transplantation versus microfracture for the treatment of single symptomatic full-thickness and osteochondral defects of the knee, in a group of 57 athletes, Gudas et al. reported significantly better results in the mosaicplasty group 3 and 10 years after the operation; however, the scores decreased from 3 to 10 years in both groups [24].

Osteochondral allograft transfer procedures provide a potential solution to overcome donor-site morbidity that limits autologous techniques for osteochondral lesions that are larger than 2 cm^2 [25]. The primary advantage is there is no restriction on the size or number of plugs that can be harvested from the donor knee, both of which are limited in autologous mosaicplasty [26]. Several studies have outlined the effectiveness of osteochondral allografts in reliably providing pain relief and return of function for activities of daily living [25]. Good clinical outcomes have been reported after osteochondral allograft transplantation in the knee, with a high satisfaction rate (86%) and a low short-term complication rate at a mean follow-up of 5 years. Furthermore, the survivorship of osteochondral allografts at 15 years' follow-up has been estimated to be 75% [27]. The return-to-sport rate after osteochondral allograft transplantation range from 75% to 82% with improvements in most patient-reported outcomes, although a high reoperation rate has been reported, with more than half of studies reporting a reoperation rate between 34%

and 53% [28]. According to a meta-analysis of return to sport after the surgical management of articular cartilage lesions in the knee, the rate of return to sport for osteochondral allograft transplantation was 88% [29].

Chondral lesions larger than 2 cm^2 with no underlying osseous defect have been successfully treated with autologous chondrocyte implantation (ACI). ACI is indicated for the treatment of medium-to-large, full-thickness cartilage defects. Due to the cost and invasiveness of the procedure, ACI is a second-line treatment for defects smaller than 2 cm^2, in which it is generally reserved for revision of prior failed cartilage repair. For larger defects, however, it can be used as a primary procedure due to the lowered efficacy of lesser procedures, such as microfracture or osteochondral autograft transfer. ACI involves the harvesting of chondrocytes from a healthy non-weight-bearing portion of the knee followed by implantation of culture-expanded autologous chondrocytes under a periosteal flap (first-generation ACI) or a collagen membrane (second-generation ACI), or onto a membrane carrier or porous scaffold prior to implantation (third-generation ACI). When performed in elite athletes, ACI resulted in a successful return to high-impact sport with excellent durability at 5 years and beyond [30]. Mithoefer et al. analyzed professional and recreational soccer players who underwent ACI and found that 33% of the players returned to soccer, including 83% of competitive-level players and 16% of recreational players. Of the returning players, 80% returned to the same competitive level and 87% maintained their level of performance [30]. The main disadvantage of ACI techniques is the long time for tissue maturation and consequent return to sport. In fact, while a meta-analysis of return to sport after the surgical management of articular cartilage lesions in the knee reported an 82% rate of return to sport after ACI [29]. Previous recommendations had been for return to activity at 18 months to allow sufficient time for tissue remodeling, but more recent accelerated protocols have athletes returning to activity at 12 months [31]. Further limitations include the requirement for multiple surgical procedures, donor-site morbidity, the expense, and potentially harmful modification of cells in culture, and the repair tissue is not hyaline cartilage.

19.5 Management of Meniscal Tears

Meniscal tears are particularly common in athletes, especially in contact sport that involves pivoting or cutting. Different types of meniscal surgery can be performed for an acute tear: meniscectomy or meniscal repair. A recently published analysis of 2004 through 2012 data from the American Board of Orthopaedic Surgery certification examination database showed an increased rate of surgeons performing meniscal repairs and a decreased rate of meniscal debridement [32]. However, meniscectomy remains one of the most frequent orthopedic procedures with a fast RTS for the athletes but with a high risk of early degenerative changes. Meniscectomy in patients with high physical demands should be used only when a meniscal repair is unworkable, evaluating factors such as tear type, location, chronicity, and potential to heal [32]. Osteoarthritis is a common consequence following meniscectomy. 56% of patients who underwent lateral meniscectomy, at a 20 years' follow-up, showed osteoarthritis. Furthermore, resection amount, age at surgery, and cartilage status are prognostic factors. 100% excellent or good results after meniscectomy for longitudinal vertical tear were obtained by Osti et al., compared to 79% for complex lesions [33]. RTP after partial meniscectomy at the pre-injury activity level in athletes occurs usually from 7 to 9 weeks, when knee pain and effusions have subsided and quadriceps/hamstring strength has returned to normal, with more adverse effects reported after partial lateral meniscectomy [34]. Few studies evaluating the return to play of athletes following meniscectomy are presented in the literature. A study by Osti et al. found that 98% of 41 athletes who underwent a partial lateral meniscectomy returned to sport at an average of 55 days, with a faster rehabilitation in patients with an isolated simple longitudinal tear than more complex tears. Moreover, Kim et al. noticed a longer RTS in recreational athletes in 88 days than in elite ones in 54 and in patients >30 years in 89 days than <30 years in 54. Different rehabilitation protocols are presented in the literature. Brelin et al. use standard method progresses in 3 phases: (1) 0–2 weeks: begin weight-bearing and range of motion as tolerated along with quadriceps, hamstring, and core strengthening; (2) 2–4 weeks: addition of sport-specific exercises and return to cardio training; and (3) 4–6 weeks: continued advancement in sport-specific training and maintenance of strengthening program [33].

In young athletes, the gold standard treatment for an unstable tear in the vascular zone is a meniscus repair to avoid early degenerative changes and alteration of the mechanism of the knee joint. For these reasons, recently, it has also been proposed to try to repair a tear into the avascular zone. All-inside, inside-out, and outside-in techniques are all effective, and indications are basically cultural: for example, use of hybrid material in Europe. The type of surgery in athletes was reported in a recent review: 625 (94%) repairs were arthroscopic surgeries, while the remaining 39 (6%) repairs were performed as open surgery via arthrotomy. An all-inside technique was used in 473 cases (71%), inside-out in 110 (17%), and a combination of outside-in and all-inside in 42 (6%) patients [35]. Functional results of meniscal repair are similar to meniscectomy, although surgical revision rates are slightly higher with repair. Moreover, meniscal repair provides long-term cartilage protection, on radiography or MRI, and failure rates are acceptable (6–28%). Although there is a high risk of failure in extended tears than small ones, lesion extension is not a prognostic factor, as the antero-posterior location of the tear. Time to surgery is probably a factor, and early repair is probably preferable: Acute-stage repair shows better prognosis than chronic repair [34]. Stein et al. compared the results of meniscus repair and partial meniscectomy in 81 patients with traumatic medial meniscal tears at midterm (mean 3.4 years) and long-term (mean 8.8 years) follow-up. Whereas the midterm examinations showed no difference between both groups, sports level at the long-term follow-up was significantly higher in the repair group with 94% being active at the pre-injury sports level compared to 44% in the partial meniscectomy group [36]. However, a meniscal preservation needs a longer rehabilitation period, delaying the RTS. After meniscal repair, 81–88.9% of athletes returned to sports on average 5.6 months. There is no consensus about postoperative rehabilitation programs in patients who underwent a meniscus repair. However, more aggressive approaches have been used to let an early postoperative weight-bearing and deep flexion with good outcomes. Kozlowski et al. published a rehabilitation protocol for athletes using a 3-phase progression based on patient abilities. The first 6 weeks of the early phase let to protect the meniscal repair. Following this, athletes begin a return-to-sport progression (static, dynamic, and ballistic phases) if they meet specified subjective and objective criteria to achieve finally full confidence in their knee [33].

Meniscal allograft transplantation (MAT) is a surgical procedure indicated for athletes with symptomatic meniscal deficiency, "the post-meniscectomy syndrome." It consists of recurrent joint effusions, pain, and symptomatic "giving way," which may develop in athletes and may limit or prevent them from returning to play after meniscal injury and surgery. The ideal patient for a MAT should have joint line pain, mild chondral changes, normal alignment, and a stable knee to achieve better outcomes. There is no consensus about the best technique about MAT. Bone-plug or soft tissue fixation and open or arthroscopic techniques are commonly used by surgeons [37]. After MAT, 67–85.7% of athletes returned to sports, and the time to RTS ranged from 7.6 to 16.5 months. No significant differences in the time to return to official competition were found between patients who underwent medial or lateral MAT, patient with none/mild or severe chondral damage, and those who underwent isolated or combined MAT [35]. Two-thirds of athletes who underwent MAT were able to participate in sports at the same pre-injury level. Graft-related reoperations were reported in

13% of patients, while the rate of joint replacement, with partial or total knee prosthesis, was 1.2%, not dramatically increased compared with the reported rates for the general population. However, high-demand sports should be discouraged to preserve the graft as long as possible until high-quality evidence becomes available on long-term safety [32]. Recently, a more aggressive rehabilitation after MAT has been proposed. Athletes are being released to full training exercises as early as 5 months postoperatively, under the guidance of the surgeon, athletic trainer, and coaches. However, there is a high risk of failure, and therefore, it is highly recommended to athletes to refrain from collision or contact sports. Usually, a more conservative rehabilitation protocol following MAT consists of protected weight-bearing for 6 weeks, immediate (or 2-week delayed) joint mobilization, and return to contact activities 6–9 months postsurgery [33].

19.6 Conclusion

In conclusion, preservation of meniscal function is the most important goal of meniscal surgery. However, when a meniscal repair is unworkable, partial meniscectomy must be performed, obtaining the shortest time to RTS and the highest RTS rate but with a high incidence of rapid chondrolysis. Although MAT is generally considered a salvage procedure and not strictly aimed at returning to physical activity, return to sport and good clinical outcomes were achieved in most recent reviews. Concurrent procedures associated with meniscal repair or meniscectomy, such as ACLR, prolonged the time to RTS, but it had no effect on the RTS rate and the level of sports activity at the time of RTS. A vast number of strategies are available in the treatment of *chondral injuries*, few of them are supported by robust clinical evidence. Depending on the chondral defect thickness and lesion size, different treatments can be considered, each of which is associated with variable success and return-to-sport rates. Nevertheless, chondroprotective measures such as stability, meniscal, and correct alignment restoration should first be considered in all patients to prevent disease progression. Finally, an early functional rehabilitation program has been implemented recently to provide a faster return to play while still minimizing the risk for re-injury.

References

1. Mitchell J, Graham W, Best TM, Collins C, Currie DW, Comstock RD, Flanigan DC. Epidemiology of meniscal injuries in US high school athletes between 2007 and 2013. Knee Surg Sports Traumatol Arthrosc. 2016;24(3):715–22.
2. Terzidis IP, Christodoulou A, Ploumis A, et al. Meniscal tear characteristics in young athletes with a stable knee: arthroscopic evaluation. Am J Sports Med. 2006;34:1170–5.
3. Flanigan DC, Harris JD, Trinh TQ, Siston RA, Brophy RH. Prevalence of chondral defects in athletes' knees: a systematic review. Med Sci Sports Exerc. 2010;42(10):1795–801.
4. Hreljac A. Impact and overuse injuries in runners. Med Sci Sports Exerc. 2004;36(5):845–9.
5. Bertelsen ML, Hulme A, Petersen J, Brund RK, Sørensen H, Finch CF, Parner ET, Nielsen RO. A framework for the etiology of running-related injuries. Scand J Med Sci Sports. 2017;27(11):1170–80.
6. Van der Worp MP, ten Haaf DS, van Cingel R, de Wijer A, Nijhuis-van der Sanden MW, Staal JB. Injuries in runners; a systematic review on risk factors and sex differences. PLoS One. 2015;23:10(2).
7. Hreljac A. Etiology, prevention, and early intervention of overuse injuries in runners: a biomechanical perspective. Phys Med Rehabil Clin N Am. 2005;16(3):651–67.
8. Snoeker BA, Bakker EW, Kegel CA, Lucas C. Risk factors for meniscal tears: a systematic review including meta-analysis. J Orthop Sports Phys Ther. 2013;43(6):352–67.
9. Abbott J, Lam KS, Srivastava VM, Moulton A. Sand pits and bucket handles: a case of bilateral traumatic bucket-handle tears of the medial menisci. Arthroscopy. 2000;16(6):12.
10. Murray IR, Benke MT, Mandelbaum BR. Management of knee articular cartilage injuries in athletes: chondroprotection, chondrofacilitation, and resurfacing. Knee Surg Sports Traumatol Arthrosc. 2016;24(5):1617–26.
11. Bannuru RR, et al. OARSI guidelines for the non-surgical management of knee, hip, and polyarticular osteoarthritis. Osteoarthr Cartil. 2019;27(11):1578–89.
12. Kon E, Mandebaum B, Buda R, Filardo G, Delcogliano M, Timoncini A, Fornasari PM, Giannini S, Marcacci M. Platelet-rich plasma intra-articular injection versus hyaluronic acid viscosupplementation as treatments for cartilage pathology: from early degeneration to osteoarthritis. Arthroscopy. 2011;27(10):1490–501.

13. Mithoefer K, Williams RJ, Warren RF, et al. High-impact athletics after knee articular cartilage repair: a prospective evaluation of the microfracture technique. Am J Sports Med. 2006;34(9):1413–8.
14. Gou-Hau G, Feng-Jen T, Sheng-Hao W, et al. Autologous chondrocyte implantation versus microfracture in the knee: a meta-analysis and systematic review. Arthroscopy. 2020;36:289–303.
15. Goyal D, Keyhani S, Lee EH, Hui JH. Evidence-based status of microfracture technique: a systematic review of level I and II studies. Arthroscopy. 2013;29:1579–88.
16. Gobbi A, Karnatzikos G, Kumar A. Long-term results after microfracture treatment for full- thickness knee chondral lesions in athletes. Knee Surg Sports Traumatol Arthrosc. 2014;22:1986–96.
17. Steadman JR, Miller BS, Karas SG, et al. The microfracture technique in the treatment of full-thickness chondral lesions of the knee in National Football League players. J Knee Surg. 2003;16:83–6.
18. Mithoefer K, McAdams T, Williams RJ, Kreuz PC, Mandelbaum BR. Clinical efficacy of the microfracture technique for articular cartilage repair in the knee: an evidence-based systematic analysis. Am J Sports Med. 2009;37:2053–63.
19. Benthien JP, Behrens P. Autologous matrix-induced chondrogenesis (AMIC): combining microfracturing and a collagen I/III matrix for articular cartilage resurfacing. Cartilage. 2010;1(1):65–8.
20. Gao L, Orth P, Cucchiarini M, Madry H. Autologous matrix-induced chondrogenesis: a systematic review of the clinical evidence. Am J Sports Med. 2019;47(1):222–31. https://doi.org/10.1177/0363546517740575.
21. Anders S, Volz M, Frick H, Gellissen J. A randomized, controlled trial comparing autologous matrix-induced chondrogenesis (AMIC) to microfracture: analysis of 1- and 2- year follow-up data of 2 centers. Open Orthop J. 2013;7:133–43.
22. Volz M, Schaumburger J, Frick H, Grifka J, Anders S. A randomized controlled trial demonstrating sustained benefit of autologous matrix-induced chondrogenesis over microfracture at five years. Int Orthop. 2017;41(4):797–804.
23. Hangody L, Dobos J, Balo E, Panics G, Hangody LR, Berkes I. Clinical experiences with autologous osteochondral mosaicplasty in an athletic population: a 17-year prospective multicenter study. Am J Sports Med. 2010;38(6):1125–33.
24. Gudas R, Gudaite A, Pocius A, Gudiene A, Cekanauskas E, Monastyreckiene E, Basevicius A. Ten-year follow-up of a prospective, randomized clinical study of mosaic osteochondral autologous transplantation versus microfracture for the treatment of osteochondral defects in the knee joint of athletes. Am J Sports Med. 2012;40(11):2499–508.
25. Krych AJ, Robertson CM, Williams RJ 3rd, Cartilage Study Group. Return to athletic activity after osteochondral allograft transplantation in the knee. Am J Sports Med. 2012;40(5):1053–9.
26. Bugbee W, Cavallo M, Giannini S. Osteochondral allograft transplantation in the knee. J Knee Surg. 2012;25(2):109–16.
27. Chahal J, Gross AE, Gross C, Mall N, Dwyer T, et al. Outcomes of osteochondral allograft transplantation in the knee. Arthroscopy. 2013;29(3):575–88.
28. Crawford ZT, Schumaier AP, Glogovac G, Grawe BM. Return to sport and sports-specific outcomes after osteochondral allograft transplantation in the knee: a systematic review of studies with at least 2 years' mean follow-up. Arthroscopy. 2019;35(6):1880–9.
29. Krych AJ, Pareek A, King AH, Johnson NR, Stuart MJ, Williams RJ 3rd. Return to sport after the surgical management of articular cartilage lesions in the knee: a meta-analysis. Knee Surg Sports Traumatol Arthrosc. 2017;25(10):3186–96. https://doi.org/10.1007/s00167-016-4262-3.
30. Mithoefer KPL, Saris DBF, Mandelbaum BR. Evolution and current role of autologous chondrocyte implantation for treatment of articular cartilage defects in the football (Soccer) player. Cartilage. 2012;3(1 Suppl):31S–6S.
31. Nho SJ, Pensak MJ, Seigerman DA, Cole BJ. Rehabilitation after autologous chondrocyte implantation in athletes. Clin Sports Med. 2010;29:267–82.
32. Grassi A, Bailey JR, Filardo G, Samuelsson K, Zaffagnini S, Amendola A. Return to sport activity after meniscal allograft transplantation: at what level and at what cost? A systematic review and meta-analysis. Sports Health. 2019;11(2):123–33.
33. Brelin AM, Rue JP. Return to play following meniscus surgery. Clin Sports Med. 2016;35(4):669–78.
34. Beaufils P, Pujol N. Management of traumatic meniscal tear and degenerative meniscal lesions. Save the meniscus. Orthop Traumatol Surg Res. 2017;103(8S):S237–44.
35. Lee YS, Lee OS, Lee SH. Return to sports after athletes undergo meniscal surgery: a systematic review. Clin J Sport Med. 2019;29(1):29–36.
36. Stein T, Mehling AP, Welsch F, von Eisenhart-Rothe R, Jäger A. Long-term outcome after arthroscopic meniscal repair versus arthroscopic partial meniscectomy for traumatic meniscal tears. Am J Sports Med. 2010;38:1542–154.
37. Chalmers PN, Karas V, Sherman SL, Cole BJ. Return to high-level sport after meniscal allograft transplantation. Arthroscopy. 2013;29(3):539–44.

20 Patellofemoral Overuse Injuries and Anterior Knee Pain

Gian Luigi Canata, Valentina Casale,
Antonio Pastrone, Alberto Vascellari,
and Davide Venturin

20.1 Introduction

Anterior knee pain is common in track and field, multifactorial and involving different anatomical structures. Peripatellar pain is usually clinically referred to as patellofemoral pain (PFP) and indicates localized pain of the anterior aspect of the knee [1], related to several different disorders quite common even among athletes.

PFP typically affects young adults, but is also common among older adults and adolescents, especially during phases of rapid growth [2].

It is usually prevalent in activities highly loading the patella, such as squatting, jumping, running, ascending, or descending stairs [1] and often affects running and jumping athletes [3] (Fig. 20.1).

PFP accounts for 33% and 18% of all chronic knee injuries among female and male athletes, respectively [4, 5].

Fig. 20.1 Triple jump

Less commonly, PFP may follow an acute trauma, especially direct blows to the patella or after patellar dislocation or subluxation [6].

Besides intra-articular pathologies, other causes of anterior knee pain are peripatellar tendinopathies or synovial syndromes, Osgood–Schlatter syndrome, Sinding–Larsen–Johansson syndrome, and neuromas [7, 8].

In several cases, the clinical presentation is chronic anterior knee pain, with a gradual onset, but the causes are other than those cited above. For these patients, the term of patellofemoral pain syndrome (PFPS) is more appropriate [9].

There is no clear consensus in the literature on the correct terminology to use: Anterior knee pain, patellar pain, patellar pain syndrome, chondromalacia patella, patellofemoral arthralgia,

G. L. Canata (✉) · V. Casale · A. Pastrone
Centre of Sports Traumatology, Koelliker Hospital, Torino, TO, Italy
e-mail: studio@ortosport.it

A. Vascellari
Center of Medicine, Villorba, TV, Italy
e-mail: info@albertovascellari.it

D. Venturin
Kinè Physiotherapic Center, San Vendemiano, TV, Italy

G. L. Canata et al. (eds.), *Management of Track and Field Injuries*,
https://doi.org/10.1007/978-3-030-60216-1_20

PFP, and PFPS may be often reported synonymously [10].

PFP prevalence is very high, affecting 11–17% of general active population [11], and leading up to 25% of recreational athletes diagnosed with PFP to quit participating in sports because of knee pain [12].

It has been largely reported that females are three times more likely to develop PFP compared to males [13].

20.2 Pathophysiology and Pathomechanics

The pathogenesis of PFP still remains a concern due to the high prevalence in athletes, and a better knowledge of the etiology of pain is advocated to guide the rationale of treatment regimens.

PFP has been related for decades solely to structural and biomechanical factors, such as chondromalacia patellae and patellofemoral malalignment, while current concepts claim a combination of anatomical, biomechanical, biologic, and psychological factors [14–17].

Many authors failed to find a connection between anterior knee pain and chondromalacia patellae [18]. Thus, PFP has been recently related to a supraphysiologic loading of anatomically normal knee components, with resulting loss of homeostasis of both osseous and soft tissues of the peripatellar region [15].

Expert consensus statements identified some biomechanical risk factors and classified them both by anatomic location to the knee and their nature [2]. Therefore, they may be correlated with proximal (upper femur, hip, and trunk), local (in and around the patella and the patellofemoral joint), and distal (lower leg, foot, and ankle) anatomical structures, as well as they may be defined as anatomical (such as enhanced femoral anteversion, trochlear dysplasia, patella alta and baja, and excessive foot pronation) and biomechanical (muscle tightness or weakness, generalized joint laxity, and gait abnormalities) risk factors [19].

Elevated PFJ loading during walking in people with PFP is the result of diminished contact area that appears to be related to the knee flexion angle [14].

Increased frontal plane motion of knee (valgus/abduction) and hip adduction with internal rotation can enhance the laterally direct component of the PFJ reaction vector [20]. In fact, people with PFP exhibit increased knee abduction during gait and single tasks such as stepping or hop landing [2, 14, 21].

Internal rotation of the femur relative to the external rotation of the tibia is associated with reduced contact area and elevated patellar cartilage stress at 15° and 45° knee flexion [2, 20, 22]. The influence of tibiofemoral rotation on contact area is less pronounced at larger knee flexion angles. A 10° change in the frontal plane alignment of the extensor mechanism increases PFJ pressures by 45% [14].

Furthermore, reduced hamstrings length and deficits in hip abduction and external rotation strength may lead to the development of PFP [23–25].

20.2.1 Malalignment

Clinical studies did not demonstrate relevant biomechanical or alignment differences between patients with or without anterior knee pain [9]. The only exception is the influence of a high Q-angle in maintaining the PFP once it has been developed [26].

Patellofemoral malalignment is described as an abnormality of the patellar tracking, leading to a lateral displacement and/or lateral tilt of the patella in extension and reducing in flexion [27]. In the past, it has been considered as a cause of anterior knee pain and patellar instability [28–31]. This theory had a great influence on orthopedic surgeons, leading to the development of several corrective surgical procedures [32]. However, today it is generally agreed that only a small percentage of patients with PFP has a true malalignment [32, 33], and there are conflicting data on the connection between patellar tracking abnormalities and PFP [9, 34–36].

Structural patellar malalignment is also influenced by the inclination of the lateral anterior femoral condyle and the height of the patella within the trochlear groove. Patella alta or patellar and trochlear dysplasia exhibit lower contact area for a given knee flexion angle and with higher patellofemoral stress during walking [37–39].

20.2.2 Muscular Imbalance

Any combination of malalignment and muscular imbalance may increase the risk of developing anterior knee pain [9]. A muscular imbalance between the medial and the lateral quadriceps muscles is frequently associated with PFP [2, 14, 16, 40, 41].

On the other hand, tightness of the iliotibial tract may cause lateral tilt of the patella, enhancing the pressure on its lateral aspect [42, 43].

Decreased knee extensor strength is usually found in patients with PFP [35, 44–47]. However, it is still unclear the significance of different strength deficits and muscular imbalances, as well as if a specific deficit in muscular activation is a cause or an effect of PFP [9].

20.2.3 Overload

It has been widely suggested a tight correlation between PFP and an increased physical activity is associated with overloading, rather than malalignment of the patellofemoral joint [10, 26, 35, 44]. A sudden rise of the activity level is a risk factor for developing PFP [35].

The relationship between increased joint loading and PFP is not fully understood.

The major hypothesis is that repetitive overloading of the PFJ may enhance patellar bone water content and/or raise patellar subchondral bone's metabolic activity [15, 17, 48].

High water content may change the intraosseous pressure within the patella, thus stimulating pressure-sensitive mechanical nociceptors [49].

The instability in patients with PFP not only depends on mechanical factors, but also depends on neural aspects, such as proprioceptive deficit both in the sense of position, and in slowing or diminution of stabilizing and protective reflexes [18].

The experience of PFP may be attributed not only to nociception. Patients with PFP exhibit abnormal nociceptive processing, altered somatosensory processing, and impaired sensorimotor function and certain psychological factors. All these characteristics complicate the pathophysiology of the syndrome and alter the perception of PFP [17, 50–54].

There is evidence supporting gender differences when considering the risk factors for developing PFP [55–58]. Moreover, it has been recently highlighted that the female characteristics of landing with decreased hip abduction and increased knee internal rotation enhance this risk [59].

20.3 Specific Pathological Patterns

The origin of PFPS could be localized in different structures, such as lateral retinaculum, medial retinaculum, infrapatellar fat pad, synovium, and subchondral bone [60].

20.3.1 Bursitis

Acute or repetitive injuries of any of the superficial bursae may result in bursitis, a common condition characterized by fluid accumulation, synovitis, and bursal wall thickening [61]. It may be associated with anterior knee pain and swelling. Superficial bursitis at the anterior knee is most commonly due to mechanical overuse [62]. It frequently happens when prolonged kneeling is required, as well as after excessive compressive or shear loads on the prepatellar tissues [62]. Nonmechanical causes of superficial bursitis include chronic glucocorticoid use, inflammatory arthritis, infection, and gout [63].

20.3.2 Tendinopathy

Quadriceps and patellar tendinopathies are typically related to overuse in athletes involved in track and field disciplines requiring repetitive eccentric contractions of the quadriceps [64, 65].

Patellar tendinopathy is one of the most common injuries, more frequently proximal posteromedial, due to a focal higher mechanical stress [66], or distal at the tibial insertion.

Cook and Purdam classified patellar tendinopathy in three stages: reactive tendinopathy, tendon disrepair, and degenerative tendinopathy, based on microstructural changes in the damaged tendon [67].

Histologically, the most common findings in overuse-related tendinopathies are noninflammatory disorders caused by repetitive tensile overloading, which results in collagen damage [68].

From a biomechanical point of view, a stiff movement pattern characterized by a small post-touchdown range of motion and a short landing time is often associated with the onset of patellar tendinopathy [69].

There is still no consensus on the proper treatment methods. Eccentric exercises are effective [68], while moderate evidence has been found for injection therapy with hyaluronic acid [70].

20.3.3 Synovial Impingement

Soft tissue impingement such as peripatellar synovitis, suprapatellar fat pad impingement, and Hoffa fat pad impingement may lead to a transitory ischemia, producing mechanical stimulation of nociceptors [71, 72]. In those cases, a peripatellar synovectomy may be an effective solution when conservative treatment has failed [33].

Focal synovial hypertrophy nearby the inferior patellar pole may be another responsible for anterior knee pain [73], and also, in this case a peripatellar synovectomy is suggested after the failure of nonsurgical management [73].

20.3.4 Hoffa Disease

Infrapatellar fat pad impingement syndrome, also known as Hoffa disease, is thought to result from mechanical irritation, which causes hemorrhage and inflammation of the adipose tissue. This process may lead to hypertrophy, mass effect, and bowing of the patellar tendon [74, 75].

These alterations are usually related to repetitive injuries, impingement, and friction-related syndromes affecting the fat pad [62].

Infrapatellar fat pad impingement syndrome in athletic runners may provoke anterior knee pain, swelling, and a sense of catching [76]. A high percentage of return to preoperative sports level after the infrapatellar fat pad arthroscopic resection has been reported [77, 78].

20.3.5 Synovial Plica Syndrome

Plica-related symptom prevalence among athletes is higher in young people and has been related to strenuous physical work or athletic activity, as well as a general increase in activity level [79].

Plicae are often seen during routine arthroscopy, usually as incidental findings of no real clinical meaning [80].

However, a primary disorder of the knee leading to transient or chronic synovitis may cause an inflammation and thickening of the plica, and this may result in a plica syndrome [80–82].

A pathologically inelastic, tight, and fibrotic plica may impinge between the quadriceps tendon and the femoral trochlea, and eventually sub-luxate over the medial or lateral femoral condyle [80]. Such a process may cause a secondary mechanical synovitis and a possible alteration of patellofemoral joint mechanics.

A recent study has reported that patients with infrapatellar fat pad syndrome and medial patellar plica syndrome may show a significantly smaller patella–patellar tendon angle than healthy controls [83]. The underlying articular cartilage then becomes soften and may go toward softening, degeneration, or even erosion [84].

20.4 Diagnosis

20.4.1 Clinical Evaluation

At the first examination, surgeons should investigate on previous knee injuries and surgeries, as well as recent changes in activity level of the patient.

Clinical examination starts by observing the patient in a static standing position, looking at axial deformities, increased femoral internal rotation/adduction, or abnormal foot pronation [85]. Patient's gait and posture evaluation are also helpful in identifying muscular imbalance, exaggerated lumbar lordosis, or asymmetric hip height [86].

The presence of pain with squatting is the most sensitive physical examination for PFPS. Patients should also be assessed with functional tasks, like the one-step squat test, looking for the presence of dynamic valgus and hip abductor weakness [87].

Patellar maltracking should be investigated from a seated position, with the patient slowly extending the knee from 90° of flexion to full extension; the presence of a lateral patellar shift during this movement outlines a positive J-sign, and it may suggest muscle imbalances or laxity [88].

Lastly, in supine position the surgeon should assess possible lateral peripatellar tissue stiffness or lateral patellar tilt that can lead to high load forces on the lateral facet [89]. In symptomatic plicae, pain is usually anteromedial and a tender cord may be palpated [90].

20.4.2 Imaging

20.4.2.1 X-ray

The initial X-ray evaluation requires standard anteroposterior (AP), and lateral and axial Merchant views [91]. Plain AP radiography of the knee can rule out osteoarthritis, osteochondral pathologies, and patellar fractures.

The lateral view is helpful for assessing patellar height, which may be quantified by the Insall–Salvati, Caton–Deschamps, and Blackburne–Peel ratio indexes [92].

Axial radiograph of the PF joint can show patellar translation and axial rotation along the trochlea. The patellofemoral (PF) angle usually opens laterally more than 8°. In the case of pathological increase in the patellar tilt, the PF angle becomes negative and can open medially. Also, lateral patellar translation more than 2 mm in axial view should be considered abnormal [93].

Suspect of trochlear dysplasia is raised by looking at the crossing sign and the trochlear bump in the lateral view [94]. Trochlear dysplasia can be confirmed on the axial view by a trochlear depth <3–5 mm and by a sulcus angle measuring >144° [95]. The lateral trochlear facet should not be more than 60% of the overall anterior trochlear articular width [96].

When the patella displaces only during active quadriceps contraction, as in the case of mild-to-moderate maltracking, static X-rays often fail to diagnose it [97]. For this reason, dynamic X-rays under quadriceps contraction provide a better understanding of patellofemoral biomechanics [41].

Long leg X-ray in monopodalic standing position is required for assessing the Q-angle of extensor apparatus and foot hyperpronation or flatfoot.

20.4.2.2 MRI

Magnetic resonance imaging is the method of choice for the diagnosis of acute dislocations and articular cartilage lesions [98]. It provides superior assessment of soft tissues, including PF cartilage focal injuries, bone marrow lesions, patellar and quadriceps tendinopathy or tears (Fig. 20.2), and retinacular assessment including MPFL integrity and deep infrapatellar bursitis plica [99] (Fig. 20.3).

Moreover, friction-related superolateral and prepatellar fat pad edema is a common finding in routine knee MRI and is suggestive of maltracking [100].

Static imaging does not evaluate the effect of active muscle contraction on the patellar position during flexion–extension of the knee [85]. Dynamic MRI has been introduced for better kinematic assessment of patellofemoral maltracking during motion and is accurate for detecting eventual soft tissue impingement or bony contact [101].

20.4.2.3 CT-Scan

Computed tomography (CT) scanning is the gold standard for measuring the rotational alignment of hip, knee, and ankle, even if it must be considered that knee joint alignment changes signifi-

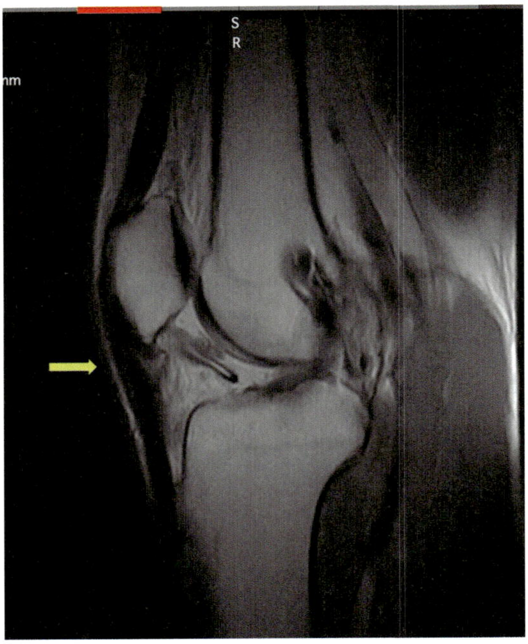

Fig. 20.2 MRI showing patellar insertional tendinopathy (yellow arrow)

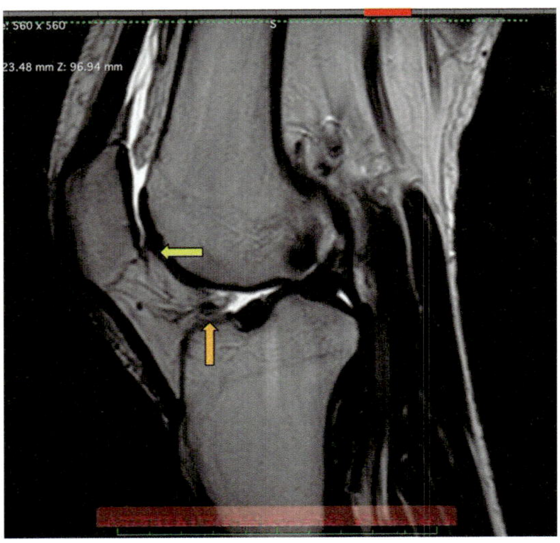

Fig. 20.3 MRI showing a fibrotic infrapatellar plica (yellow arrow) and anterior synovial hypertrophy (orange arrow)

cantly in the upright weightbearing, when compared to supine non-weightbearing CT [102].

Rotational deformities may be a predisposing factor of anterior knee pain rather than a direct etiology [103].

The tibial tubercle–trochlear groove (TT-TG) distance, which is a surrogate marker of tibial tuberosity lateralization and Q-angle, may be measured on both axial CT and MRI. TT-TG distance value >20 mm is considered indicative of pathological lateralization of the tibial tubercle, and in this case, a correction osteotomy can be considered [104].

20.5 Treatment

PFP therapy is challenging as there is a lack of evidence-based clinical guidelines.

Basically, international consensus and evidence recommend exercise therapy focused on hip and knee strengthening [11].

Conservative treatment options for patients with PFP also include pain control, enhancing flexibility and improving the lower extremity biomechanics by correcting gait and retraining with proper techniques and adequate rest [105].

High-quality studies showing pain reduction with NSAIDs are lacking and results are conflicting [68]. If analgesics are used, a short course of NSAIDs is preferred. In one small double-blind randomized trial, 1 week of naproxen improved pain compared with placebo [106].

Surgery should be considered only in case of symptoms after 6 months of conservative treatment [107].

20.5.1 Injections

There is no consensus on the effectiveness of hyaluronic acid (HA) injections; recent systematic reviews report no improvement on pain relief or activity recovery in patients with PFP [108, 109]. HA injections for patellar tendinopathy showed pain relief and improvement of knee function after short-term follow-up and could be applied during treatment with eccentric exercises [110].

Further investigations are required for testing the effectiveness of therapy with injection of mesenchymal stem cells (MSCs) and platelet-rich plasma (PRP). A preliminary pilot study showed benefit in clinical scores at short–medium-term follow-up, but no significant improvements in chondral lesions detected with MRI [111].

20.5.2 Exercise

Quadriceps strengthening program is a common rehabilitation technique that has been shown to be effective both in isolation and when paired with other treatment modalities.

Strong evidence recommends a combined exercise therapy, targeting both the hip and quadriceps muscle, as the therapy of choice for improving pain and function in patients with PFP, especially women [112, 113].

Closed kinetic chain exercises are usually well-tolerated and are generally recommended as initial treatment [114].

There is some evidence that selective muscle strengthening of the vastus medialis obliquus (VMOs) reduces pain and improves knee function, by its role in the medial patellar stabilization [115].

Despite this, it has not been clearly proven if exercises can selectively contract VMO [115].

Gait retraining and core muscle strengthening reduce pressure on the patellofemoral joint by stabilizing muscle recruitment and reducing pain; movement retraining in patellofemoral pain may be effective, but its short- and long-term benefits remain uncertain [116, 117].

Patellar taping aims to control the patellar tilt, leading to wider distribution of forces and improving patellar maltracking in athletes [118]. Its use is partially supported by literature, but only when combined with traditional exercise therapy and not in isolation; currently, the overall evidence is insufficient to recommend its routine use [119, 120].

Eccentric exercises have evidence for patellar tendinopathy, with better results than treatment with concentric exercises [121–123]. Eccentric exercise improves the elasticity and tensile strength of the patellar tendon by increasing crosslinking among collagen fibers [124]. Nevertheless, there is still no consensus on the most effective treatment protocol [125, 126].

Combining exercise with foot orthotics is likely more beneficial than either treatment alone. Semi-rigid foot orthotics absorb shock and provide medial longitudinal arch support, correcting dynamic valgus due to flatfoot and rearfoot eversion [127, 128].

Knee braces have not demonstrated benefit over exercise [129].

20.5.3 Surgery

Surgical treatment can be taken into consideration only when there is a detectable organic lesion of the knee, as well as if the patient shows no improvement after strict adherence to conservative therapy for after 6 months [107, 130].

Knee arthroscopy is particularly useful to treat articular pathologies like chondral lesions, anterior synovial impingement, and patellar tendinopathies [73, 131]. Treatment of cartilage lesions is challenging because of its incapability to regenerate or repair. Little evidence does exist of better results after surgical treatment of cartilage lesions [132].

Clinical scores 5 years after surgery do not show any differences between no treatment, debridement, or microfractures in full-thickness cartilage lesions. Furthermore, small asymptomatic lesions may not necessitate surgical treatment [133].

Large full-thickness defects in young patients may be treated by attempting autologous chondrocyte implantation and scaffold-based repair [132].

The concomitant treatment of associated pathology, including patellar malalignment, is recommended as it showed to improve the success of cartilage restoration procedures [134]. Irrespective of the surgical technique used, outcomes are generally worse in the patellofemoral compartment than in the tibiofemoral joint [134].

Lateral release is an accessory and technically simple procedure, which does not produce lasting effects when executed in isolation [135]. It is indicated only in the case of truly tight and symptomatic lateral patellar retinaculum after MPFL reconstruction or joint-preserving osteotomies [135].

20.6 Prevention

The identification of modifiable risk factors for PFP is an effective strategy to prevent a new onset of symptoms. Although several studies demonstrated that lower limb strengthening and stretching programs do not significantly reduce the risk of PFP in military and sporting population, further research is recommended [103].

Neuromuscular training programs aimed to correct known risk factors, such as quadriceps weakness, have in fact proven effective in preventing ACL injuries [136].

Another valid prevention strategy is the training load optimization, avoiding overload of the PFJ [52].

It is widely assumed that training errors may predispose to the development of PFP. They include improper warm-up or cool-down, a rapid increase in frequency or intensity of activity, changes in training pattern, and training on hard, slippery, or slanting surfaces [137–139].

Considering the runners' category, training errors have been reported as present in 60–80% of running injuries [140]. The most common errors are too long a distance, as well as too fast a progression and too much hill work.

In most sports, the risk factors to be corrected and avoided should be monotony, asymmetry, and too much specialization [141].

Poor technique plays a role in the development of anterior knee pain too. Even the least technical fault, if constantly repeated, may lead to an overuse injury [141].

20.7 Conclusions

Anterior knee pain in track and field is frequent and may jeopardize the career of an athlete. Prevention correcting several risk factors and avoiding overloads is of utmost importance. When symptoms arise, a careful clinical, biomechanical, and radiological evaluation allows the planning of a proper treatment, not surgical in most cases. When symptoms persist notwithstanding the conservative efforts, a well-planned targeted surgery can be effective.

References

1. Crossley KM, Stefanik JJ, Selfe J, Collins NJ, Davis IS, Powers CM, et al. 2016 patellofemoral pain consensus statement from the 4th international patellofemoral pain research retreat, Manchester. Part 1: terminology, definitions, clinical examination, natural history, patellofemoral osteoarthritis and patient-reported outcome measures. Br J Sports Med. 2016;50:839–43.
2. Witvrouw E, Callaghan MJ, Stefanik JJ, Noehren B, Bazett-Jones DM, Willson JD, et al. Patellofemoral pain: consensus statement from the 3rd international patellofemoral pain research retreat held in Vancouver, September 2013. Br J Sports Med. 2014;48:411–4.
3. Petersen W, Rembitzki I, Liebau C. Patellofemoral pain in athletes. Open Access J Sports Med. 2017;8:143–54.
4. LaBella C. Patellofemoral pain syndrome: evaluation and treatment. Prim Care. 2004;31:977–1003.
5. DeHaven KE, Lintner DM. Athletic injuries: comparison by age, sport, and gender. Am J Sports Med. 1986;14:218–24.
6. Crossley KM, Callaghan MJ, van Linschoten R. Patellofemoral pain. Br J Sports Med. 2016;50:247–50.
7. Aysin IK, Askin A, Mete BD, Guvendi E, Aysin M, Kocyigit H. Investigation of the relationship between anterior knee pain and chondromalacia patellae and patellofemoral malalignment. Eurasian J Med. 2018;50:28–33.
8. Houghton KM. Review for the generalist: evaluation of anterior knee pain. Pediatr Rheumatol Online J. 2007;5:8.
9. Thomeé R, Augustsson J, Karlsson J. Patellofemoral pain syndrome: a review of current issues. Sports Med. 1999;28:245–62.
10. Reid DC. The myth, Mystic, and frustration of anterior knee pain. Clin J Sport Med. 1993;3:139–43.
11. Crossley KM, van Middelkoop M, Callaghan MJ, Collins NJ, Rathleff MS, Barton CJ. 2016 patellofemoral pain consensus statement from the 4th international patellofemoral pain research retreat, Manchester. Part 2: recommended physical interventions (exercise, taping, bracing, foot orthoses and combined interventions). Br J Sports Med. 2016;50:844–52.
12. Rathleff MS, Rasmussen S, Olesen JL. [Unsatisfactory long-term prognosis of conservative

treatment of patellofemoral pain syndrome]. Ugeskr Laeger. 2012;174:1008–1013.
13. Boling M, Padua D, Marshall S, Guskiewicz K, Pyne S, Beutler A. Gender differences in the incidence and prevalence of patellofemoral pain syndrome. Scand J Med Sci Sports. 2010;20:725–30.
14. Powers CM, Witvrouw E, Davis IS, Crossley KM. Evidence-based framework for a pathomechanical model of patellofemoral pain: 2017 patellofemoral pain consensus statement from the 4th international patellofemoral pain research retreat, Manchester, UK: part 3. Br J Sports Med. 2017;51:1713–23.
15. Dye SF. The pathophysiology of patellofemoral pain: a tissue homeostasis perspective. Clin Orthop. 2005:100–10. https://doi.org/10.1097/01.blo.0000172303.74414.7d.
16. Sisk D, Fredericson M. Update of risk factors, diagnosis, and Management of Patellofemoral Pain. Curr Rev Musculoskelet Med. 2019;12:534–41.
17. Willy RW, Hoglund LT, Barton CJ, Bolgla LA, Scalzitti DA, Logerstedt DS, et al. Patellofemoral pain. J Orthop Sports Phys Ther. 2019;49:CPG1–95.
18. Sanchis-Alfonso V. Pathophysiology of anterior knee pain. In: Patellofemoral pain instability arthritis clinical presentation and imaging treatment. New York: Springer; 2010. p. 1–16.
19. Davis IS, Powers CM. Patellofemoral pain syndrome: proximal, distal, and local factors, an international retreat, April 30-may 2, 2009, fells point, Baltimore, MD. J Orthop Sports Phys Ther. 2010;40:A1–16.
20. Powers CM, Heino JG, Rao S, Perry J. The influence of patellofemoral pain on lower limb loading during gait. Clin Biomech (Bristol, Avon). 1999;14:722–8.
21. Nakagawa TH, Moriya ETU, Maciel CD, Serrão FV. Trunk, pelvis, hip, and knee kinematics, hip strength, and gluteal muscle activation during a single-leg squat in males and females with and without patellofemoral pain syndrome. J Orthop Sports Phys Ther. 2012;42:491–501.
22. Liao T-C, Yang N, Ho K-Y, Farrokhi S, Powers CM. Femur rotation increases patella cartilage stress in females with patellofemoral pain. Med Sci Sports Exerc. 2015;47:1775–80.
23. White LC, Dolphin P, Dixon J. Hamstring length in patellofemoral pain syndrome. Physiotherapy. 2009;95:24–8.
24. Boling MC, Padua DA, Alexander CR. Concentric and eccentric torque of the hip musculature in individuals with and without patellofemoral pain. J Athl Train. 2009;44:7–13.
25. Lankhorst NE, Bierma-Zeinstra SMA, van Middelkoop M. Factors associated with patellofemoral pain syndrome: a systematic review. Br J Sports Med. 2013;47:193–206.
26. Messier SP, Davis SE, Curl WW, Lowery RB, Pack RJ. Etiologic factors associated with patellofemoral pain in runners. Med Sci Sports Exerc. 1991;23:1008–15.
27. Sanchis-Alfonso V, Ordono F, Subias-Lopez A, Monserrat C. Pathogenesis of anterior knee pain and patellar instability in the active young. What have we learned from realignment surgery? In: Anterior knee pain patellar instability. New York: Springer; 2006. p. 21–54.
28. Hughston JC. Subluxation of the patella. J Bone Joint Surg Am. 1968;50:1003–26.
29. Insall J. "Chondromalacia patellae": patellar malalignment syndrome. Orthop Clin North Am. 1979;10:117–27.
30. Merchant AC, Mercer RL, Jacobsen RH, Cool CR. Roentgenographic analysis of patellofemoral congruence. J Bone Joint Surg Am. 1974;56:1391–6.
31. Sanchis-Alfonso V, Rosello-Sastre E, Martinez-Sanjuan V. Pathogenesis of anterior knee pain syndrome and functional patellofemoral instability in the active young. Am J Knee Surg. 1999;12:29–40.
32. Sanchis-Alfonso V. Background: patellofemoral malalignment versus tissue homeostasis. Myths and truths about patellofemoral disease. In: Anterior knee pain patellar instability. New York: Springer; 2006. p. 3–20.
33. Sanchis-Alfonso V, Dye SF. How to Deal with anterior knee pain in the active young patient. Sports Health. 2017;9:346–51.
34. Insall JN, Aglietti P, Tria AJ. Patellar pain and incongruence. II: clinical application. Clin Orthop Relat Res. 1983;176:225–32.
35. Thomeé R, Renström P, Karlsson J, Grimby G. Patellofemoral pain syndrome in young women. I. a clinical analysis of alignment, pain parameters, common symptoms and functional activity level. Scand J Med Sci Sports. 1995;5:237–44.
36. Maffulli N. Anterior knee pain: an overview of management options. In: Rehabilitation of sports injuries - current concepts. New York: Springer; 2013. p. 148–53.
37. Lankhorst NE, Bierma-Zeinstra SMA, van Middelkoop M. Risk factors for patellofemoral pain syndrome: a systematic review. J Orthop Sports Phys Ther. 2012;42:81–94.
38. Stefanik JJ, Zhu Y, Zumwalt AC, Gross KD, Clancy M, Lynch JA, et al. Association between patella Alta and the prevalence and worsening of structural features of patellofemoral joint osteoarthritis: the multicenter osteoarthritis study. Arthritis Care Res. 2010;62:1258–65.
39. Rothermich MA, Glaviano NR, Li J, Hart JM. Patellofemoral pain: epidemiology, pathophysiology, and treatment options. Clin Sports Med. 2015;34:313–27.
40. Van Tiggelen D, Cowan S, Coorevits P, Duvigneaud N, Witvrouw E. Delayed vastus medialis obliquus to vastus lateralis onset timing contributes to the development of patellofemoral pain in previously healthy men: a prospective study. Am J Sports Med. 2009;37:1099–105.

41. Canata G, Guazzotti L, Ferrero G. Functional evaluation of patellar subluxation with extended knee. Ital J Sports Traumatol. 1988;10:25–31.
42. Fredericson M, Powers CM. Practical management of patellofemoral pain. Clin J Sport Med. 2002;12:36–8.
43. Collado H, Fredericson M. Patellofemoral pain syndrome. Clin Sports Med. 2010;29:379–98.
44. Fairbank JC, Pynsent PB, van Poortvliet JA, Phillips H. Mechanical factors in the incidence of knee pain in adolescents and young adults. J Bone Joint Surg Br. 1984;66:685–93.
45. Kujala UM, Osterman K, Kvist M, Aalto T, Friberg O. Factors predisposing to patellar chondropathy and patellar apicitis in athletes. Int Orthop. 1986;10:195–200.
46. Bennett JG, Stauber WT. Evaluation and treatment of anterior knee pain using eccentric exercise. Med Sci Sports Exerc. 1986;18:526–30.
47. Werner S. An evaluation of knee extensor and knee flexor torques and EMGs in patients with patellofemoral pain syndrome in comparison with matched controls. Knee Surg Sports Traumatol Arthrosc. 1995;3:89–94.
48. Draper CE, Fredericson M, Gold GE, Besier TF, Delp SL, Beaupre GS, et al. Patients with patellofemoral pain exhibit elevated bone metabolic activity at the patellofemoral joint. J Orthop Res. 2012;30:209–13.
49. Ho K-Y, Hu HH, Colletti PM, Powers CM. Recreational runners with patellofemoral pain exhibit elevated patella water content. Magn Reson Imaging. 2014;32:965–8.
50. Rathleff MS, Samani A, Olesen JL, Roos EM, Rasmussen S, Christensen BH, et al. Neuromuscular activity and knee kinematics in adolescents with patellofemoral pain. Med Sci Sports Exerc. 2013;45:1730–9.
51. Maclachlan LR, Collins NJ, Matthews MLG, Hodges PW, Vicenzino B. The psychological features of patellofemoral pain: a systematic review. Br J Sports Med. 2017;51:732–42.
52. Barton CJ, Rathleff MS. "Managing my patellofemoral pain": the creation of an education leaflet for patients. BMJ Open Sport Exerc Med. 2016;2:e000086.
53. Noehren B, Shuping L, Jones A, Akers DA, Bush HM, Sluka KA. Somatosensory and biomechanical abnormalities in females with patellofemoral pain. Clin J Pain. 2016;32:915–9.
54. Doménech J, Sanchis-Alfonso V, Espejo B. Changes in catastrophizing and kinesiophobia are predictive of changes in disability and pain after treatment in patients with anterior knee pain. Knee Surg Sports Traumatol Arthrosc. 2014;22:2295–300.
55. Kernozek TW, Torry MR, VAN Hoof H, Cowley H, Tanner S. Gender differences in frontal and sagittal plane biomechanics during drop landings. Med Sci Sports Exerc. 2005;37:1003–12. discussion 1013
56. Lephart SM, Ferris CM, Riemann BL, Myers JB, Fu FH. Gender differences in strength and lower extremity kinematics during landing. Clin Orthop. 2002:162–9. https://doi.org/10.1097/00003086-200208000-00019.
57. Jacobs CA, Uhl TL, Mattacola CG, Shapiro R, Rayens WS. Hip abductor function and lower extremity landing kinematics: sex differences. J Athl Train. 2007;42:76–83.
58. Ferber R, Davis IM, Williams DS. Gender differences in lower extremity mechanics during running. Clin Biomech (Bristol, Avon). 2003;18:350–7.
59. Boling MC, Nguyen A-D, Padua DA, Cameron KL, Beutler A, Marshall SW. Gender-specific risk factor profiles for patellofemoral pain. Clin J Sport Med. 2019; https://doi.org/10.1097/JSM.0000000000000719.
60. Dye SF, Vaupel GL, Dye CC. Conscious neurosensory mapping of the internal structures of the human knee without intraarticular anesthesia. Am J Sports Med. 1998;26:773–7.
61. Hudson K, Delasobera B. Bursae. In: Musculoskelet sports medicine primary care practice. 4th ed. Boca Raton, FL: CRC Press; 2016. p. 111–6.
62. Flores DV, Mejía Gómez C, Pathria MN. Layered approach to the anterior knee: normal anatomy and disorders associated with anterior knee pain. Radiographics. 2018;38:2069–101.
63. Steinbach LS, Stevens KJ. Imaging of cysts and bursae about the knee. Radiol Clin N Am. 2013;51:433–54.
64. Mascaró A, Cos MÀ, Morral A, Roig A, Purdam C, Cook J. Load management in tendinopathy: clinical progression for Achilles and patellar tendinopathy. Apunts Med Esport. 2018;53:19–27.
65. McLoughlin RF, Raber EL, Vellet AD, Wiley JP, Bray RC. Patellar tendinitis: MR imaging features, with suggested pathogenesis and proposed classification. Radiology. 1995;197:843–8.
66. Toumi H, Higashiyama I, Suzuki D, Kumai T, Bydder G, McGonagle D, et al. Regional variations in human patellar trabecular architecture and the structure of the proximal patellar tendon enthesis. J Anat. 2006;208:47–57.
67. Cook JL, Purdam CR. Is tendon pathology a continuum? A pathology model to explain the clinical presentation of load-induced tendinopathy. Br J Sports Med. 2009;43:409–16.
68. Loiacono C, Palermi S, Massa B, Belviso I, Romano V, Gregorio AD, et al. Tendinopathy: pathophysiology, therapeutic options, and role of nutraceutics. a narrative literature review. Medicina (Kaunas Lith). 2019;55:447.
69. Van der Worp H, de Poel HJ, Diercks RL, van den Akker-Scheek I, Zwerver J. Jumper's knee or lander's knee? A systematic review of the relation between jump biomechanics and patellar tendinopathy. Int J Sports Med. 2014;35:714–22.
70. Muneta T, Koga H, Ju Y-J, Mochizuki T, Sekiya I. Hyaluronan injection therapy for athletic

patients with patellar tendinopathy. J Orthop Sci. 2012;17:425–31.
71. Sanchis-Alfonso V. Holistic approach to understanding anterior knee pain. Clinical implications. Knee Surg Sports Traumatol Arthrosc. 2014;22:2275–85.
72. Woolf CJ, American College of Physicians, American Physiological Society. Pain: moving from symptom control toward mechanism-specific pharmacologic management. Ann Intern Med. 2004;140:441–51.
73. Maier D, Bornebusch L, Salzmann GM, Südkamp NP, Ogon P. Mid- and long-term efficacy of the arthroscopic patellar release for treatment of patellar tendinopathy unresponsive to nonoperative management. Arthroscopy. 2013;29:1338–45.
74. Eymard F, Chevalier X. Inflammation of the infrapatellar fat pad. Joint Bone Spine. 2016;83:389–93.
75. Mace J, Bhatti W, Anand S. Infrapatellar fat pad syndrome: a review of anatomy, function, treatment and dynamics. Acta Orthop Belg. 2016;82:94–101.
76. Emad Y, Ragab Y. Liposynovitis prepatellaris in athletic runner (Hoffa's syndrome): case report and review of the literature. Clin Rheumatol. 2007;26:1201–3.
77. Canata G. The anterior knee impingement syndrome. Ital J Sports Traumatology. 1989;11:239–46.
78. Kumar D, Alvand A, Beacon JP. Impingement of infrapatellar fat pad (Hoffa's disease): results of high-portal arthroscopic resection. Arthroscopy. 2007;23:1180–1186.e1.
79. Ogata S, Uhthoff HK. The development of synovial plicae in human knee joints: an embryologic study. Arthroscopy. 1990;6:315–21.
80. Schindler OS. "The sneaky plica" revisited: morphology, pathophysiology and treatment of synovial plicae of the knee. Knee Surg Sports Traumatol Arthrosc. 2014;22:247–62.
81. Canata G, Pugliese M. The synovial plica in the athlete's knee. Med Dello Sport Riv Fisiopatol Dello Sport. 1988;41:183–6.
82. O'Dwyer KJ, Peace PK. The plica syndrome. Injury. 1988;19:350–2.
83. Kim YM, Joo YB, Lee WY, Park IY, Park YC. Patella-patellar tendon angle decreases in patients with infrapatellar fat pad syndrome and medial patellar plica syndrome. Knee Surg Sports Traumatol Arthrosc. 2020;28(8):2609–18.
84. Dupont JY. Synovial plicae of the knee. Controversies and review. Clin Sports Med. 1997;16:87–122.
85. Gulati A, McElrath C, Wadhwa V, Shah JP, Chhabra A. Current clinical, radiological and treatment perspectives of patellofemoral pain syndrome. Br J Radiol. 2018;91:20170456.
86. Patel DR, Villalobos A. Evaluation and management of knee pain in young athletes: overuse injuries of the knee. Transl Pediatr. 2017;6:190–8.
87. Lopes Ferreira C, Barton G, Delgado Borges L, Dos Anjos Rabelo ND, Politti F, Garcia Lucareli PR. Step down tests are the tasks that most differentiate the kinematics of women with patellofemoral pain compared to asymptomatic controls. Gait Posture. 2019;72:129–34.
88. Fredericson M, Yoon K. Physical examination and patellofemoral pain syndrome. Am J Phys Med Rehabil. 2006;85:234–43.
89. Nunes GS, Stapait EL, Kirsten MH, de Noronha M, Santos GM. Clinical test for diagnosis of patellofemoral pain syndrome: systematic review with meta-analysis. Phys Ther Sport. 2013;14:54–9.
90. Yilmaz C, Golpinar A, Vurucu A, Ozturk H, Eskandari MM. Retinacular band excision improves outcome in treatment of plica syndrome. Int Orthop. 2005;29:291–5.
91. van der Heijden RA, de Kanter JLM, Bierma-Zeinstra SMA, Verhaar JAN, van Veldhoven PLJ, Krestin GP, et al. Structural abnormalities on magnetic resonance imaging in patients with patellofemoral pain: a cross-sectional case-control study. Am J Sports Med. 2016;44:2339–46.
92. Berg EE, Mason SL, Lucas MJ. Patellar height ratios. A comparison of four measurement methods. Am J Sports Med. 1996;24:218–21.
93. Kim T-H, Sobti A, Lee S-H, Lee J-S, Oh K-J. The effects of weight-bearing conditions on patellofemoral indices in individuals without and with patellofemoral pain syndrome. Skelet Radiol. 2014;43:157–64.
94. Paiva M, Blønd L, Hölmich P, Steensen RN, Diederichs G, Feller JA, et al. Quality assessment of radiological measurements of trochlear dysplasia; a literature review. Knee Surg Sports Traumatol Arthrosc. 2018;26:746–55.
95. Davies AP, Costa ML, Shepstone L, Glasgow MM, Donell S, Donnell ST. The sulcus angle and malalignment of the extensor mechanism of the knee. J Bone Joint Surg Br. 2000;82:1162–6.
96. Elias DA, White LM. Imaging of patellofemoral disorders. Clin Radiol. 2004;59:543–57.
97. Muhle C, Brossmann J, Heller M. [Functional MRI of the femoropatellar joint]. Radiology. 1995;35:117–124.
98. Diederichs G, Issever AS, Scheffler S. MR imaging of patellar instability: injury patterns and assessment of risk factors. Radiographics. 2010;30:961–81.
99. Conway WF, Hayes CW, Loughran T, Totty WG, Griffeth LK, el-Khoury GY, et al. Cross-sectional imaging of the patellofemoral joint and surrounding structures. Radiographics. 1991;11:195–217.
100. Jarraya M, Diaz LE, Roemer FW, Arndt WF, Goud AR, Guermazi A. MRI findings consistent with Peripatellar fat pad impingement: how much related to patellofemoral Maltracking? Magn Reson Med Sci. 2018;17:195–202.
101. McNally EG, Ostlere SJ, Pal C, Phillips A, Reid H, Dodd C. Assessment of patellar maltracking using combined static and dynamic MRI. Eur Radiol. 2000;10:1051–5.
102. Hirschmann A, Buck FM, Fucentese SF, Pfirrmann CWA. Upright CT of the knee: the effect of

weight-bearing on joint alignment. Eur Radiol. 2015;25:3398–404.
103. Crossley KM, van Middelkoop M, Barton CJ, Culvenor AG. Rethinking patellofemoral pain: prevention, management and long-term consequences. Best Pract Res Clin Rheumatol. 2019;33:48–65.
104. Saudan M, Fritschy D. [AT-TG (anterior tuberosity-trochlear groove): interobserver variability in CT measurements in subjects with patellar instability]. Rev Chir Orthop Reparatrice Appar Mot. 2000;86:250–255.
105. Dutton RA, Khadavi MJ, Fredericson M. Update on rehabilitation of patellofemoral pain. Curr Sports Med Rep. 2014;13:172–8.
106. Heintjes E, Berger MY, Bierma-Zeinstra SMA, Bernsen RMD, Verhaar JAN, Koes BW. Pharmacotherapy for patellofemoral pain syndrome. Cochrane Database Syst Rev. 2004;CD003470.
107. Krieves M, Ramakrishnan R, Lavelle W, Lavelle E. Knee pain. In: Current therapy pain. Amsterdam: Elsevier; 2009.
108. Hart JM, Kuenze C, Norte G, Bodkin S, Patrie J, Denny C, et al. Prospective, randomized, double-blind evaluation of the efficacy of a single-dose hyaluronic acid for the treatment of patellofemoral chondromalacia. Orthop J Sports Med. 2019;7:2325967119854192.
109. Jones BQ, Covey CJ, Sineath MH. Nonsurgical management of knee pain in adults. Am Fam Physician. 2015;92:875–83.
110. Uchida R, Nakamura N, Horibe S. Pathogenesis and treatment of patellar tendinopathy. In: Muscle tendon injection evaluation management. New York: Springer; 2017. p. 295–304.
111. Pintat J, Silvestre A, Magalon G, Gadeau AP, Pesquer L, Perozziello A, et al. Intra-articular injection of mesenchymal stem cells and platelet-rich plasma to treat patellofemoral osteoarthritis: preliminary results of a long-term pilot study. J Vasc Interv Radiol. 2017;28:1708–13.
112. Collins NJ, Barton CJ, van Middelkoop M, Callaghan MJ, Rathleff MS, Vicenzino BT, et al. 2018 consensus statement on exercise therapy and physical interventions (orthoses, taping and manual therapy) to treat patellofemoral pain: recommendations from the 5th international patellofemoral pain research retreat, Gold Coast, Australia, 2017. Br J Sports Med. 2018;52:1170–8.
113. Lack S, Barton C, Sohan O, Crossley K, Morrissey D. Proximal muscle rehabilitation is effective for patellofemoral pain: a systematic review with meta-analysis. Br J Sports Med. 2015;49:1365–76.
114. Kooiker L, Van De Port IGL, Weir A, Moen MH. Effects of physical therapist-guided quadriceps-strengthening exercises for the treatment of patellofemoral pain syndrome: a systematic review. J Orthop Sports Phys Ther. 2014;44:391–402. B1
115. Syme G, Rowe P, Martin D, Daly G. Disability in patients with chronic patellofemoral pain syndrome: a randomised controlled trial of VMO selective training versus general quadriceps strengthening. Man Ther. 2009;14:252–63.
116. Chevidikunnan MF, Al Saif A, Gaowgzeh RA, Mamdouh KA. Effectiveness of core muscle strengthening for improving pain and dynamic balance among female patients with patellofemoral pain syndrome. J Phys Ther Sci. 2016;28:1518–23.
117. Roper JL, Harding EM, Doerfler D, Dexter JG, Kravitz L, Dufek JS, et al. The effects of gait retraining in runners with patellofemoral pain: a randomized trial. Clin Biomech (Bristol, Avon). 2016;35:14–22.
118. Derasari A, Brindle TJ, Alter KE, Sheehan FT. McConnell taping shifts the patella inferiorly in patients with patellofemoral pain: a dynamic magnetic resonance imaging study. Phys Ther. 2010;90:411–9.
119. Callaghan MJ, Selfe J. Patellar taping for patellofemoral pain syndrome in adults. Cochrane Database Syst Rev. 2012;CD006717.
120. Logan CA, Bhashyam AR, Tisosky AJ, Haber DB, Jorgensen A, Roy A, et al. Systematic review of the effect of taping techniques on patellofemoral pain syndrome. Sports Health. 2017;9:456–61.
121. Jonsson P, Alfredson H. Superior results with eccentric compared to concentric quadriceps training in patients with jumper's knee: a prospective randomised study. Br J Sports Med. 2005;39:847–50.
122. Stasinopoulos D, Stasinopoulos I. Comparison of effects of exercise programme, pulsed ultrasound and transverse friction in the treatment of chronic patellar tendinopathy. Clin Rehabil. 2004;18:347–52.
123. Cannell LJ, Taunton JE, Clement DB, Smith C, Khan KM. A randomised clinical trial of the efficacy of drop squats or leg extension/leg curl exercises to treat clinically diagnosed jumper's knee in athletes: pilot study. Br J Sports Med. 2001;35:60–4.
124. Graham HK, Holmes DF, Watson RB, Kadler KE. Identification of collagen fibril fusion during vertebrate tendon morphogenesis. The process relies on unipolar fibrils and is regulated by collagen-proteoglycan interaction. J Mol Biol. 2000;295:891–902.
125. Wasielewski NJ, Kotsko KM. Does eccentric exercise reduce pain and improve strength in physically active adults with symptomatic lower extremity tendinosis? A systematic review. J Athl Train. 2007;42:409–21.
126. Pearson SJ, Hussain SR. Region-specific tendon properties and patellar tendinopathy: a wider understanding. Sports Med. 2014;44:1101–12.
127. Hossain M, Alexander P, Burls A, Jobanputra P. Foot orthoses for patellofemoral pain in adults. Cochrane Database Syst Rev. 2011;CD008402.

128. Barton CJ, Menz HB, Crossley KM. Effects of prefabricated foot orthoses on pain and function in individuals with patellofemoral pain syndrome: a cohort study. Phys Ther Sport. 2011;12:70–5.
129. Swart NM, van Linschoten R, Bierma-Zeinstra SMA, van Middelkoop M. The additional effect of orthotic devices on exercise therapy for patients with patellofemoral pain syndrome: a systematic review. Br J Sports Med. 2012;46:570–7.
130. Cucurulo T, Louis M-L, Thaunat M, Franceschi J-P. Surgical treatment of patellar tendinopathy in athletes. A retrospective multicentric study. Orthop Traumatol Surg Res. 2009;95:S78–84.
131. Figueroa D, Figueroa F, Calvo R. Patellar tendinopathy: diagnosis and treatment. J Am Acad Orthop Surg. 2016;24:e184–92.
132. Falah M, Nierenberg G, Soudry M, Hayden M, Volpin G. Treatment of articular cartilage lesions of the knee. Int Orthop. 2010;34:621–30.
133. Dewan AK, Gibson MA, Elisseeff JH, Trice ME. Evolution of autologous chondrocyte repair and comparison to other cartilage repair techniques. Biomed Res Int. 2014;2014:272481.
134. Brophy RH, Wojahn RD, Lamplot JD. Cartilage restoration techniques for the patellofemoral joint. J Am Acad Orthop Surg. 2017;25:321–9.
135. Clifton R, Ng CY, Nutton RW. What is the role of lateral retinacular release? J Bone Joint Surg Br. 2010;92:1–6.
136. Webster KE, Hewett TE. Meta-analysis of meta-analyses of anterior cruciate ligament injury reduction training programs. J Orthop Res. 2018;36:2696–708.
137. Duri ZA, Aichroth PM, Wilkins R, Jones J. Patellar tendonitis and anterior knee pain. Am J Knee Surg. 1999;12:99–108.
138. Fredberg U, Bolvig L. Jumper's knee. Review of the literature. Scand J Med Sci Sports. 1999;9:66–73.
139. Renström P, Johnson RJ. Overuse injuries in sports. A review. Sports Med. 1985;2:316–33.
140. James SL, Bates BT, Osternig LR. Injuries to runners. Am J Sports Med. 1978;6:40–50.
141. Renström PAFH, Kannus P. Prevention of injuries in endurance athletes. In: Endura sport. Hoboken, NJ: John Wiley & Sons; 2008. p. 458–85.

21 Knee Ligament Injuries in Track and Field Athletes

Christopher M. Gibbs, Jonathan D. Hughes, Giacomo Dal Fabbro, Margaret L. Hankins, Khalid Alkhelaifi, Stefano Zaffagnini, and Volker Musahl

21.1 Introduction

Knee ligamentous injuries are one of the most common injuries in sport and result in loss of articular stability leading to significant functional impairment. Athletes who injure knee ligaments often miss extended periods of participation in sport and competition, with potential long-term disability, particularly post-traumatic osteoarthritis [1]. Athletes in track and field compete in a variety of events, with the likelihood of sustaining a ligamentous injury varying with each event.

C. M. Gibbs (✉) · J. D. Hughes · M. L. Hankins
V. Musahl
Department of Orthopaedic Surgery, University of Pittsburgh, Pittsburgh, PA, USA
e-mail: gibbscm2@upmc.edu; hughesjd3@upmc.edu; hankinsml@upmc.edu; musahlv@upmc.edu

G. D. Fabbro · S. Zaffagnini
IRCCS - Istituto Ortopedico Rizzoli - Clinica Ortopedica e Traumatologica II, Bologna, Italy
e-mail: giacomo.dalfabbro@studio.unibo.it; stefano.zaffagnini@unibo.it

K. Alkhelaifi
Orthopaedic Surgery Department, Aspetar Orthopaedic and Sports Medicine Hospital, Doha, Qatar
e-mail: Khalid.Alkhelaifi@aspetar.com

21.2 Epidemiology of Knee Ligamentous Injuries in Track and Field

Although a detailed description of the prevalence of each knee ligament injury is lacking, a high prevalence of total knee injuries has been reported among track and field athletes, with the majority of these injuries being overuse-related conditions such as tendinopathies and stress fractures [2, 3]. At the elite level, knee sprains represented 4% of all injuries and 2% of injuries resulting in loss-of-time participating in sport [4]. In high school athletes, knee sprain or strain is the most common type of season-ending injury in track and field (13% in male, 22% in female) [5]. Amongst high school track and field athletes, the incidence of ACL injury was found to be 0.05 and 0.16 per 10,000 athlete exposures and the rate of MCL injury was 0.05 and 0.11 per 10,000 athlete exposures in male and in female athletes, respectively [6].

Knee ligament injuries in track and field athletes are more commonly sustained in competition than during practice, with the injury rate higher for females than males [6, 7]. Injury frequency and characteristics among elite athletes show substantial differences between disciplines with the highest number of reported knee ligament injuries occurring in marathons, combined events, and throwing and jumping events [3].

G. L. Canata et al. (eds.), *Management of Track and Field Injuries*,
https://doi.org/10.1007/978-3-030-60216-1_21

In a survey of elite athletes from the United Kingdom, 7%, 17%, 10%, and 20% of sprinters, hurdlers, long-distance runners, and middle-distance runners, respectively, reported injuries about the knee which lasted more than 1 week [2]. Sprinters typically experience more acute injuries compared to events where endurance is emphasized to a greater degree, such as long-distance running, in which more gradual or chronic-use injuries typically occur [8].

Although runners demonstrate a fairly high rate of injury, serious knee ligamentous injuries are relatively uncommon in track and field athletes competing in running events [3, 9–13]. The literature lacks detailed descriptions of the individual knee ligaments injured as well as the severity of injury. Knee injuries are often grouped into categories which are typically either "sprain" or "tendinopathy." Thus, ligamentous injuries of the knee in track runners seem to be relatively rare compared to other sports, and when injury does occur, it is more likely to be a sprain rather than a rupture.

Although few epidemiologic studies have been reported on lower extremity injuries in throwing track and field athletes, one study demonstrated that the most common body part injured among throwers was the ankle, followed by the back, which shows the importance of considering injuries other than those to the upper extremities in throwing athletes [2]. While upper extremity injuries are well-described in javelin throwers, lower extremity injury in these throwers should not be overlooked.

Lower extremity injuries in jumping or vaulting track and field athletes have been reported in a few epidemiologic studies. The high jump has historically been associated with a high incidence of "jumper's knee," or patellar tendinitis. "Knee sprain" was found to be the most common injury of the knee in a collegiate group of pole vault athletes [14].

The decathlon and heptathlon are high-intensity events that require a combination of speed, strength, power, and endurance [15]. Injuries occur more frequently during the decathlon and heptathlon than other disciplines. Additionally, the dropout rate for international combined events remains high, with one group demonstrating athlete dropout in international combined events to be 22% for decathletes and 13% for heptathletes [16]. Injury has been shown to be the reason for dropout in up to 36% of cases, with younger age being a higher risk factor for injury [17]. Explosive events on the first day, such as the 100 m dash and long jump, were the highest risk events for musculoskeletal injury [16, 18]. For the combined events, the incidence of injuries requiring time away from sport during international championships ranges from 115 to 200 per 1000 registered athletes [4, 16, 17, 19, 20]. Approximately 77% of injuries involve the lower extremity, with knee tendinopathy (14%) being the most common diagnosis [20].

21.3 Mechanism, Diagnosis, and Management of Knee Ligament Injuries

While knee ligamentous injuries occur infrequently in track and field athletes, an understanding of the nature of knee ligament injuries is helpful to properly prevent, diagnose, and manage these injuries when they do occur. Additionally, as track and field athletes by nature are fast, explosive, and strong, it may be necessary to provide care to athletes who have sustained knee ligament injuries while participating in other sporting events. This section will begin with a discussion of the general principles regarding the mechanism of injury, diagnosis, and management of knee ligament injuries.

The anterior cruciate ligament (**ACL**) is the most commonly injured ligament of the knee in sports and is frequently associated with other ligamentous or meniscal injuries. The majority of ACL ruptures happen in a noncontact trauma, often during a quick deceleration or landing maneuver immediately after initial foot contact with the ground, particularly when the knee is at or near full extension [21–25].

The Lachman and anterior drawer tests must be performed to assess for pathologic, excessive anterior tibial translation, and the pivot shift used to assess rotatory stability. Radiographs should always be performed to exclude osseous lesions and avulsions. Magnetic resonance imaging (MRI) is invaluable for the diagnosis of an ACL injury with 95% specificity and 86% sensitivity [26].

There is a limited role for nonsurgical treatment of ACL injuries in pivoting sport athletes [27, 28]. Sports activity with an ACL-deficient knee can cause significant functional impairment and predisposes to early damage of the articular surface and periarticular structures of the knee, such as the menisci. The three most commonly used grafts for ACL reconstruction (ACLR) in athletes are the autologous central third of patellar tendon (BPTB), the autologous four strand hamstring (HS), and the autologous quadriceps tendon (QT).

The posterior cruciate ligament (**PCL**) is the primary restraint to posterior translation of the tibia in relation to the femur. Ninety percent of PCL injuries occurring during sport result from a typical "dashboard mechanism" characterized by a posterior force on the anterior tibia with the knee in a flexed position.

On examination, loss of medial and lateral tibial eminence prominence (Clancy sign) or a posterior sag sign representing posterior subluxation of the tibia, at 90° of knee flexion, may be seen. A posterior drawer test to assess for posterior translation of the tibia relative to the femur with the knee flexed to 90° should also be performed. Measurement of posterior tibial translation during the posterior drawer test permits PCL injury classification with grade I <0.5 cm, grade II 0.5 to 1 cm, and grade III >1 cm of posterior tibial translation [29].

Radiographs are necessary in acute trauma to exclude osseous injury such as a fracture or tibial spine avulsion injury. MRI represents the gold standard for acute injuries. Stress radiographs are useful in the diagnosis of a chronic PCL injury, in which MRI can be normal [30].

Nonoperative treatment is indicated in grade I and II PCL tears [31]. Management of PCL grade III injuries is controversial. A possible approach consists of 2–4 weeks of immobilization with a brace locked in extension followed by a rehabilitation program. In the event of persistent pain, instability, and swelling, surgery may be indicated. Surgical treatment is indicated in multiligament PCL-based injuries and with displaced avulsions of the tibial spine [32].

The medial collateral ligament (**MCL**) is the principle restraining structure of the medial knee. MCL injuries account for up to 8% of sport-related knee injuries [33]. The most common MCL injury mechanism is a direct blow on the lateral side of the flexed knee with the foot planted on the ground.

Examination of medial joint line gapping under a valgus stress applied between 0° and 30° of knee flexion compared to the healthy knee should be performed. Grade I tears consist of tenderness without instability, grade II tears consist of broad tenderness with partially torn medial knee structures, and grade III tears are characterized by complete disruption of the medial knee structures without an endpoint.

X-rays are required to evaluate for potential fractures or chondral damage. Valgus stress radiographs can objectively identify a medial knee injury. Greater than 10 mm of increased medial compartment gapping at 20° of knee flexion represents a complete tear of the medial knee structures [34]. Finally, MRI is a fundamental tool in the diagnosis of medial sided injuries and any associated lesions [35].

In the case of a grade I to II lesion of the MCL and for isolated, acute grade III injuries, nonoperative treatment is the first line of therapy. Surgical treatment is indicated in multiligament injuries or knee dislocation involving the MCL and in the presence of a tear involving both the midsubstance and tibial insertion. In such circumstances, direct repair with sutures, repair augmentation with a hamstring graft, or acute reconstruction with auto- or allograft may be indicated [35, 36]. In chronic grade III medial knee injuries, surgery is indicated for patients with instability.

The lateral collateral ligament (**LCL**) is the primary varus stabilizer of the knee, with the most common mechanism of injury being a direct blow to the medial aspect of the knee [37]. On examination, varus laxity at 30° of flexion indicates isolated LCL damage, while varus laxity in full extension is associated with additional injury to one or both of the cruciate ligaments.

As with other knee ligament injuries, X-rays should be obtained, but MRI is considered the gold standard when evaluating for an LCL injury. MRI also permits classification of LCL tears based on interstitial injury from grades I to III [37].

Grade I and II LCL lesions are generally treated conservatively with knee immobilization. In grade III injuries, the risk of developing chronic instability is very high, and thus surgical treatment must be considered.

21.4 Knee Ligamentous Injury Considerations by Event

There are important considerations regarding knee ligament injuries for athletes in the various events. The nature of the event may place the athlete at high risk of injury by creating a scenario in which a mechanism which causes a knee ligament injury is more likely to occur. Additionally, prevention and management of knee ligament injuries may be optimized based upon the type of event an athlete competes in.

21.4.1 Injury Mechanisms by Event

Theoretically, runners are at relatively low risk of sustaining a cruciate ligament injury as they are unlikely to experience the mechanisms leading to knee ligament injury. Hurdlers are at higher risk as a wrong step over the hurdle may cause the athlete's limb to be positioned in a pattern causing injury. Additionally, PCL injury may result from a posteriorly directed force on the tibia by a hurdle or the ground.

Although relatively rare, contact between runners, which can occur in any running event, could result in cruciate or collateral ligament injury. As relays involve passing of a baton from one team member to another, often occurring alongside multiple teams simultaneously, athletes in a relay race are likely at higher risk than other runners.

Running events are held with runners traveling in a counterclockwise direction around the track. This has been thought to account for the tendency to have medial or posterolateral pain in the right (outer) knee and medial pain in the left (inner) knee [38]. These effects would theoretically be further accentuated when running on indoor tracks as these typically have a greater angle of track embankment than outdoor surfaces.

Throwing events involve the generation of energy beginning through the legs and exiting out of the arms during the throw. In shot put, javelin, and discus, the rotational motion about the throwing circle puts these athletes at high risk of rotational knee injury.

The throwing circle itself can also be a threat to an athlete's knee. Many shotput throwers plant their nondominant foot under the toe bar to stop movement at the end of their throw, causing their body's momentum to exit through this extremity [38]. This is a dangerous maneuver that can lead to ACL tear or meniscal injury due to internal rotation of the planted leg.

In the hammer throw, the athlete balances his or her center of gravity and leverage to generate maximum energy for the throw. Elevating the hammer too quickly during the rotational approach is associated with an increased risk of LCL strain [38].

The javelin throw involves less rotational energy than the other throwing events but involves a similar high-intensity approach and generation of momentum. If the approach to a javelin throw is executed too fast or if conditions are poor, desynchronization of upper and lower extremity motion can occur and cause knee injury due to loss of control [38].

During the take-off or landing phase of jumping and vaulting events, displacement of the center of gravity can create unexpected stress on the knees. Specifically, during landing, athletes can sustain ligamentous or meniscal injury depending on the position of the leg in relation to the body and center of gravity during impact [38]. In the long jump and triple jump, misstep during the end of the approach upon reaching the ramp can result in a twisting injury to the knee, and improper acceleration can lead to acute or chronic strains about the knee.

The high jump and pole vault are associated with a unique set of injury patterns given the nature of the events, with athletes reaching a height of more than 6 ft., or up to 15 ft. for the average collegiate pole vaulter. Highest risk of injury to the knee occurs during take-off and landing. Energy generated during the horizontal approach is converted into vertical lift during take-off, which requires braking; this is thought to place these athletes at high risk of patellar tendinitis and chronic extensor mechanism pathology [38]. Improper landing technique can understandably increase the risk for acute traumatic knee injury, including any variety of ligamentous rupture or meniscal tear.

Due to the 2-day length as well as the multiple high-intensity events of decathlons and heptathlons, athletes are more susceptible to injury in this discipline. The majority of reported knee ligamentous injuries occur during a noncontact traumatic injury or a direct blow during competition [19, 20]. As the decathlon and heptathlon combine the different types of events previously described, the mechanism of knee ligamentous injury depends on the particular event being performed.

21.4.2 Treatment of Knee Ligament Injury by Event

Ligamentous injuries of the knee in runners, throwers, jumping and vaulting athletes, and athletes participating in combined events are treated in a similar manner with consideration of the demands of each athlete's event.

21.4.3 Prevention of Knee Ligament Injury by Event

To prevent injuries in all events, it is important that the athlete be adequately conditioned and perform an adequate warm-up and cool-down to prepare the musculotendinous unit for rapid elongation and contraction prior to activity [39–41]. Additionally, due to the nature of each event, additional factors may be particularly important in preventing injury.

Evidence regarding the prevention of knee ligamentous injuries in runners is lacking with mixed reports as to the effectiveness of a proper training regimen to reduce injury in long-distance recreational athletes [42, 43]. However, runners have been shown to a have a significantly lower flexor to extensor strength ratio which has been shown to be a risk factor for ACL rupture and failure of ACLR [44–46]. Therefore, strengthening the knee flexors to create a better flexor to extensor ratio would be beneficial for preventing injury.

For throwing athletes, the throwing circle, ambient conditions, and footwear can all contribute to knee injury due to mechanical disruption in motion or loss of control or balance. The athlete should always be aware of the conditions in which they are performing and attend to the location of any obstacles. Most importantly, proper technique must be taught, practiced, and executed. The rotational movement of the trunk during throwing events must be in perfect coordination with the rotation of the lower extremities to avoid excessive rotational force on the knees [47]. Proper follow-through can help avoid deceleration injury [38, 47].

Prior to jumping and vaulting events, proper facilities must be provided to ensure the safety of athletes in every event, but particularly in the high jump and pole vault. The introduction of adequate padding has dramatically reduced the rate of injury in these events [38]. Proper jumping, vaulting, falling, and stopping technique must be taught and practiced.

The combined events require stamina, skill, speed, determination, endurance, and concentration which emphasize the importance of the physiological condition of the athletes [48]. Significant technical, mental, and physical demands are required from the athletes. The athletes must remain concentrated through all the events and cannot be distracted by their performance in a previous event; thus, training on mental awareness and fortitude cannot be understated.

21.5 The Case of a Skeletally Immature Female Athlete

History: A 12-year-old female track and field throwing athlete presented with an acute right knee sprain following a noncontact injury during competition. She reported an acutely swollen right knee, medial knee pain, and the sensation of right knee instability, especially with pivoting and cutting activities.

Diagnosis: Clinical examination showed a slightly swollen right knee with a complete, normal range of motion. Mild pain was evoked by palpation of the joint space. Both the anterior drawer and Lachman tests were positive, with a high-grade pivot shift.

Right Knee MRI confirmed the suspected diagnosis of ACL rupture (Fig. 21.1). Moreover, the growth plates of the patient remained open, demonstrating skeletal immaturity.

An instrumental PS examination using a triaxial accelerometer device (KiRA, Orthokey, Florence, Italy) was used to assess the acceleration value of the lateral tibial compartment during the pivot shift test. The difference between the injured knee and contralateral limb was more than 3 meters/sec^2, confirming a high-grade pivot shift injury (Fig. 21.2) [49, 50].

Treatment: The authors performed a physeal-sparing ACLR with HS autograft [51]. First, an arthroscopic repair of the posterior horn of the medial meniscus was performed. Next, the ipsilateral gracilis and semitendinosus tendons were harvested. The tibial insertion of both tendons was preserved to maintain their neurovascular supply. An all-epiphyseal tibial tunnel was drilled above the tibial growth plate of the patient, with the aid of intraoperative X-ray imaging (Fig. 21.3). Subsequently the graft was retrieved through the knee joint from a lateral incision and

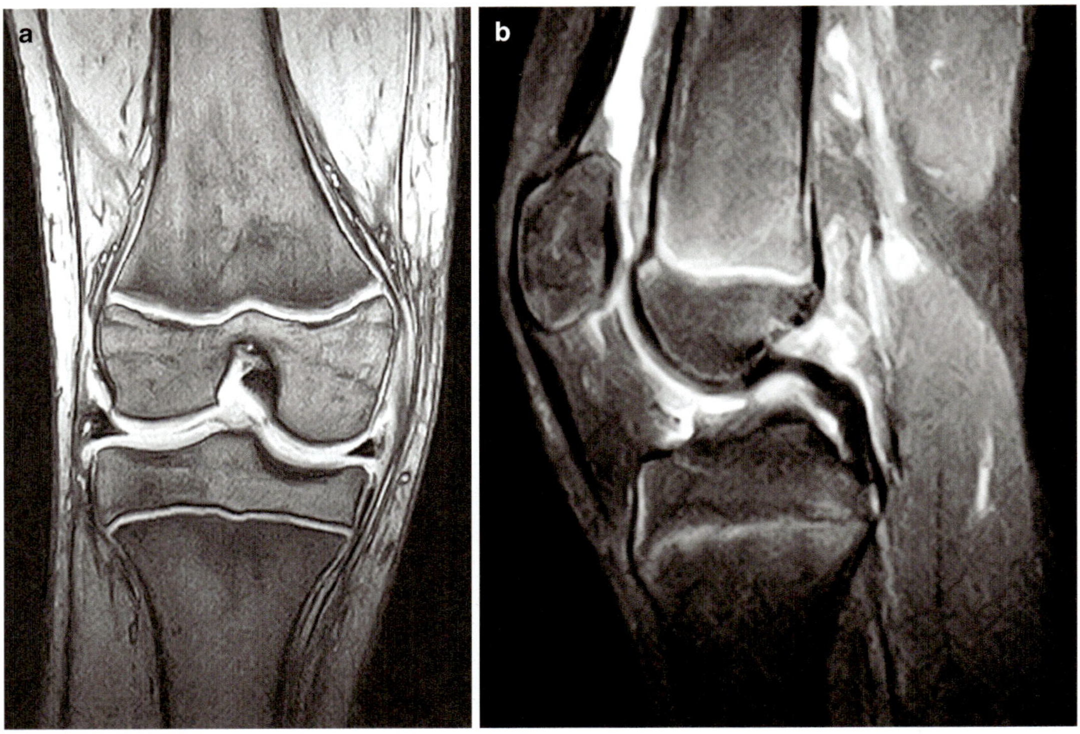

Fig. 21.1 Coronal (**a**) and sagittal (**b**) MRI images of the knee obtained preoperatively demonstrate acute ACL rupture; also note the open physes indicating skeletal immaturity

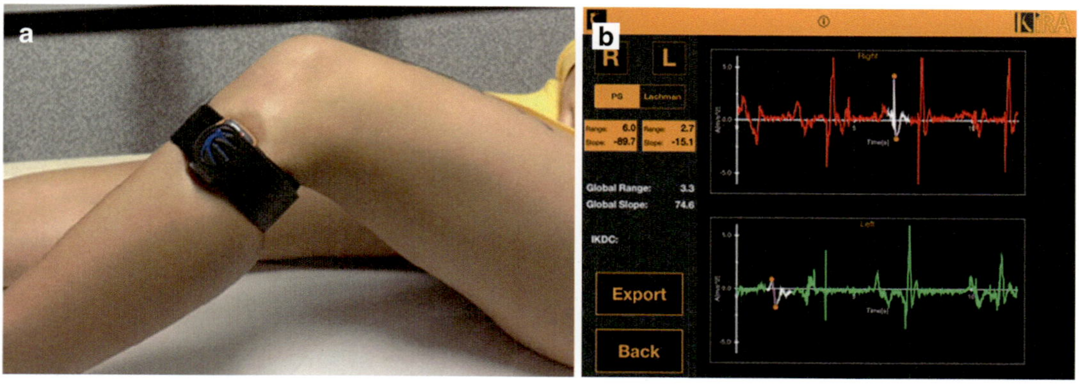

Fig. 21.2 Triaxial accelerometer device (KiRA, Orthokey, Florence, Italy) used to quantitatively evaluate the pivot shift (**a**); the preoperative side-to-side difference between the injured knee and contralateral limb was more than 3 m/s^2 (**b**)

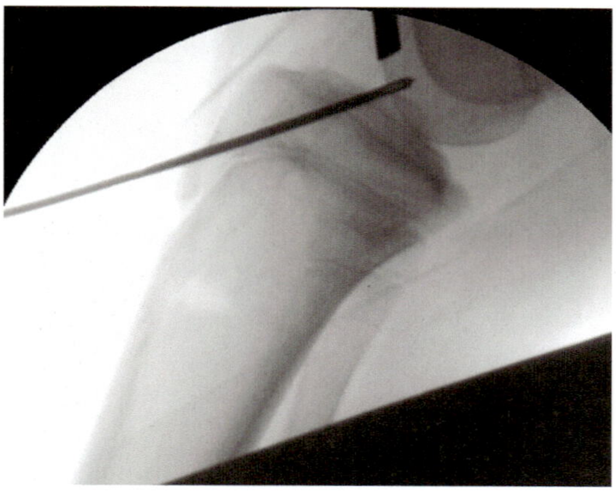

Fig. 21.3 ACL reconstruction was performed by drilling the tibial tunnel above the growth cartilage under fluoroscopic guidance

fixed with two staples to the cortex of the lateral femoral condyle, in the over-the-top position. The remaining part of the graft was fixed below Gerdy's tubercle to the lateral aspect of the tibia with one staple (Fig. 21.4).

Rehabilitation Protocol and Follow-Up: The patient was nonweightbearing with a brace locked in extension for the first 2 weeks to protect the meniscal repair during the healing process. Following this, a progressive return to sport protocol was followed. At 10 months follow-up, objective pivot shift measurement showed a difference between the injured and contralateral knee less than 0.4 m/s^2. The patient returned to track and field competition 11 months following surgery. X-ray evaluation at 4 years follow-up demonstrated normal alignment (Fig. 21.5). She was still engaged in track and field competition without pain or functional impairment.

21.6 Conclusion

Knee ligament injuries are one of the most common injuries in sport, resulting in loss of joint stability and significant functional impairment. Although track and field athletes sustain knee ligamentous injury at a rate lower than athletes in pivoting, contact sports, injuries still may occur. Thus, healthcare providers must be knowledgeable of knee ligament injuries. The treatment of knee ligament injuries is largely the same for track and field athletes who compete in various events. However, track and field athletes may be at higher or lower risk for injury based on the potential movement patterns of the lower limb inherent to the events in which they participate. The specific nature of an athlete's event must be considered when selecting the proper individualized treatment for the athlete.

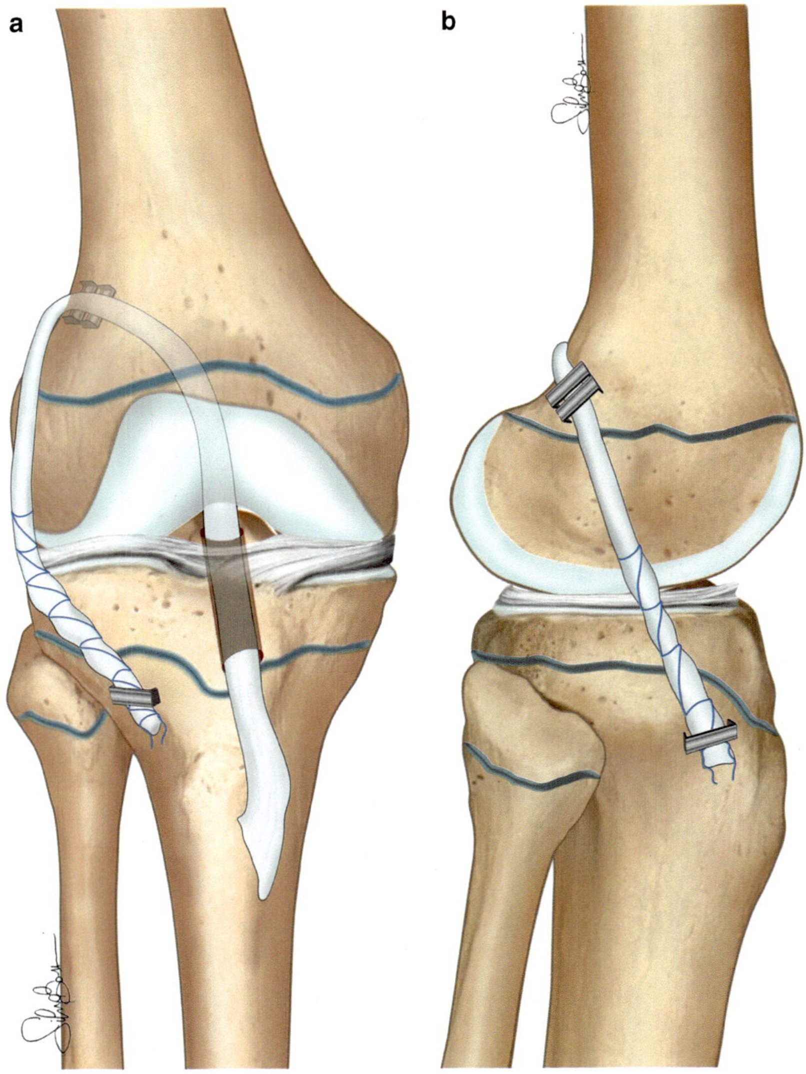

Fig. 21.4 Illustration of the over-the-top physeal sparing ACL-reconstruction technique; (**a**) Antero-posterior view, (**b**) Lateral view

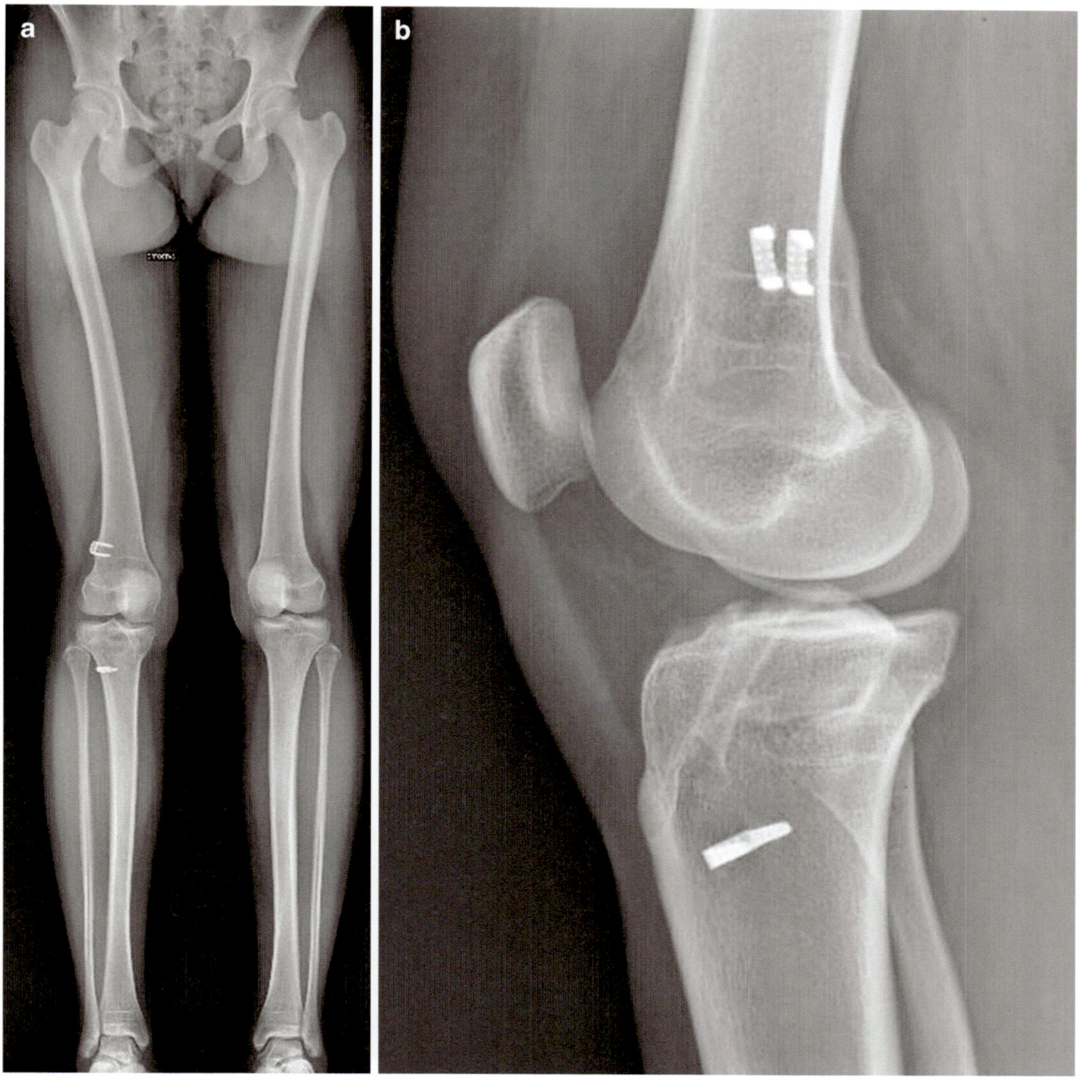

Fig. 21.5 Full standing (**a**) and lateral (**b**) X-ray at 4-years postoperative follow-up

References

1. Wiggins AJ, Grandhi RK, Schneider DK, Stanfield D, Webster KE, Myer GD. Risk of secondary injury in younger athletes after anterior cruciate ligament reconstruction: a systematic review and meta-analysis. Am J Sports Med. 2016;44(7):1861–76. https://doi.org/10.1177/0363546515621554.
2. D'Souza D. Track and field athletics injuries – a one-year survey. Br J Sports Med. 1994;28(3):197–202. https://doi.org/10.1136/bjsm.28.3.197.
3. Edouard P, Navarro L, Branco P, Gremeaux V, Timpka T, Junge A. Injury frequency and characteristics (location, type, cause and severity) differed significantly among athletics ('track and field') disciplines during 14 international championships (2007–2018): implications for medical service planning. Br J Sports Med. 2020;54(3):159–67. https://doi.org/10.1136/bjsports-2019-100,717.
4. Alonso JM, Edouard P, Fischetto G, Adams B, Depiesse F, Mountjoy M. Determination of future prevention strategies in elite track and field: analysis of Daegu 2011 IAAF Championships injuries and illnesses surveillance. Br J Sports Med. 2012;46(7):505–14. https://doi.org/10.1136/bjsports-2012-091008.
5. Tirabassi J, Brou L, Khodaee M, Lefort R, Fields SK, Comstock RD. Epidemiology of high school sports-related injuries resulting in medical disqualification: 2005–2006 through 2013–2014 academic years. Am J Sports Med. 2016;44(11):2925–32. https://doi.org/10.1177/0363546516644604.

6. Swenson DM, Collins CL, Best TM, Flanigan DC, Fields SK, Comstock RD. Epidemiology of knee injuries among U.S. high school athletes, 2005/2006–2010/2011. Med Sci Sports Exerc. 2013;45(3):462–9. https://doi.org/10.1249/MSS.0b013e318277acca.
7. Jacobsson J, Timpka T, Kowalski J, Nilsson S, Ekberg J, Renstrom P. Prevalence of musculoskeletal injuries in Swedish elite track and field athletes. Am J Sports Med. 2012;40(1):163–9. https://doi.org/10.1177/0363546511425467.
8. Bennell KL, Crossley K. Musculoskeletal injuries in track and field: incidence, distribution and risk factors. Aust J Sci Med Sport. 1996;28(3):69–75.
9. Gornitzky AL, Lott A, Yellin JL, Fabricant PD, Lawrence JT, Ganley TJ. Sport-specific yearly risk and incidence of anterior cruciate ligament tears in high school athletes: a systematic review and meta-analysis. Am J Sports Med. 2015;44(10):2716–23. https://doi.org/10.1177/0363546515617742.
10. Pierpoint LA, Williams CM, Fields SK, Comstock RD. Epidemiology of injuries in united states high school track and field: 2008–2009 through 2013–2014. Am J Sports Med. 2016;44(6):1463–8. https://doi.org/10.1177/0363546516629950.
11. Prodromos CC, Han Y, Rogowski J, Joyce B, Shi K. A meta-analysis of the incidence of anterior cruciate ligament tears as a function of gender, sport, and a knee injury-reduction regimen. Arthroscopy. 2007;23(12):1320–5.e6. https://doi.org/10.1016/j.arthro.2007.07.003.
12. Kerr ZY, Kroshus E, Grant J, Parsons JT, Folger D, Hayden R, et al. Epidemiology of national collegiate athletic association men's and women's cross-country injuries, 2009–2010 through 2013–2014. J Athl Train. 2016;51(1):57–64. https://doi.org/10.4085/1062-6050-51.1.10.
13. Jacobsson J, Timpka T, Kowalski J, Nilsson S, Ekberg J, Renström P. Prevalence of musculoskeletal injuries in swedish elite track and field athletes. Am J Sports Med. 2011;40(1):163–9. https://doi.org/10.1177/0363546511425467.
14. Rebella G. A prospective study of injury patterns in collegiate pole vaulters. Am J Sports Med. 2015;43(4):808–15. https://doi.org/10.1177/0363546514564542.
15. Linden M. Factor analytical study of olympic decathlon data. Res Q. 1977;48(3):562–8.
16. Edouard P, Pruvost J, Edouard J-L, Morin J-B. Causes of dropouts in decathlon. A pilot study. Phys Ther Sport. 2010;11(4):133–5.
17. Edouard P, Samozino P, Escudier G, Baldini A, Morin J-B. Injuries in youth and national combined events championships. Int J Sports Med. 2012;33(10):824–8.
18. Edouard P, Morin J, Celli F, Celli Y, Edouard J. Dropout in international combined events competitions. New Stud Athlet. 2009;24(4):63–8.
19. Edouard P, Kerspern A, Pruvost J, Morin J. Four-year injury survey in heptathlon and decathlon athletes. Sci Sport. 2012;27(6):345–50.
20. Edouard P, Morin J, Pruvost J, Kerspern A. Injuries in high-level heptathlon and decathlon. Br J Sports Med. 2011;45(4):346.
21. Koga H, Nakamae A, Shima Y, Iwasa J, Myklebust G, Engebretsen L, et al. Mechanisms for noncontact anterior cruciate ligament injuries: knee joint kinematics in 10 injury situations from female team handball and basketball. Am J Sports Med. 2010;38(11):2218–25. https://doi.org/10.1177/0363546510373570.
22. Krosshaug T, Nakamae A, Boden BP, Engebretsen L, Smith G, Slauterbeck JR, et al. Mechanisms of anterior cruciate ligament injury in basketball: video analysis of 39 cases. Am J Sports Med. 2007;35(3):359–67. https://doi.org/10.1177/0363546506293899.
23. Taylor KA, Cutcliffe HC, Queen RM, Utturkar GM, Spritzer CE, Garrett WE, et al. In vivo measurement of ACL length and relative strain during walking. J Biomech. 2013;46(3):478–83. https://doi.org/10.1016/j.jbiomech.2012.10.031.
24. Taylor KA, Terry ME, Utturkar GM, Spritzer CE, Queen RM, Irribarra LA, et al. Measurement of in vivo anterior cruciate ligament strain during dynamic jump landing. J Biomech. 2011;44(3):365–71. https://doi.org/10.1016/j.jbiomech.2010.10.028.
25. Li G, Defrate LE, Rubash HE, Gill TJ. In vivo kinematics of the ACL during weight-bearing knee flexion. J Orthop Res. 2005;23(2):340–4. https://doi.org/10.1016/j.orthres.2004.08.006.
26. Musahl V, Karlsson J. Anterior cruciate ligament tear. N Engl J Med. 2019;380(24):2341–8. https://doi.org/10.1056/NEJMcp1805931.
27. Walden M, Hagglund M, Magnusson H, Ekstrand J. ACL injuries in men's professional football: a 15-year prospective study on time trends and return-to-play rates reveals only 65% of players still play at the top level 3 years after ACL rupture. Br J Sports Med. 2016;50(12):744–50. https://doi.org/10.1136/bjsports-2015-095952.
28. Weiler R, Monte-Colombo M, Mitchell A, Haddad F. Non-operative management of a complete anterior cruciate ligament injury in an English Premier League football player with return to play in less than 8 weeks: applying common sense in the absence of evidence. BMJ Case Rep. 2015;2015. https://doi.org/10.1136/bcr-2014-208012.
29. Wang D, Graziano J, Williams RJ 3rd, Jones KJ. Nonoperative treatment of PCL injuries: goals of rehabilitation and the natural history of conservative care. Curr Rev Musculoskelet Med. 2018;11(2):290–7. https://doi.org/10.1007/s12178-018-9487-y.
30. Pache S, Aman ZS, Kennedy M, Nakama GY, Moatshe G, Ziegler C, et al. Posterior cruciate ligament: current concepts review. Arch Bone Jt Surg. 2018;6(1):8–18.
31. Johnson D. Posterior cruciate ligament injuries: my approach. Oper Tech Sports Med. 2009;17(3):167–74. https://doi.org/10.1053/j.otsm.2009.06.004.
32. Pierce CM, O'Brien L, Griffin LW, Laprade RF. Posterior cruciate ligament tears: functional and postoperative rehabilitation. Knee Surg Sports

Traumatol Arthrosc. 2013;21(5):1071–84. https://doi.org/10.1007/s00167-012-1970-1.
33. Majewski M, Susanne H, Klaus S. Epidemiology of athletic knee injuries: a 10-year study. Knee. 2006;13(3):184–8. https://doi.org/10.1016/j.knee.2006.01.005.
34. Laprade RF, Bernhardson AS, Griffith CJ, Macalena JA, Wijdicks CA. Correlation of valgus stress radiographs with medial knee ligament injuries: an in vitro biomechanical study. Am J Sports Med. 2010;38(2):330–8. https://doi.org/10.1177/0363546509349347.
35. LaPrade MD, Kennedy MI, Wijdicks CA, LaPrade RF. Anatomy and biomechanics of the medial side of the knee and their surgical implications. Sports Med Arthrosc Rev. 2015;23(2):63–70. https://doi.org/10.1097/JSA.0000000000000054.
36. Owens BD, Neault M, Benson E, Busconi BD. Primary repair of knee dislocations: results in 25 patients (28 knees) at a mean follow-up of four years. J Orthop Trauma. 2007;21(2):92–6. https://doi.org/10.1097/BOT.0b013e3180321318.
37. Grawe B, Schroeder AJ, Kakazu R, Messer MS. Lateral collateral ligament injury about the knee: anatomy, evaluation, and management. J Am Acad Orthop Surg. 2018;26(6):e120–e7. https://doi.org/10.5435/jaaos-d-16-00028.
38. Ciullo JV, Ciullo JR. Sports injuries: mechanisms, prevention, treatment. 2nd ed. Philadelphia: Lippincott Williams and Wilkins; 2001.
39. Schiff MA, Caine DJ, O'Halloran R. Injury prevention in sports. Am J Lifestyle Med. 2009;4(1):42–64. https://doi.org/10.1177/1559827609348446.
40. Petersen J, Hölmich P. Evidence based prevention of hamstring injuries in sport. Br J Sports Med. 2005;39(6):319. https://doi.org/10.1136/bjsm.2005.018549.
41. Edouard P, Mori J-B, Samozino B. Maximal lower extremity power output changes during a decathlon. New Stud Athlet. 2013;28(3/4):19–37.
42. van Mechelen W, Hlobil H, Kemper HCG, Voorn WJ, de Jongh HR. Prevention of running injuries by warm-up, cool-down, and stretching exercises. Am J Sports Med. 1993;21(5):711–9. https://doi.org/10.1177/036354659302100513.
43. Jakobsen BW, Króner K, Schmidt SA, Kjeldsen A. Prevention of injuries in long-distance runners. Knee Surg Sports Traumatol Arthrosc. 1994;2(4):245–9. https://doi.org/10.1007/BF01845597.
44. Deli CK, Paschalis V, Theodorou AA, Nikolaidis MG, Jamurtas AZ, Koutedakis Y. Isokinetic knee joint evaluation in track and field events. J Strength Cond Res. 2011;25(9):2528–36. https://doi.org/10.1519/JSC.0b013e3182023a7a.
45. Myer GD, Ford KR, Barber Foss KD, Liu C, Nick TG, Hewett TE. The relationship of hamstrings and quadriceps strength to anterior cruciate ligament injury in female athletes. Clin J Sports Med. 2009;19(1):3–8. https://doi.org/10.1097/JSM.0b013e318190bddb.
46. Yamanashi Y, Mutsuzaki H, Iwai K, Ikeda K, Kinugasa T. Failure risks in anatomic single-bundle anterior cruciate ligament reconstruction via the outside-in tunnel technique using a hamstring autograft. J Orthop. 2019;16(6):504–7. https://doi.org/10.1016/j.jor.2019.04.015.
47. Meron A, Saint-Phard D. Track and field throwing sports: injuries and prevention. Curr Sports Med Rep. 2017;16(6):391–6. https://doi.org/10.1249/jsr.0000000000000416.
48. Zarnowski F. The nature of decathlon. A basic guide to decathlon. Torrance, CA: Griffin Publishing Group; 2001. p. 27–37.
49. Marcheggiani Muccioli GM, Signorelli C, Grassi A, TRD S, Raggi F, Carbone G, et al. In-vivo pivot-shift test measured with inertial sensors correlates with the IKDC grade. J ISAKOS. 2018;3(2):89–93. https://doi.org/10.1136/jisakos-2017-000167.
50. Lopomo N, Zaffagnini S, Signorelli C, Bignozzi S, Giordano G, Marcheggiani Muccioli GM, et al. An original clinical methodology for non-invasive assessment of pivot-shift test. Comput Methods Biomech Biomed Eng. 2012;15(12):1323–8. https://doi.org/10.1080/10255842.2011.591788.
51. Roberti di Sarsina T, Macchiarola L, Signorelli C, Grassi A, Raggi F, Marcheggiani Muccioli GM, et al. Anterior cruciate ligament reconstruction with an all-epiphyseal "over-the-top" technique is safe and shows low rate of failure in skeletally immature athletes. Knee Surg Sports Traumatol Arthrosc. 2019;27(2):498–506. https://doi.org/10.1007/s00167-018-5132-y.

Part VI

Common Foot and Ankle Injuries

Achilles Tendon, Calf, and Peroneal Tendon Injuries

22

Pim A. D. van Dijk, Guilherme França, Jari Dahmen, Gino M. M. J. Kerkhoffs, Pieter D'Hooghe, and Jon Karlsson

22.1 Introduction

In track and field, the suddenness of motion combined with running and jumping on uneven grounds requires great stability and power of the foot and ankle. The calf and peroneal muscles play an important role in both static and dynamic support of the foot and ankle and thus provide both stability and power during running and jumping. In this manner, track and field exposes these muscles to high mechanical loads, putting them at higher risk for injuries. In fact, the majority of injuries (approximately 30%) in track and field are located within the foot and ankle [1, 2].

This chapter provides an overview of track and field injuries related to the calf, Achilles tendon, and peroneal tendons, including anatomy, epidemiology, sports dynamics, and physical demands. Moreover, it provides a framework for management and return to sport guidelines of the most common pathologies related to the muscles and tendons of the lower leg.

22.2 Anatomy

22.2.1 Anatomy of the Calf and the Achilles Tendon

The calf muscle, *or triceps surae,* is the primary plantiflexor of the foot and is formed by the gastrocnemius and the soleus muscles (Fig. 22.1). The gastrocnemius muscle is located most superficially and contains a medial head, originating posterior at the medial femoral condyle, and a lateral head, originating from the lateral femoral condyle. In this way, it bridges over three joints; the knee, ankle, and subtalar joints. Deep to the gastrocnemius the soleus muscle is located, originating posterior at the proximal fibula and middle third of the medial border of the tibia. The soleus bridges over the ankle and the subtalar joints. Distal, both muscles converge into one tendon,

P. A. D. van Dijk (✉) · J. Dahmen
G. M. M. J. Kerkhoffs
Department of Orthopaedic Surgery, Amsterdam University Hospital - Academic Medical Centre, University of Amsterdam, Amsterdam, The Netherlands

Academic Center for Evidence Based Sports Medicine, Amsterdam, The Netherlands

Amsterdam Collaboration on Health and Safety in Sports, Amsterdam, The Netherlands

G. França
Orthopaedic Surgery Department, Hospital de Braga, Braga, Portugal

P. D'Hooghe
Orthopedic Surgery, Aspire Zone, ASPETAR Qatar Orthopaedic and Sports Medicine Hospital, Doha, Qatar

J. Karlsson
Department of Orthopaedics, Sahlgrenska University Hospital, Sahlgrenska Academy, Gothenburg University, Gothenburg, Sweden

G. L. Canata et al. (eds.), *Management of Track and Field Injuries*,
https://doi.org/10.1007/978-3-030-60216-1_22

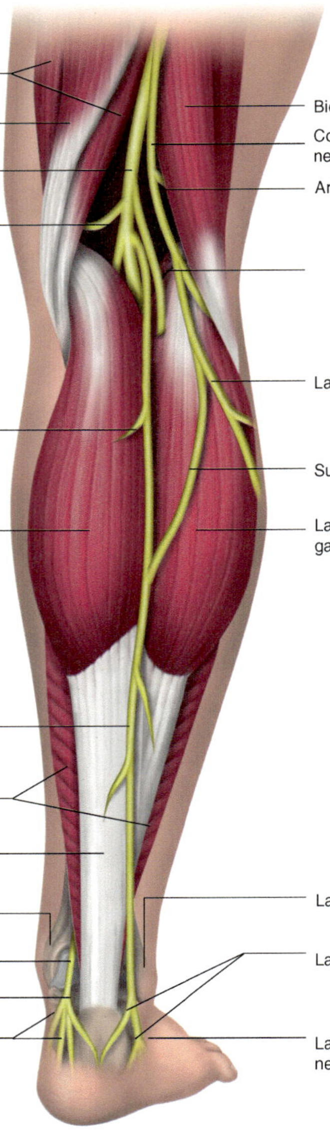

Fig. 22.1 Anatomy of theposterior lower leg including the calf muscles and Achilles tendon

the Achilles tendon, which inserts on the posterior tuberosity of the calcaneus [3].

The Achilles tendon is a round, fibro-elastic structure which spirals approximately 90° along its course. In this way, an area of concentrated stress arises which gives the tendon the possibility to produce forceful elastic recoil and elongation which is indispensable in track and field [4]. Blood is supplied by the muscular-tendon junction (proximal), surrounding highly vascularized endo- and paratenon (central) and bone-tendon junction (distal) [5]. A zone of hypo-vascularity is present, 2–6 cm above the insertion, leading to a relatively poor healing capacity. Innervation arises from three main sources: cutaneous, muscular, and peritendinous nerves [6].

In up to 92–94% of the population, a plantaris muscle is present, originating from the lateral supracondylar line of the femur and is located in between the gastrocnemius and soleus muscles [3]. It is a relatively small muscle with an appreciably long tendinous portion, not to be mistaken for a nerve. The insertion is mostly found at the calcaneus, just medio-anterior to the Achilles tendon. In 6–8% of the population, the tendon inserts into the flexor retinaculum [3].

22.2.2 Anatomy of the Peroneal Tendons

In general, two peroneal muscles are identified: the peroneus brevis (PB) and peroneus longus (PL) (Figs. 22.2 and 22.3), together acting as the primary evertors and abductors of the foot. Moreover, they play an important role in active lateral ankle stability and stabilization of the lateral column of the foot, especially during stance.

The PL originates at the lateral tibial condyle, lateral aspect of the proximal fibular head, intramuscular septa, and adjacent fascia. The PB originates more distally, on the fibular shaft and interosseous membrane. The PL muscle becomes tendinous 3–4 cm proximal to the distal fibular tip, while the PB muscle usually runs up to 2 cm more distally [7]. In some cases, the musculotendinous junction runs beyond the fibular tip, known as a low-lying muscle belly [8]. In literature, it is argued whether this variation possibly predisposes the tendons to pathology [8].

Around the fibular tip, the PB lays anteromedially to the PL and is flattened against the bone within the fibular groove. The superior peroneal retinaculum provides stability of the tendons within the groove and is therefore critical in preventing dislocation. Distal to the fibular tip, the tendons are separated by the calcaneal peroneal tubercle and each tendon enters an individual fibrous tunnel. A cadaveric study found a prominent peroneal tubercle in 29% of specimens and this may lead to pain [8]. The PB inserts at the fifth metatarsal base. The PL, after turning plantarly at the cuboid groove, inserts at the medial cuneiform and first metatarsal base. Within the cuboid groove, an os peroneum is found in up to 4–30% [9, 10]. It protects the PL from damage at the level where it redirects medially, but has also been associated with pathology [9, 10].

The superficial peroneal nerve innervates both tendons and blood is supplied by branches of the peroneal artery and anterior tibial artery running through common vincula [11].

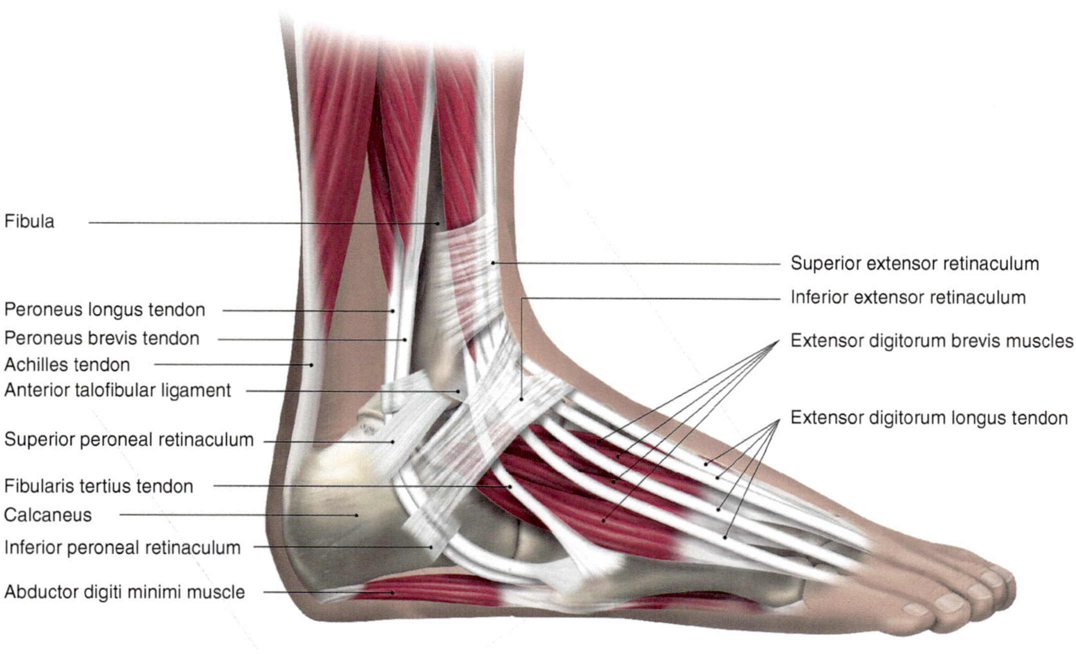

Fig. 22.2 Anatomy of the lateral ankle including the peroneal tendons

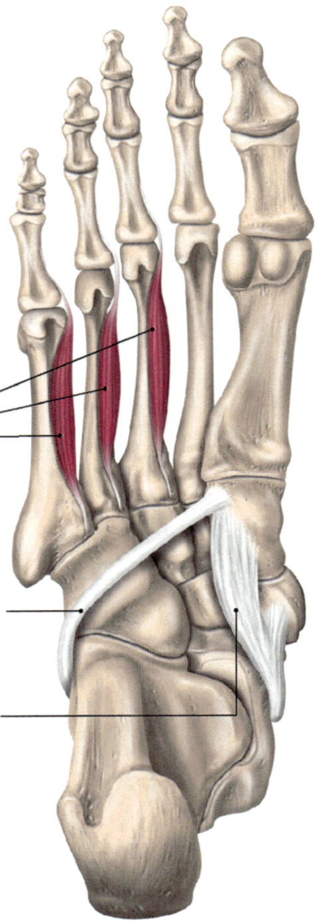

Fig. 22.3 Anatomy of the plantar side of the foot including the insertion of the peroneus longus tendon

22.3 Epidemiology

In track and field athletes, incidence rates between 43% and 76% have been reported, with a strong dominance of overuse-related conditions [1, 12]. A study among 321 Swedish track and field athletes found an overall 1-year injury prevalence of 43% with 12% occurring in the foot and ankle [12].

Achilles tendinopathy is one of the most frequent foot and ankle overuse injuries in the active population. It is known to affect 9% of recreational runners and has a cumulative lifetime incidence of 24% in the athletic population and 52% in runners. In up to 5% of professional athletes, it can lead to a career-ending injury [13–15].

22.4 Sports Dynamics and Related Physical Demands of the Calf and Peroneal Tendons

Track and field injuries of the calf, Achilles tendon, and peroneal tendons, mostly chronic overuse injuries, can often be related to sports specifics biodynamical aspects such as excessive loading, rapid transitions on uneven ground, acceleration, and middle to long-distance running, which will be discussed below.

22.4.1 Excessive Loading

The stretch-shortening cycle is commonly observed during running and jumping (Fig. 22.4). The stretch-shortening cycle refers to the prestretch or countermovement action and allows the athlete to produce more force and move quicker due to a combination of active state and storage of elastic energy within the tendon [16]. During the stretch-shortening cycle, the Achilles tendon is subjected to tensile loads up to ten times its body weight [17]. Moreover, running and jumping (on uneven grounds) require high-energy storage loading within the Achilles and peroneal tendons. With excessive loading above the tendon's capacity—a phenomenon being associated with tendinopathy—the tendons are at higher risk of injury [18].

22.4.2 Rapid, Repeated Transitions on Uneven Ground

In running, rapid and repeated transitions from pronation to supination cause the Achilles tendon to undergo a "whipping" action. This whipping action creates shear forces across the tendon, exerting a particularly high eccentric stress on the medial side of the Achilles tendon [19]. Moreover, the peroneal tendons are exposed to high mechanical loads when jumping or running on uneven ground. They remain under significant pressure within the retromalleolar groove, predisposing them to (repetitive micro) trauma.

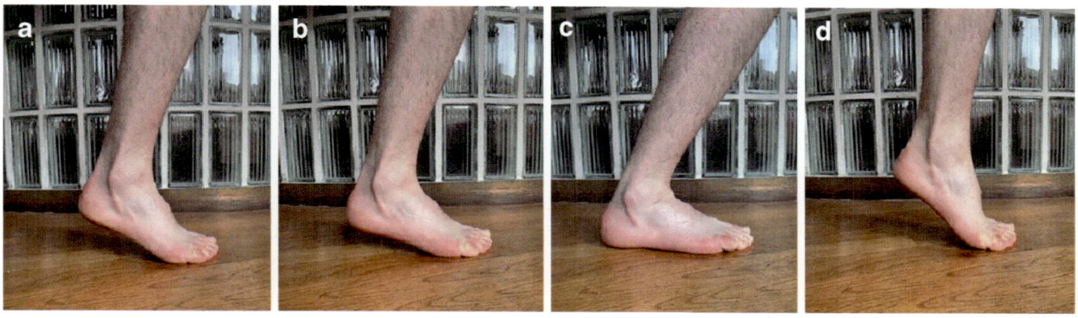

Fig. 22.4 Stretch-shortening cycle: (**a**, **b**) eccentric phase, (**c**) amortisation phase, (**d**) concentric phase

22.4.3 Acceleration

As runners accelerate, they move toward a more forefoot strike which increases the loading on the Achilles tendon. Especially in sprinters, rapid acceleration is required to achieve and maintain a very high pace. During acceleration, the calf muscle is exposed to powerful eccentric contraction and thereby prone to stretching past its capacity. This may potentially lead to partial or even full muscle tears.

22.4.4 Middle- and Long-distance Running

Middle- and long-distance runners sustain repetitive microtrauma of the Achilles tendon for longer periods. In this way, they have a great susceptibility to noninsertional tendinopathy [20].

22.5 Calf Injuries

22.5.1 Pathologies

In the calf, the gastrocnemius muscle is most prone to injury because it bridges over three joints and has a larger musculotendinous junction in comparison to the soleus muscle. Most gastrocnemius injuries occur distally, near the musculotendinous junction and happen during sudden eccentric contraction with the knee in full extension and the ankle dorsiflexed. Injuries associated with the plantaris muscle, while less common, occur in a similar way. Soleus muscle injures are far less common as the muscle solely bridges the ankle and subtalar joint. Typical trauma mechanism includes passive dorsiflexion of the ankle with a flexed knee.

For a long period of time, calf muscle injuries were classified as either muscle strains or full muscle tears. The Munich consensus statement, however, stated that a muscle strain is a biomechanical term which is not properly defined and used indiscriminately for anatomically and functionally different muscle injuries [21]. Since the use of the term strain is not recommended anymore, soleus injuries are being classified as either partial or full muscle tears.

22.5.2 Clinical Signs and Diagnostics

Typical clinical signs include sharp pain or cramping at the level of the tear, often during stretching of the calf. Moreover, swelling and ecchymosis may be visible. On palpation, possible tenderness, swelling, thickening, defects, and masses can be observed. In case of retraction, the actual rupture may be palpable although it can be difficult to differentiate from a total Achilles tendon rupture [22].

Based on the degree of knee flexion when testing ankle plantar flexion strength, injuries of the different calf muscles can be differentiated. When the knee is maximally flexed, the soleus acts as the primary plantiflexor, while the gastrocnemius is the stronger plantiflexor with the knee in full extension. Moreover, in injuries of the plantaris and the soleus, pain may be exacerbated upon weight-bearing and with passive dorsiflexion.

Although ultrasound (US) is a user-dependent diagnostic method, it is less expensive than Magnetic Resonance Imaging (MRI) and it can be employed in the outpatient clinic. Furthermore, USA has the ability to dynamically evaluate the muscle groups and differentiate partial from full ruptures. Signs of a rupture include discontinuity of the muscle, edema, hematoma, and an intramuscular fluid collection [23]. Moreover, Doppler ultrasonography can be used to evaluate hyperaemia and possible deep venous thrombosis [22]. A large hematoma can be drained during ultrasound. In professional athletes or when ultrasound is inconclusive, MRI is recommended. Also, MRI is useful during follow-up [24]. In case of a muscle strain, MRI often reveals discontinuity or rupture of the muscle, retraction of the damaged muscle fibers or a hematoma or hemorrhage within the musculotendinous junction [22].

22.5.3 Treatment

In general, calf injuries are treated nonsurgically with recovery time being highly patient-specific. Factors defining (time) to return to sports include the (transverse) location of the muscle tear (injuries of the central aponeurosis need a significantly longer recovery period than injuries in the medial or lateral aponeurosis and myofascial sites), gap or retraction length, weight, and age [25]. To prevent reinjury, complete muscle flexibility and strength should be restored before return to sports.

In the acute phase, treatment should focus on hemorrhage, pain, and prevention of complications such as a compartment syndrome. This includes a period of rest, ice, compression, and elevation. The use of NonSteroidal Anti-Inflammatory Drugs (NSAIDs) is relatively contraindicated due to the antiplatelet effects and thus possibly increasing bleeding and thereby hampering healing. Moreover, the Cyclo-oxygenase-2 (COX-2) inhibitors negatively affect the muscle's healing tendency.

22.6 Achilles Tendon Injuries

22.6.1 Pathologies

Nomenclature of Achilles tendon pathology has been much debated and is still controversial. Ever since histopathological studies have demonstrated a lack of inflammatory cells, Achilles tendinopathy is the most consensual term to describe this type of pathology resulting from a failed healing response [26]. In general, two anatomic categories can be distinguished: insertional and noninsertional tendinopathy. A third category, total Achilles tendon rupture, will not be discussed in this chapter since it is rare in track and field.

Insertional tendinopathy occurs at the level of the calcaneal-tendon junction. It sometimes involves a Haglund's deformity, retrocalcaneal bursitis or calcifications within the Achilles tendon. Noninsertional tendinopathy is located more proximal, at the hypovascular zone, 2–6 cm proximal to the insertion. It involves the tendon's substance, with or without inflammation of the paratenon.

Achilles tendinopathy is considered a multifactorial condition. Known extrinsic risk factors include excessively hard, slippery or uneven weight-bearing surfaces, inappropriate footwear, training errors, use of fluoroquinolone, and the type of exercise activity (the stretch-shortening cycle is known to increase the risk). Most relevant and correctable intrinsic risk factors include previous injury, low flexibility of the calf, and altered lower limb biomechanics [27].

22.6.2 Clinical Signs and Diagnostics

Patients typically present with pain at the level of the tendinopathy (2–6 cm proximal to the insertion vs. at the calcaneal tuberosity), swelling, and stiffness of the Achilles tendon. Exercise, climbing stairs, and running on hard surfaces may exacerbate pain. As the tendinopathy progresses, walking on flat ground and even rest may provoke pain. Some patients report pain over the posterior heel, which may cause them to struggle with shoe wear.

Physical examination is important to rule out other injuries such as a (total) tendon rupture, retrocalcaneal bursitis or stress fractures. Tenderness, pain, swelling, thickening, and crepitus may be felt at the involved portion of the tendon. In case of pain on the lateral or medial border of the insertion without tendon thickening, retrocalcaneal bursitis is more likely. Active plantarflexion against resistance may provoke pain.

Lateral weight-bearing radiographs of the foot can be used to evaluate enthesophytes, intratendinous calcifications, and a possible Haglund's deformity (suggestive for insertional tendinopathy). Moreover, the width of the Achilles's shadow and Kager's fat pad triangle can be evaluated. MRI may show tendon thickening, degenerative changes, retrocalcaneal bursitis, and the impact of the Haglund's deformity. Recent studies have shown equal or even better accuracy using (Doppler) ultrasonography when compared to MRI, as the pain in Achilles tendinopathy seems to be related to areas of neovascularization [28]. Computed tomography (CT) or conventional radiographs can additionally be used in case of suspicion of Haglund's deformity and retrocalcaneal bursitis.

22.6.3 Treatment

In the early phase, 3–6 months of conservative treatment is the first step in management of Achilles tendinopathy. Precipitating factors are controlled by modifying training regimes or even complete rest. A systematic review by Rowe et al. showed strong evidence for the use of eccentric exercises and the use of low-energy shock-wave therapy to improve healing [29]. No strong evidence for use of platelet-rich plasma (PRP) was found [30]. In case conservative treatment fails, which happens in around 25–33% of the patients, surgery can be considered to remove degenerative tissue and stimulate tendon healing (Fig. 22.5).

Surgical treatment for noninsertional Achilles tendinopathy results in a success rate exceeding the 80% [31]. Open surgery with excision of the degenerative tissue and repair of normal tissue is commonly performed. In case of removing more than 50% of the tendon thickness, augmentation or reconstruction, often with the flexor hallucis longus, is recommended to minimize the risk of rupture and optimize strength. Minimally invasive procedures have been growing in popularity with reduced complication rates and faster recovery time [32]. Tendoscopy allows striping the paratenon from the tendon.

Surgical treatment of insertional tendinopathy is usually more complex. Degenerative tissue is removed, combined with excision of intratendinous calcification or degenerative tendon above the insertion, excision of the inflamed retrocalcaneal bursa or resection of the Haglund's deformity. In general, an open approach is required to remove all unhealthy tissue, which also allows easier augmentation when needed. Detachment up to 50% of the insertion can be safely performed. When >50% of the insertion is detached, reinsertion has been recommended [33]. For patients with a gastrocnemius contracture, a gastrocnemius release may be advised. A Haglund's deformity or retrocalcaneal bursitis is amenable to an endoscopic calcaneoplasty.

22.7 Peroneal Tendon Injuries

22.7.1 Pathologies

Due to its anatomical position within the retromalleolar groove, the PB is most prone to pathology. Pathology of the peroneal tendons may occur anywhere along their course but is most often found within areas of greatest stress: around the lateral malleolus (PB), the peroneal tubercle (PB and PL), or within the cuboid groove (PL). In general, peroneal pathology is categorized into three types: (1) tendinopathy (tendinitis, tenosynovitis, tendinosis, and stenosis), (2) partial or complete ("rupture") tears, and (3) subluxation or dislocation [34].

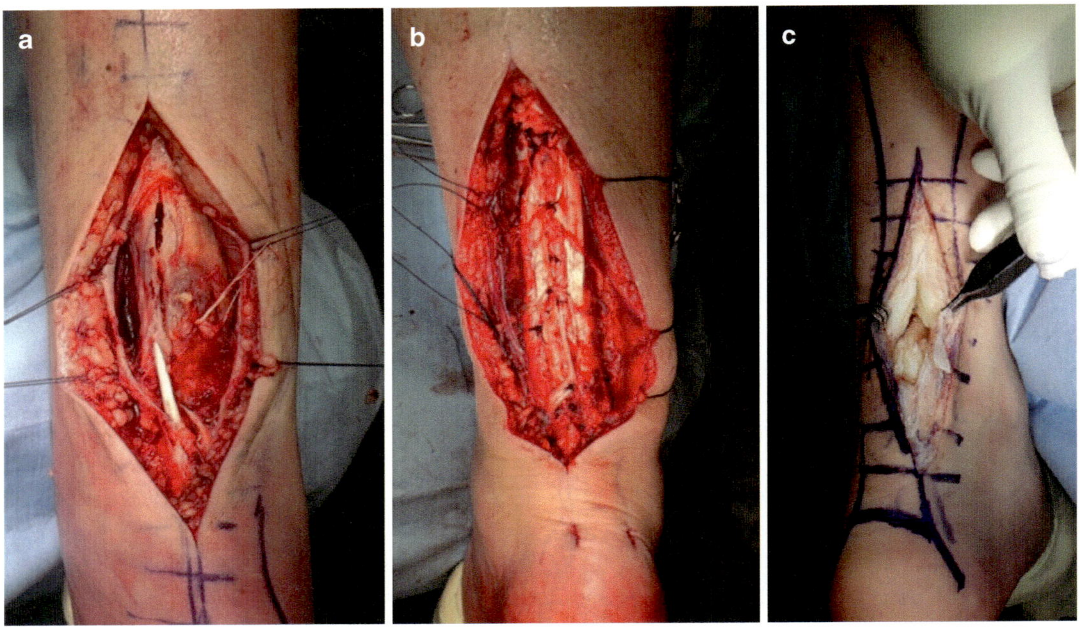

Fig. 22.5 Achilles tendinopathy. (**a**) Achilles tendon thickened with severe tendinosis. (**b**) The same tendon after extensive debridement, tabularization, proximal release, and distal reinforcement with suture anchors in the calcaneus. (**c**) A different Achilles tendon with severe tendinosis including fibrous and lipoid degeneration. *All images were kindly provided and its use authorized by Dr. Bruno Pereira*

Predisposing factors for peroneal tendon injuries include anatomical variations (i.e., low-lying muscle belly, prominent peroneal tubercle or flat retromalleolar groove), rheumatoid or psoriatic arthritis, diabetic neuropathy, calcaneal fractures, use of fluoroquinolone, and local steroid injections [35–38].

22.7.2 Clinical Signs and Diagnostics

Careful patient history and clinical examination are keys in diagnosing peroneal tendon injuries. Acute injuries are often described by the patient as "an ankle sprain that never resolved," while chronic disorders occur after a gross ankle inversion in the medical history or in patients with chronic lateral ankle ligament instability. The patient typically presents with pain along the course of the tendons that worsens upon activity. Other symptoms include swelling, tenderness, giving way, and lateral ankle instability. In case of dislocation, the patient may report a popping or snapping sensation.

Physical examination may reveal tenderness, crepitus, and swelling. Passive plantarflexion and inversion or active dorsiflexion and eversion often exacerbate pain. Muscle strength may be weaker when compared to the contralateral side. In case of tendon rupture, pain is exacerbated on acute loosening of resistance during the provocation test [34]. Dislocation of the tendons can often be provoked by combined active dorsiflexion and eversion. During physical examination, differentiation between tendinopathy and a tendon's tear is challenging; a tendon tear may appear with less pain but more weakness and swelling.

To rule out osseous pathologies such as fractures, spurs or calcifications, weight-bearing radiographs in anteroposterior and lateral direction are recommended. Moreover, in case of peroneal tendon dislocation, a small avulsion fracture of the lateral malleolus ("fleck sign") may be visible on the anteroposterior view [39]. MRI remains the standard diagnostic test [40]. Abnormalities include a C-shaped tendon, clefts, irregularity of the tendon's contour, and increased signal intensity due to fluid within the tendon sheath [40, 41]. An increased signal intensity, however, can also be seen in asymptomatic patients due to the so called magic angle effect [41]. While this effect only appears on T1-weighted images, in tears these signal abnormalities are found on both T1- and T2-weighted images. This underscores the importance of evaluating the tendons in both settings. Ultrasound is especially useful in detecting dynamic injuries such as (episodic) subluxation, dislocation, and tears that are not seen on MRI. Ultrasonic abnormalities include tendon thickening, peritendinous fluid within the tendon sheath, ruptures, and dislocation of the tendons over the fibular tip.

22.7.3 Treatment

With only limited evidence, nonsurgical management is the first step in treatment of peroneal tendon injuries, including a period of rest, activity modification or immobilization to reduce symptoms [34]. Physical therapy is recommended to strengthen the peroneal- and surrounding muscles. When symptoms persist longer than 3 months, there is, at least, some evidence for the use of shock-wave therapy [34]. If nonsurgical treatment fails, surgery should be considered.

Especially in tears and dislocation, surgery is required in most cases since these pathologies rarely heal themselves [42]. According to ESSKA-AFAS's peroneal tendon consensus statement, first choice in surgical treatment of peroneal tendon tears includes debridement and tubularization of one or both tendons. In cases this is clinically not feasible, single-stage autograft with the hamstrings, or tenodesis is recommended. If one of the tendons is deemed irreparable, it is recommended to perform debridement and tubularization on the reparable tendon and autograft or tenodesis of the irreparable tendon. If neither of the tendons can be repaired and the proximal muscle tissue is healthy, single-stage autograft is recommended [34]. Inadequate management of anatomical abnormalities may lead to persistent pain and dysfunction so additional predisposing factors should simultaneously be assessed.

In treatment of dislocation within athletes, evidence showed that the combination of retinaculum repair and retromalleolar groove deepening provides significant higher return to sports rates as compared to retinaculum repair alone [43].

Over the last years, peroneal tendoscopy has become more appreciated as diagnostic and treatment modality [44, 45]. It should be reserved for patients with a high clinical suspicion of peroneal pathology, though with absence of positive findings or inconclusive abnormalities on imaging [46]. Peroneal tendoscopy is highly sensitive and specific for both static and dynamic injuries and provides easy transition to (minimally invasive) treatment [46], with a relatively low complication rate, low costs, and earlier recovery when compared to open procedures [46, 47].

22.7.4 Rehabilitation

Adequate rehabilitation is a key for optimal management of peroneal tendon injuries and should be individualized for each patient [48]. Importantly, the surgeon must distinguish whether or not the SPR was repaired during surgical treatment. When the retinaculum was not repaired, rehabilitation should be goal-based with the promotion of early mobilization, rather than time-based. In case surgery included repair of the SPR, rehabilitation should start with 2 weeks of nonweight-bearing in a lower leg cast, followed by active range of motion and 4 weeks of weight-bearing in a cast or walker boot. It is important that the tendons are not loaded until 6 weeks after repair of the SPR [34].

22.8 Injury Prevention

As track and field injuries often result from training errors, it is important to identify and modify them [49]. A training routine record can be kept to identify recent changes susceptible to have caused the injury, such as a sudden increase in load, different training surface or equipment, and change in intensity and frequency.

Repetitive microtrauma caused by impact, whether by prolonged running or higher but shorter jumping loads, is key in most common athletics injuries. Moreover, tendon injuries are more likely to occur with an increase in training pace rather than volume [49]. The type of surface has an impact on injury pattern and incidence. While asphalt decreases the incidence of (overuse) noninsertional Achilles tendinopathy, sand increases it [20]. Throughout the years, much time and money has been invested to improve performance while reducing injury risk by incorporating improved shock absorption mechanisms into sportswear and surfaces. New foot-wear materials such as Kevlar, foam-blown polyurethane or thermoplastic polyurethane allow more comfort and lower weight. Moreover, different designs have been adapted for sports-specific performance improvement like small spikes on heel and front for high jump, just in front for long jump and sprinting or high flexibility for sprinting. Also comfort is a real priority. Cushioning is developed for maximal shock absorption. Most advanced systems have built in a spring-like mechanism at the base of the heel, allowing for a large portion of the impact to be transferred into the spring, putting much lower strain on the joints.

Malalignment of the foot increases the incidence of injuries and may be compensated by orthotics. Insoles to slightly elevate the heel and individualized footwear might be helpful in order to maintain sports activity in athletes with insertional tendinopathy. Decreased ankle flexibility and muscle weakness may be treated by appropriate physiotherapy.

Before starting an event or training, stretching is important to ensure sufficient range of motion to perform optimally and decrease muscle stiffness (or increase muscle compliance). Theoretically, the risk of injury is thereby decreased. Stretching is therefore intended to enhance performance while decreasing the risk of injury.

22.9 Conclusion

Track and field puts high mechanical loads on the calf and peroneal tendons, making them prone to (overuse) injuries. Patient history and physical examination are the keys to accurate diagnosis. In general, adequate conservative treatment should be attempted before surgery. Since track and field injuries often occur as a result of training errors, the most important step in prevention is to identify and modify these errors.

References

1. D'Souza D. Track and field athletics injuries--a one-year survey. Br J Sports Med. 1994;28(3):197–202.
2. Sanchis-Gimeno JA, Casas-Roman E, Garcia-Campero C, Hurtado-Fernandez R, Aparicio-Bellver L. Anatomical location of athletic injuries during training: a prospective two year study in 2701 athletes. Br J Sports Med. 2005;39(7):467.
3. Kelikian AS, Sarrafian SK. Sarrafian's anatomy of the foot and ankle. 3rd ed. Philadelphia: Wolters Kluwer; 2011.
4. Edama M, Kubo M, Onishi H, Takabayashi T, Inai T, Yokoyama E, et al. The twisted structure of the human Achilles tendon. Scand J Med Sci Sports. 2015;25(5):e497–503.
5. Carr AJ, Norris SH. The blood supply of the calcaneal tendon. J Bone Joint Surg Br. 1989;71(1):100–1.
6. van Sterkenburg MN, van Dijk CN. Mid-portion Achilles tendinopathy: why painful? An evidence-based philosophy. Knee Surg Sports Traumatol Arthrosc. 2011;19(8):1367–75.
7. Edwards M. The relations of the peroneal tendons to the fibula, calcaneus,and cuboideum. Am J Anat. 1928;42:213–53.
8. Saupe N, Mengiardi B, Pfirrmann CW, Vienne P, Seifert B, Zanetti M. Anatomic variants associated with peroneal tendon disorders: MR imaging findings in volunteers with asymptomatic ankles. Radiology. 2007;242(2):509–17.
9. Sobel M, Pavlov H, Geppert MJ, Thompson FM, DiCarlo EF, Davis WH. Painful os peroneum syndrome: a spectrum of conditions responsible for plantar lateral foot pain. Foot Ankle Int. 1994;15(3):112–24.
10. Stockton KG, Brodsky JW. Peroneus longus tears associated with pathology of the os peroneum. Foot Ankle Int. 2014;35(4):346–52.
11. van Dijk PA, Madirolas FX, Carrera A, Kerkhoffs GM, Reina F. Peroneal tendons well vascularized: results from a cadaveric study. Knee Surg Sports Traumatol Arthrosc. 2016;24:1140–7.
12. Jacobsson J, Bergin D, Timpka T, Nyce JM, Dahlstrom O. Injuries in youth track and field are

perceived to have multiple-level causes that call for ecological [holistic-developmental] interventions: a national sporting community perceptions and experiences. Scand J Med Sci Sports. 2018;28(1):348–55.
13. Kujala UM, Sarna S, Kaprio J. Cumulative incidence of achilles tendon rupture and tendinopathy in male former elite athletes. Clin J Sport Med. 2005;15(3):133–5.
14. Lysholm J, Wiklander J. Injuries in runners. Am J Sports Med. 1987;15(2):168–71.
15. Maffulli N, Wong J, Almekinders LC. Types and epidemiology of tendinopathy. Clin Sports Med. 2003;22(4):675–92.
16. Nicol C, Avela J, Komi PV. The stretch-shortening cycle: a model to study naturally occurring neuromuscular fatigue. Sports Med. 2006;36(11):977–99.
17. O'Brien M. The anatomy of the Achilles tendon. Foot Ankle Clin. 2005;10(2):225–38.
18. Girdwood M, Docking S, Rio E, Cook J. Pathophysiologie of tendinopathy. In: Canata GL, d'Hooghe P, Hunt KJ, editors. Muscle and tendon injuries. New York: Springer; 2017.
19. Van Ginckel A, Thijs Y, Hesar NG, Mahieu N, De Clercq D, Roosen P, et al. Intrinsic gait-related risk factors for Achilles tendinopathy in novice runners: a prospective study. Gait Posture. 2009;29(3):387–91.
20. Knobloch K, Yoon U, Vogt PM. Acute and overuse injuries correlated to hours of training in master running athletes. Foot Ankle Int. 2008;29(7):671–6.
21. Mueller-Wohlfahrt HW, Haensel L, Mithoefer K, Ekstrand J, English B, McNally S, et al. Terminology and classification of muscle injuries in sport: the Munich consensus statement. Br J Sports Med. 2013;47(6):342–50.
22. Campbell JT. Posterior calf injury. Foot Ankle Clin. 2009;14(4):761–71.
23. Hayashi D, Hamilton B, Guermazi A, de Villiers R, Crema MD, Roemer FW. Traumatic injuries of thigh and calf muscles in athletes: role and clinical relevance of MR imaging and ultrasound. Insights Imaging. 2012;3(6):591–601.
24. Radice F, Velasquez F, Orizola A. Triceps surae injuries. In: Canata GL, d'Hooghe P, Hunt KJ, editors. Muscle and tendon injuries. New York: Springer; 2017.
25. Pedret C, Rodas G, Balius R, Capdevila L, Bossy M, Vernooij RW, et al. Return to play after soleus muscle injuries. Orthop J Sports Med. 2015;3(7):2325967115595802.
26. Maffulli N, Khan KM, Puddu G. Overuse tendon conditions: time to change a confusing terminology. Arthroscopy. 1998;14(8):840–3.
27. Munteanu SE, Barton CJ. Lower limb biomechanics during running in individuals with achilles tendinopathy: a systematic review. J Foot Ankle Res. 2011;4:15.
28. Khan KM, Forster BB, Robinson J, Cheong Y, Louis L, Maclean L, et al. Are ultrasound and magnetic resonance imaging of value in assessment of Achilles tendon disorders? A two year prospective study. Br J Sports Med. 2003;37(2):149–53.
29. Rowe V, Hemmings S, Barton C, Malliaras P, Maffulli N, Morrissey D. Conservative management of midportion Achilles tendinopathy: a mixed methods study, integrating systematic review and clinical reasoning. Sports Med. 2012;42(11):941–67.
30. Abat F, Alfredson H, Cucchiarini M, Madry H, Marmotti A, Mouton C, et al. Current trends in tendinopathy: consensus of the ESSKA basic science committee. Part II: treatment options. J Exp Orthop. 2018;5(1):38.
31. Khan WS, Malvankar S, Bhamra JS, Pengas I. Analysing the outcome of surgery for chronic Achilles tendinopathy over the last 50 years. World J Orthop. 2015;6(6):491–7.
32. Maquirriain J. Surgical treatment of chronic achilles tendinopathy: long-term results of the endoscopic technique. J Foot Ankle Surg. 2013;52(4):451–5.
33. Maffulli N, Testa V, Capasso G, Sullo A. Calcific insertional Achilles tendinopathy: reattachment with bone anchors. Am J Sports Med. 2004;32(1):174–82.
34. van Dijk PA, Miller D, Calder J, DiGiovanni CW, Kennedy JG, Kerkhoffs GM, et al. The ESSKA-AFAS international consensus statement on peroneal tendon pathologies. Knee Surg Sports Traumatol Arthrosc. 2018;26(10):3096–107.
35. Truong DT, Dussault RG, Kaplan PA. Fracture of the os peroneum and rupture of the peroneus longus tendon as a complication of diabetic neuropathy. Skelet Radiol. 1995;24(8):626–8.
36. Vainio K. The rheumatoid foot. A clinical study with pathological and roentgenological comments. 1956. Clin Orthop Relat Res. 1991;(265):4–8.
37. Wright DG, Sangeorzan BJ. Calcaneal fracture with peroneal impingement and tendon dysfunction. Foot Ankle Int. 1996;17(10):650.
38. Younger AS, Hansen ST Jr. Adult cavovarus foot. J Am Acad Orthop Surg. 2005;13(5):302–15.
39. Church CC. Radiographic diagnosis of acute peroneal tendon dislocation. AJR Am J Roentgenol. 1977;129(6):1065–8.
40. Park HJ, Cha SD, Kim HS, Chung ST, Park NH, Yoo JH, et al. Reliability of MRI findings of peroneal tendinopathy in patients with lateral chronic ankle instability. Clin Orthop Surg. 2010;2(4):237–43.
41. Wang XT, Rosenberg ZS, Mechlin MB, Schweitzer ME. Normal variants and diseases of the peroneal tendons and superior peroneal retinaculum: MR imaging features. Radiographics. 2005;25(3):587–602.
42. Redfern D, Myerson M. The management of concomitant tears of the peroneus longus and brevis tendons. Foot Ankle Int. 2004;25(10):695–707.
43. van Dijk PA, Gianakos AL, Kerkhoffs GM, Kennedy JG. Return to sports and clinical outcomes in patients treated for peroneal tendon dislocation: a systematic review. Knee Surg Sports Traumatol Arthrosc. 2016;24(4):1155–64.
44. Sammarco VJ. Peroneal tendoscopy: indications and techniques. Sports Med Arthrosc Rev. 2009;17(2):94–9.

45. Scholten PE, van Dijk CN. Tendoscopy of the peroneal tendons. Foot Ankle Clin. 2006;11(2):415–20.
46. Kennedy JG, van Dijk PA, Murawski CD, Duke G, Newman H, DiGiovanni CW, et al. Functional outcomes after peroneal tendoscopy in the treatment of peroneal tendon disorders. Knee Surg Sports Traumatol Arthrosc. 2016;24(4):1148–54.
47. Jerosch J, Aldawoudy A. Tendoscopic management of peroneal tendon disorders. Knee Surg Sports Traumatol Arthrosc. 2007;15(6):806–10.
48. van Dijk PA, Lubberts B, Verheul C, DiGiovanni CW, Kerkhoffs GM. Rehabilitation after surgical treatment of peroneal tendon tears and ruptures. Knee Surg Sports Traumatol Arthrosc. 2016;24(4):1165–74.
49. Nielsen RO, Buist I, Sørensen H, Lind M, Rasmussen S. Training errors and running related injuries: a systematic review. Int J Sports Phys Ther. 2012;7(1):58–75.

23 Bunions, Hallux Rigidus, Turf Toe, and Sesamoid Injury in the Track and Field Athlete

Kenneth J. Hunt and Mark W. Bowers

23.1 Introduction

Injuries to the hallux metatarsophalangeal joint complex are common in the track and field athlete and can result in deformity, chronic pain, and a decline in performance. In addition, underlying alignment-related conditions can place athletes at risk for pain, weakness, and functional deficits impacting sport. A thorough understanding of the anatomy and pathomechanics of first metatarsophalangeal (MTP) joint is crucial for managing athletes with great toe injuries and disorders. Early injury recognition and implementation of appropriate management strategies can help these athletes return to their selected activities safely and expeditiously.

K. J. Hunt (✉)
Department of Orthopedics, University of Colorado School of Medicine, Aurora, CO, USA

UCHealth Steadman Hawkins Clinic, Denver, CO, USA
e-mail: Kenneth.J.Hunt@cuanschutz.edu

M. W. Bowers
Department of Orthopedics, University of Colorado School of Medicine, Aurora, CO, USA

23.1.1 Anatomy and Biomechanics

23.1.1.1 Anatomy of the Hallux MTP Joint Complex

The round metatarsal head articulates with the concave elliptical base of the proximal phalanx and allows for plantarflexion, dorsiflexion, and to a limited degree abduction and adduction. Unlike the lesser toes, the first MTP joint has a sesamoid mechanism running on the plantar aspect of the joint. The bony articulation overall provides little stability to the joint. Instead the majority of the joint stability comes from the capsular-ligamentous-sesamoid complex [1]. The medial and lateral collateral ligaments help stabilize the metatarsal (MT) head and proximal phalanx articulation. These fan-shaped ligaments originate from the medial and lateral epicondyle of the MT head and run distal and plantar, interdigitating with the metatarsosesamoid ligaments which fan out plantarly to the margin of the sesamoids and plantar plate [2].

The strong, fibrous plantar plate is a confluence of structures including the two tendons of the flexor hallucis brevis (FHB), the abductor and adductor hallucis, the plantar aponeurosis, and the joint capsule. This structure is firmly attached to the proximal phalanx and only loosely attached to the neck of the MT via the capsule [2]. The FHB tendons run along the plantar aspect of the first MTP joint and encase the sesamoids prior to inserting on the proximal

G. L. Canata et al. (eds.), *Management of Track and Field Injuries*,
https://doi.org/10.1007/978-3-030-60216-1_23

phalanx. The intersesamoid ligament tethers the two sesamoids together and helps to maintain the course of the flexor hallucis longus (FHL) tendon. In addition, the extensor hallucis brevis tendon, and the adductor and abductor hallucis tendons insert and blend into the capsular-ligamentous-sesamoid complex and contribute to the overall stability of the hallux [3].

23.1.1.2 Biomechanics

The first MTP joints support approximately twice the load of the lesser toes and can see forces up to 40–60% of body weight during normal gait [4]. During activities such as jogging or running, peak forces can reach two to three times body weight and up to eight times body weight when a running jump is performed [5].

Range of motion of the first MTP is highly variable and decreases as we age. Normal motion of the hallux MTP joint is 85° dorsiflexion and 40° plantarflexion [6]. During the pushoff phase of the gait cycle, the hallux has been found to dorsiflex from 60° to 84° [7, 8]. A study by Bowman showed that athletes can accommodate up to 50% reduction in MTP joint motion after an acute injury through various gait adjustments [9].

23.2 Bunions and Hallux Valgus

Hallux valgus and metatarsus primus varus are common in running athletes and can be progressive. Symptoms related to hallux valgus deformity range from simple pain with shoe wear, to loss of pushoff strength, transfer metatarsalgia or callus from abnormal weightbearing distribution, and resultant decreased athletic performance. The hallux valgus deformity in the athlete is no different than the deformity in the nonathletic population with likely causes including a genetic predisposition which may be worsened with improper shoe wear. The deformity is characterized by a lateral deviation of the hallux and adduction of the first MT with a resultant increase of the first intermetatarsal (IM) angle. As the deformity progresses, the sesamoid complex no longer sits beneath the metatarsal head, resulting in a less functional windlass mechanism, decreased medial longitudinal arch stability, and diminished pushoff strength. Subsequently, the weightbearing distribution of the foot transfers from the first metatarsal to the second, which can cause metatarsalgia with callus formation underneath the lesser metatarsal heads [3]. In athletes, the primary disability is typically related to the prominent medial eminence, which rubs on their shoe resulting in skin irritation or callus formation. With advanced deformity, the hallux pronates leading to concentrated weightbearing on the medial aspect of the metatarsal head and potential compression of the dorsal cutaneous nerve, causing pain and a loss of functional pushoff strength.

23.2.1 Conservative Management

Initial management of athletes with symptomatic hallux valgus begins with conservative management. Identifying the specific area of pain is necessary to help guide treatment. The running athlete often complains of pain directly over the medial eminence as well as symptoms related to compression of the dorsal medial cutaneous nerve. The athlete's shoes should be carefully evaluated and preferably would have a wide-toe box. Increasing the shoe size to ½ to 1 size larger may be required to accommodate for the medial eminence. In general, there should be approximately 1 cm of space beyond the toes to allow them to move freely [10]. The seams of the shoe should be evaluated as they may cross the medial eminence and could cause increased pressure [11]. This may necessitate changing to shoes with a different seam pattern or altering the seam configuration on the current shoe. If the shoes are found to be of adequate size, the shoes may be stretched or a balloon patch may be utilized to help alleviate pressure over the prominence. Shoe stiffeners, such as a carbon fiber footplate, can be used to help decrease the forces across the first MTP joint as long as it does not affect the athlete's performance [11].

In addition to shoe modifications, accommodative or corrective foot orthotics can be prescribed to correct any malalignment of the foot and to help distribute the concentration of pressure that causes pain. The main goals of the orthotic should be to support the medial arch, correct the forefoot pronation, and offload the lesser metatarsal heads with a metatarsal pad or bar [10]. However, care should be taken when prescribing an orthotic, as adding an orthotic will take up space within the shoe, potentially exacerbating tightness and pressure over the bunion. In addition, elite runners can be sensitive to modifications of their footwear and other issue may arise with improper orthotic fit. A cutout pad may also be used to help offload the medial eminence, but it is important that the pad is not placed directly over the eminence as this may result in increased pressure and pain. The use of a toe spacer between the first and second toes may also be helpful.

Stretching exercises should be emphasized to the athlete and can be incorporated into their warmup routine. Since hallux valgus is associated with gastrocnemius and Achilles contractures, stretching of the Achilles tendon and plantar fascia has been shown to reduce the strain across the forefoot [11]. Improving toe function can help alleviate midfoot pain and can be achieved with intrinsic toe strengthening exercises such as towel gathering, toe splaying, and purposely flexing toes while walking.

23.2.2 Surgical Treatment

The decision to operate on athletes with symptomatic hallux valgus should be made with caution and the operative procedure should be selected carefully. Surgery may reduce the first MTP joint ROM, which can significantly decrease the competitiveness of athletes that requires extreme dorsiflexion of the first MTP joint, such as with sprinters. The need for excessive dorsiflexion is less important in the middle- and long-distance runners. It is important to keep in mind the significant demands placed on the foot during running and jumping activities. The increased forces can result in tremendous strain across the forefoot. When arthrodesis procedures are performed, such as a Lapidus, the stress transfer to the surrounding joints of the midfoot and forefoot is greatly exaggerated [12].

In the setting of an acute post-traumatic hallux valgus, athletes may undergo repair of the medial collateral ligament and capsule [13]. In some cases, partial release of the lateral structures is also indicated in order to restore the normal hallux alignment. An untreated traumatic hallux valgus injury could result in alterations of the first MTP mechanics and joint reactive forces can limit power and lead to early-onset arthritis and functional decline [14].

With chronic hallux valgus, as in the nonathletic population, radiographs should be analyzed to determine the hallux valgus angle, intermetatarsal angle, joint congruency, distal metatarsal articular angle, and evidence of arthrosis. The degree of deformity is used to guide the selection of the bunion surgery. For athletes with mild to moderate deformity that is affecting performance and/or competitiveness, a distal chevron osteotomy is recommended. In patients with a subluxated metatarsophalangeal joint, a distal soft tissue procedure is required. Lillich et al. described two world class middle-distance and marathon runners who underwent distal chevron bunionectomies and neuroma removals who were able to return to world class caliber running [12].

A proximal osteotomy is more effective in reducing larger angular deformity as well as sagittal plane deformity (i.e., elevated first ray) and is recommended in cases of moderate to severe deformity. However, traditional proximal osteotomies come with additional concerns and complication rates. These osteotomies have been considered to be unstable and have been associated with delayed healing, malunion, shortening of the first metatarsal, and necessitate longer postoperative immobilization [15, 16]. The use of a proximal rotational metatarsal osteotomy (PROMO) is a new technique that corrects the first metatarsal adductus and pronation through a single oblique osteotomy (Fig. 23.1). The angle of the osteotomy is determined through radiographic measurement of the intermetatarsal angle and the metatarsal rotation angle [17]. In addition to correction of the rotational deformity, this osteotomy limits the amount of shortening of the first metatarsal. With all of these procedures, it is crucial to adequately correct both the bony and soft tissue components of the deformity to lessen the chances of recurrence.

There have been mixed results in the literature regarding first metatarsophalangeal arthrodesis in the athletic population. This procedure, also known as a Lapidus, is typically reserved for patients with severe hallux valgus deformity with a hypermobile first tarsometatarsal joint (TMT) or a degenerative TMT joint. As discussed earlier, fusion of this joint increases the stress across the midfoot and forefoot. However, MacMahon et al. reported promising results in 48 athletes who underwent a Lapidus procedure with a mean follow-up of 2.8 years. The study included only subjective findings, and reported 81% of the patients being satisfied with their return to activities and 80% being able to participate in their previous sports [18]. In contrast, McInnes and Bouche published a retrospective study on outcomes of the Lapidus procedure and had less favorable outcomes. Thirty-two feet in 25 patients were included with a mean follow-up time of 3.3 years with athletes demonstrating a lower return to activity with only 30% returning to their previous level of activity [19]. Mann does not recommend this procedure be performed in the active athletic population [11].

Postoperatively, patients are placed into a bunion dressing holding the hallux in proper alignment for a period of 6 to 8 weeks. For distal osteotomies, patients are placed into a postoperative shoe and begin hallux ROM at 3–5 days postop. In patients undergoing a proximal osteotomy, the foot is protected in a nonweightbearing splint for 2 weeks before transitioning to a sandal or boot.

Regardless of which bunion procedure is performed, the athlete will require a period of physical therapy to regain ROM as well as adequate time to allow the osteotomies to heal and the soft tissues to mature. Return to activity varies depending on the procedure performed and the physical demands of the athlete. Saxena reported the average return to activity for athletes (RTA) time with a distal Chevron procedure. Athletes were defined as those engaged in 6 or more hours of sports-specific activity/week, running 25 miles/week, varsity high school, college, or professional sports, and averaged 8.9 weeks before returning to the athletes desired sport [20]. Giotis et al. published a prospective analysis measuring both subjective and objective out-

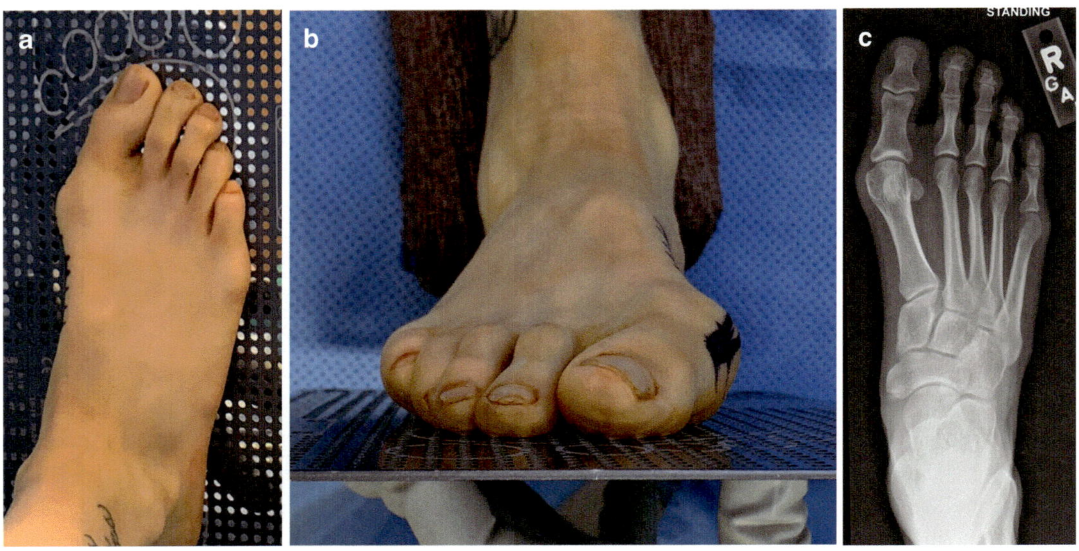

Fig. 23.1 (**a** and **b**) Photographs of foot in 22-year-old collegiate track athlete with hallux valgus impacting performance. Note the valgus alignment and pronated toe position. (**c**) Anteroposterior foot radiograph demonstrating an increased hallux valgus angle (HVA) and intermetatarsal angle (IMA)

comes of the modified Chevron osteotomy for the treatment of mild to moderate hallux valgus deformity in the female athlete. The athletes were allowed to bear weight at 2 weeks postoperatively and returned to their desired level of activities at 12 weeks [21].

In general, it is best to avoid surgery for hallux valgus in a competitive athlete. However, if the deformity creates functional deficits and pain that are not corrected with conservative measures, correcting the deformity can be a reasonable approach. It is important to explain to the athlete that not everyone is able to return to their previous level of competition following surgery.

23.3 Hallux Rigidus

Dorsal impingement of the first metatarsophalangeal joint is referred to as Hallux rigidus and is the most common pathology affecting the first MTP joint in the athletic population [22–25]. In the early stages of the condition, athletes typically complain of pain only at extremes of motion; however, with progression of degenerative changes, midrange of motion becomes painful [26]. The natural history of the condition involves cartilage degeneration with dorsal osteophyte formation followed by progressive degenerative changes throughout the entire first MTP joint [26–28]. Inability to dorsiflex the great toe leads to a decreased ability to rise onto the toes, roll through the toes, and can make running difficult and painful [27].

The exact etiology for the development of hallux rigidus remains in question; however, there are numerus potential causes. There are several anatomic and structural factors that may lead to hallux rigidus including a flat or pronated foot, a long first metatarsal or hallux, a flat or chevron-shaped metatarsal head, hallux valgus, hypermobility of the first ray, and metatarsus adductus [24, 29]. The condition may also be a result of a traumatic injury, such as a turf toe injury, or an osteochondral lesion [24]. In the running athlete, overuse and repetitive dorsiflexion forces may lead to chondral lesions and other occult injuries [30].

Hallux rigidus is graded radiographically based on the degenerative changes in the first MTP joint. Grade I hallux rigidus is characterized by mild to moderate dorsal osteophyte formation with preservation of the joint space. Grade II hallux rigidus involves moderate osteophyte formation with evidence of joint space narrowing and subchondral sclerosis. Grade III changes demonstrate significant osteophyte formation with severe loss of the first MTP joint space and subchondral cyst formation [26].

Normal range of motion (ROM) of the first MTP joint is approximately 40° of plantarflexion and 85° of dorsiflexion. The track and field athlete requires greater dorsiflexion (~80–100°) due to increased stride length while running, a prolonged propulsive phase of gait, and the greater ROM required for pushoff during activities such as jumping [23, 31]. Limitations in range-of motion can cause significant disability in athletes. This is particularly true in sprinters who require extreme ROM and to a lesser degree in middle- to long-distance runners who require less ROM. With progression of the condition, the interphalangeal (IP) joint will often compensate with hyperextension. This hyperextension may force the toe nail into the toe box of the shoe resulting in nail changes or subungual hematoma [31].

23.3.1 Conservative Management

Initial nonoperative management of hallux rigidus should be aimed at pain relief. Nonsteroidal anti-inflammatory drugs (NSAIDs) may be used to help alleviate acute episodes of pain. Similarly, injections of corticosteroids may provide temporary relief, but should be avoided, if possible [22]. The role of injectable viscosupplementation and biological agents has not been demonstrated in the literature [32]. Activity modification is often not a practical option for high-level athletes.

Shoes with wide and deep toe boxes are helpful in preventing compression of dorsal osteophytes. Additional modifications with a balloon patch over the bony prominence can be made to the shoes to further alleviate pressure on the toe. The use of a rigid Morton's extension footplate can be used to limit dorsiflexion and subsequent first MTP dorsal impingement [33]. Although these may improve pain symptoms, the reduced ROM may limit performance in the elite runner. Similarly, rigid shoes with a rocker bottom sole can limit ROM of the first MTP and improve pain, but may not be tolerated by the track and field athlete due to the added weight and excessive stiffness. Taping techniques to limit motion at the first MTP joint can help to provide pain relief, but may cause skin problems such as blistering [1].

23.3.2 Surgical Treatment

There are various surgical options for symptomatic athletes who have failed conservative management. The most common surgical interventions include cheilectomy, arthroscopic cheilectomy, interposition arthroplasty, synthetic cartilage implant (SCI), and arthrodesis.

The cheilectomy procedure was first described by Mann in 1979 and involves resection of both the dorsal osteophyte and the dorsal third of the metatarsal head as well as removal of any loose bodies or synovitis [34]. The procedure increases dorsiflexion by removing the bony impingement lesions and, additionally, removes the prominence associated with painful shoe pressure. Indications for cheilectomy are early stage (grade I and II) hallux rigidus. A lateral radiograph should demonstrate preserved joint space of the plantar half of the MTP joint and there should be an absence of pain through mid-range of motion, and a negative grind test [35].

There are multiple techniques for performing a cheilectomy including open, percutaneous, and arthroscopic. Selection of the correct technique depends on the size of the dorsal osteophyte, as well as the presence of loose bodies, lateral osteophytes, or chondral injury. A dorsal or dorsomedial incision is utilized for the open cheilectomy technique. After the extensor hallucis longus (EHL) tendon and dorsomedial cutaneous nerve are identified and protected, a synovectomy is performed as well as release of plantar adhesions, although this may not typically be necessary in athletes. The cheilectomy is performed with a goal of achieving at least 80° of dorsiflexion intraoperatively as dorsal scar formation can limit ROM in the postoperative period. This scar formation can be mitigated to some degree with early ROM in the postoperative period.

Arthroscopic and percutaneous cheilectomies are minimally invasive procedures and have been found to be associated with decreased postoperative swelling and improved postop motion [36] (Figs. 23.2 and 23.3). An arthroscopic cheilectomy is useful as it allows complete joint visualization including cartilage loss and the health of the sesamoid articulations. It also minimizes disruption to the soft tissues that can occur with open procedures, allowing early range-of-motion and resulting in less scar.

A Moberg osteotomy can be used as an adjunct to a cheilectomy and involves a dorsal closing wedge osteotomy of the proximal phalanx. The procedure translates the first MTP joint arc of motion plantarly, increasing the functional ROM and in turn, decreasing the stress on the hallux with pushoff [1]. This procedure may be helpful in the running athlete who requires increased ROM; however, decreased pushoff power can occur and should be used with caution in athletes who require increased pushoff strength, such as sprinters or jumpers.

The Valenti procedure has been shown to allow athletes increased hallux ROM. The procedure was first described in 1987 and involves a cheilectomy of the metatarsal head as well as removal of the proximal aspect of the proximal phalanx in a "V"-shaped osteotomy [37]. The procedure was later modified with less bony resection to allow for future arthrodesis or arthroplasty, if necessary [37–39]. Saxena et al. reported that the modified Valenti procedure is highly effective in the running and jumping athlete allowing 94% of athletes within the study to return to their desired level of activity [40].

Multiple reports describe successful results following cheilectomy. In 1999, Mulier et al. reported on 22 open cheilectomies in high-level

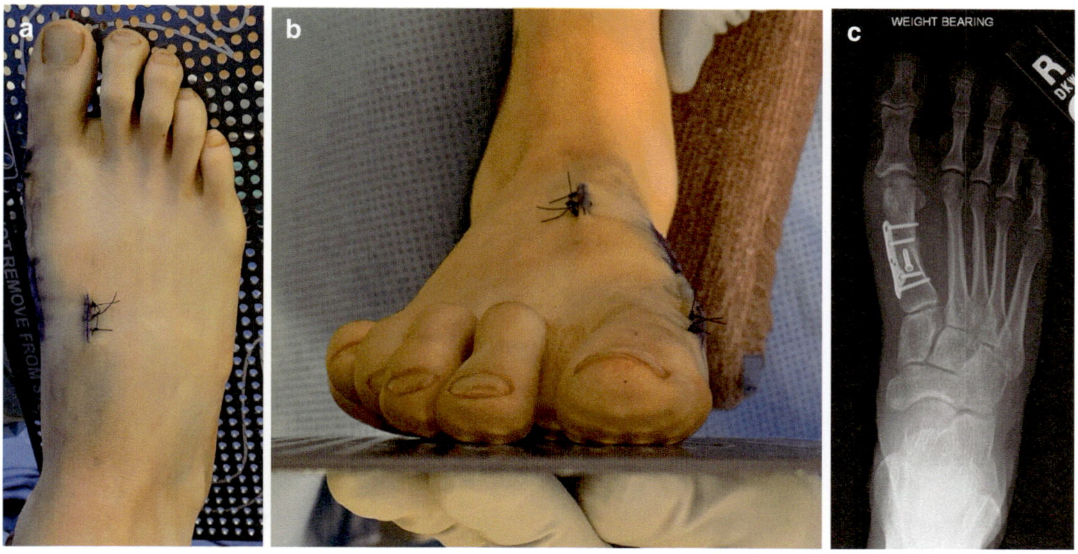

Fig. 23.2 (**a** and **b**) Immediate postoperative photographs following proximal rotational metatarsal osteotomy procedure. (**c**) Postoperative anteroposterior radiograph

Fig. 23.3 (**a**) Preoperative and (**b**) postoperative anteroposterior radiographs of patient with hallux valgus and first TMT instability undergoing a Lapidus procedure

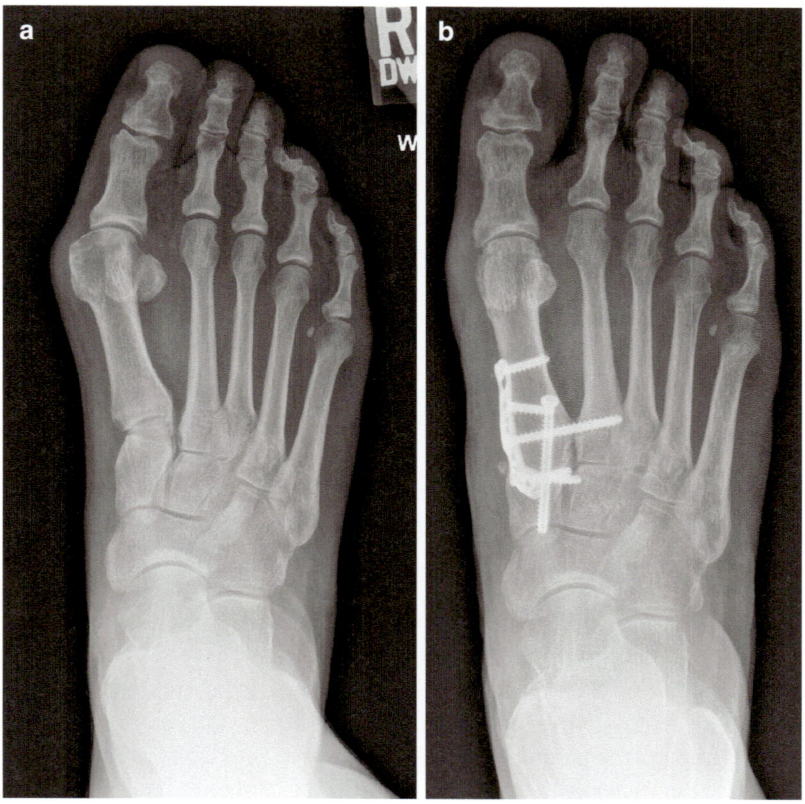

athletes with a mean follow-up of greater than 5 years achieving 90% good and excellent results [23]. Two studies examining the results of arthroscopic cheilectomy found 67% good to excellent outcomes; however, these studies both had small sample sizes [41, 42] and were conducted prior to current arthroscopic equipment and low torque bur systems. Teoh et al. concluded that minimally invasive dorsal cheilectomy resulted in improved patient-reported outcomes with minimal complications, however, 10% of patients with higher grade hallux rigidus went on to an arthrodesis [43].

The postoperative course and return to activity is fairly rapid after a cheilectomy. Saxena found that RTA in athletes who had undergone a cheilectomy or Valenti procedure were, 5.5 weeks and 6.5 weeks, respectively [20]. Mulier allowed athletes to RTA at 6 weeks postoperatively following a cheilectomy [23]. After the wounds have healed, athletes can return to training by engaging in activities that avoid significant stresses or impact to the MTP joint, such as cycling, swimming, or running in water.

A variety of MTP joint arthroplasty procedures are available for end-stage hallux rigidus with a common goal of persevering ROM and relieving pain. These include metallic implants, interposition arthroplasty, and polyvinyl alcohol hemiarthroplasty. At this time, there is limited evidence of effectiveness and longevity for MTP implants in the athletic population. The magnitude of the shear forces across the MTP joint required during running and jumping activities would put the implant at high risk for early failure and potentially leading to progressive degenerative changes and decreased athletic performance. For these reasons, arthroplasty should generally be avoided in the track and field athlete.

First MTP arthrodesis is usually considered for end-stage degenerative changes (grade III or IV) or after failure of joint-sparing procedures. This procedure should rarely be considered for first-line treatment of hallux rigidus in athletes and is best to be avoided in the sprinting athlete. If an arthrodesis must be performed in an athlete, the hallux should be fused in a position that is at least 10 mm off the ground, to help decrease the stress on the distal hallux and IP joint during activity. In addition, slight shortening of the hallux may also be of benefit as this will lessen the potential of the athlete having to vault over the hallux during running [1]. Da Cunha et al. investigated return to sports in younger patients (age range 23–55 years) following first MTP arthrodesis with a mean follow-up of 5.1 years. They found that 96% of patients in the study were satisfied with the procedure regarding return to sports and activities [44]. Similarly, in a study by DeFrino et al., nine patients with a mean age of 56 years underwent first MTP arthrodesis with six of the nine patients able to return to activity without limitations, and all patients who participated in running preoperatively returned to it postoperatively [45]. It is important to note that the participants and these studies were not high-level athletes.

23.4 Sesamoid Disorders and Turf Toe Injuries

Running and jumping activities create substantial forces across the plantar aspect of the first MTP joint. When the force is excessive or repetitive, inflammation or injury to the sesamoid complex can occur. There are numerous causes of sesamoid pain in the athlete. The term "sesamoiditis" implies pain in the sesamoid region with negative radiographs and an equivocal magnetic resonance imaging (MRI) and is considered a diagnosis of exclusion. A history of overuse or mild trauma is common and can result in bursitis or flexor tendinitis [46, 47].

Sesamoid fractures are another cause of pain and can be either an acute fracture or a result of a stress injury, which is common in the running athlete with repetitive impact through the forefoot. Fractures typically involve the tibial sesamoid due to its larger size and the resultant increased contact stresses seen with weightbearing (Fig. 23.4). The fracture line is most often transverse and located at the mid-waist region [1]. Acute fractures generally occur as a result of a forceful impact to the forefoot. Additional causes of sesamoid pain include degenerative etiologies such as chondromalacia, impingement, or osteophyte formation. These pathologies can result from a chondral injury or repetitive damage. Sesamoid avascular necrosis (AVN) can occur as a sequela from a crush injury or a stress

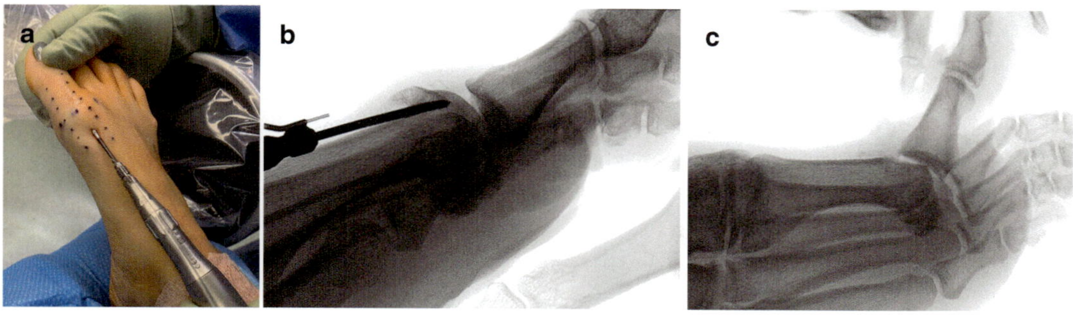

Fig. 23.4 (**a**) Intraoperative photograph depicting Bur placement for percutaneous cheilectomy. (**b**) Intraoperative lateral fluoroscopic image demonstrating planned dorsal osteophyte resection with bur. (**c**) Postresection lateral fluoroscopic image with improved dorsiflexion at least 80° after cheilectomy

fracture. Sesamoid AVN is most commonly seen in women between ages 18 and 29 years [48]. Patients typically have pain with direct palpation of the affected sesamoid with worsened symptoms during resisted plantar flexion of the first MTP joint. AVN often results in flattening of the sesamoid, with cyst formation and fragmentation. The fibular sesamoid is more frequently affected by AVN [1]. Prominent sesamoids can result in bursitis or intractable plantar keratosis (IPK), which often can be seen in the long-distance runner. In the absence of acute or repetitive trauma, sesamoiditis can also be caused by conditions such as infection, inflammatory arthropathies, and rarely tumors.

Patients typically report pain along the plantar aspect of the first MTP joint that is worsened with weightbearing or any athletic activity. Oftentimes there is not a single inciting event, but rather a gradual or insidious onset of pain. A thorough physical exam is important to localize the specific area of maximal tenderness as well as to assess for any anatomic variations such as a cavus foot position, hindfoot varus, or equinus. These variations may create increased stress across the base of the first metatarsal head and predispose the athlete to overload injuries to the sesamoid complex.

23.4.1 Turf Toe Injuries

Although not commonly seen in the track and field athlete, turf toe injuries can occur. A turf toe injury is defined as a sprain or tear of the capsular ligamentous structure of the first MTP joint. These injuries more commonly occur in football players participating on artificial surfaces. The typical mechanism of injury is an axial load applied to a foot fixed in equinus resulting in injury to the plantar plate. However, injury to the capsular ligamentous structure can occur with repetitive hyper-dorsiflexion of the first MTP joint. The use of a more flexible or lighter shoe may predispose an athlete to injury. Although the injury is typically a result of a hyper-dorsiflexion mechanism, injury to the capsular ligamentous structures can also occur through a hyper-plantarflexion mechanism. Clanton et al. described two track and field athletes who sustained a turf toe injury plantar flexion mechanism while participating on a Tartan track [49]. When evaluating a turf toe injury, the hallux MTP joint should be assessed for ecchymosis or swelling. Range of motion and stability of the toe should be examined and compared to the contralateral hallux. Decreased resistance to dorsiflexion suggests plantar plate injury, and a dorsoplantar drawer test should be performed to evaluate the integrity of the joint capsule.

The workup of every patient should include weightbearing AP and lateral foot radiographs with axial or tangential views of the sesamoid articulation. These views can be helpful in detecting arthrosis, osteophyte, or fracture. It is important to keep in mind that approximately 33% of the population have a bipartite sesamoid, which typically will have smooth cortical edges [14].

Contralateral AP radiographs can be helpful in this determination as it has been reported that there is a 90% incidence of bilateral bipartite sesamoids [50]. In contrast, fractures will have sharp, irregular borders on both sides of the fracture line. If there is concern for a plantar plate injury, a forced dorsiflexion lateral radiograph can be obtained (Fig. 23.5).

Computed tomography (CT), MRI, or three-phase bone scan can be helpful in patients with normal radiographs. CT scan can be used to assess the bony anatomy, evaluate fracture or fracture healing, as well as can help define the degree of arthritis at the sesamoid articulation. MRI is the most sensitive diagnostic tool to assess for AVN and can also be helpful in differentiating between bone and soft tissue abnormality (Fig. 23.6). Although there is a relatively high false-positive rate, a bone scan is a sensitive and inexpensive tool that can be used to detect increased areas of inflammation or stress fracture to the sesamoid.

23.4.1.1 Conservative Management

Initial treatment of sesamoid problems begins with rest, ice, compression, elevation, and activity modification. Anti-inflammatory medication can be a useful adjunct. Methods to unload the first MTP joint can be utilized and include a metatarsal pad or dancer's pad, arch support or an orthotic with a first MTP cutout. Furthermore, the athlete's shoe can be stiffened with the use of a full-length, carbon fiber foot plate or Morton's extension plate. In athletes with severe pain, a period of nonweightbearing with a boot or a cast may be warranted. The duration of immobilization is variable, but is usually continued until pain and tenderness have improved. Nondisplaced stress fractures should be treated with nonweightbearing for 6 weeks as these fractures are at higher risk for nonunion [51]. In patients with milder symptoms, taping of the hallux can provide compression and increased stability to the joint. Corticosteroid and/or anesthetic intra-articular injection is not recommended for any

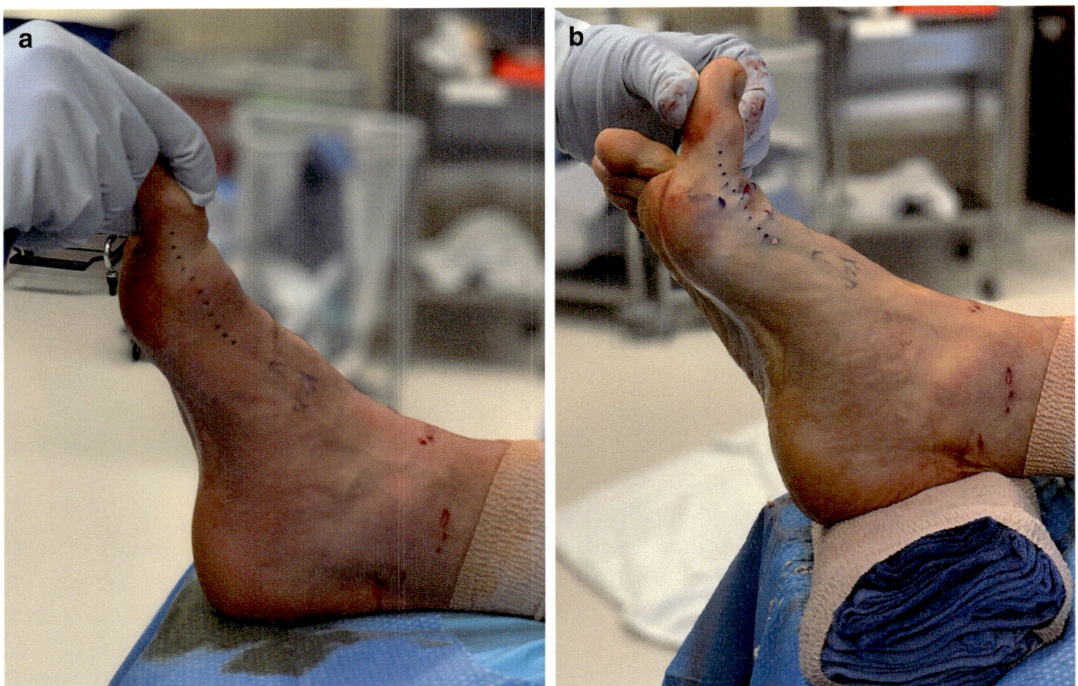

Fig. 23.5 (**a**) Preoperative and (**b**) postoperative photographs of first MTP dorsiflexion following percutanous cheilectomy

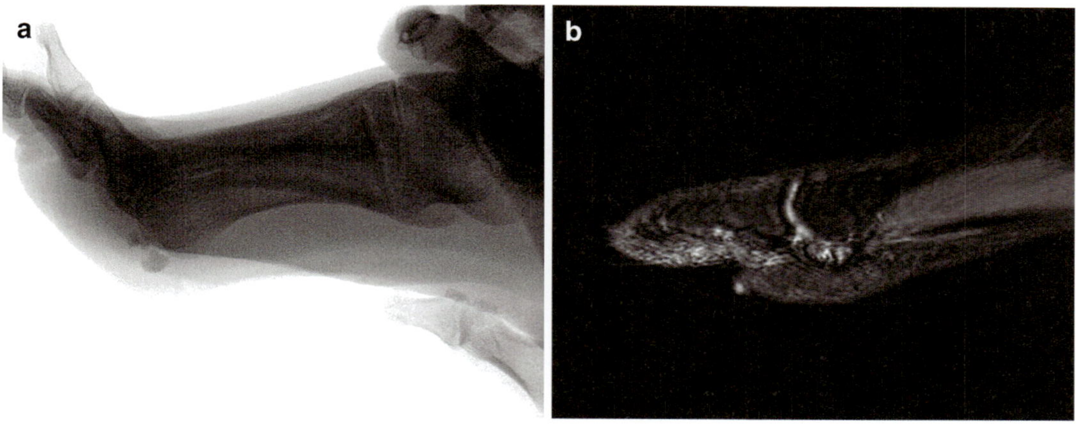

Fig. 23.6 (**a**) Lateral foot fluoroscopic image demonstrating fracture of the tibial sesamoid with displacement upon dorsiflexion, (**b**) sagittal T2 MRI images showing avascular necrosis of the sesamoid

injury; however, an anesthetic injection alone in a single nerve distribution can be used for pain [1].

23.4.1.2 Surgical Treatment

Surgical intervention should be reserved for patients with persistent pain despite appropriate conservative management. The specific surgical treatment is directed by the etiology of the sesamoid pathology. In patients with painful IPK due to a bony prominence or plantar exostosis, a sesamoid shaving procedure is indicated. The tibial sesamoid is most commonly involved and as such a plantarmedial approach to the sesamoid is utilized. Great care must be taken to identify and protect the plantarmedial digital nerve during exposure. After the sesamoid is exposed, the plantar half of the sesamoid is resected with the use of a microsagittal saw. The overlying soft tissues are then meticulously repaired. Weightbearing as tolerated in a hard-sole shoe is allowed immediately with a gradual return to normal shoe wear and activities as pain and swelling allow over 6–8 weeks [14].

Acute fractures of the sesamoid typically heal with nonsurgical management consisting of a period of nonweightbearing in a cast with a toe spica extension, or in a boot. Internal fixation has been described as a treatment option; however, it is unclear whether surgical intervention provides any benefit over traditional treatment methods [52–54]. Stress fractures of the sesamoid are often diagnosed several months after the onset of symptoms and at that time have likely progressed to a nonunion [1]. Successful treatment of sesamoid fracture nonunions with bone grafting has been described and is indicated for mid-waist fractures with displacement of less than 2 mm [55]. It is important that there is no significant injury to the articular surfaces and that the fracture fragments are stable. The tibial sesamoid is most commonly involved and as such an extra-articular plantarmedial approach is performed to expose the sesamoid. Autogenous bone graft is harvested from the metatarsal head through the capsulotomy. The nonunion site is thoroughly debrided, removing all fibrous tissue and exposing the underlying bone. The nonunion site is then packed with the bone graft followed by closure of the periosteum and soft tissues with absorbable suture. The capsulotomy is repaired and the wound closed. Patients are then placed into a nonweightbearing splint that goes past the toes. At the 2-week postop mark, the splint is removed and the patient is placed into a nonweightbearing short-leg cast with a toe spic extension. Transition to a walking cast or boot occurs at 6 weeks with initiation of gradual weightbearing. At 8 weeks, the patient advances to a regular shoe with a turf toe plate. A CT scan is obtained 3 months postoperatively, and if bony union has been achieved the patient is allowed to resume running activities. This technique was described by Anderson and McByde in series of 21 patients, 19 of which went on to bony union and return to their prior level of activity [55].

Sesamoidectomy is considered a viable treatment option for patients with osteochondrosis of the sesamoid or degenerative disease that has failed extensive nonoperative management (Fig. 23.7). Other indications for sesamoidectomy are infection or tumor. Careful consideration of the procedure must be made in the running or jumping athlete as removal of the sesamoid has been shown to reduce pushoff strength which could result in decreased athletic performance. Aper et al. reported a 10% loss of pushoff strength with removal of the tibial sesamoid, 16% loss with fibular resection, and 30% loss of strength with simultaneous removal of both sesamoids [56, 57].

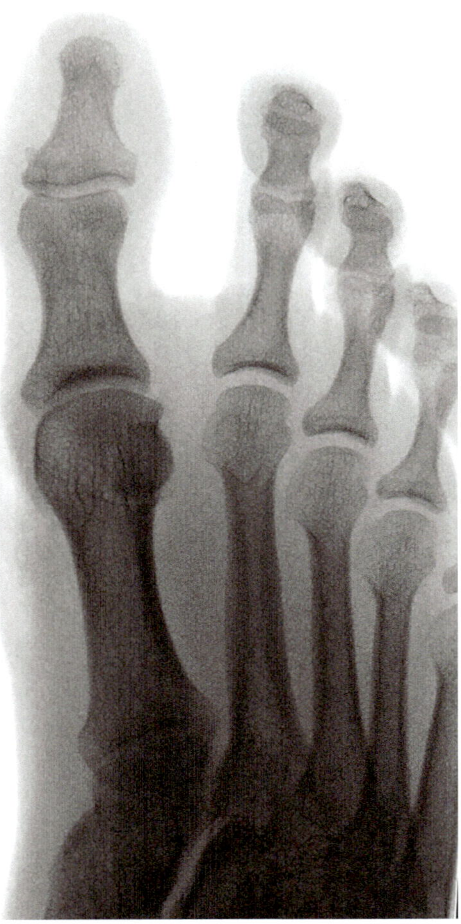

Fig. 23.7 Intraoperative anteroposterior fluoroscopic image following sesamoid excision

A tibial sesamoidectomy is performed through a plantarmedial approach, as described earlier. The sesamoid is shelled out from the FHB tendon, with care taken to avoid injury to the FHL tendon. After removal of the sesamoid, the defect is then repaired side to side with absorbable suture. If there is a large defect, the abductor hallucis tendon can be transferred from its distal insertion into the soft tissue defect. In addition to filling the plantar defect, the tendon transfer also acts by supplementing plantar flexor strength [1].

A curvilinear plantar incision lateral to the weightbearing surface is utilized for a fibular sesamoidectomy. During this approach, the plantarlateral digital nerve should be identified and protected. After the sesamoid is removed, the FHB tendon is directly repaired and the skin is meticulously closed with careful approximation of the dermal edges to minimize the risk of hypertrophic scar formation.

Removal of both sesamoids is not recommended in the running or jumping athlete due to the significant loss of pushoff strength, and the potential for the development of a cock-up toe deformity.

Postoperatively, the toe is maintained in plantarflexion with slight varus for tibial sesamoidectomy or slight valgus for fibular sesamoidectomy. Following tibial sesamoidectomy, the patient may weightbear as tolerated in a boot or hard-soled shoe and should wear a bunion dressing or splint for 6 weeks to allow for healing of the soft tissue repair. Weightbearing should be restricted for 2 weeks follow fibular sesamoidectomy to allow the plantar incision to heal. Afterwards, weightbearing can be initiated. Around 6 weeks postoperatively, the patient can transition to a regular shoe with a carbon fiber foot plate or a Morton's extension. Good results have been published following sesamoidectomy [58–62]. Saxena looked at return to activity in 24 athletic patients with a mean follow-up of 7.2 years. Eleven athletes (defined as professional or varsity sports level) returned to activity at a mean of 7.5 weeks, while the "active" individual had a slower mean RTA of 12 weeks. Of the 10 patients who underwent fibular sesamoidectomy, there was one case of hallux varus and two cases of a painful plantar scar. There was one case of hallux valgus deformity in the patients that underwent tibial sesamoidectomy [62]. Bichara et al. reported on 24 athletic patients who underwent sesamoidectomy with a mean follow-up of 35 months. Ninety-two percent of patients returned to activity with a mean RTA of 11.6 months. One patient did develop a hallux valgus deformity following a tibial sesamoidectomy [61].

Operative management of turf toe injuries is rarely necessary. Indications for surgical intervention include failure of extensive nonoperative management, retracted sesamoids, large capsular avulsion, diastasis of a bipartite sesamoid, and traumatic hallux valgus [14]. The plantar plate is approached through either a medial "J" incision or utilizing a two-incision technique. Through these incisions, the capsular disruption is identified and is typically found just distal to the fibular sesamoid. The plantar plate is then repaired using nonabsorbable sutures or with suture anchors of need due to inadequate soft tissue. The sutures are then tied with the MTP joint placed into approximately 15° of plantar flexion. Postoperatively, the toe is immobilized in 5–10° of plantar flexion with a toe spica splint. Gentle passive plantar flexion ROM can be initiated approximately 1 week postoperatively. Dorsiflexion of the MTP joint should be avoided. The patient will remain non-weightbearing with a removable splint or boot until 4 weeks postop. At that time, protected weightbearing in a boot can begin with initiation of active ROM. The patient can transition into a regular shoe with a carbon fiber footplate or Morton's extension at 2 months postop. Activities can then gradually be advanced as tolerated with protective taping. Return to activity typically occurs at 3–4 months; however, full recovery can take 6 months to a year. There have been multiple studies reporting satisfactory results with operative fixation of turf toe injuries [49, 63, 64]. Common complications include MTP joint stiffness and persistent pain with athletic activity. The vast majority of the literature describes turf toe injuries in football players; however, Lohrer described a case study of an elite level female sprinter who sustained an acute injury to the plantar plate and medial capsular tissue. The athlete underwent surgical repair and was able to return to full activity 6 months postoperatively [65].

These injuries can lead to significant functional disability. Short-term sequalae include decreased pushoff strength, stiffness, and difficulties with running. In the long-term, athletes may have troubles returning to preinjury performance due to pain, and may develop hallux rigidus.

23.5 Conclusions

Injuries and disorders of the hallux MTP joint complex commonly impact the track and field athlete. A thorough understanding of anatomy and pathophysiology are critical for the medical team to identify and appropriately manage these injuries. While a large majority of these injuries and conditions can be managed nonoperatively, it is important to be aware of surgical indications and current techniques. It is important to have a clear understanding of the goals and risks of any treatment, especially those surgical. With this knowledge, the medical team can help optimize the performance and safety of the track and field athlete.

References

1. Anderson RB, Shawen SB. Great-toe disorders. In: Baxter's the foot and ankle in sport. 2nd ed. Philadelphia: Mosby; 2008. p. 411–33.
2. Martin BD, McGuigan FX. Foot and ankle. Orthopaedic surgery: principles of diagnosis and treatment. Alphen aan den Rijn.: Wolters Kluwer; 2011.
3. Porter D, Schon L, Marymont J. Baxter's the foot and ankle in sport. Baxter's the foot and ankle in sport. Maryland Heights, MO: Mosby Elsevier; 2008. p. 636.
4. Stokes IA, Hutton WC, Stott JR, Lowe LW. Forces under the hallux valgus foot before and after surgery. Clin Orthop Relat Res. 1979;142:64–72.

5. Nigg B. Biomechanical aspects of running. In: Biomechanics of running shoes, vol. 3. Champaign, IL: Humen Kinetics Publiser; 1986. p. 1–25.
6. Nordin M, Frankel VH. Basic biomechanics of the musculoskeletal system. Basic biomechanics of the musculoskeletal system. Philadelphia: Wolters Kluwer Health/Lippincott Williams & Wilkins; 2012. p. 1–454.
7. Joseph J. Range of movement of the great toe in men. J Bone Joint Surg Br. 1954;36-B(3):450–7.
8. Bojsen-Møller F, Lamoreux L. Significance of free dorsiflexion of the toes in walking. Acta Orthop. 1979;50(4):471–9.
9. Bowman M. Athletic injuries to the great toe MP joint. In: disorders of the great toe. Rosemont, IL: AAOS; 1997.
10. Tanaka Y. Hallux valgus for athletes. In: Sports injuries of the foot and ankle: a focus on advanced surgical techniques. Berlin: Springer; 2019. p. 265–71.
11. Mann R. Bunion deformity in elite athletes. In: Baxter's the foot and ankle in sport. Amsterdam: Elsevier; 2008. p. 435–43.
12. Lillich JS, Baxter DE. Bunionectomies and related surgery in the elite female middle-distance and marathon runner. Am J Sports Med. 1986;14(6):491–3.
13. Covell DJ, Lareau CR, Anderson RB. Operative treatment of traumatic hallux valgus in elite athletes. Foot Ankle Int. 2017;38(6):590–5.
14. Hunt KJ, McCormick JJ, Anderson RB. Management of forefoot injuries in the athlete. Oper Tech Sports Med. 2010;18(1):34–45.
15. Thordarson DB, Leventen EO. Hallux valgus correction with proximal metatarsal osteotomy: two-year follow-up. Foot Ankle Int. 1992;13(6):321–6.
16. Mann RA, Rudicel S, Graves SC. Repair of hallux valgus with a distal soft-tissue procedure and proximal metatarsal osteotomy. A long-term follow-up. J Bone Jt Surg Ser A. 1992;74(1):124–9.
17. Wagner P, Wagner E. The use of a triplanar metatarsal rotational osteotomy to correct hallux valgus deformities. JBJS Essent Surg Tech. 2019;9(4):e43.1–2.
18. Macmahon A, Karbassi J, Burket JC, Elliott AJ, Levine DS, Roberts MM, et al. Return to sports and physical activities after the modified Lapidus procedure for hallux valgus in young patients. Foot Ankle Int. 2016;37(4):378–85.
19. McInnes BD, Bouché RT. Critical evaluation of the modified Lapidus procedure. J Foot Ankle Surg. 2001;40(2):71–90.
20. Saxena A. Return to athletic activity after foot and ankle surgery: a preliminary report on select procedures. J Foot Ankle Surg. 2000;39(2):114–9.
21. Giotis D, Paschos NK, Zampeli F, Giannoulis D, Gantsos A, Mantellos G. Modified Chevron osteotomy for hallux valgus deformity in female athletes. A 2-year follow-up study. Foot Ankle Surg. 2016;22(3):181–5.
22. Yee G, Lau J. Current concepts review: hallux rigidus. Foot Ankle Int. 2008;29:637–46.
23. Mulier T, Steenwerckx A, Thienpont E, Sioen W, Hoore KD, Peeraer L, et al. Results after cheilectomy in athletes with hallux rigidus. Foot Ankle Int. 1999;20(4):232–7.
24. Shurnas PS. Hallux rigidus: etiology, biomechanics, and nonoperative treatment. Foot Ankle Clin. 2009;14:1–8.
25. Coughlin MJ, Mann RA, Saltzman CL. Arthritic conditions of the foot. Mann's Surg Foot Ankle. 2007:805–922.
26. Coughlin MJ, Shurnas PS. Hallux rigidus. Grading and long-term results of operative treatment. J Bone Jt Surg Ser A. 2003;85(11):2072–88.
27. Lucas DE, Hunt KJ. Hallux rigidus relevant anatomy and pathophysiology. Foot Ankle Clin. 2015;20(3):381–9.
28. Hunt KJ, Githens M, Riley GM, Kim M, Gold GE. Foot and ankle injuries in sport. Imaging correlation with arthroscopic and surgical findings. Clin Sports Med. 2013;32(3):525–57.
29. Goodfellow J. Aetiology of hallux rigidus. Proc R Soc Med. 1966;59(9):821–4.
30. McMaster MJ. The pathogenesis of hallux rigidus. J Bone Jt Surg Ser B. 1978;60B(1):82–7.
31. Lichniak JE. Hallux limitus in the athlete. Clin Podiatr Med Surg. 1997;14:407–26.
32. Tenforde AS, Yin A, Hunt KJ. Foot and Ankle Injuries in Runners. Phys Med Rehabil Clin N Am. 2016;27(1):121–37.
33. Sammarco VJ, Nichols R. Orthotic management for disorders of the hallux. Foot Ankle Clin. 2005;10:191–209.
34. Mann RA, Coughlin MJ, DuVries HL. Hallux rigidus: a review of the literature and a method of treatment. Clin Orthop Relat Res. 1979;142:57–63.
35. Walter R, Perera A. Open, arthroscopic, and percutaneous cheilectomy for hallux rigidus. Foot Ankle Clin. 2015;20:421–31.
36. Hunt KJ. Hallux metatarsophalangeal (MTP) joint arthroscopy for hallux rigidus. Foot Ankle Int. 2015;36(1):113–9.
37. Grady JF, Smith AM, Boumendjel Y, Saxena A. Hallux rigidus: the valenti arthroplasty. In: International advances in foot and ankle surgery. London: Springer; 2012. p. 37–43.
38. Grady JF, Axe TM. The modified Valenti procedure for the treatment of hallux limitus. J Foot Ankle Surg. 1994;33(4):365–7.
39. Saxena A. The valenti procedure for hallux limitus/rigidus. J Foot Ankle Surg. 1995;34(5):485–8.
40. Saxena A, Valerio DL, Behan SA, Hofer D. Modified Valenti arthroplasty in running and jumping athletes with hallux Limitus/Rigidus: analysis of one hundred procedures. J Foot Ankle Surg. 2019;58(4):609–16.
41. Iqbal MJ, Chana GS. Arthroscopic cheilectomy for hallux rigidus. Arthroscopy. 1998;14(3):307–10.
42. Van Dijk CN, Veenstra KM, Nuesch BC. Arthroscopic surgery of the metatarsophalangeal first joint. Arthroscopy. 1998;14(8):851–5.

43. Teoh KH, Tan WT, Atiyah Z, Ahmad A, Tanaka H, Hariharan K. Clinical outcomes following minimally invasive dorsal cheilectomy for hallux rigidus. Foot Ankle Int. 2019;40(2):195–201.
44. Da Cunha RJ, MacMahon A, Jones MT, Savenkov A, Deland J, Roberts M, et al. Return to sports and physical activities after first metatarsophalangeal joint arthrodesis in young patients. Foot Ankle Int. 2019;40(7):745–52.
45. DeFrino PF, Brodsky JW, Polio FE, Crenshaw SJ, Beischer AD. First metatarsophalangeal arthrodesis: a clinical, pedobarographic and gait analysis study. Foot Ankle Int. 2002;23(6):496–502.
46. Coughlin MJ. Sesamoid pain: causes and surgical treatment. Instr Course Lect. 1990;39:23–35.
47. McBryde AM, Anderson RB. Sesamoid foot problems in the athlete. Clin Sports Med. 1988;7:51–60.
48. Ogata K, Sugioka Y, Urano Y, Chikama H. Idiopathic osteonecrosis of the first metatarsal sesamoid. Skelet Radiol. 1986;15(2):141–5.
49. Clanton TO, Butler JE, Eggert A. Injuries to the metatarsophalangeal joints in athletes. Foot Ankle Int. 1986;7(3):162–78.
50. Zinman H, Keret D, Reis ND, Reis ND. Fracture of the medial sesamoid bone of the hallux. J Trauma Inj Infect Crit Care. 1981;21(7):581–2.
51. Kahanov L, Eberman L, Games K, Wasik M. Diagnosis, treatment, and rehabilitation of stress fractures in the lower extremity in runners. Open Access J Sport Med. 2015;6:87–95.
52. Blundell CM, Nicholson P, Blackney MW. Percutaneous screw fixation for fractures of the sesamoid bones of the hallux. J Bone Jt Surg Ser B. 2002;84(8):1138–41.
53. Riley J, Selner M. Internal fixation of a displaced tibial sesamoid fracture. J Am Podiatr Med Assoc. 2001;91(10):536–9.
54. Pagenstert GI, Hintermann B, Valderrabano V. Percutaneous fixation of hallux sesamoid fractures. Tech Foot Ankle Surg. 2008;7(2):107–14.
55. Anderson RB, McBryde AM. Autogenous bone grafting of hallux sesamoid nonunions. Foot Ankle Int. 1997;18(5):293–6.
56. Aper RL, Saltzman CL, Brown TD. The effect of hallux sesamoid excision on the flexor hallucis longus moment arm. Clin Orthop Relat Res. 1996;325:209–17.
57. Aper RL, Saltzman CL, Brown TD. The effect of hallux sesamoid resection on the effective moment of the flexor hallucis brevis. Foot Ankle Int. 1994;15(9):462–70.
58. Inge GAL. Surgery if the sesamoid bones of the great toe. Arch Surg. 1933;27(3):466.
59. Leventen EO. Sesamoid disorders and treatment: an update. Clin Orthop Relat Res. 1991;(269):236–40.
60. Shimozono Y, Hurley ET, Brown AJ, Kennedy JG. Sesamoidectomy for hallux sesamoid disorders: a systematic review. J Foot Ankle Surg. 2018;57:1186–90.
61. Bichara DA, Henn RF, Theodore GH. Sesamoidectomy for hallux sesamoid fractures. Foot Ankle Int. 2012;33(9):704–6.
62. Saxena A, Krisdakumtorn T. Return to activity after sesamoidectomy in athletically active individuals. Foot Ankle Int. 2003;24(5):415–9.
63. Anderson RB. Turf toe injuries of the hallux metatarsophalangeal joint. Tech Foot Ankle Surg. 2002;1(2):102–11.
64. Coker TP, Arnold JA, Weber DL. Traumatic lesions of the metatarsophalangeal joint of the great toe in athletes. Am J Sports Med. 1978;6(6):326–34.
65. Lohrer H. MP I joint giving way - A case study. Foot Ankle Int. 2001;22(2):153–7.

Ankle Sprains and Instability

24

J. Nienke Altink, Liam D. A. Paget, Robin P. Blom, Jari Dahmen, Miki Dalmau-Pastor, and Gino M. M. J. Kerkhoffs

24.1 Introduction

Ankle sprains and instability are among the most common musculoskeletal disorders in the general and athletic population. Sustaining an ankle sprain can lead to a variety of disabling symptoms. Pain and loss of function are well-known symptoms in the acute phase after an initial ankle sprain. However, ankle sprains are also associated with severe long-term consequences such as persistent instability and eventually degeneration of the articular cartilage. In both the general and athletic population, these long-term consequences can severely impact the quality of life. In track and field athletes, ankle sprains can cause long-term inability to sport or loss of performance. To prevent these long-term consequences and facilitate quick return to performance, adequate management of ankle sprains is extremely important. This chapter provides an overview of the most important aspects on diagnosis and treatment of ankle sprains with a special consideration for this type of injury in track and field athletes.

24.2 Epidemiology

In the general population, the incidence of ankle sprains ranges from 2.15 to 6.97 per 1000 person years [1, 2]. Half of all these ankle sprains occur during sport activities resulting in a much higher incidence of ankle sprains in the athletic population [3, 4]. In 2010, Waterman et al. studied the incidence of ankle sprains in a cohort of active duty military personnel (e.g., an athletic population) which was 58.3 ankle sprains per 1000 person years [4]. Ankle sprains account for approximately 12% of all injuries in intercollegiate athletes and for approximately 16% in high school athletes [5, 6]. However, incidence rates and prevalence of ankle sprains differ significantly per type of sport, with higher incidence

J. N. Altink (✉) · L. D. A. Paget · R. P. Blom
J. Dahmen · G. M. M. J. Kerkhoffs
Department of Orthopaedic Surgery, Amsterdam Movement Sciences, Amsterdam UMC, Location AMC, University of Amsterdam, Amsterdam, The Netherlands

Academic Center for Evidence Based Sports Medicine (ACES), Amsterdam, The Netherlands

Amsterdam Collaboration for Health and Safety in Sports (ACHSS), AMC/VUmc IOC Research Center, Amsterdam, The Netherlands
e-mail: j.n.altink@amsterdamumc.nl

M. Dalmau-Pastor
Human Anatomy Unit, Department of Pathology and Experimental Therapeutics, School of Medicine and Health Sciences. University of Barcelona, Barcelona, Spain

GRECMIP - MIFAS (Groupe de Recherche et d'Etude en Chirurgie Mini-Invasive du Pied - Minimally Invasive Foot and Ankle Society), Merignac, France

G. L. Canata et al. (eds.), *Management of Track and Field Injuries*,
https://doi.org/10.1007/978-3-030-60216-1_24

rates in court and team sports. The incidence rate of ankle injuries in track and field athletes is approximately 29 per 1000 person years [7]. Ankle injuries contribute significantly to the total number of injuries in track and field athletes. In track events, ankle injuries account for 4–9% of all injuries, and in field events the ankle is the most commonly injured body site accounting for 39% of all injuries [5, 7] Lateral ankle sprains are the most frequently occurring type of ankle sprain, followed by high ankle sprains and medial ankle sprains. The reported incidence rates are 4.95 for lateral ankle sprains, 1 for high ankle sprains, and 0.7 for medial ankle sprains per 10,000 athlete exposures [5, 8, 9].

24.3 Ankle Joint Anatomy

Knowledge of the ankle anatomy is vital in order to understand the trauma mechanism of ankle injuries, the symptoms that occur after an ankle injury, the effective treatment strategies, and preventive measures necessary to be taken after an ankle injury. The most important structures associated with lateral ankle ligament injuries are described in the following section.

The ankle or talocrural joint, formed by the talus and distal tibia and fibula, is stabilized by three main ligament complexes that can be identified based on their anatomical location, respectively:

1. The lateral collateral ligament (LCL) complex is located on the lateral side of the ankle and consists of three ligaments originating from the distal fibula and inserting on the talus or calcaneus,
   (a) *The anterior talofibular ligament (ATFL)*
   The ATFL originates approximately 1 cm proximal from the tip of the anterior lateral malleolus and runs toward the neck of the talus [10]. The function of the ATFL is to limit anterior displacement of the talus and plantarflexion of the ankle [10]. Tension of the ATFL occurs when the ankle is in maximum plantarflexion, and it is the first ligament to be injured during an ankle sprain, thus being the most frequently injured ankle ligament [11]. Although diverse morphologies of ATFL have been described, the most recent evidence reports that the ATFL is a ligament formed by two fascicles, one superior and one inferior [12] (Fig. 24.1). Of these two fascicles, ATFL's superior fascicle has been described as an intra-articular structure; while ATFL's inferior fascicle has connections with the calcaneofibular ligament [12, 13].
   (b) *The calcaneofibular ligament (CFL)*
   The CFL originates from the tip of the lateral malleolus and inserts at the lateral side of the calcaneus. It is connected to ATFL's inferior fascicle through arciform fibers forming an isometric ankle stabilizing structure called the lateral fibulotalocalcaneal ligament

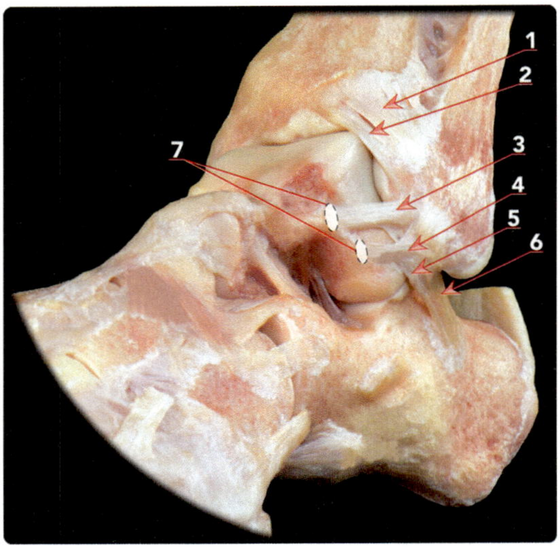

Fig. 24.1 Lateral view of an osteoarticular dissection demonstrating the anatomy of the LFTCL Complex. (1) Anterior tibiofibular ligament. (2) Distal fascicle of the anterior tibiofibular ligament. (3) ATFL superior fascicle. (4) ATFL inferior fascicle. (5) Arciform fibers of the LFTCL Complex. (6) CFL. (7) Note the different talar insertion points of ATFL fascicles. Figure reproduced with permission from: Vega J, Malagelada F, Manzanares Céspedes MC, Dalmau-Pastor M. The lateral fibulotalocalcaneal ligament complex: an ankle stabilizing isometric structure. Knee Surg Sports Traumatol Arthrosc. 2020;28(1):8–17. https://doi.org/10.1007/s00167-018-5188-8

complex [12] (Fig. 24.2). The insertion at the calcaneus is slightly posterior from its origin at the lateral malleolus [10]. In addition to the talocrural joint, the CFL also bridges the subtalar joint. As an isometric structure, tension of the CFL occurs in all ankle positions (neutral position, dorsiflexion, and plantarflexion), being more vertical in dorsiflexion and running in a posterior to anterior direction in plantarflexion (Fig. 24.3). This ligament is injured in approximately 20% of all ankle sprains [11]. Isolated injury of the CFL is rare because it is practically always in combination with damage of the ATFL.

(c) *The posterior talofibular ligament (PTFL)*

The PTFL originates from the medial posterior surface of the lateral malleolus and runs almost horizontally toward its insertion at the posterolateral tubercle and body of talus [10]. Tension of the PTFL occurs when the ankle is in dorsiflexion. The PTFL is least frequently involved in ankle sprains and damage of the PTFL practically only occurs in combination

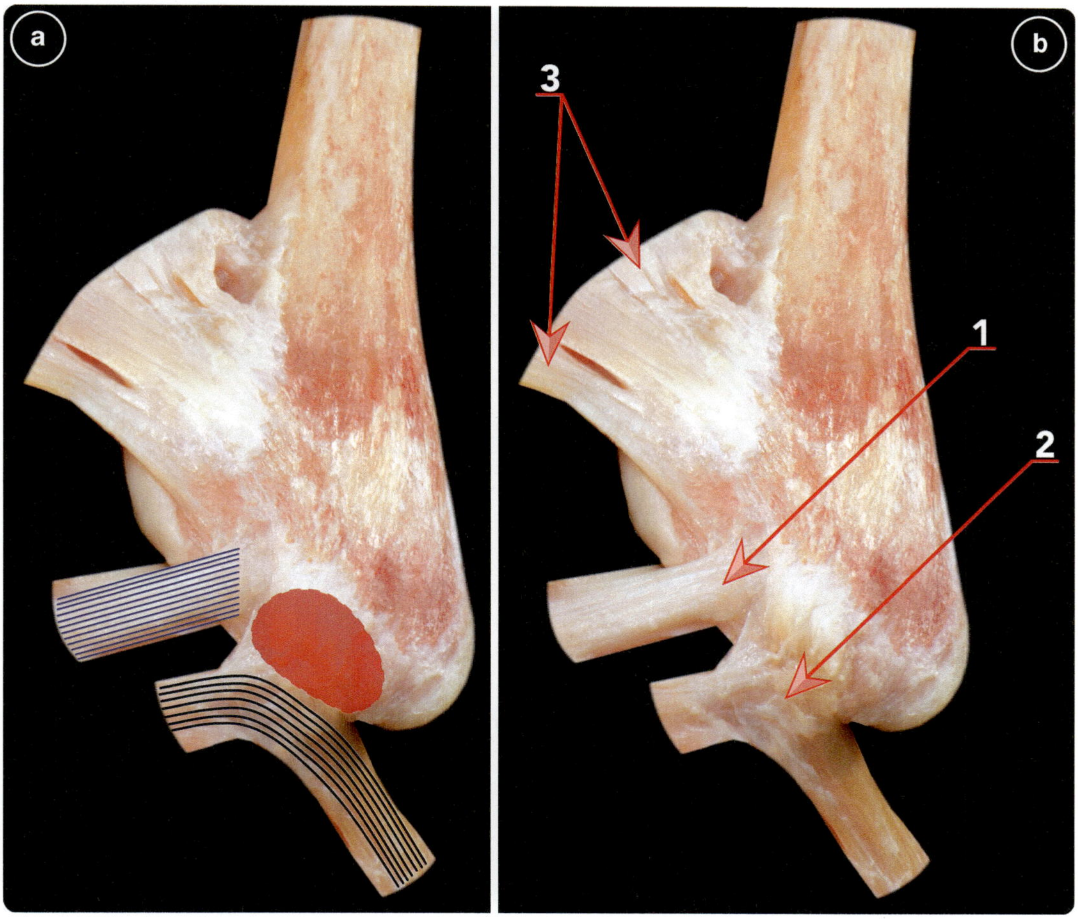

Fig. 24.2 Schematic view of the LFTCL Complex with the lateral malleolus disarticulated from the ankle. (**a**) View with the lateral ankle ligaments highlighted: ATFL superior fascicle (blue lines), LFTCL Complex (black lines), and area showing the common origin of the LFTCL Complex (red area). (**b**) Classic view of the LFTCL Complex. (1) ATFL superior fascicle. (2) LFTCL Complex. (3) Anterior tibiofibular ligament and distal fascicle. Figure reproduced with permission from: Vega J, Malagelada F, Manzanares Céspedes MC, Dalmau-Pastor M. The lateral fibulotalocalcaneal ligament complex: an ankle stabilizing isometric structure. Knee Surg Sports Traumatol Arthrosc. 2020;28(1):8–17. https://doi.org/10.1007/s00167-018-5188-8

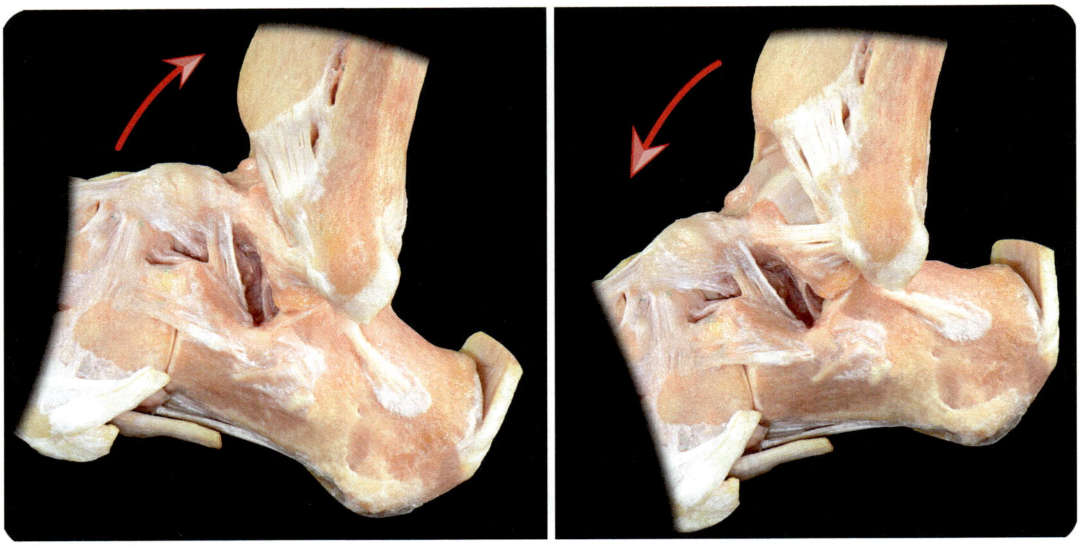

Fig. 24.3 Lateral view of an osteoarticular dissection showing the unchanged length of CFL during dorsiflexion and plantarflexion

with other ligamentous ankle injuries. Recently, intra-articular connections to the rest of the lateral ankle ligaments have been described [14] (*Dalmau-Pastor M, Malagelada F, Calder J, Manzanares MC, Vega J. The lateral ankle ligaments are interconnected: the medial connecting fibers between the anterior talofibular, calcaneofibular, and posterior talofibular ligaments. Doi:* https://doi.org/10.1007/s00167-019-05794-8) (Fig. 24.4).

2. The medial collateral ligament (MCL) complex is located on the medial side of the ankle joint. The ligaments of the MCL originate from the distal part of the medial malleolus and insert at the talus, calcaneus, and navicular bone. The function of the MCL is to prevent anterior and lateral translation and valgus tilting of the talus [15]. Injury to the MCL occurs in approximately 10% of all ligamentous ankle injuries [11]. Its anterior fibers are tense in plantarflexion while its posterior fibers are tense in dorsiflexion.
3. The distal tibiofibular ligaments or distal tibiofibular syndesmosis is located between the distal parts of the tibia and fibula. The function of the syndesmosis is to stabilize the fibula and distal tibia by limiting axial, rotational, and translational forces that attempt to separate the fibula from the tibia [10]. The syndesmosis consists of the anterior tibiofibular ligament (AITFL), the posterior tibiofibular ligament (PITFL), and the interosseus ligament (IOL) [16].

24.4 Trauma Mechanism of Ankle Sprains

The most commonly reported trauma mechanism of an ankle sprain is a combination of supination and adduction (inversion) of the foot. Inversion injuries account for approximately 77% of all ankle sprains [17] and result in damage to the ATFL. Although the LCL consists of three ligaments, the ATFL is the first ligament to be damaged [18]. The high incidence of injuries to the LCL complex, specifically the ATFL, can be explained by various anatomical and biomechanical factors. One of these factors is the extendible strength of the different ligaments around the ankle joint. The strength needed to stretch or rupture a ligament is lower in ligaments with less extendible strength. When the individual ligaments of the LCL are assessed,

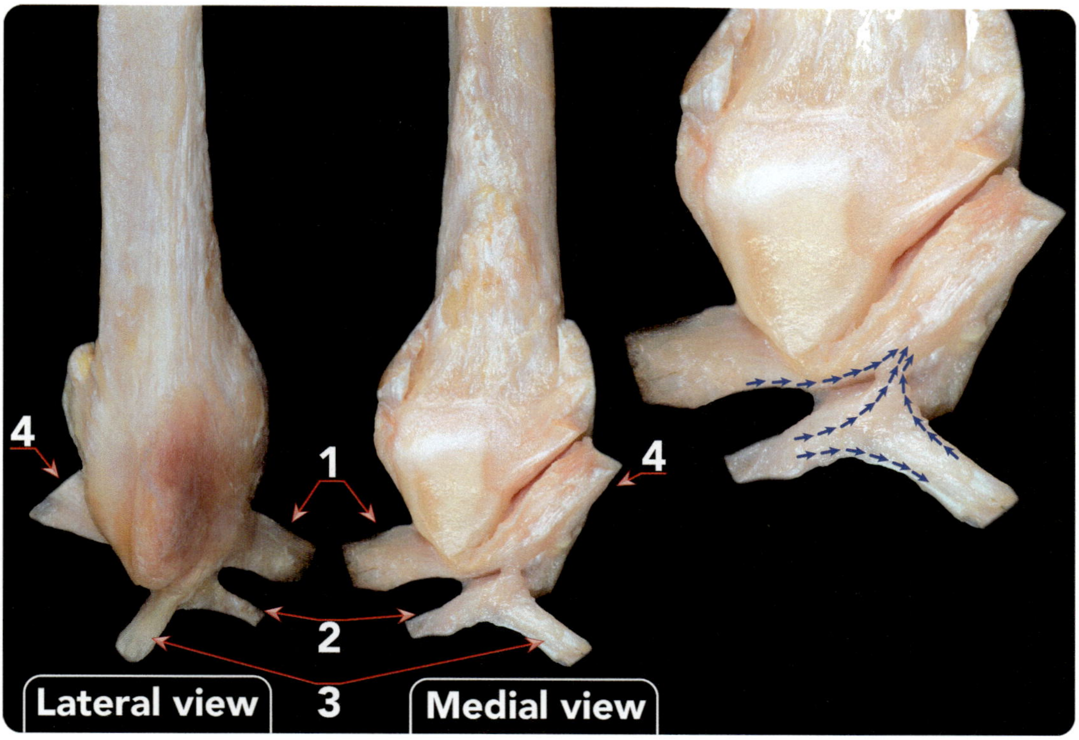

Fig. 24.4 Lateral and medial view of the fibular malleolus in a specimen with ATFLsf, ATFLif, CFL, and PTFL connections. (1) ATFLsf. (2) ATFLif. (3) CFL. (4) PTFL. Figure reproduced with permission from: Dalmau-Pastor M, Malagelada F, Calder J, Manzanares MC, Vega J. The lateral ankle ligaments are interconnected: the medial connecting fibres between the anterior talofibular, calcaneofibular and posterior talofibular ligaments. Knee Surg Sports Traumatol Arthrosc. 2020;28(1):34–39. https://doi.org/10.1007/s00167-019-05794-8

the ATFL has the least extendible strength, followed by the CFL, and the PTFL [19, 20]. Apart from strength in the individual ligaments, the position of the foot also contributes to the stability of the ankle joint. Due to the saddle shape of the talus, i.e., a broad anterior aspect and narrow posterior aspect, the contact area of the articulating surface is smaller in plantarflexion resulting in a less stable ankle during plantarflexion. Additionally, inversion of the talus is more likely to occur because eversion of the talus is blocked by the distal fibula, extending further distally compared to the medial malleolus of the tibia. The combination of an unstable joint in plantarflexion, the tendency of the ankle to move in inversion rather than eversion, and the low extendible strength of the ATFL results in a much higher incidence of ATFL damage compared to damage to the remaining ankle ligaments.

Ligamentous injuries of the MCL complex and syndesmosis are most likely to occur in impact sports with direct player contact such as soccer, basketball, ice hockey or rugby. The trauma mechanism associated with ligamentous injuries of the MCL complex is excessive pronation and abduction (eversion) and the mechanism associated with syndesmotic injury is forced dorsiflexion and external rotation. Due to the low incidence of MCL and syndesmotic injuries in track and field sports, the present chapter will focus on inversion injuries leading to ligamentous injuries of the LCL complex.

24.5 Diagnosis

Initial assessment of an athlete directly after an ankle sprain is performed in order to assess the severity of the injury and to exclude fractures.

The trauma mechanism and the (in)ability to bear weight on the injured leg are important factors to assess directly after an ankle sprain. Inability to bear weight after an ankle sprain might be suspicious for an ankle fracture. In athletes where an ankle fracture is suspected, the Ottawa ankle rules can be used as a clinical decision aid. Due to the high sensitivity of >97%, the Ottawa ankle rules are a valid tool to exclude fractures of the ankle and mid-foot [21]. Radiographic imaging is indicated in case of positive Ottawa ankle rules, as the sensitivity of the Ottawa ankle rules is approximately 32% [21].

In case of negative Ottawa ankle rules, ankle ligament damage should be assessed. Physical examination immediately after trauma to assess ankle ligament damage is unreliable. Diffuse pain and swelling make it difficult to localize the exact location of the pain and an anterior drawer test can often not be performed due to pain in the ankle [22]. As the pain and swelling rapidly decrease in the first days after trauma, delayed physical examination 4–5 days after trauma should be performed to assess rupture of the ATFL. Hematoma, pain on palpation of the ATFL, and the anterior drawer test are important aspects of physical examination of athletes following an ankle sprain. Complete rupture of the ATFL is likely in case of a positive anterior drawer test. A positive anterior drawer test has a sensitivity of 73% and a specificity of 97% for rupture of the ATFL, and in combination with hematoma and localized pain on palpation of the ATFL; a sensitivity of 96% and a specificity of 84% [22]. If localized pain on palpation of the ATFL is absent, rupture of the ATFL is unlikely [22]. Additional imaging is only indicated in case of severely unstable ankles or in case of persistent symptoms [23]. Magnetic Resonance Imaging (MRI) has a sensitivity of 93–96% and specificity of approximately 100% to assess ligament, tendon, bone, and chondral injuries [23]. Although MRI is a reliable method to assess these injuries, it should only be used in professional athletes and if concomitant injuries or multiple ligament involvements are suspected. If osteochondral damage is suspected, an additional Computed Tomography (CT) scan should be performed as MRI tends to overestimate the size of these osteochondral lesions due to bone marrow edema [24]. Other diagnostic tools provide limited additional information and are not indicated. Needle arthroscopy may be a promising new tool to assess ligamentous injuries, (osteo)chondral damage, and tendon pathology after an initial ankle sprain [25, 26].

24.6 Treatment

In the acute phase, the goal of the treatment of an ankle sprain is to prevent further damage, and to provide optimal healing circumstances by decreasing pain, swelling, and hematoma. Rest, Ice, Compression, and Elevation (RICE) are widely used in the acute phase despite the limited evidence on the efficacy of RICE [23]. Nonsteroidal anti-inflammatory drugs (NSAIDs) are effective in reducing pain and swelling [23]. Additionally, transcutaneous electrical nerve stimulation (TENS) may be applied as pain relief [27]. Functional therapy should start as soon as weightbearing is possible and is associated with shorter rehabilitation and superior functional outcomes compared to immobilization [23]. Plaster cast immobilization to reduce pain should only be applied in athletes who are unable to bear weight and never for a longer period than the initial 5–10 days after the trauma. Following a phase-sensitive approach, the athlete should start with restoring range of motion (ROM) and basic neuromuscular control [27]. Manual mobilization of the ankle joint can provide a short-term increase in dorsiflexion [28]. Manual therapy in combination with exercise therapy has been shown to be superior to exercise therapy alone [23]. Once the phase of restoring function is possible, exercise therapy can improve ankle instability and is associated with a shorter convalescent period and superior functional outcomes [23]. Contradicting results have been reported in the literature regarding the need for supervision of the rehabilitation program [23]. This is most likely due to differences in compliance and should therefore be assessed on individual patient level. The exercise therapy should aim to restore neuromuscular function through basic strength, endurance, plyometric, and in particular proprioceptive exercises. Optionally, electrical muscle stimulation (EMS) can be applied throughout the exercise therapy [27]. The athlete can then progress to more dynamic function exercises such as running, jumping, and cutting before continuing/starting sport-specific exercises.

Ligamentous ankle injuries can also be treated with surgical therapy. In professional athletes, the need for surgical treatment should be assessed on an individual level keeping in mind that surgical therapy leads to lower reinjury rates and quicker return to sports but is also associated with an increased risk of complications, decreased ankle ROM, and higher costs compared to conservative therapy [29, 30]. In professional track and field athletes, surgical therapy can be considered in professional athletes with rupture of all three lateral ankle ligaments or combined rupture of ligaments of the LCL and MCL or syndesmosis to ensure quick return to play [30]. In other patients, surgical therapy for lateral ligament injuries is only recommended in chronic cases in patients with persistent mechanical laxity after extensive functional therapy [23].

24.7 Return to Performance/Sport

Setting goals is an integral part of rehabilitation, as it is more often than not a gruelling process with mental and physical setbacks. Furthermore, it allows both the athlete as well as the medical staff to manage the anticipated setbacks, thereby fast-tracking the road to recovery whilst reducing reinjury risk. Convalescence should therefore be phase-sensitive as opposed to time contingency based and tailored to the athlete [31]. Currently 44.4% of the athletes return to sport (RTS) within 24 hours (hrs) following a lateral ankle sprain. The average RTS is 3 days for a first and <24 h for a recurrent ankle sprain. Around 95% of the athletes achieve RTS within 10 days after sustaining a first ankle sprain [31]. However, the fact that 40% retain residual symptoms, more caution is needed during rehabilitation and RTS clearance of ankle sprain injuries [32].

Throughout the rehabilitation, clinicians should be aware of certain risk factors for reinjury and act accordingly. Possible risk factors for reinjury include high body mass index (BMI), reduced range of motion (dorsiflexion), poor proprioception (static or dynamic), reduced muscle strength, and increased ankle laxity [33, 34]. Furthermore, it must be taken into account that ligament healing has been distinctly divided into three phases: the inflammatory phase (3–5 days), the proliferative phase (3–21 days), and the remodeling phase (14–28 days). However, mechanical stability does not coincide with the end of the remodeling phase or typical return to sport time frame.

Before returning to play, it is important to assess the current physical and mental limitations. In the anterior cross-ligament domain, fear of reinjury and lack of confidence in the injured limb are considered major factors for a successful RTS. Following consensus among the medical staff, the athlete can begin phasing in to RTS. Due to the repetitive trauma and overuse injuries in track and field athletes, load monitoring is essential. Clinicians should not only be aware of overloading but also inadequate loading, i.e., insufficiently preparing the athlete for RTS [31].

24.8 Long-Term Consequences

Besides discontinuation of sport participation and decreased athletic performance, ankle sprains are also associated with more severe long-term consequences such as chronic ankle instability, osteochondral lesions, and finally ankle osteoarthritis.

24.8.1 Chronic Ankle Instability (CAI)

Chronic ankle instability (CAI) is defined as the presence of perceived instability in combination with a history of (recurrent) sprains for a period of at least 12 months [35]. Recurrence rates of 34% have been reported in literature and many athletes do not recover within 3 years [36]. Specifically for track and field, approximately 18% of elite athletes who sustained an ankle sprain, had a recurrent sprain within 24 months after the initial ankle sprain [37]. Over 40% of elite track and field athletes report perceived ankle instability after an initial ankle sprain [37, 38]. The recent findings of ATFL's superior fascicle being an intra-articular ligament, added to the fact that this fascicle is the first one to be injured in inversion ankle sprains could explain the high index of CAI and perceived instability in these patients, as intra-articular ligaments are not expected to heal in the same way that an extra-articular ligament does [12, 13, 39].

24.8.2 Articular Cartilage Degeneration

Excessive articular loading (i.e., when the talus impacts the distal tibia such as during an ankle sprain) could result in acute articular cartilage damage within the ankle joint. A recent study by Blom et al. showed that single high impact loads did not induce osteochondral damage visualized with microcomputed tomography (microCT); however, the observed changes in biomechanical behavior imply that the ankle joints were compromised by the impacts [40]. These changes were found directly after the initial impact and could be the first step in the process toward articular cartilage damage and finally post-traumatic osteoarthritis (PTOA) of the ankle [40, 41]. These findings could explain why in 66% of patients with chronic ankle instability damage to the articular cartilage is observed [42]. Be aware that when track and field athletes may present with persisting pain after an ankle sprain, one should think of cartilaginous damage to the ankle joint, and additional imaging through an MRI or CT is the next step in the treatment algorithm and therefore indicated.

24.9 Primary and Secondary Prevention

Prevention of ankle injuries in the track and field athlete is important not only for the continued sport participation or performance, but also for reducing long-term consequences. Sufficient hydration, nutrition, and sleep are general key factors to be considered. Traditionally track and field athletes suffer from overuse due to the non-contact nature of the sport. However, sprains are among the most frequent ankle injuries with a recurrence rate of 3–34% and a previously sustained ankle injury is the primary predisposing factor for ankle injuries in general [23]. Additionally, chronic instability or osteoarthritis is more likely to develop as a result of multiple sustained ankle injuries.

Several methods have been proposed for primary and secondary prevention of ankle sprains, including a warm-up program, footwear, bracing, taping, and exercise therapy [23, 43]. Despite the lack of clear evidence regarding the effect of a warm-up program on ankle sprain incidence, the 32% reduction in ankle injuries following implementation of the FIFA 11+ program in football suggests its efficacy [44]. Evidence for shoe type or height is inconclusive [23]. The use of an ankle brace or tape reduces the risk of primary ankle injuries, but even more so in athletes with a history of a previously sustained ankle injury. Athletes suffering from recurrent ankle injuries are 50–70% less likely to sustain a recurrent sprain [23, 45]. Currently, there is no difference between the use of an ankle brace or taping. There is also no evidence for a superior taping technique [46]. However, nonelastic tape was found to be superior to elastic tape [47]. ROM-restriction, reduced comfort, and stability were the most frequent athlete-reported factors contributing to their choice in functional support. As there are currently no significant differences between and among braces and taping (techniques), the choice should be an individualized one [45]. However, a brace is considered the most cost-effective method compared to taping [23]. Additionally, both tape and a brace lose their mechanical function during exercise, with most of the mechanical support of tape being lost during the first 20 min of exercise [46]. Exercise therapy is another intervention shown to significantly reduce ankle sprain recurrence, specifically proprioceptive exercises, being effective up to 12 months after the initial ankle sprain [23]. In fact, the longer the exercise therapy is carried out, the longer and greater the prophylactic benefit [43]. However, compliance to the training program is a main issue.

Based on the current evidence, it is therefore advised to implement a warm-up program such as the FIFA 11+. Additionally, a structured rehabilitation program including proprioceptive exercises should be advocated to all athletes who sustained an ankle injury. Athletes should be advised to wear some form of functional support; at least until normal ankle function is restored.

References

1. Waterman BR, et al. The epidemiology of ankle sprains in the United States. JBJS. 2010;92(13):2279–84.
2. Bridgman S, et al. Population based epidemiology of ankle sprains attending accident and emergency units in the West Midlands of England, and a survey of UK practice for severe ankle sprains. Emerg Med J. 2003;20(6):508–10.
3. Doherty C, et al. The incidence and prevalence of ankle sprain injury: a systematic review and meta-analysis of prospective epidemiological studies. Sports Med. 2014;44(1):123–40.
4. Waterman BR, et al. Epidemiology of ankle sprain at the United States Military Academy. Am J Sports Med. 2010;38(4):797–803.
5. Roos KG, et al. The epidemiology of lateral ligament complex ankle sprains in National Collegiate Athletic Association sports. Am J Sports Med. 2017;45(1):201–9.
6. Pierpoint LA, et al. Epidemiology of injuries in United States High School Track and Field: 2008-2009 through 2013-2014. Am J Sports Med. 2016;44(6):1463–8.
7. Fong DT, et al. A systematic review on ankle injury and ankle sprain in sports. Sports Med. 2007;37(1):73–94.
8. Mauntel TC, et al. The epidemiology of high ankle sprains in National Collegiate Athletic Association Sports. Am J Sports Med. 2017;45(9):2156–63.
9. Kopec TJ, et al. The epidemiology of deltoid ligament sprains in 25 National Collegiate Athletic Association Sports, 2009-2010 through 2014-2015 academic years. J Athl Train. 2017;52(4):350–9.
10. Golano P, et al. Anatomy of the ankle ligaments: a pictorial essay. Knee Surg Sports Traumatol Arthrosc. 2016;24(4):944–56.
11. Brostrom L. Sprained ankles. 3. Clinical observations in recent ligament ruptures. Acta Chir Scand. 1965;130(6):560–9.
12. Vega J, et al. The lateral fibulotalocalcaneal ligament complex: an ankle stabilizing isometric structure. Knee Surg Sports Traumatol Arthrosc. 2020;28(1):8–17.
13. Dalmau-Pastor M, et al. Redefining anterior ankle arthroscopic anatomy: medial and lateral ankle collateral ligaments are visible through dorsiflexion and non-distraction anterior ankle arthroscopy. Knee Surg Sports Traumatol Arthrosc. 2020;28(1):18–23.
14. Dalmau-Pastor M, et al. The lateral ankle ligaments are interconnected: the medial connecting fibres between the anterior talofibular, calcaneofibular and posterior talofibular ligaments. Knee Surg Sports Traumatol Arthrosc. 2020;28(1):34–9.
15. Michelson J, et al. The effect of ankle injury on subtalar motion. Foot Ankle Int. 2004;25(9):639–46.
16. McCollum GA, et al. Syndesmosis and deltoid ligament injuries in the athlete. Knee Surg Sports Traumatol Arthrosc. 2013;21(6):1328–37.
17. Woods C, et al. The football association medical research Programme: an audit of injuries in professional football: an analysis of ankle sprains. Br J Sports Med. 2003;37(3):233–8.
18. Brostrom L. Sprained ankles. V. Treatment and prognosis in recent ligament ruptures. Acta Chir Scand. 1966;132(5):537–50.
19. Siegler S, Block J, Schneck CD. The mechanical characteristics of the collateral ligaments of the human ankle joint. Foot Ankle. 1988;8(5):234–42.
20. Beumer A, et al. A biomechanical evaluation of the tibiofibular and tibiotalar ligaments of the ankle. Foot Ankle Int. 2003;24(5):426–9.
21. Bachmann LM, et al. Accuracy of Ottawa ankle rules to exclude fractures of the ankle and mid-foot: systematic review. BMJ. 2003;326(7386):417.
22. van Dijk CN, et al. Physical examination is sufficient for the diagnosis of sprained ankles. J Bone Joint Surg Br. 1996;78(6):958–62.
23. Vuurberg G, et al. Diagnosis, treatment and prevention of ankle sprains: update of an evidence-based clinical guideline. Br J Sports Med. 2018;52(15):956.
24. Yasui Y, et al. Lesion size measured on MRI does not accurately reflect arthroscopic measurement in talar osteochondral lesions. Orthop J Sports Med. 2019;7(2):2325967118825261.
25. Stornebrink T, et al. 2-mm diameter operative tendoscopy of the tibialis posterior, peroneal, and achilles tendons: a cadaveric study. Foot Ankle Int. 2020;41(4):473–8.
26. Stornebrink T, et al. Two-millimetre diameter operative arthroscopy of the ankle is safe and effective. Knee Surg Sports Traumatol Arthrosc. 2020; https://doi.org/10.1007/s00167-020-05889-7.

27. Feger M, Hertel J. Rehabilitation after ankle football injuries. In: The ankle in football. New York: Springer; 2014. p. 269–85.
28. van der Wees PJ, et al. Effectiveness of exercise therapy and manual mobilisation in ankle sprain and functional instability: a systematic review. Aust J Physiother. 2006;52(1):27–37.
29. Kerkhoffs GM, et al. Surgical versus conservative treatment for acute injuries of the lateral ligament complex of the ankle in adults. Cochrane Database Syst Rev. 2007;2:CD000380.
30. Petersen W, et al. Treatment of acute ankle ligament injuries: a systematic review. Arch Orthop Trauma Surg. 2013;133(8):1129–41.
31. Tassignon B, et al. Criteria-based return to sport decision-making following lateral ankle sprain injury: a systematic review and narrative synthesis. Sports Med. 2019;49(4):601–19.
32. Wikstrom EA, Mueller C, Cain MS. Lack of consensus on return-to-sport criteria following lateral ankle sprain: a systematic review of expert opinions. J Sport Rehabil. 2020;29(2):231–7.
33. Vuurberg G, et al. Weight, BMI and stability are risk factors associated with lateral ankle sprains and chronic ankle instability: a meta-analysis. J ISAKOS. 2019;4(6):313–27.
34. Beynnon BD, Murphy DF, Alosa DM. Predictive factors for lateral ankle sprains: a literature review. J Athl Train. 2002;37(4):376–80.
35. Gribble PA, et al. 2016 consensus statement of the international ankle consortium: prevalence, impact and long-term consequences of lateral ankle sprains. Br J Sports Med. 2016;50(24):1493–5.
36. van Rijn RM, et al. What is the clinical course of acute ankle sprains? A systematic literature review. Am J Med. 2008;121(4):324–331.e6.
37. Malliaropoulos N, et al. Reinjury after acute lateral ankle sprains in elite track and field athletes. Am J Sports Med. 2009;37(9):1755–61.
38. Edouard P, et al. Invertor and evertor strength in track and field athletes with functional ankle instability. Isokinet Exerc Sci. 2011;19(2):91–6.
39. Vega J, et al. Ankle arthroscopy: the wave that's coming. Knee Surg Sports Traumatol Arthrosc. 2020;28(1):5–7.
40. Blom RP, et al. A single axial impact load causes articular damage that is not visible with micro-computed tomography: an ex vivo study on caprine tibiotalar joints. Cartilage. 2019:1947603519876353. https://doi.org/10.1177/1947603519876353.
41. Griffin TM, Guilak F. The role of mechanical loading in the onset and progression of osteoarthritis. Exerc Sport Sci Rev. 2005;33(4):195–200.
42. Hintermann B, Boss A, Schafer D. Arthroscopic findings in patients with chronic ankle instability. Am J Sports Med. 2002;30(3):402–9.
43. McKeon PO, Mattacola CG. Interventions for the prevention of first time and recurrent ankle sprains. Clin Sports Med. 2008;27(3):371–82. viii
44. Thorborg K, et al. Effect of specific exercise-based football injury prevention programmes on the overall injury rate in football: a systematic review and meta-analysis of the FIFA 11 and 11+ programmes. Br J Sports Med. 2017;51(7):562–71.
45. Barelds I, van den Broek AG, Huisstede BMA. Ankle bracing is effective for primary and secondary prevention of acute ankle injuries in athletes: a systematic review and meta-analyses. Sports Med. 2018;48(12):2775–84.
46. Zwiers R, et al. Taping and bracing in the prevention of ankle sprains: current concepts. J ISAKOS. 2016;1(6):304–10.
47. Vaes P, et al. Comparative radiological study of the influence of ankle joint strapping and taping on ankle stability. J Orthop Sports Phys Ther. 1985;7(3):110–4.

25 Osteochondral Lesions of the Ankle: An Evidence-Based Approach for Track and Field Athletes

Quinten G. H. Rikken, Jari Dahmen, J. Nienke Altink, Gian Luigi Canata, Pieter D'Hooghe, and Gino M. M. J. Kerkhoffs

25.1 Introduction

An osteochondral lesion (OCL) of the ankle is characterized by damage to the subchondral bone and the overlying cartilage. This lesion can occur after trauma, such as an inversion sprain or an ankle fracture [1, 2]. Ankle OCLs typically cause deep ankle pain during weight-bearing activities, subsequently impacting the patient's quality of life [3]. Although OCLs in the ankle can be considered a frequent entity among sports injuries of the lower extremity, a definite treatment paradigm has yet to achieve consensus among experts in the field [4, 5]. Therefore, an individualized evidence-based approach is best suitable as a treatment algorithm. This chapter serves as a practical guideline for the diagnosis, management, and rehabilitation of ankle OCLs in the track and field athlete.

Q. G. H. Rikken (✉) · J. Dahmen · J. N. Altink
G. M. M. J. Kerkhoffs
Department of Orthopaedic Surgery, Amsterdam Movement Sciences, Amsterdam UMC, Location AMC, University of Amsterdam, Amsterdam, The Netherlands

Academic Center for Evidence Based Sports Medicine (ACES), Amsterdam, The Netherlands

Amsterdam Collaboration for Health and Safety in Sports (ACHSS), AMC/VUmc IOC Research Center, Amsterdam, The Netherlands
e-mail: q.rikken@amsterdamumc.nl; j.dahmen@amsterdamumc.nl; j.n.altink@amsterdamumc.nl; g.m.kerkhoffs@amsterdamumc.nl

G. L. Canata
Centre of Sports Traumatology, Koelliker Hospital, Torino, Italy
e-mail: studio@ortosport.it

P. D'Hooghe
Orthopedic Surgery, Aspire Zone, ASPETAR Qatar Orthopaedic and Sports Medicine Hospital, Doha, Qatar

25.2 Incidence and Pathogenesis

The incidence of ankle OCLs, specifically in the talus, has been estimated to be around 27 per 100.000 person years in an athletic population [6]. OCLs are considered to have a strong relationship with traumatic events, as they occur in up to 70% of ankle fractures and sprains [2, 7, 8]. One can consider two essential theories concerning the pathogenesis of ankle OCLs. Firstly, during an ankle sprain or fracture, the talus impacts on the distal tibia, damaging the articular cartilage of the talar dome through microfractures ("cartilage cracks"). Blom et al. [9] showed that a single axial impact load leads to changes in the whole-joint biomechanics while no osteochondral damage was observed on micro-CT (computed tomography). Damage to

G. L. Canata et al. (eds.), *Management of Track and Field Injuries*,
https://doi.org/10.1007/978-3-030-60216-1_25

the subchondral bone occurs as a result of infiltration of synovial fluid through the damaged cartilage in the subchondral bone, leading to osteonecrosis [10]. Furthermore, weight-bearing then accelerates this cyclic process by increasing synovial pressure in the ankle joint and increasing the lesion size and/or depth (i.e., cysts formation).

Another theory is as follows: the OCL may be present as an osteochondritis dissecans in the ankle joint. This lesion has a fragmentous morphology and its exact origin is unknown. After a trauma, the fibrinous tissue attaching the dissecans to the surrounding dome may become loose and unstable, thereby inducing a symptomatic phase for the patient (Fig. 25.1). These lesions seem to be present since childhood and can occur in a bilateral fashion [11].

25.3 Clinical Presentation

The anamnesis in a patient potentially presenting with an OCL in the ankle is key. Patients typically present 6–12 months after an ankle sprain or fracture with deep ankle pain during weight-bearing. Track and field athletes may present sooner due to an increased physical self-awareness and proprioception of the ankle during high-load activities. Other symptoms can include stiffness, a catching or locking sensation, swelling after activities or an impaired range of motion (ROM) [3]. Track and field athletes may typically experience these complaints with explosive plantar flexion of the ankle while running, jumping or landing. Dependent on the location, a symptomatic OCL may be painful on palpation of the ankle mortise when the ankle is in full plantar

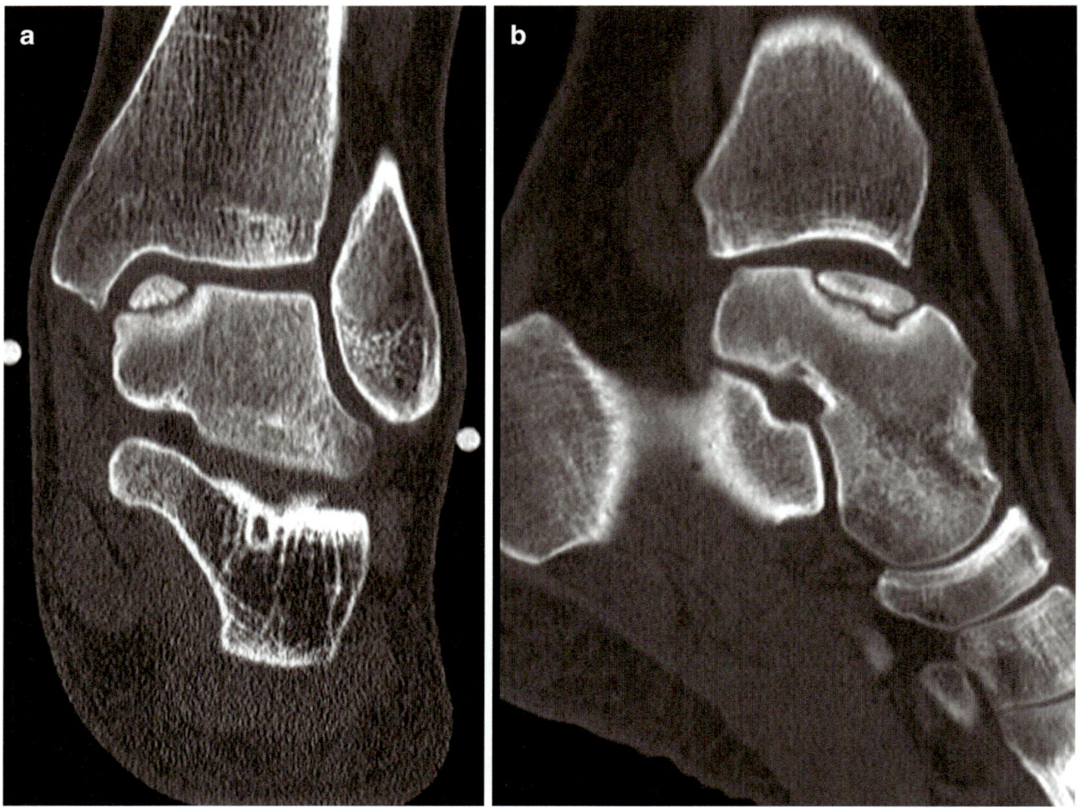

Fig. 25.1 Coronal- (**a**) and sagittal- (**b**) computed tomography scan of an ankle in plantar flexion with an osteochondritis dissecans lesion

flexion [12]. However, the recognizable pain cannot always be induced, especially when lesions are located on the posterior talar dome [12]. Van Diepen et al. [13] found that the majority of talar osteochondral lesions are located on posteromedial and centromedial dome. Suspicion of an OCL justifies further imaging for diagnosis and treatment planning.

25.4 Imaging Strategies

Imaging is crucial for the diagnosis of OCLs of the ankle. Radiographs only allow up to 60% of OCLs to be detected and should not be used as a decision tool for treatment choice [14]. Computed Tomography (CT) scans are the preferred modality to assess bony morphology including the subchondral bone plate [15]. Lesion size should be measured in three planes (anterior-posterior, medial-lateral, and depth). Additionally, the morphological aspects of the lesion should be carefully assessed (e.g., fragmentous-, cystic-, and sclerotic-morphology). The sensitivity and specificity of CT-scans for OCLs are 81% and 99%, respectively [14]. CT scans with the ankle in maximum plantar flexion can be obtained to determine arthroscopic accessibility (Fig. 25.1) [12]. An alternative to CT is the application of magnetic resonance imaging (MRI). Lesion size tends to be overestimated on MRI due to subchondral edema and is therefore less suited for determining lesion dimensions [16]. MRI has been reported to have a sensitivity and specificity of 96% for the diagnosis of an OCL making it a suitable imaging modality for OCLs [14].

25.5 Treatment

Choosing the right treatment option for the patient is an individualized, evidence-based process, guided by patient and surgeon preference, and individual patient characteristics such as lesion morphology, size, and primary or nonprimary nature of the lesion (i.e., failed prior surgical intervention(s)), as well as preoperative level of activity, hindfoot alignment, and presence of concomitant injuries.

25.5.1 Conservative Treatment

The first-line treatment is conservative, which can consist of one or a combination of the following treatment options: restriction of physical activities and/or sports, (cast)immobilization, injection therapy, insoles, physiotherapy, and nonsteroidal anti-inflammatory drugs (NSAIDs) [3]. By unloading the ankle joint, the goal of conservative therapy is to reduce symptoms through a reduction of joint edema and prevention of damage to the subchondral bone. Furthermore, natural healing of the articular cartilage can occur by offloading the joint. Conservative treatment has been advocated for asymptomatic lesions, nondisplaced lesions, patients with joint arthritis, and older patients with a low functional status [17].

A decrease in pain and lesion size is observed after conservative treatment at mid-term follow-up [18, 19]. Seo et al. [19] reported no progression of joint arthritis in a cohort of 142 patients at a mean follow-up of 5.7 years, with 84% of patients showing no limitations in sporting activities. However, other studies showed that the outcome of conservative treatment of OCLs can be regarded unsatisfactory, as up to 55% of cases fail [20]. In case of persistence of symptoms in athletes, surgical treatment may be considered.

25.5.2 Surgical Treatment

Surgical interventions can be considered between 3 and 6 months after the start of conservative therapy in the absence of clinical improvement. Earlier intervention is advocated in case of unstable fragmentous lesions potentially requiring immediate fixation. A wide variety of treatment options are available.

25.5.2.1 Bone Marrow Stimulation (BMS)

Arthroscopic bone marrow stimulation (BMS) is the most common surgical procedure for primary small OCLs (<15millimeter (mm) diameter) [21]. The purpose of BMS is to facilitate the growth of fibrocartilaginous tissue through revascularization [3]. Damaged cartilage tissue is removed from the lesion site until healthy bone is observed after which the subchondral bone is perforated. This results in the infiltration of multipotent mesenchymal stem cells and the formation of a fibrin clot, stimulating the growth of fibrocartilaginous tissue [22]. When there is a relatively healthy cartilage layer though a damaged subchondral bone, retrograde drilling can be considered [3]. This treatment option allows for perforation of the subchondral bone and revascularization, aiming at the formation of novel supportive bone [3].

BMS is successful in up to 82% of primary lesions, and up to 75% of secondary lesions [4, 5]. Clinical results at mid-term follow-up are considered good; however, repair tissue surface damage was found in 74% of the patients [21]. An eight- to 20-year follow-up study [23] showed similar clinical outcomes, though the presence of osteoarthritic changes were observed in 33% of patients. The observed osteoarthritic changes after BMS treatment can be explained by the deterioration of fibrocartilage, as fibrocartilage shows inferior wear characteristics compared to native hyaline cartilage [24]. Treatment failure of BMS, seen in up to 20% of patients, can partially be explained by this condition. Another essential factor for successful BMS treatment is the critical defect size as recent studies found a lesion diameter of 11-15 mm to be the optimal upper limit lesion size for a successful outcome [25, 26].

The return to sport (RTS) rate following BMS at any level of sports is 88% and to preinjury level of sports is 79% [27]. Mean time to RTS ranges from 15 to 26 weeks [27]. Hurley et al. [28] found a RTS (at any level) of 87%, and a mean time to RTS of 4.5 months. When compared to other surgical treatment options BMS is relatively less invasive, and allows for a shorter rehabilitation time and faster return to sports.

Additional therapeutics could also aid future BMS treatment by optimizing its effects. Bone Marrow Aspirate Concentrate (BMAC) and Platelet-rich Plasma (PRP) are adjunct therapies assisting in the growth of novel cartilage by the regenerative effect of growth factors from highly-concentrated stem cells or plasma from the blood. These techniques show promising results but need to be thoroughly investigated in future randomized controlled trials [29, 30].

25.5.2.2 Fixation Techniques

A fixation procedure is indicated for fragmentous primary lesions with a diameter of >15 mm and with a bony fragment of at least 3 mm on preoperative CT [31]. The treatment goal is to achieve subchondral bone healing, preserve the hyaline cartilage, and restore the natural joint congruency [32]. This fixation procedure can be performed through an open or arthroscopic technique using standard portals. During the lift, drill, fill, and fix (LDFF) procedure, an osteochondral bone flap is created with a blade and lifted while leaving the posterior side of the lesion intact (lift), analogous to lifting the hood of a car [32, 33]. Revascularization is promoted by drilling the sclerotic bone of the osteochondral bed of the talus (drill). Healthy, cancellous bone is harvested from the distal tibia and used to fill the lesion (fill) [33]. Because the osteochondral flap is still attached to the talus, it will automatically return to its original position when fixated with a (bio-)compression screw (fix).

Success rates range from 89% to 100% at short-term to mid-term follow-up [33, 34]. Advantages of fixation are the preservation of the hyaline cartilage, which shows better wear characteristics than fibrocartilaginous tissue [24]. Subchondral bone healing is found to be superior after fixation as compared to BMS [35]. Due to the important role of the subchondral bone in OCL restoration and the development of osteoarthritis, the rate of osteoarthritis development may be lower from a theoretical point of view.

Return to sport rates reported in the literature range from 87% to 93% [33, 34]. Lambers et al. [33] found the Foot and Ankle Outcome Score (FAOS) sports subscale improved significantly from 40 points preoperatively to 70 points postoperatively. Pain during running improved from 7.8 points preoperatively to 2.9 points postoperatively (on a 0–10 point scale).

25.5.2.3 Cartilage Transplantation and Chondrogenesis Inducing Techniques

Autologous cartilage implantation (ACI) and matrix-associated chondrocyte implantation (MACI) are cartilage transplantation techniques while autologous matrix-induced Chondrogenesis (AMIC) and bone marrow-derived cells transplantation (BMDCT) are chondrogenesis-inducing techniques (CIT). The techniques are used for larger (>15 mm diameter) primary and failed primary lesions, including cystic lesions. ACI and MACI are both two-stage procedures aiming to restore the natural hyaline cartilage layer. For ACI, autologous chondrocytes are harvested from nonweight-bearing areas and cultured, after which the culture expansion is implanted with an autologous periosteal membrane. During MACI, the harvested chondrocytes are embedded onto a scaffold and thereafter implanted. AMIC and BMDCT are both in essence one-stage procedures. In AMIC, first microfracturing is performed, after which the site is covered by a biodegradable collagen type I/III membrane. With BMDCT, platelet-rich fibrin (PRF) from the blood and bone marrow from the iliac crest are extracted and concentrated. This product is later injected onto a collagen scaffold which is placed over the arthroscopically cleaned lesion site.

Systematic reviews found ACI, MACI, and AMIC to result in a treatment success rate around 80% for primary and nonprimary lesions [4, 5]. A pooled RTS (preinjury level) rate of 69% for ACI was found by a systematic review by Steman and Dahmen et al. [36]. In cohort studies, MACI and BMDCT were found to have an RTS (preinjury level) rate of 81% and 73%, respectively [37, 38]. Vannini et al. [38] found a mean time to return to preinjury level of sport of 18.5 (±15.7) months for patients treated with BMDCT and observed no difference in return to sports rates for high- or low-impact sports. Rehabilitation can be elongated compared to BMS due to the need for strict immobilization in the initial phase but is similar to osteochondral transplantation and fixation.

25.5.2.4 Osteo(Chondral) Transplantation

Osteo(chondral) transplantation can be used in larger lesions (>15 mm diameter), secondary lesions, and cystic lesions. The most commonly described technique is autologous osteochondral transplantation (AOT). During this technique, the area containing the lesion is excised and replaced with an osteochondral autograft which is most commonly harvested from a nonweight-bearing part of the ipsilateral femoral condyle. Good to excellent clinical results are reported in 87 to 90% of patients after AOT [39]. However, approximately up to 11% of patients develop some sort of donor site morbidity following AOT [40]. An alternative osteo(chondral) transplantation technique is autologous osteoperiosteal cylinder grafting during which an osteoperiosteal graft is harvested from the iliac crest to replace the area of the talus containing the lesion. Success rates up to 94% have been reported [41]. Recently Kerkhoffs et al. [42] developed a new osteochondral transplantation technique which utilizes an osteoperiosteal graft from the iliac crest. The technique is called Talar OsteoPeriostic grafting from the Iliac Crest (TOPIC). In this technique, the graft is shaped exactly in the preferred shape, matching the curvature, size, and depth of the talus and additionally minimalizing the need for removal of healthy tissue [42]. The TOPIC procedure is a promising, simple, (cost-)effective, one-stage technique.

The return to any level of sports and preinjury level of sports rate was assessed to be 90% and 72%, respectively after osteochondral transplantation [36]. The average return to sport time ranges from 13 to 26 weeks [36].

Treatment guidelines	
Commence with conservative therapy, when complaints do not improve 3–6 months after starting conservative therapy, surgery can be considered	
Surgical indication	Treatment
• Small (<15 mm diameter) lesion	BMS (±PRP or BMAC)
• Larger lesion (>15 mm diameter), without cysts, or secondary lesions	1. TOPIC [42] 2. AOT 3. ACI, MACI, AMIC, BMDCT

Treatment guidelines	
• Large lesion (>15 mm diameter), with (massive) cysts, or secondary lesions	1. TOPIC [42] 2. AOT 3. ACI, MACI, ACI, BMDCT
• Good cartilage layer with cysts or arthroscopically unreachable	Retrograde drilling (±cancellous graft)
• Fixable lesion	LDFF [32], or other fixation technique

Abbreviations: *BMS* bone marrow stimulation, *PRP* platelet-rich plasma, *BMAC* bone marrow aspirate concentrate, *ACI* autologous cartilage implantation, *MACI* matrix-associated chondrocyte implantation, *AMIC* autologous matrix-induced chondrogenesis, *BMDCT* bone marrow-derived cells transplantation, TOPIC: Talar OsteoPeriostal grafting from the Iliac Crest [42], *AOT* autologous osteochondral transplantation, *LDFF* Lift-Drill-Fill-Fix [32]

25.6 Rehabilitation and Return to Sports

A uniform rehabilitation protocol has not yet been established for optimal postoperative recovery [28, 43]. Athletes progress through the stages of the "return to sports continuum," defined by three elements (i.e., return to participation, –pre-injury sports, and -performance) [44]. Track and field athletes should focus on event-based sport-specific rehabilitation with phased rehabilitation goals. Pain (during and after activities), joint swelling, proprioception, and stability are key clinical indicators on which the temporary limitation of shear forces and the progression of ankle activity are based.

The general phased protocol is shown in Fig. 25.2. In phase 1, athletes should perform ROM exercises (in between casts or boot) and focus on the neurologic "mind-muscle" connection to limit atrophy of the muscles supporting the ankle, electric muscle stimulation can aid this process [45]. Treatments which restrict postoperative ROM exercises such as scaffold therapies and postoperatively casted patients should adhere to these specific instructions and can resume ROM exercises when allowed. Icing of the ankle to limit joint swelling after activities throughout the rehabilitation process is encouraged. During phase 2, weight-bearing is gradually increased to full-load bearing. In this phase, the track and field athlete should focus on regaining normal gait. Additionally, athletes can start exercises which increase overall fitness, without stressing the ankle joint. Phase 3 incorporates the treatment of a physiotherapist to aid recovery. Strength and balance exercises and low-load exercises can be started, keeping in mind that no axial peak forces

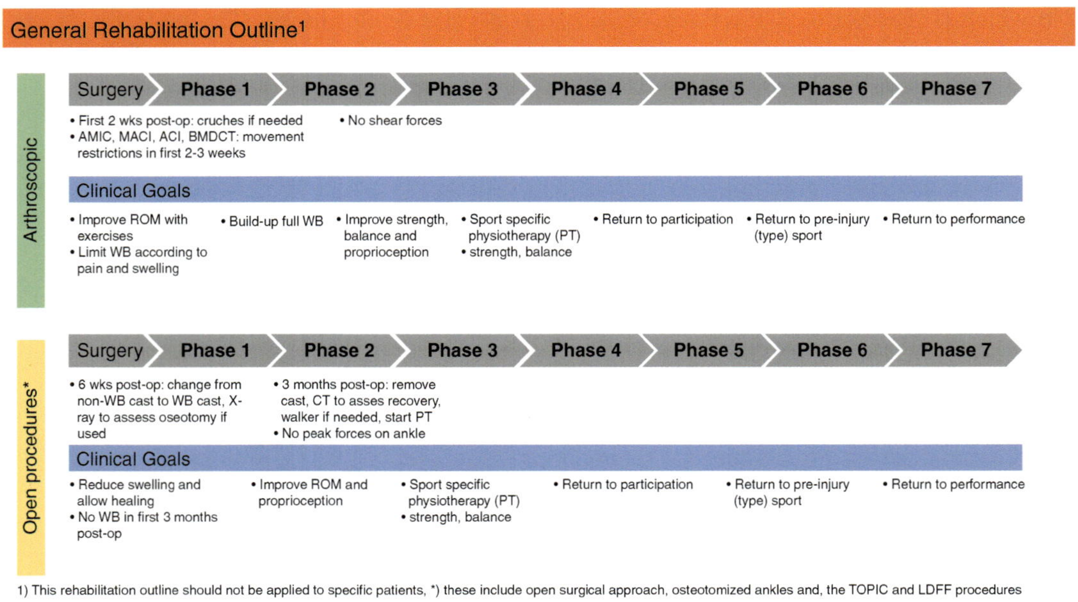

Fig. 25.2 Phased general rehabilitation protocol

are allowed. In phase 4, sport-specific training is started. Progression to more dynamic exercises only occurs when the ROM and strength in the ankle are sufficient to perform these exercises in phase 4 safely. For track and field athletes, this means running and jumping event specialized athletes should focus more on explosive strength, and athletes primarily competing in throwing events should pay attention to ankle stability and proprioception training. Phase 4 overlaps with phase 5 which is defined by return to participation. However, the exercises performed in phase 4 are muscle group-specific and led by (or supervised by) a (team) physiotherapist. In phase 5, track and field athletes can focus on sport-specific exercises which increase strength and technique, without stressing the ankle joint as in preinjury sports. The ROM, balance, proprioception, and strength needed to progress to phase 6 are dependent on the specialization of the athlete. For example, more plantarflexion strength is needed for athletes specialized in high jump compared to athletes specialized in throwing events. In phase 6, athletes can return to their preinjury sport and gradually increase training load, and return to their preinjury level or improve performance (phase 7). The "Fit-to-play" prognosis is highly individualized. It is important to recognize that field events have a relatively higher ankle activity score compared to track events, meaning the ankle experiences higher load, which can require a longer rehabilitation time [46]. Biomechanics, treatment of choice, lesion size, psychological factors (especially fear of reinjury), level of preinjury sports, and age affect the rehabilitation time and should be taken into account by the treating medical team [43]. Objective measures are available to aid this decision by testing recovery progression and determine if an athlete is fit for play.

References

1. Hintermann B, Boss A, Schäfer D. Arthroscopic findings in patients with chronic ankle instability. Am J Sports Med. 2002;30:402–9.
2. Martijn HA, Lambers KTA, Dahmen J, Stufkens SAS, Kerkhoffs GMMJ. High incidence of (osteo) chondral lesions in ankle fractures. Knee Surg Sports Traumatol Arthrosc. 2021;29:1523–34.
3. Reilingh M, van Bergen C, van Dijk C. Diagnosis and treatment of osteochondral defects of the ankle. SA Orthop J. 2009;8:44–50.
4. Dahmen J, Lambers KTA, Reilingh ML, Van Bergen CJA, Stufkens SAS, Kerkhoffs GMMJ. No superior treatment for primary osteochondral defects of the talus. Knee Surg Sport Traumatol Arthrosc. 2018;26:2142–57.
5. Lambers KTA, Dahmen J, Reilingh ML, van Bergen CJA, Stufkens SAS, Kerkhoffs GMMJ. No superior surgical treatment for secondary osteochondral defects of the talus. Knee Surg Sport Traumatol Arthrosc. 2018;26:2158–70.
6. Orr JD, Dawson LK, Garcia EJ, Kirk KL. Incidence of osteochondral lesions of the talus in the United States military. Foot Ankle Int. 2011;32:948–54.
7. Hintermann B, Regazzoni P, Lampert C, Stutz G, Gächter A. Arthroscopic findings in acute fractures of the ankle. J Bone Joint Surg Br. 2000;82:345–51.
8. Saxena A, Eakin C. Articular Talar injuries in athletes. Am J Sports Med. 2007;35:1680–7.
9. Blom RP, Mol D, van Ruijven LJ, Kerkhoffs GMMJ, Smit TH. A single axial impact load causes articular damage that is not visible with micro-computed tomography: an ex vivo study on caprine Tibiotalar joints. Cartilage. 2019; https://doi.org/10.1177/1947603519876353.
10. Van der Vis HM, Aspenberg P, Marti RK, Tigchelaar W, Van Noorden CJ. Fluid pressure causes bone resorption in a rabbit model of prosthetic loosening. Clin Orthop Relat Res. 1998;350:201–8.
11. Reilingh ML, Kerkhoffs GMMJ, Telkamp CJA, Struijs PAA, van Dijk CN. Treatment of osteochondral defects of the talus in children. Knee Surg Sports Traumatol Arthrosc. 2014;22:2243–9.
12. Van Bergen CJA, Tuijthof GJM, Maas M, Sierevelt IN, Van Dijk CN. Arthroscopic accessibility of the talus quantified by computed tomography simulation. Am J Sports Med. 2012;40:2318–24.
13. Van Diepen PR, Dahmen J, Altink JN, Stufkens SAS, Kerkhoffs GMMJ. Location distribution of 2,087 osteochondral lesions of the talus. Cartilage. 2020;1947603520954510. https://doi.org/10.1177/1947603520954510.
14. Verhagen RAW, Maas M, Dijkgraaf MGW, Tol JL, Krips R, van Dijk CN. Prospective study on diagnostic strategies in osteochondral lesions of the talus. Is MRI superior to helical CT? J Bone Jt Surg Br. 2005;87:41–6.
15. van Bergen CJ, Gerards RM, Opdam KT, Terra MP, van Bergen CJ, Kerkhoffs GM. Diagnosing, planning and evaluating osteochondral ankle defects with imaging modalities. World J Orthop. 2015; 6:944–53.
16. Yasui Y, Hannon CP, Fraser EJ, Ackermann J, Boakye L, Ross KA, et al. Lesion size measured on MRI does not accurately reflect arthroscopic measurement

in Talar osteochondral lesions. Orthop J Sport Med. 2019;7:2325967118825261.

17. Dombrowski ME, Yasui Y, Murawski CD, Fortier LA, Giza E, Haleem AM, et al. Conservative management and biological treatment strategies: proceedings of the international consensus meeting on cartilage repair of the ankle. Foot Ankle Int. 2018;39:9S–15S.
18. Klammer G, Maquieira GJ, Spahn S, Vigfusson V, Zanetti M, Espinosa N. Natural history of nonoperatively treated osteochondral lesions of the talus. Foot Ankle Int. 2015;36:24–31.
19. Seo SG, Kim JS, Seo D-K, Kim YK, Lee S-H, Lee HS. Osteochondral lesions of the talus. Few patients require surgery. Acta Orthop. 2018;89:462–7.
20. Zengerink M, Struijs PAA, Tol JLC, Van Dijk N. Treatment of osteochondral lesions of the talus: a systematic review. Knee Surg Sports Traumatol Arthrosc. 2010;18:238–46.
21. Toale J, Shimozono Y, Mulvin C, Dahmen J, Kerkhoffs GMMJ, Kennedy JG. Midterm outcomes of bone marrow stimulation for primary osteochondral lesions of the talus: a systematic review. Orthop J Sport Med. 2019;7:1–8.
22. O'Driscoll SW. The healing and regeneration of articular cartilage. J Bone Joint Surg Am. 1998;80:1795–812.
23. van Bergen CJA, Kox LS, Maas M, Sierevelt IN, Kerkhoffs GMMJ, van Dijk CN. Arthroscopic treatment of osteochondral defects of the talus: outcomes at eight to twenty years of follow-up. J Bone Joint Surg Am. 2013;95:519–25.
24. Lynn AK, Brooks RA, Bonfield W, Rushton N. Repair of defects in articular joints. Prospects for material-based solutions in tissue engineering. J Bone Jt Surg Br. 2004;86:1093–9.
25. Ramponi L, Yasui Y, Murawski CD, Ferkel RD, Digiovanni CW, Kerkhoffs GMMJ, et al. Lesion size is a predictor of clinical outcomes after bone marrow stimulation for osteochondral lesions of the talus: a systematic review. Am J Sports Med. 2017;45:1698–705.
26. Choi WJ, Park KK, Kim BS, Lee JW. Osteochondral lesion of the talus: is there a critical defect size for poor outcome? Am J Sports Med. 2009;37:1974–80.
27. Steman JAH, Dahmen J, Lambers KTA, Kerkhoffs GMMJ. Return to sports after surgical treatment of osteochondral defects of the talus: a systematic review of 2347 cases. Orthop J Sport Med. 2019;7:1–15.
28. Hurley ET, Shimozono Y, McGoldrick NP, Myerson CL, Yasui Y, Kennedy JG. High reported rate of return to play following bone marrow stimulation for osteochondral lesions of the talus. Knee Surg Sport Traumatol Arthrosc. 2019;27:2721–30.
29. Guney A, Akar M, Karaman I, Oner M, Guney B. Clinical outcomes of platelet-rich plasma (PRP) as an adjunct to microfracture surgery in osteochondral lesions of the talus. Knee Surg Sport Traumatol Arthrosc. 2015;23:2384–9.
30. Kim YS, Park EH, Kim YC, Koh YG. Clinical outcomes of mesenchymal stem cell injection with arthroscopic treatment in older patients with osteochondral lesions of the talus. Am J Sports Med. 2013;41:1090–9.
31. Reilingh ML, Murawski CD, DiGiovanni CW, Dahmen J, Ferrao PNF, Lambers KTA, et al. Fixation techniques: proceedings of the international consensus meeting on cartilage repair of the ankle. Foot Ankle Int. 2018;39:23S–7S.
32. Kerkhoffs GMMJ, Reilingh ML, Gerards RM, de Leeuw PAJ. Lift, drill, fill and fix (LDFF): a new arthroscopic treatment for talar osteochondral defects. Knee Surg Sport Traumatol Arthrosc. 2016;24:1265–71.
33. Lambers KTA, Dahmen J, Reilingh ML, van Bergen CJA, Stufkens SAS, Kerkhoffs GMMJ. Arthroscopic lift, drill, fill and fix (LDFF) is an effective treatment option for primary talar osteochondral defects. Knee Surg Sport Traumatol Arthrosc. 2020;28:141–7.
34. Kumai T, Takakura Y, Kitada C, Tanaka Y, Hayashi K. Fixation of osteochondral lesions of the talus using cortical bone pegs. J Bone Jt Surg Br. 2002;84:369–74.
35. Reilingh ML, Lambers KTA, Dahmen J, Opdam KTM, Kerkhoffs GMMJ. The subchondral bone healing after fixation of an osteochondral talar defect is superior in comparison with microfracture. Knee Surg Sport Traumatol Arthrosc. 2018;26:2177–82.
36. Steman JA, Dahmen J, Lambers KT, MMJ Kerkhoffs G. Return to sports after surgical treatment of osteochondral defects of the talus: a systematic review of 2347 cases. Orthop J Sports Med. 2019;7:1–15.
37. D'Ambrosi R, Villafañe JH, Indino C, Liuni FM, Berjano P, Usuelli FG. Return to sport after arthroscopic autologous matrix-induced Chondrogenesis for patients with osteochondral lesion of the talus. Clin J Sport Med. 2019;29:470–5.
38. Vannini F, Cavallo M, Ramponi L, Castagnini F, Massimi S, Giannini S, et al. Return to sports after bone marrow–derived cell transplantation for osteochondral lesions of the talus. Cartilage. 2017;8:80–7.
39. Shimozono Y, Hurley ET, Myerson CL, Kennedy JG. Good clinical and functional outcomes at mid-term following autologous osteochondral transplantation for osteochondral lesions of the talus. Knee Surg Sport Traumatol Arthrosc. 2018;26:3055–62.
40. Shimozono Y, Seow D, Yasui Y, Fields K, Kennedy JG. Knee-to-talus donor-site morbidity following autologous osteochondral transplantation: a meta-analysis with best-case and worst-case analysis. Clin Orthop Relat Res. 2019;477:1915–31.
41. Hu Y, Guo Q, Jiao C, Mei Y, Jiang D, Wang J, et al. Treatment of large cystic medial osteochondral lesions of the talus with autologous osteoperiosteal cylinder grafts. Arthroscopy. 2013;29:1372–9.

42. Kerkhoffs G, Altink J, Stufkens S, Dahmen J. Talar osteoperiostic grafting from the iliac crest (TOPIC) for large medial talar osteochondral defects: operative technique. Oper Orthop Traumatol. 2021; 33:160–9.
43. D'Hooghe P, Murawski CD, Boakye LAT, Osei-Hwedieh DO, Drakos MC, Hertel J, et al. Rehabilitation and return to sports: proceedings of the international consensus meeting on cartilage repair of the ankle. Foot Ankle Int. 2018;39:61S–7S.
44. Ardern CL, Glasgow P, Schneiders A, Witvrouw E, Clarsen B, Cools A, et al. 2016 consensus statement on return to sport from the first world congress in sports physical therapy, Bern. Br J Sports Med. 2016;50:853–64.
45. Wainwright TW, Burgess LC, Middleton RG. Does neuromuscular electrical stimulation improve recovery following acute ankle sprain? A pilot randomised controlled trial. Clin Med Insights Arthritis Musculoskelet Disord. 2019;12:1179544119849024. https://doi.org/10.1177/1179544119849024.
46. Halasi T, Kynsburg Á, Tállay A, Berkes I. Development of a new activity score for the evaluation of ankle instability. Am J Sports Med. 2004;32:899–908.

Heel Spurs and Plantar Fasciitis in Runners

26

Masato Takao, Kosui Iwashita, Yasuyuki Jujo, Mai Katakura, and Yoshiharu Shimozono

26.1 Etiology and Epidemiology

Plantar fasciitis commonly causes inferior heel pain and occurs in up to 10% of the population. The condition accounts for more than 600,000 annual outpatient visits in the United States [1] and is also one of the most common disorders in runners occurring in 4.5–31% of runners [2, 3]. In running, the ground reaction force at the time of the midstance phase ranges from 1.5 to 5 times body weight [4, 5]. At the pace of 7 min per mile, running implies approximately 5000 contacts per hour of running [4, 5]. Considering the huge loads on the tissues, it is clear that even small abnormalities can result in a significant load concentration on the foot [2]. Foot and lower limbs muscles also play a pivotal role in movement patterns of gait and run cycle and, as expected, in the onset and progression of plantar fasciitis [6]. It has been highlighted that a difference in rearfoot load in recreational runners with plantar fasciitis, with respect to the stage of disease and with respect to the healthy runners [7], may be related to plantar fascia stiffness [8].

M. Takao (✉) · K. Iwashita · Y. Jujo
Clinical and Research Institute for Foot and Ankle Surgery, Jujo Hospital, Chiba, Japan
e-mail: m.takao@carifas.com

M. Katakura
Fortius Clinic, London, UK

Y. Shimozono
New York University, New York, NY, USA

The plantar fascia is attached proximally to the calcaneus at the anterior medial calcaneal tubercle, the site of attachment for the digitorum brevis, and abductor halluces. Lemont et al. [9] reviewed the histological findings of 50 cases of plantar fasciitis and clarified that all included myxoid degeneration with fragmentation and degeneration of the plantar fascia and bone marrow vascular ectasia. Accordingly, plantar fasciitis is defined as degenerative fasciosis without inflammation. However, most surgeons consider the inciting inflammation to be local or systemic and that the inflammation may stem from the plantar fascia proper or may be secondary to inflammation in surrounding tissue [10]. The subcalcaneal bursa and medial tibial branch of the tibial nerve may be involved in what is seen as the general symptom complex of plantar fasciitis, especially in chronic cases.

26.2 Patient Evaluation

26.2.1 History

Patients typically report a gradual onset of pain in the inferior heel that is usually worse with the first steps in the morning or after a period of inactivity. Patients may also describe limping with the heel off the ground. The pain tends to lessen with gradually increased activity but worsens toward the end of the day with increased duration

G. L. Canata et al. (eds.), *Management of Track and Field Injuries*,
https://doi.org/10.1007/978-3-030-60216-1_26

of weight-bearing activity. Patients sometimes report that before the onset of their symptoms, they increased the amount or intensity of their regular walking or running regimen, changed footwear, or exercised on a different surface.

26.2.2 Physical Examination

Diagnosis of plantar fasciitis can be made with reasonable certainly on the basis of clinical assessment alone. Pain is usually localized to a small area of maximal tenderness over the anteromedial aspect of the inferior heel which is the proximal insertion of plantar fascia into the medial tubercle of the calcaneus. The pain response to palpation over this small area involves considerable apprehension, and evasive action may be taken by the patient to avoid further investigation. A small percentage of cases are positive for the windlass test [10], which is generally regarded as a clinical test with high specificity and low sensitivity for diagnosis of plantar fasciitis [11].

26.2.3 Imaging

Imaging modalities include ultrasonography and magnetic resonance imaging (MRI) for investigation of soft tissue structures and plane radiography for bone abnormalities, which help to elucidate the underlying pathology of the disorder and assist in the formation of an accurate diagnosis and targeted treatment plan [12].

The thickness of the proximal plantar fascia is considered to reflect the pathology of plantar fasciitis [12]. Some reports using ultrasonography have shown that patients with plantar fasciitis had a 2.16 mm thicker plantar fascia than controls, and are more likely to have plantar fascia thickness >4 mm. Similarly, MRI has revealed that patients with plantar fasciitis have proximal plantar fascia 3.35 mm thicker than controls [12]. With the development of imaging technology and progress in equipment, ultrasonography can be a reliable method for the measurement of plantar fascia thickness [8]. Therefore, the authors recommend ultrasonography as a simple, reliable, and cost-effective tool for diagnosis of plantar fasciitis (Fig. 26.1).

Lateral plain radiography can show the presence of a plantar calcaneal spur in many cases of plantar fasciitis. Spurs are closely associated with the abductor halluces and the flexor digitorum brevis origin [13] and most commonly occur close to the plantar fascia enthesis [14]. The formation of plantar calcaneal spurs has traditionally been attributed to repetitive longitudinal traction of the fascia [15] with subsequent

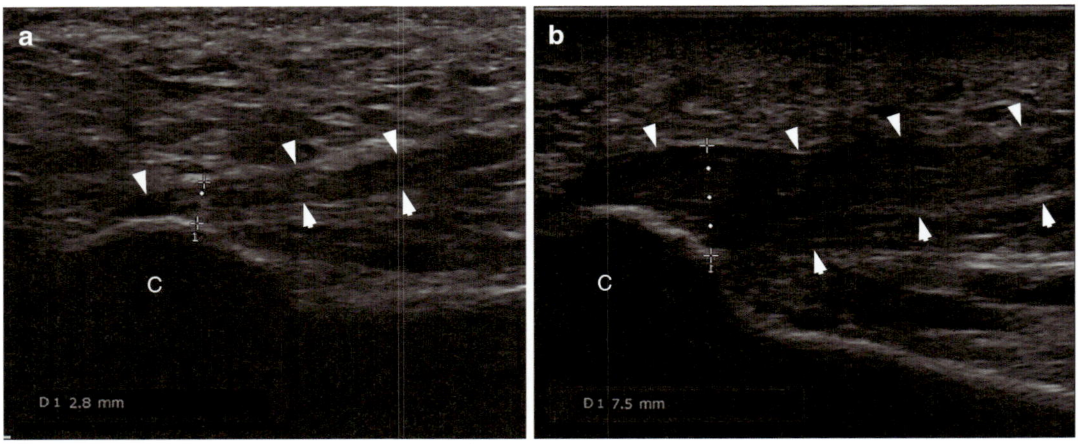

Fig. 26.1 Ultrasonography of normal (**a**) and abnormal case (**b**) **c** indicates a calcaneus and the arrow heads indicate a plantar fascia. The thickness of the proximal plantar fascia in abnormal case (7.5 mm, **b**) is thicker than normal case (2.8 mm, **a**)

inflammation and reactive ossification [16]. However, Li and Murhleman [14] performed a histological study and clarified that a spur tubercle commonly forms perpendicular to its long axis. Furthermore, Menz et al. [16] reported that spur development is unrelated to medial arch height. These reports suggest that vertical compression may play an important role in the spur development. The role of the plantar calcaneal spur in the pathogenesis of plantar fasciitis has been questioned for several decades [15, 17]. The basis of this uncertainty was the reportedly high prevalence of the calcaneal spur in the asymptomatic population [18], leading to an emerging view that the finding has limited diagnostic value [19]. On the other hand, a previous study reported evidence of the plantar calcaneal spur by ultrasonography and clarified that the presence of this structure was found in 45% of chronic plantar heel pain participants and in only 2% of controls [20]. To conclude this question, McMillan et al. [12] conducted a systematic review of 23 studies and performed a meta-analysis, and they concluded that plantar calcaneal spur formation is strongly associated with pain beneath the heel. An anatomical dissection study showed that there are rich vascular and nerve structures around the plantar calcaneal spur [21]. Accordingly, we recommend excision of the plantar calcaneal spurs in cases treated surgically.

Other causes of pain in the inferior heel include rupture of the plantar fascia, subcalcaneal bursitis, calcaneal stress fracture, infection, fat-pad atrophy, medial calcaneal nerve entrapment, tarsal tunnel syndrome, seronegative arthropathy, Reiter's syndrome, Paget's disease, psoriatic arthritis, Sever's disease, and tumors, which are usually distinguishable by assessment of history, physical examination, and imaging [10, 22].

26.3 Management

26.3.1 Conservative Treatments

Most patients with plantar fasciitis respond to conservative modalities, which are considered as the first line treatments. Lutter [23] reported that 85% of patients with symptomatic plantar fasciitis responded to conservative management, with surgery indicated for the remaining 15%. A long-term follow-up study [24] showed that 80% of patients with plantar fasciitis treated conservatively had complete resolution of pain after 4 years. Several conservative treatments have been reported including corticosteroid local injection [25, 26], Botulinum toxin local injection [25], platelet-rich plasma (PRP) local injection [27], autologous blood local injection [28], extracorporeal shockwave therapy [29], orthosis [30], manipulation [30], stretching [30], bipolar radiofrequency therapy [31], low-frequency electrical stimulation [32], acupuncture [33], taping [34], laser therapy [35], custom made footwear [30], and trigger point block of the gastrocnemius [36]. Although most patients improve in response to these conservative treatments, there is a lack of data from high-quality, randomized, controlled trials that support the efficacy of these therapies.

In runners, treatments that can degrade performance should be avoided. In addition, runners tend to dislike orthoses, which cause foot discomfort during running. Special care should be taken for local injection of corticosteroids. There is a risk that the plantar aponeurosis may rupture, causing unbearable pain in the heel for long periods of time [37]. Accordingly, the authors recommend manipulation and stretching (Fig. 26.2), extracorporeal shockwave therapy (Fig. 26.3) and local injection of PRP under ultrasonography (Fig. 26.4) for the treatment of plantar fasciitis in runners.

Extracorporeal shockwave therapy (ESWT) has been applied to orthopedic surgery since 2000 and widely used in the treatment of plantar fasciitis due to its noninvasive nature and fast recovery time. This therapy not only promotes the destruction of nerve endings [38] and suppresses conduction of neurotransmitters to alleviate chronic pain [39], but also stimulates production of various growth factors and differentiation/migration factors of cells by stimulating cells and the extracellular matrix, inducing tissue repair and regeneration [40, 41]. In addition, it is thought to directly affect inflammation by suppressing the production of inflammatory

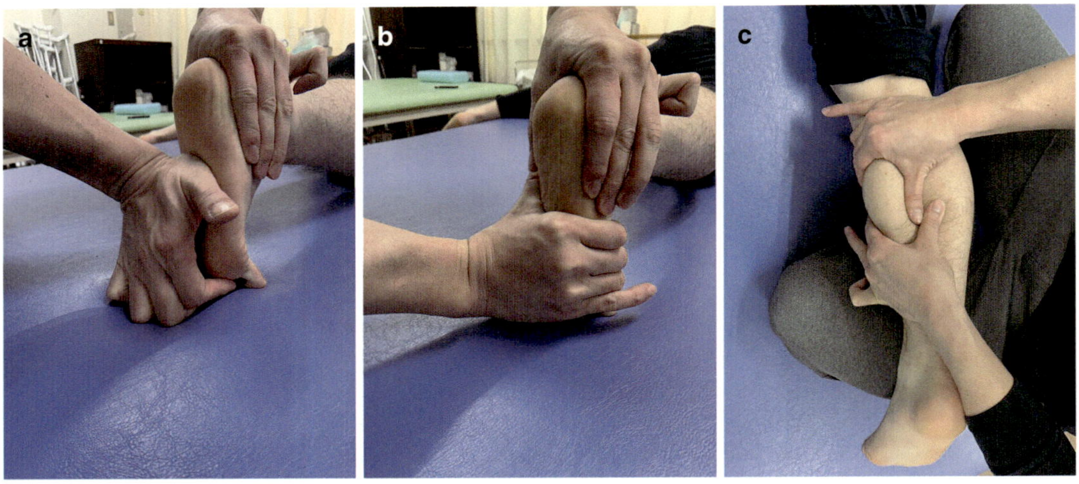

Fig. 26.2 Stretching (**a**), plantar fascia manipulation (**b**), and calf muscle manipulation (**c**)

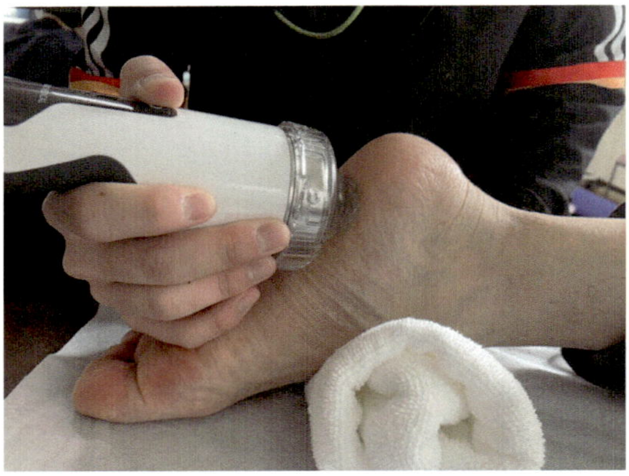

Fig. 26.3 Extracorporeal shock wave therapy

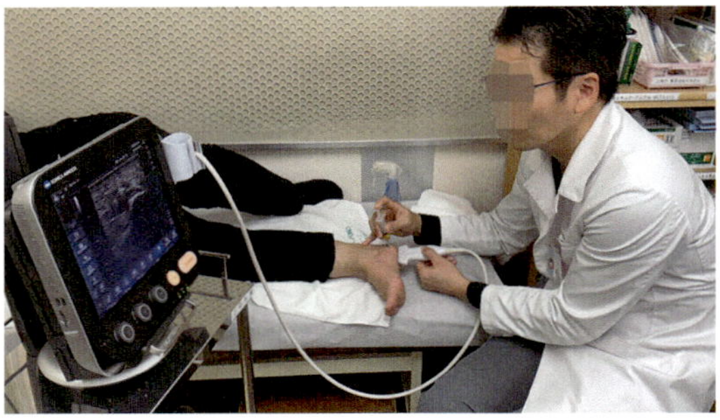

Fig. 26.4 Local injection of platelet-rich plasma under ultrasonography

cytokines [42]. Currently, both focused and radial shockwave therapies are available as treatment options and numerous studies reported their effectiveness in the treatment of plantar fasciitis. However, a recent meta-analysis revealed that focused shockwave therapy can result in a higher success rate and greater pain reduction [42]. Further studies are warranted due to limitations of studies included in the meta-analysis.

PRP is an autologous biological product with increased concentration of platelets suspended in a small amount of plasma after centrifugation. The utility of PRP for plantar fasciitis has been demonstrated in a meta-analysis of randomized controlled trials with high levels of evidence [27, 43–48], and PRP is a safe and effective opinion. PRP contains abundant growth factors and bioactive cytokines, which are believed to promote tissue healing, although corticosteroids have no such regenerative capacity. Therefore, while both PRP and corticosteroids can decrease inflammation, PRP is advantageous over corticosteroids. However, the composition of PRP is different among the preparation devices, although detailed composition including platelets and leukocyte is a critical factor for the treatment. For instance, leukocyte-rich or leukocyte-poor PRP can have differing effects on various pathologies; however, no study has investigated this in plantar fasciitis. Furthermore, it is known that the blood component before PRP generation varies depending on the timing of collection, and it has been pointed out that it affects the therapeutic effect [49]. In our experience, the effectiveness of PRP therapy for plantar fasciitis in runners is about 50%. To improve patient outcomes, it is an important research topic to clarify which components of PRP are most effective for the treatment of plantar fasciitis.

26.3.2 Surgical Treatment

When conservative treatment has failed, open or endoscopic partial fasciotomy is considered. There are many causes of heel pain including calcaneal stress fracture, heel pad atrophy, systemic inflammation disease, nerve compression, neoplasia, and infection, which must be excluded before surgical treatment of plantar fasciitis. In plantar fasciotomy, resection of the plantar fascia more than medial 60% may lead to progressive pes planus and/or lateral foot pain [50, 51], and no such complications were observed after medial one-third to one-half plantar fascia release in the report of 19 plantar fascia endoscopic release [52]. Accordingly, releasing the medial one-third to one-half of the plantar fascia is recommended.

There are only four studies in the literature that directly compare open and endoscopic surgery [53–56]. Tomczak and Haverstock [54] reviewed 34 cases of endoscopic plantar fasciotomy and 34 cases of open plantar fasciotomy with calcaneal spur resection and showed that the time between surgery and return to work was 34 days for the endoscopic surgery group and 84 days for open surgery group. Kinley et al. [53] compared the results of 66 endoscopic and 26 open plantar fascia releases. Eighty percent of patients had pain resolution and returned to activity in 6.3 weeks in the endoscopic surgery group and 10.3 weeks in the open surgery group. Pain was 45% less for endoscopic surgery compared with open surgery. Serious and total complications were seen in 17% and 41% of endoscopic surgery subjects and 35% and 58% of open surgery subjects. Accordingly, endoscopic surgery for plantar fasciitis is generally considered less invasive than open surgery.

Endoscopic surgery for plantar fasciitis was firstly described by Barrett and Day in 1991 [57]. This technique has the advantage of no exposure of the nerve to the abductor digiti quinti. Following Barrett's first report [57], most surgeons have used the superficial fascial approach, in which surgical devices are inserted from inferior to the plantar fascia, to release the medial one-third to one-half of the plantar fascia using the same type of hook knife used for endoscopic carpal tunnel release [52, 55, 58–69]. Endoscopic techniques have potential risks of damaging relevant structures in the operation field; however, a cross-sectional anatomic study by Reeve et al. [70] investigating the structures at risk during endoscopic plantar fascia release showed that the average distance between the cannula margin to the nerve to the abductor digiti minimi was 6 mm at the medial border of the plantar fascia, and no damage of the nerve was observed after endoscopic plantar fascia release. According to previous reports, 68–100% of patients showed good to excellent clinical results [52, 55, 58–69]. However, the superficial fascial approach has some disadvantages, including insufficient field of vision and narrow working space because the operative field of view is between the skin and the plantar fascia and is filled with adipose tissue. Another disadvantage is that it is difficult to remove calcaneal spurs through this approach because they typically exist deep underneath the plantar fascia [71].

The deep fascial approach was developed to resolve the issues of the superficial fascial approach [70–72] by facilitating a larger space in the dorsal side of the plantar fascia than in the plantar side. Therefore, the deep fascial approach enables a wider field of vision and a larger working space. Usually, two portals, the medial and lateral portals, are adopted [72]. The authors recommend this approach, especially in cases with calcaneal spurs, because of easier access to the spurs.

In the deep fascial approach, the patient is placed in the supine position to elevate the affected foot by approximately 15 cm using a leg holder (Fig. 26.5). A pneumotourniquet is applied to the thigh and inflated to a pressure of systolic blood pressure plus 100–150 mmHg. A medial portal is then made. Under fluoroscopy, a needle is inserted 5 mm superior to the plantar fascia and

10 mm anterior to its origin on the calcaneus (Fig. 26.6a). A 5-mm vertical incision is made only in the skin (Fig. 26.6b), and blunt dissection is performed with Pean's mosquito forceps to only the supra-medial aspect of the plantar fascia (Fig. 26.6c). During blunt dissection, it is important to touch the anterior calcaneal tubercle and calcaneal spurs with the tip of the forceps in order to dissect bluntly enough around them to ensure a large working space with minor excision of the flexor digitorum brevis. Next, a lateral portal is established by passing a blunt troche through the medial portal superior and perpendicular to the plantar fascia and across to the lateral aspect of the foot. A vertical skin incision is created in a tent of skin, which is pushed up by the troche (Fig. 26.7a), and a blunt troche is penetrated (Fig. 26.7b). Then, a 4.0-mm diameter (30°) arthroscope is inserted through the lateral portal, while the surgical devices are inserted through the medial portal (Fig. 26.8). A motorized shaver

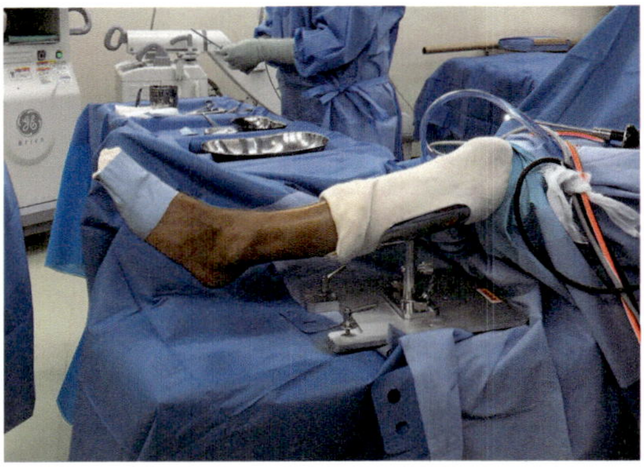

Fig. 26.5 Position of the patients. The patient is placed in the supine position to elevate the affected foot by approximately 15 cm with a leg holder

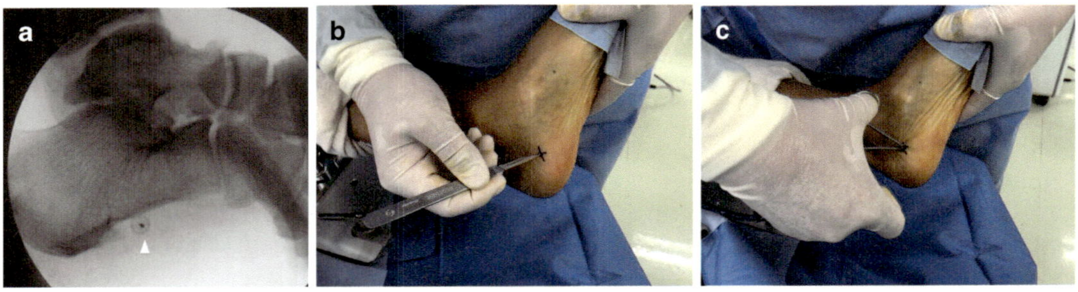

Fig. 26.6 Making a medial portal. Under fluoroscopy, a needle (arrowhead) is inserted 5 mm superior to the plantar fascia and 10 mm anterior to its origin on the calcaneus (**a**). A 5-mm vertical incision is made only in the skin (**b**), and blunt dissection is done with Pean's mosquito forceps to only the supra-medial aspect of the plantar fascia (**c**)

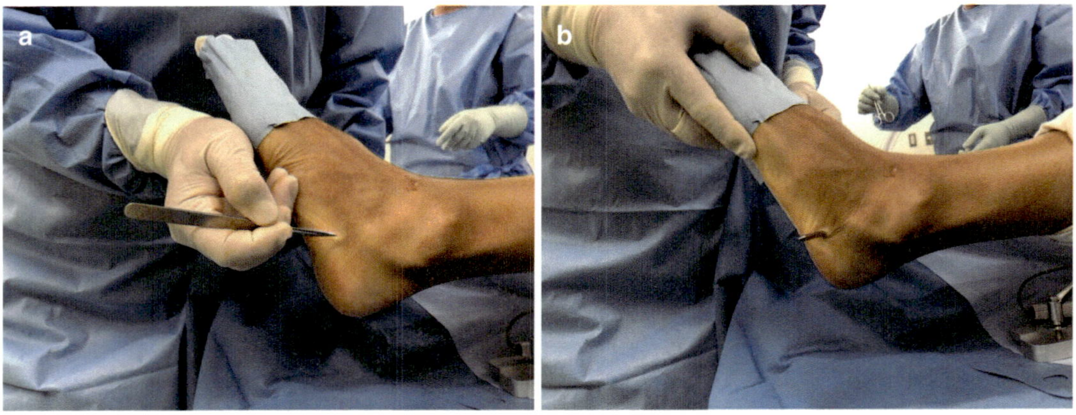

Fig. 26.7 Making a lateral portal. A blunt troche through the medial portal superior and perpendicular to the plantar fascia and across to the lateral aspect of the foot. A vertical skin incision is created in a tent of skin which is pushed up by the troche (**a**), and a blunt troche is penetrated (**b**)

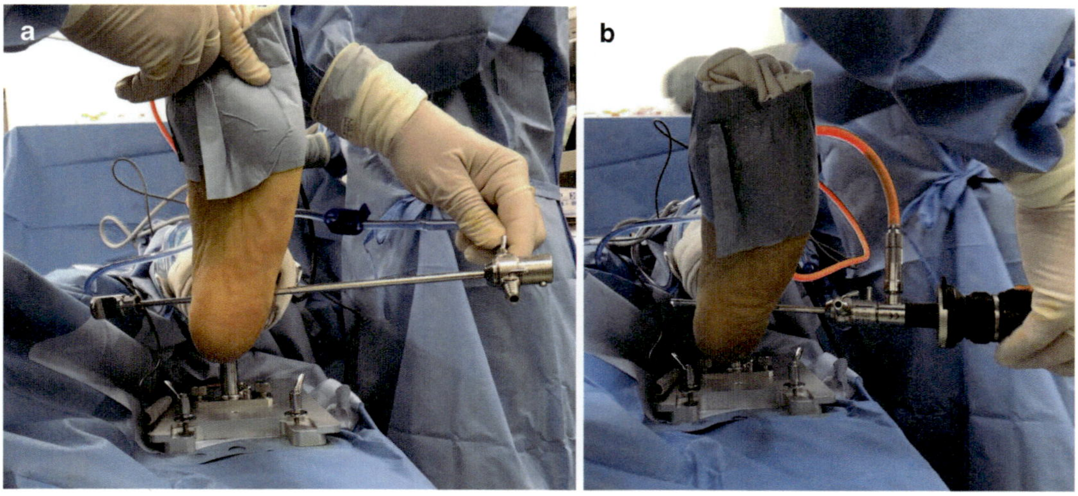

Fig. 26.8 Inserting an arthroscope through the lateral portal. A 4.0-mm diameter (30°) arthroscope is inserted through the lateral portal, while the surgical devices are inserted through the medial portal

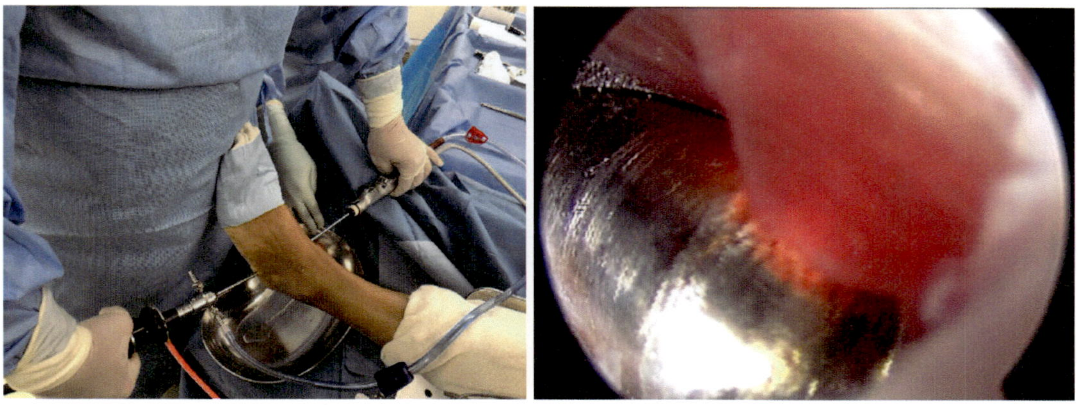

Fig. 26.9 Making a working space. A motorized shaver with a diameter of 3.5 mm is inserted via medial portal to resect a part of flexor digitorum muscle for making a working space

with a diameter of 3.5 mm is used for making a working space to excise the adipose tissue and a plantar part of the flexor digitorum brevis, as minimally as possible, to obtain a good view (Fig. 26.9). For leading the shaver into the field of vision of the endoscopy, it is helpful that the tip of the shaver is outside of the medial portal and afterwards moves into the working space. In making the working space, the anterior wall of the calcaneus and the calcaneal attachment of the plantar fascia should be identified as landmarks. In most cases with a calcaneal spur, the upper side of the calcaneal spur is covered with the flexor digitorum brevis and the lower side is covered with the plantar fascia. After detaching these structures from the spur using the arthroknife (Fig. 26.10a), the calcaneal spur is resected using an abrader burr (Fig. 26.10b). The plantar fascia can be observed after removing the calcaneal spur. A width of plantar fascia is measured with a probe, and an area less than the medial one-third of the plantar fascia is resected using an arthroknife (Fig. 26.11). Care should be taken to remove all layers of the plantar fascia to ensure no residual plantar pain after surgery. The plantar fascia should be removed until the plantar adipose

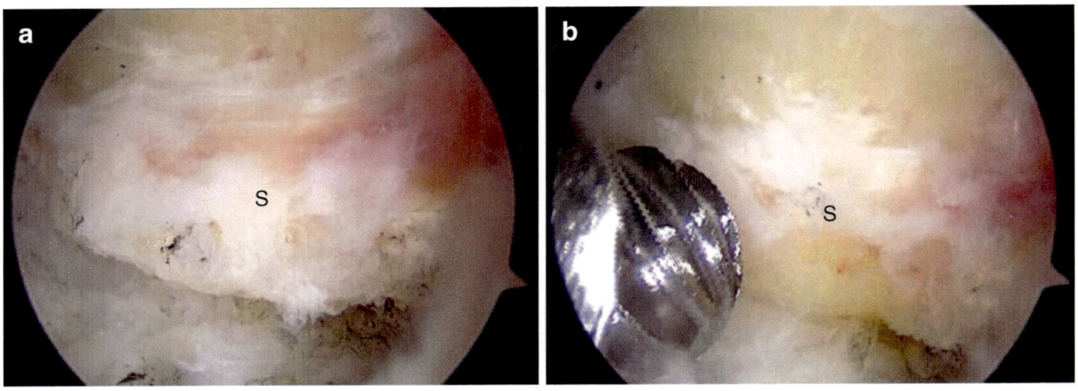

Fig. 26.10 Resection of the calcaneal spurs. S indicates a calcaneal spur. After detaching these structures from the spur using the arthroknife (**a**), the calcaneal spur is resected using an abrader burr (**b**)

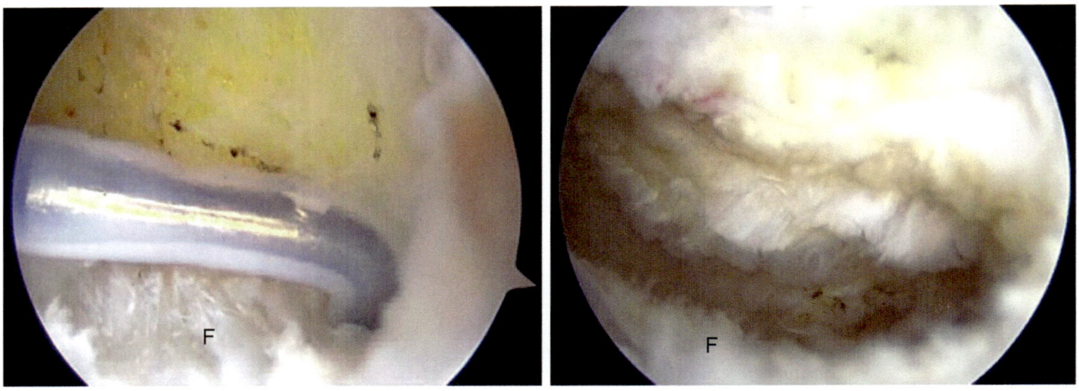

Fig. 26.11 Partial resection of the plantar fascia. An area less than the medial half of the plantar fascia is resected using an arthroknife

tissue is exposed, which is a sign that the plantar fascia has been completely resected toward its deeper layer.

Active range-of-motion exercises of the foot and ankle are performed 1 day after surgery. Partial weight-bearing is allowed 3 days after surgery and gradually increases to full weight-bearing in accordance with patient tolerance.

A previous report of endoscopic surgery using the deep fascial approach for 10 ft. of eight patients [23] showed that the mean AOFAS score was 64.2 ± 6.3 points before surgery and 92.6 ± 7.1 points at 2 years after surgery ($p < 0.001$). In a recent study conducted by authors for 33 ft. of 33 runners, the mean AOFAS score was 65.8 ± 8.8 points before surgery and 90.4 ± 7.6 points at 1 years after surgery ($p < 0.001$). The duration to full weight-bearing after surgery was a mean 4.2 ± 6.3 days (range 1–14 days), and the duration to jogging after surgery was a mean 4.5 ± 8.3 weeks (range 2–6 weeks). All patients had returned to full athletic activities by a mean of 11.7 ± 5.6 weeks (range 6–18 weeks). There were no serious complications, but four patients showed dysfunction of abductor digiti minimi. The dominant nerve of the abductor digiti minimi is the lateral plantar nerve (Baxter's nerve). It runs about 9 mm away from the incision of the plantar aponeurosis, and it can be damaged during surgery. Although this is a mild complication that does not interfere with daily life and sports activities, it is necessary to fully explain the possibility of this disorder to patients before surgery.

References

1. Riddle DL, Schappert SM. Volume of ambulatory care visits and patterns of care for patients diagnosed with plantar fasciitis: a national study of medical doctors. Foot Ankle Int. 2004;25:303–10.
2. Di Caprio F, Buda R, Mosca M, Calabro' A, Giannini S. Foot and lower limb diseases in runners: assessment of risk factors. J Sports Sci Med. 2010;9:587–96.
3. Lopes AD, Hespanhol Júnior LC, Yeung SS, Costa LOP. What are the main running-related musculoskeletal injuries? A systematic review. Sports Med. 2012;42:891–905.
4. Birrer RB, Buzermanis S, DellaCorte MP, Grisalfi PJ. Biomechanics of running. Textb Run Med. New York: McGraw-Hill, Medical Publishing Division; 2001. p. 11–19.
5. Cavanagh PR, Lafortune MA. Ground reaction forces in distance running. J Biomech. 1980;13:397–406.
6. Petraglia F, Ramazzina I, Costantino C. Plantar fasciitis in athletes: diagnostic and treatment strategies. A systematic review. Muscles Ligaments Tendons J. 2017;7:107–18.
7. Ribeiro AP, João SMA, Dinato RC, Tessutti VD, Sacco ICN. Dynamic patterns of forces and loading rate in runners with unilateral plantar fasciitis: a cross-sectional study. PLoS One. 2015;10:e0136971.
8. Wu C-H, Chang K-V, Mio S, Chen W-S, Wang T-G. Sonoelastography of the plantar fascia. Radiology. 2011;259:502–7.
9. Lemont H, Ammirati KM, Usen N. Plantar fasciitis: a degenerative process (fasciosis) without inflammation. J Am Podiatr Med Assoc. 2003;93:234–7.
10. Bartold SJ. The plantar fascia as a source of pain—biomechanics, presentation and treatment. J Bodyw Mov Ther. 2004;8:214–26.
11. De Garceau D, Dean D, Requejo SM, Thordarson DB. The association between diagnosis of plantar fasciitis and windlass test results. Foot Ankle Int. 2003;24:251–5.
12. McMillan AM, Landorf KB, Barrett JT, Menz HB, Bird AR. Diagnostic imaging for chronic plantar heel pain: a systematic review and meta-analysis. J Foot Ankle Res. 2009;2:32.
13. Osborne HR, Breidahl WH, Allison GT. Critical differences in lateral X-rays with and without a diagnosis of plantar fasciitis. J Sci Med Sport. 2006;9:231–7.
14. Li J, Muehleman C. Anatomic relationship of heel spur to surrounding soft tissues: greater variability than previously reported. Clin Anat. 2007;20:950–5.
15. Wainwright AM, Kelly AJ, Winson IG. Calcaneal spurs and plantar fasciitis. Foot. 1995;5:123–6.
16. Menz HB, Zammit GV, Landorf KB, Munteanu SE. Plantar calcaneal spurs in older people: longitudinal traction or vertical compression? J Foot Ankle Res. 2008;1:7.
17. Wearing SC, Smeathers JE, Urry SR, Hennig EM, Hills AP. The pathomechanics of plantar fasciitis. Sports Med. 2006;36:585–611.
18. Prichasuk S, Subhadrabandhu T. The relationship of pes planus and calcaneal spur to plantar heel pain. Clin Orthop. 1994;306:192–6.
19. Singh D, Angel J, Bentley G, Trevino SG. Fortnightly review. Plantar fasciitis. BMJ. 1997;315:172–5.
20. Gibbon WW, Long G. Ultrasound of the plantar aponeurosis (fascia). Skelet Radiol. 1999;28:21–6.
21. Kumai T, Benjamin M. Heel spur formation and the subcalcaneal enthesis of the plantar fascia. J Rheumatol. 2002;29:1957–64.
22. Buchbinder R. Clinical practice. Plantar fasciitis. N Engl J Med. 2004;350:2159–66.
23. Lutter LD. Surgical decisions in athletes' subcalcaneal pain. Am J Sports Med. 1986;14:481–5.
24. Wolgin M, Cook C, Graham C, Mauldin D. Conservative treatment of plantar heel pain: long-term follow-up. Foot Ankle Int. 1994;15:97–102.
25. Díaz-Llopis IV, Rodríguez-Ruíz CM, Mulet-Perry S, Mondéjar-Gómez FJ, Climent-Barberá JM, Cholbi-Llobel F. Randomized controlled study of the efficacy of the injection of botulinum toxin type A versus corticosteroids in chronic plantar fasciitis: results at one and six months. Clin Rehabil. 2012;26:594–606.
26. Moustafa AMA, Hassanein E, Foti C. Objective assessment of corticosteroid effect in plantar fasciitis: additional utility of ultrasound. Muscles Ligaments Tendons J. 2015;5:289–96.
27. Akşahin E, Doğruyol D, Yüksel HY, Hapa O, Doğan O, Celebi L, et al. The comparison of the effect of corticosteroids and platelet-rich plasma (PRP) for the treatment of plantar fasciitis. Arch Orthop Trauma Surg. 2012;132:781–5.
28. Lee TG, Ahmad TS. Intralesional autologous blood injection compared to corticosteroid injection for treatment of chronic plantar fasciitis. A prospective, randomized, controlled trial. Foot Ankle Int. 2007;28:984–90.
29. Rompe JD, Cacchio A, Weil L, Furia JP, Haist J, Reiners V, et al. Plantar fascia-specific stretching versus radial shock-wave therapy as initial treatment of plantar fasciopathy. J Bone Joint Surg Am. 2010;92:2514–22.
30. Drake M, Bittenbender C, Boyles RE. The short-term effects of treating plantar fasciitis with a temporary custom foot orthosis and stretching. J Orthop Sports Phys Ther. 2011;41:221–31.
31. Weil L, Glover JP, Weil LS. A new minimally invasive technique for treating plantar fasciosis using bipolar radiofrequency: a prospective analysis. Foot Ankle Spec. 2008;1:13–8.
32. Stratton M, McPoil TG, Cornwall MW, Patrick K. Use of low-frequency electrical stimulation for the treatment of plantar fasciitis. J Am Podiatr Med Assoc. 2009;99:481–8.
33. Cotchett MP, Landorf KB, Munteanu SE, Raspovic AM. Consensus for dry needling for plantar heel pain (plantar fasciitis): a modified Delphi study. Acupunct Med. 2011;29:193–202.
34. Abd El Salam MS, Abd Elhafz YN. Low-dye taping versus medial arch support in managing pain and

pain-related disability in patients with plantar fasciitis. Foot Ankle Spec. 2011;4:86–91.
35. Kiritsi O, Tsitas K, Malliaropoulos N, Mikroulis G. Ultrasonographic evaluation of plantar fasciitis after low-level laser therapy: results of a double-blind, randomized, placebo-controlled trial. Lasers Med Sci. 2010;25:275–81.
36. Renan-Ordine R, Alburquerque-Sendín F, de Souza DPR, Cleland JA, Fernández-de-Las-Peñas C. Effectiveness of myofascial trigger point manual therapy combined with a self-stretching protocol for the management of plantar heel pain: a randomized controlled trial. J Orthop Sports Phys Ther. 2011;41:43–50.
37. Acevedo JI, Beskin JL. Complications of plantar fascia rupture associated with corticosteroid injection. Foot Ankle Int. 1998;19:91–7.
38. Ohtori S, Inoue G, Mannoji C, Saisu T, Takahashi K, Mitsuhashi S, et al. Shock wave application to rat skin induces degeneration and reinnervation of sensory nerve fibres. Neurosci Lett. 2001;315:57–60.
39. Takahashi N, Wada Y, Ohtori S, Saisu T, Moriya H. Application of shock waves to rat skin decreases calcitonin gene-related peptide immunoreactivity in dorsal root ganglion neurons. Auton Neurosci. 2003;107:81–4.
40. Wang C-J, Wang F-S, Yang KD, Weng L-H, Hsu C-C, Huang C-S, et al. Shock wave therapy induces neovascularization at the tendon-bone junction. A study in rabbits. J Orthop Res. 2003;21:984–9.
41. Han SH, Lee JW, Guyton GP, Parks BG, Courneya J-P, Schon LC. J.Leonard Goldner award 2008. Effect of extracorporeal shock wave therapy on cultured tenocytes. Foot Ankle Int. 2009;30:93–8.
42. Sun J, Gao F, Wang Y, Sun W, Jiang B, Li Z. Extracorporeal shock wave therapy is effective in treating chronic plantar fasciitis: a meta-analysis of RCTs. Medicine (Baltimore). 2017;96:e6621.
43. Martinelli N, Marinozzi A, Carnì S, Trovato U, Bianchi A, Denaro V. Platelet-rich plasma injections for chronic plantar fasciitis. Int Orthop. 2013;37:839–42.
44. Ragab EMS, Othman AMA. Platelets-rich plasma for treatment of chronic plantar fasciitis. Arch Orthop Trauma Surg. 2012;132:1065–70.
45. Jain K, Murphy PN, Clough TM. Platelet-rich plasma versus corticosteroid injection for plantar fasciitis: a comparative study. Foot Edinb Scotl. 2015;25:235–7.
46. Mahindra P, Yamin M, Selhi HS, Singla S, Soni A. Chronic plantar fasciitis: effect of platelet-rich plasma, corticosteroid, and placebo. Orthopedics. 2016;39:e285–9.
47. Monto RR. Platelet-rich plasma efficacy versus corticosteroid injection treatment for chronic severe plantar fasciitis. Foot Ankle Int. 2014;35:313–8.
48. Hurley ET. Platelet-rich plasma versus corticosteroids for plantar fasciitis: a meta-analysis of randomized control trials. Barcelona, Spain: 19th EFORT Annual Congress; 2018 Jun.
49. Chahla J, Cinque ME, Piuzzi NS, Mannava S, Geeslin AG, Murray IR, et al. A call for standardization in platelet-rich plasma preparation protocols and composition reporting: a systematic review of the clinical Orthopaedic literature. J Bone Joint Surg Am. 2017;99:1769–79.
50. Brugh AM, Fallat LM, Savoy-Moore RT. Lateral column symptomatology following plantar fascial release: a prospective study. J Foot Ankle Surg. 2002;41:365–71.
51. Sharkey NA, Ferris L, Donahue SW. Biomechanical consequences of plantar fascial release or rupture during gait: part I – disruptions in longitudinal arch conformation. Foot Ankle Int. 1998;19:812–20.
52. Bazaz R, Ferkel RD. Results of endoscopic plantar fascia release. Foot Ankle Int. 2007;28:549–56.
53. Kinley S, Frascone S, Calderone D, Wertheimer SJ, Squire MA, Wiseman FA. Endoscopic plantar fasciotomy versus traditional heel spur surgery: a prospective study. J Foot Ankle Surg. 1993;32:595–603.
54. Tomczak RL, Haverstock BD. A retrospective comparison of endoscopic plantar fasciotomy to open plantar fasciotomy with heel spur resection for chronic plantar fasciitis/heel spur syndrome. J Foot Ankle Surg. 1995;34:305–11.
55. Brekke MK, Green DR. Retrospective analysis of minimal-incision, endoscopic, and open procedures for heel spur syndrome. J Am Podiatr Med Assoc. 1998;88:64–72.
56. Wander DS. A retrospective comparison of endoscopic plantar fasciotomy to open plantar fasciotomy with heel spur resection for chronic plantar fasciitis/heel spur syndrome. J Foot Ankle Surg. 1996;35:183–4.
57. Barrett SL, Day SV. Endoscopic plantar fasciotomy for chronic plantar fasciitis/heel spur syndrome: surgical technique – early clinical results. J Foot Surg. 1991;30:568–70.
58. Bader L, Park K, Gu Y, O'Malley MJ. Functional outcome of endoscopic plantar fasciotomy. Foot Ankle Int. 2012;33:37–43.
59. Barrett SL, Day SV. Endoscopic plantar fasciotomy: two portal endoscopic surgical techniques – clinical results of 65 procedures. J Foot Ankle Surg. 1993;32:248–56.
60. Barrett SL. Endoscopic plantar fasciotomy. Clin Podiatr Med Surg. 1994;11:469–81.
61. Hake DH. Endoscopic plantar fasciotomy: a minimally traumatic procedure for chronic plantar fasciitis. Ochsner J. 2000;2:175–8.
62. Hogan KA, Webb D, Shereff M. Endoscopic plantar fascia release. Foot Ankle Int. 2004;25:875–81.
63. Lundeen RO, Aziz S, Burks JB, Rose JM. Endoscopic plantar fasciotomy: a retrospective analysis of results in 53 patients. J Foot Ankle Surg. 2000;39:208–17.
64. Ogilvie-Harris DJ, Lobo J. Endoscopic plantar fascia release. Arthroscopy. 2000;16:290–8.
65. Othman AMA, Ragab EM. Endoscopic plantar fasciotomy versus extracorporeal shock wave therapy for treatment of chronic plantar fasciitis. Arch Orthop Trauma Surg. 2010;130:1343–7.

66. Saxena A. Uniportal endoscopic plantar fasciotomy: a prospective study on athletic patients. Foot Ankle Int. 2004;25:882–9.
67. El Shazly O, El Hilaly RA, Abou El Soud MM, El Sayed MNMA. Endoscopic plantar fascia release by hooked soft-tissue electrode after failed shock wave therapy. Arthroscopy. 2010;26:1241–5.
68. Stone PA, Davies JL. Retrospective review of endoscopic plantar fasciotomy – 1992 through 1994. J Am Podiatr Med Assoc. 1996;86:414–20.
69. Urovitz EP, Birk-Urovitz A, Birk-Urovitz E. Endoscopic plantar fasciotomy in the treatment of chronic heel pain. Can J Surg. 2008;51:281–3.
70. Reeve F, Laughlin RT, Wright DG. Endoscopic plantar fascia release: a cross-sectional anatomic study. Foot Ankle Int. 1997;18:398–401.
71. Komatsu F, Takao M, Innami K, Miyamoto W, Matsushita T. Endoscopic surgery for plantar fasciitis: application of a deep-fascial approach. Arthroscopy. 2011;27:1105–9.
72. Blanco CE, Leon HO, Guthrie TB. Endoscopic treatment of calcaneal spur syndrome: a comprehensive technique. Arthroscopy. 2001;17:517–22.

27 Nerve Injuries in the Foot and Ankle: Neuromas, Neuropathy, Entrapments, and Tarsal Tunnel Syndrome

Lorraine Boakye, Nia A. James, Cortez L. Brown, Stephen P. Canton, Devon M. Scott, Alan Y. Yan, and MaCalus V. Hogan

27.1 Morton's Neuroma

Morton's neuroma results in metatarsalgia from the entrapment of the third interdigital nerve between the transverse intermetatarsal ligament and fascia [1]. Interdigital nerves arise from the medial and lateral plantar nerves. They course along the metatarsals and cross the deep transverse metatarsal ligaments. Symptoms are thought to be due to mechanical compression of the nerve, with incidence high in sports requiring relatively higher strain on the foot. The compression is thought to result in demyelination of the nerve as well as a fibrotic nodule with "perineural fibrosis" [2]. Patients will describe pain in the forefoot with shooting pain to the toes upon compression. Diagnosis is made via a combination of clinical evaluation and imaging; typically, ultrasonography or magnetic resonance imaging (MRI). Hallmarks of nonoperative treatment include use of oral nonsteroidal anti-inflammatory drugs (NSAIDs), cessation of athletic activity, change in footwear and corticosteroid injection [1]. Surgical treatment is considered only if symptoms are recurrent or persistent. The goal of surgery is to decompress the nerve with or without neuroma excision. Both dorsal and plantar approaches have been described; however, there

L. Boakye
Department of Orthopaedic Surgery, University of Pittsburgh Medical Center, Pittsburgh, PA, USA
e-mail: boakyela@upmc.edu

N. A. James · S. P. Canton
Department of Orthopaedic Surgery, University of Pittsburgh Medical Center, Pittsburgh, PA, USA

The University of Pittsburgh School of Medicine, Pittsburgh, PA, USA
e-mail: james.nia@medstudent.pitt.edu; canton.stephen@medstudent.pitt.edu

C. L. Brown
Department of Orthopaedic Surgery, University of Pittsburgh Medical Center, Pittsburgh, PA, USA

Florida State University College of Medicine, Tallahassee, FL, USA
e-mail: cortezbl@upmc.edu

D. M. Scott
Department of Orthopaedic Surgery, University of Pittsburgh Medical Center, Pittsburgh, PA, USA

Loma Linda University School of Medicine, Loma Linda, CA, USA
e-mail: dmscott@students.llu.edu

A. Y. Yan · M. V. Hogan (✉)
Department of Orthopaedic Surgery, University of Pittsburgh Medical Center, Pittsburgh, PA, USA

The University of Pittsburgh School of Medicine, Pittsburgh, PA, USA

The Foot and Ankle Injury [F.A.I.R] Group, University of Pittsburgh Medical Center, Pittsburgh, PA, USA
e-mail: yanay@upmc.edu; hoganmv@upmc.edu

G. L. Canata et al. (eds.), *Management of Track and Field Injuries*,
https://doi.org/10.1007/978-3-030-60216-1_27

is little data in regards to the ideal surgical approach [2]. Konstantine et al. describe a plantar approach in which sharp dissection is carried out to identify the common digital nerve and traced to its bifurcation. Branches are isolated and the common digital nerve is released proximal to the head of the metatarsal. A study by Konstantine et al. [1] found the majority of their patients had significant improvement of symptoms postoperatively.

27.2 Superficial Peroneal Nerve Entrapment

Superficial peroneal nerve entrapment (SPNE) is most commonly found in running athletes in their late twenties and early thirties. However, the general population can succumb to this injury as well. The incidence of men and women presenting with superficial peroneal nerve entrapment is fairly equal [3]. It is important to note that both the superficial and deep branches of the peroneal nerve may be at risk from forceful, repetitive movements such as those required to carry out the act of running [4]. However, we found no studies specific to superficial peroneal nerve dysfunction and running athletes.

The superficial peroneal nerve provides both motor and sensory innervation throughout its course. From a motor standpoint, this nerve allows the peroneus longus and brevis muscles to evert and plantarflex the foot and ankle, respectively. It also sends sensory information to the dorsum of the foot [5].

There are several possible etiologies that can lead to SPNE. Chronic inversion ankle sprains account for the majority of cases. However, this condition can also result from fibula fractures, exertional compartment syndrome, and potentially unknown etiologies in some cases [3].

The diagnosis of a superficial peroneal nerve entrapment can be quite elusive. Styf and Morberg reported only 3.5% of patients presenting with chronic leg pain to have entrapment of said nerve [6].The neuropathy is oftentimes a clinical diagnosis. Matsumoto et al. found that all of their patients reported pain and paresthesia at the lateral leg and dorsum foot [7]. Additionally, a positive Tinel's sign was present. Brown et al. reported tenderness to palpation at the fascial exit point for 87% (40/46) of their patients and a positive Tinel's sign for 84% [8]. Bregman and Schuenke introduced the use of a diagnostic nerve block to better identify this neuropathy [9]. Injecting lidocaine into the subcutaneous tissue at the point of maximum tenderness has shown to be therapeutic and a successful indicator for postoperative outcomes [8, 9]. Brown et al. also found a nerve block clinically useful at diagnosing this neuropathy. In their study, 31/44 (70%) of their patients reported pain relief after the nerve block. Many nonsurgical treatments are available for nerve dysfunction: without intervention, various medication classes, physical therapy, and psychosocial therapy [9, 10].

Surgical management of nerve entrapment involves decompression and neurolysis [10]. To do so, the surgeon must release the entrapped nerve circumferentially. Additionally, care must be taken to ensure the blood supply to the freed nerve is not compromised [10]. Bregman and Schuenke stated surgeons should locate the point of maximum tenderness preoperatively. This increases postoperative symptom relief. Brown et al. supported such claims [8, 9]. Eighty-four percent of their patients undergoing an isolated decompression reported symptom improvement postoperatively.

27.3 Sural Nerve Entrapment

Sural nerve entrapment is not very common in comparison to other pathologies involving the lower limb. However, it is important to be aware of because diagnosis can be difficult and if delayed, can lead to poor long-term outcomes for patients. It typically affects running athletes but can occur in anyone who maintains an active lifestyle [5].

The sural nerve is purely sensory, and its journey begins midway down the posterior aspect of the gastrocnemius muscle. It then terminates at the base of the fifth metatarsal [3]. Unfortunately, the sural nerve can become entrapped anywhere along its course, but it most commonly occurs along the lateral portion of the heel.

There are several causes that can contribute to this pathology; however, a large number of cases are due to some sort of trauma to the area, ultimately affecting the nerve. An acute or recurrent ankle injury precedes patient complaints involving the distribution of the sural nerve [3]. One can imagine how edema, scar tissue, or nerve stretching secondary to an ankle injury can lead to symptomatic nerve compression and irritation.

Most patients will present with loss of sensation or neuropathic pain (numbness, tingling, burning) along the sural nerve pathway: between the medial and lateral heads of the gastrocnemius muscles and posterolateral to Achilles tendon. Additionally, patients can present with calf pain that becomes exacerbated at night or with physical activity [5, 11].

Similar to superficial peroneal nerve entrapment, sural nerve entrapment is also a clinical diagnosis. Although if necessary, imaging can be obtained to rule out bony malformations and vascular or soft tissue pathologies that may be related to patient symptoms [5].

Prior to surgical treatment, trying supplementation with Vitamin B_6, gabapentin, and nonsteroidal anti-inflammatory drugs (NAIDS) may prove to be beneficial when treating isolated sural neuralgia [5]. Several studies discussed Vitamin C supplementation and its possible benefits for orthopedic pain and complex regional pain syndrome [12, 13]. Although no studies specifically report pain improvement for nerve entrapment, several reported success after wrist fractures using high dose Vitamin C [14, 15]. Sural nerve entrapment cannot always be treated with conservative management. Fabre et al. reported sural nerve decompression and neurolysis as the procedures of choice to relieve patient symptoms when pursuing surgical intervention [11].

Regarding return to play outcomes, Fabre et al. found that 12 out of 13 athletes returned to the same level of activity after nerve decompression. The same athletes returned to play within 2–25 weeks with a mean of 8 weeks after surgery [11]. Fabre also reported the following complications postoperatively: superficial hematoma on postoperative day 3, persistent lateral knee pain with radiation to anterolateral side of lower leg during physical activity, and persistent focal pain along the sural nerve pathway leading to a bilateral neurectomy [11].

It should be noted that there was a paucity of literature relating to both superficial peroneal and sural nerve entrapment specifically associated with runners and track and field athletes. Therefore, peripheral nerve entrapment in these populations should be an area of future research.

27.4 Tarsal Tunnel Syndrome

Briefly, the tarsal tunnel is a continuation of the deep posterior compartment of the leg that is bounded by the medial malleolus anterosuperiorly, by the posterior talus and calcaneus laterally, and held against the bone by the flexor retinaculum. The tarsal tunnel contains many important structures: the tendons of the posterior tibialis, flexor digitorum longus, flexor hallucis longus muscles, posterior tibial artery/vein, and the posterior tibial nerve. Notably, compression or entrapment of the tibial nerve occurs in the region where the nerve passes under the transverse tarsal ligament, leading to the primary symptoms of tarsal tunnel syndrome.

Tarsal tunnel syndrome has a number intrinsic and extrinsic causes [16, 17]. Extrinsic causes include anatomic and biomechanical abnormalities (tarsal coalition, valgus/varus hindfoot), poorly fitting shoes, post-traumatic or postsurgical scarring, diabetes, and inflammatory diseases. Intrinsic causes include osteophytes, (perineural) fibrosis, tendinopathy, tenosynovitis, hypertrophic retinaculum, and space-occupying or mass effect lesions. Trauma, with incidence of up to 43%, is the most common cause—specifically fracture or dislocation involving the talus, calcaneus, or medial malleolus [18–20].

Tarsal tunnel syndrome is a rare and underdiagnosed disease. The incidence is unknown, but there is a higher rate in women than men [18]. Tarsal tunnel syndrome is most commonly diagnosed in patients with prior foot trauma, whereas "idiopathic" tarsal tunnel syndrome, unlike other nerve entrapments such as carpal tunnel syndrome, is quite rare. Based on clinical data from 1986 to 2020, Kinoshita et al. reported that an average of 2.7 patients (3.4 ft.) were treated annually, and relatively large percentage were sport-related cases (39%) [21].

The diagnosis of tarsal tunnel syndrome is usually made with a detailed history and clinical examination. The general population and athletes alike typically present with aching and concomitant paresthesia [22]. The pain and tenderness usually localize to the location of the tarsal tunnel and radiate to the arch or to the plantar foot. There may be associated radiation up to the calf or higher, mimicking sciatica. The patient may also note weakness in the muscles of the foot. Generally, the symptoms are worse at night, with standing and walking, and get better with rest; nocturnal dysesthesias are reported to be the most irritating. However, these symptoms are less common in the athletic population [22]. Chronic cases can lead to lower motor neuron pathology signs (atrophy, weakness of the intrinsic foot muscles, and contractures of the toes). Patients may also have diminished plantar sensation in the distribution of the tibial nerve (either the medial or lateral plantar nerve). Overall, the symptoms of tarsal tunnel syndrome can be quite vague, making diagnosis very difficult.

The physical exam may be relatively benign, but recreation of the symptoms can be elicited in some patients via repetitive tapping over the tarsal tunnel, also called the Tinel sign. Pain or paresthesia in the distribution of the tibial nerve indicates a positive test.

Plain radiographs of foot and ankle are the preferred initial study to identify structural abnormalities. MRI is not sensitive for the diagnosis but may be helpful in excluding alternate diagnoses. Both, ultrasound and MRI, may be helpful in the evaluation of other soft tissue abnormalities such as tenosynovitis, tendonitis, or space-occupying lesions (e.g., lipomas) [23, 24]. Electromyography and nerve conduction studies have been used in some cases; however, the sensitivity and specificity are suboptimal and false-negatives are not uncommon [18].

Treatment initially includes nonoperative measures. Corrections of overpronation (e.g., pes planus deformity), with accommodative orthotics, arch support, and medial wedge are useful first steps in management. Physical therapy should be instituted with focus on strengthening medial flexors. Immobilization is also an option to rest the irritated tibial nerve. Nonsteroidal anti-inflammatory medications are first line, but topical compound creams (e.g., lidocaine) have also been used. A steroid injection can also be considered with failure of the previously noted less invasive methods. Surgery may be indicated in a patient who does not respond to these treatments. Decompression is effective in some patients—a retrospective study with 47 patients over a 10-year period, 72% of patients reported improvement of their symptoms [25]. Other studies have reported more variable results [26, 27].

There is a paucity of literature regarding tarsal tunnel syndrome in general. Of the literature available, much of it is outdated and not relevant for track and field athletes. Despite the low incidence, tarsal tunnel syndrome should be a future area of study to improve clinical and performance outcomes in these patients.

27.5 Baxter's Neuropathy

For the track and field athlete, especially those in running disciplines, who report plantar heel pain, Baxter's neuropathy should be a key part of the differential diagnosis [28]. The Baxter nerve, otherwise known as the inferior calcaneal nerve or the motor branch of the abductor digiti quinti, is the first branch of the lateral plantar nerve. It originates near the bifurcation of the tibial nerve or may arise before the bifurcation. At this level, the nerve courses close to the superior border of the abductor hallucis and quadratus plantae muscles. Here, there is a thicker layer of fascia laterally due to interfascicular ligament traveling with the medial intermuscular septum [29]. There is supported evidence for multiple areas of possible entrapment of the nerve. One of which is the point at which the nerve runs laterally between the thick fascia of the abductor hallucis and the medial border of the quadratus plantae [30]. Another possible point is anterior to the medial calcaneal tuberosity, in this case, hypertrophy of the muscle, pronation of the midfoot can increase contact area and lead to impingement [31]. Other considerations at this area are local trauma and venous engorgement [32].

Heel pain is a common foot and ankle complaint for the track and athlete. Many different pathologies can lead to pain in this area, some of which include plantar fasciitis, heel pad atrophy, tarsal tunnel syndrome, and calcaneal stress fractures. Reaching the distinct diagnosis of Baxter's neuropathy can prove challenging due to the overlapping signs and symptoms from similar diagnoses. Symptoms of this nerve entrapment include tenderness over the area of the origin of the abductor hallucis, other areas of the heal just proximal to the plantar fascia [33]. Pain may be brought on by performing the Phalen's maneuver and Tinel sign may be elicited [33, 34]. In some chronic cases, abduction of the fifth digit may be limited. It is important to compare with the contralateral side as some patients may lack this ability inherently [35].

Imaging can aid in diagnosis. Plain films can demonstrate osseous pathologies like calcaneal enthesophyte, and MRIs may show hypertrophy of the surrounding musculature and reveal possible inflammation. Lack of significant inflammation can support the diagnosis of entrapment as the cause of heel pain. If there is fatty replacement and increased water signal of the abductor digiti minimi, this may indicate atrophy caused by nerve entrapment [36]. Meadows et al. suggest that a nerve block administered between the abductor hallucis and quadratus plantae that results in relief of pain is diagnostic of the condition [33]. Treatment for Baxter's neuropathy begins with conservative measures akin to those prescribed for plantar fasciitis. The use of heel gel cups, soft sole shoes, night splints, physical therapy may prove beneficial although not as effective as in treatment for plantar fasciitis [33, 34]. Corticosteroid injections can also be utilized. If this option is taken, it should be noted that the injection should only be directed to the Baxter's nerve. If pain is decreased but recurs in short duration, surgical nerve decompression, and fascial release can provide good results [36, 37]. Hendrix et al. reported that a majority of athletes were asymptomatic after decompression and had a return to sport time of 5–8 weeks [38].

References

1. Konstantine B, Anastasia T, Catherine B, George T, Pavlos K. The treatment of Morton's neuroma, a significant cause of metatarsalgia for people who exercise. Int J Clin Med. 2013;4:19–24.
2. McCrory P, Bell S, Bradshaw C. Nerve entrapments of the lower leg, ankle and foot in sport. Sports Med. 2002;32:371–91.
3. Coughlin MJ, Saltzman CL, Mann RA. Mann's surgery of the foot and ankle E-book: expert consult - online. Amsterdam: Elsevier; 2013.
4. Leach RE, Purnell MB, Saito A. Peroneal nerve entrapment in runners. Am J Sports Med. 1989;17(2):287–91. https://doi.org/10.1177/036354658901700224.
5. Miller MD, Thompson SR. DeLee & Drez's orthopaedic sports medicine E-Book. Amsterdam: Elsevier; 2018.
6. Styf J, Morberg P. The superficial peroneal tunnel syndrome. Results of treatment by decompression. J Bone Joint Surg Br. 1997;79(5):801–3.
7. Matsumoto J, Isu T, Kim K, Iwamoto N, Yamazaki K, Isobe M. Clinical features and surgical treatment of superficial peroneal nerve entrapment neuropathy. Neurol Med Chir (Tokyo). 2018;58(7):320–5. https://doi.org/10.2176/nmc.oa.2018-0039.
8. Brown CL, Worts P, Borom A. Superficial peroneal nerve entrapment: diagnosis and surgical management. a retrospective study of 46 patients. Foot Ankle Orthop. 2019;4(4):2473011419S00120. https://doi.org/10.1177/2473011419s00120.
9. Bregman PJ, Schuenke MJ. A commentary on the diagnosis and treatment of superficial peroneal (fibular) nerve injury and entrapment. J Foot Ankle Surg. 2016;55(3):668–74. https://doi.org/10.1053/j.jfas.2015.11.005.
10. Lipinski LJ, Spinner RJ. Neurolysis, neurectomy, and nerve repair/reconstruction for chronic pain. Neurosurg Clin N Am. 2014;25(4):777–87. https://doi.org/10.1016/j.nec.2014.07.002.
11. Fabre T, Montero C, Gaujard E, Gervais-Dellion F, Durandeau A. Chronic calf pain in athletes due to sural nerve entrapment. A report of 18 cases. Am J Sports Med. 2000;28(5):679–82. https://doi.org/10.1177/03635465000280051001.
12. Carr AC, McCall C. The role of vitamin C in the treatment of pain: new insights. J Transl Med. 2017;15(1):77. https://doi.org/10.1186/s12967-017-1179-7.

13. Bruehl S. Complex regional pain syndrome. BMJ. 2015;351:h2730. https://doi.org/10.1136/bmj.h2730.
14. Zollinger PE, Tuinebreijer WE, Breederveld RS, Kreis RW. Can vitamin C prevent complex regional pain syndrome in patients with wrist fractures? A randomized, controlled, multicenter dose-response study. J Bone Joint Surg Am. 2007;89(7):1424–31. https://doi.org/10.2106/JBJS.F.01147.
15. Zollinger PE, Tuinebreijer WE, Kreis RW, Breederveld RS. Effect of vitamin C on frequency of reflex sympathetic dystrophy in wrist fractures: a randomised trial. Lancet. 1999;354(9195):2025–8. https://doi.org/10.1016/S0140-6736(99)03059-7.
16. Hong CH, Lee YK, Won SH, Lee DW, Moon SI, Kim WJ. Tarsal tunnel syndrome caused by an uncommon ossicle of the talus: a case report. Medicine (Baltimore). 2018;97(25):e11008. https://doi.org/10.1097/MD.0000000000011008.
17. Komagamine J. Bilateral tarsal tunnel syndrome. Am J Med. 2018;131(7):e319. https://doi.org/10.1016/j.amjmed.2017.10.028.
18. Kiel JKK. Tarsal tunnel syndrome. Treasure Island, FL: StatPearls Publishing; 2020.
19. Francis H, March L, Terenty T, Webb J. Benign joint hypermobility with neuropathy: documentation and mechanism of tarsal tunnel syndrome. J Rheumatol. 1987;14(3):577–81.
20. Lau JT, Daniels TR. Tarsal tunnel syndrome: a review of the literature. Foot Ankle Int. 1999;20(3):201–9. https://doi.org/10.1177/107110079902000312.
21. Kinoshita M, Okuda R, Yasuda T, Abe M. Tarsal tunnel syndrome in athletes. Am J Sports Med. 2006;34(8):1307–12. https://doi.org/10.1177/0363546506286344.
22. Raikin SMSL. Nerve entrapment in the foot and ankle of an athlete. Sports Med Arthrosc Rev. 2000; https://doi.org/10.1097/00132585-200008040-00010.
23. Erickson SJ, Quinn SF, Kneeland JB, Smith JW, Johnson JE, Carrera GF, et al. MR imaging of the tarsal tunnel and related spaces: normal and abnormal findings with anatomic correlation. AJR Am J Roentgenol. 1990;155(2):323–8. https://doi.org/10.2214/ajr.155.2.2115260.
24. Kerr R, Frey C. MR imaging in tarsal tunnel syndrome. J Comput Assist Tomogr. 1991;15(2):280–6. https://doi.org/10.1097/00004728-199103000-00018.
25. Bailie DS, Kelikian AS. Tarsal tunnel syndrome: diagnosis, surgical technique, and functional outcome. Foot Ankle Int. 1998;19(2):65–72. https://doi.org/10.1177/107110079801900203.
26. Macaré van Maurik JF, Schouten ME, ten Katen I, van Hal M, Peters EJ, Kon M. Ultrasound findings after surgical decompression of the tarsal tunnel in patients with painful diabetic polyneuropathy: a prospective randomized study. Diabetes Care. 2014;37(3):767–72. https://doi.org/10.2337/dc13-1787.
27. Turan I, Rivero-Melián C, Guntner P, Rolf C. Tarsal tunnel syndrome. Outcome of surgery in longstanding cases. Clin Orthop Relat Res. 1997;343:151–6.
28. Schon LC, Baxter DE. Neuropathies of the foot and ankle in athletes. Clin Sports Med. 1990;9(2):489–509.
29. Kelikian AS, Sarrafian SK. Sarrafian's anatomy of the foot and ankle: descriptive, topographic, functional. Philadelphia: Wolters Kluwer Health/Lippincott Williams & Wilkins; 2011.
30. Rondhuis JJ, Huson A. The first branch of the lateral plantar nerve and heel pain. Acta Morphol Neerl Scand. 1986;24(4):269–79.
31. Baxter DE, Thigpen CM. Heel pain--operative results. Foot Ankle. 1984;5(1):16–25. https://doi.org/10.1177/107110078400500103.
32. Schon LC. Nerve entrapment, neuropathy, and nerve dysfunction in athletes. Orthop Clin North Am. 1994;25(1):47–59.
33. Meadows JR, Finnoff JT. Lower extremity nerve entrapments in athletes. Curr Sports Med Rep. 2014;13(5):299–306. https://doi.org/10.1249/JSR.0000000000000083.
34. Offutt SDP. How to address baxter's nerve entrapment. Podiatry Today. 2004;17:52–8.
35. Seidelmann, FE. MRI evaluation of tibialis posterior tendinopathy. International Foot and Ankle Congress. New Orleans, LA; 2004.
36. Goecker RM, Banks AS. Analysis of release of the first branch of the lateral plantar nerve. J Am Podiatr Med Assoc. 2000;90(6):281–6. https://doi.org/10.7547/87507315-90-6-281.
37. Baxter DE. Functional nerve disorders in the athlete's foot, ankle, and leg. Instr Course Lect. 1993;42:185–94.
38. Hendrix CL, Jolly GP, Garbalosa JC, Blume P, DosRemedios E. Entrapment neuropathy: the etiology of intractable chronic heel pain syndrome. J Foot Ankle Surg. 1998;37(4):273–9. https://doi.org/10.1016/s1067-2516(98)80062-8.

Foot and Ankle Stress Fractures in Athletics

28

Silvio Caravelli, Simone Massimi, Thomas P. A. Baltes, Jari Dahmen, Pieter D'Hooghe, and Gino M. M. J. Kerkhoffs

28.1 Introduction

Although stress fractures are rare, they present a significant burden for athletes as they are associated with prolonged absence from sports and high rates of reoccurrence [1–3]. The aim of this chapter is to outline the most common stress fractures of the lower extremity and provide specific guidelines for the diagnosis, treatment, and return to sport in the (elite) track and field athlete.

28.2 Epidemiology

In elite sports, stress fractures most commonly occur in the lower extremity. As observed during the Rio de Janeiro Olympics, stress injuries were most common among Track and Field athletes (44%) and affected the lower limb in 84% of the cases [4]. In nonelite athletes, a similar pattern is observed, with stress fractures primarily affecting the foot (34.9%) [5]. As established in collegiate student-athletes (NCAA), the incidence of stress fractures ranges from 16.23 per 100.000 Athlete-Exposures for indoor Track and Field sports to 29.46 per 100.000 Athlete-Exposures

S. Caravelli (✉) · S. Massimi
II Clinic of Orthopaedics and Traumatology, IRCCS Istituto Ortopedico Rizzoli, Alma Mater Studiorum, Bologna, Italy

T. P. A. Baltes
Department of Orthopaedic Surgery, Aspetar Sports Medicine and Orthopedic Surgery Hospital, Doha, Qatar

Department of Orthopaedic Surgery, Amsterdam Movement Sciences, Amsterdam UMC, Location AMC, University of Amsterdam, Amsterdam, The Netherlands

Academic Center for Evidence-based Sports Medicine (ACES), Amsterdam, The Netherlands

Amsterdam Collaboration for Health and Safety in Sports (ACHSS), AMC/VUmc IOC Research Center, Amsterdam, The Netherlands
e-mail: t.p.baltes@amsterdamumc.nl

J. Dahmen · G. M. M. J. Kerkhoffs
Department of Orthopaedic Surgery, Amsterdam Movement Sciences, Amsterdam UMC, Location AMC, University of Amsterdam, Amsterdam, The Netherlands

Academic Center for Evidence-based Sports Medicine (ACES), Amsterdam, The Netherlands

Amsterdam Collaboration for Health and Safety in Sports (ACHSS), AMC/VUmc IOC Research Center, Amsterdam, The Netherlands
e-mail: j.dahmen@amsterdamumc.nl; g.m.kerkhoffs@amsterdamumc.nl

P. D'Hooghe
Orthopedic Surgery, Aspire Zone, ASPETAR Qatar Orthopaedic and Sports Medicine Hospital, Doha, Qatar

G. L. Canata et al. (eds.), *Management of Track and Field Injuries*,
https://doi.org/10.1007/978-3-030-60216-1_28

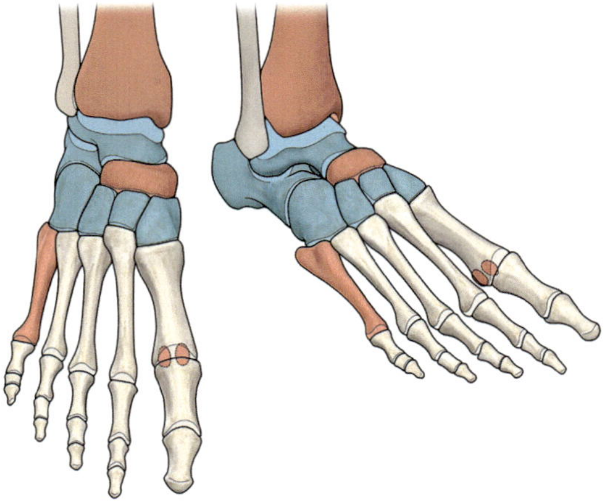

Fig. 28.1 Higher (red) and lower (blue) risk anatomical sites for stress fractures, in foot and ankle, are shown

for outdoor Track and Field sports, predominantly affecting female athletes [5]. Stress fractures of the lower extremity are most prevalent in the tibia, the navicular bone, and the (fifth) metatarsals [1, 5] (Fig. 28.1).

28.3 Etiopathogenesis

Load applied to the bone during sports activities or normal weight-bearing activities results in external forces (strain) and internal forces (stress), both of which are vital for the maintenance of normal bone strength [6]. Stress fractures occur when the mechanical forces (e.g., repetitive cyclic loads) exceed the physiological forces that result in normal bone remodeling.

In the event of persistent overload (i.e., mechanical forces exceeding physiological forces), the regenerative and reparative capacities of the involved bone are insufficient to manage the resulting microtrauma. Failure to repair microtrauma leads to bone fatigue and loss of structural strength due to the predominance of osteoclastic activity (stress reaction). When overload persists, formation and propagation of microscopic "cracks" inside the bone may further affect bone strength [7–9]. Finally, the areas of fragility may accumulate to form a frank fracture pattern [1].

Table 28.1 Risk factors for stress fractures of foot

Extrinsic factors	Intrinsic factors
Footwear/insole/ orthotics	Foot morphology and lower limb alignment
Type of activity	Bone turnover
External loadings	Bone geometry
Field surface	Hormonal factors
Improper technique	Recovery periods
New excessive training regimen	Genetic predisposition
	Nutritional aspects
	Age, sex, BMI

The odds of an athlete sustaining a stress fracture is correlated with the presence of intrinsic and extrinsic risk factors (Table 28.1) [1]. An important risk factor associated with stress fractures is "Relative Energy Deficiency in Sport (RED-S)", which is discussed in this Track and Field ISAKOS book as well.

28.4 Clinical Assessment and Radiological Evaluation

The presence of a stress fracture should be suspected in athletes with a gradual onset of atraumatic pain. Symptoms are often associated with an increase in workload. When a stress injury has evolved into a complete stress fracture, pain may be present continuously and affect athletic performance and daily activities [10].

Clinical examination should aim to localize the point of tenderness [11]. In some stress fractures, periosteal thickening (sign of inadequate callus formation) or the presence of swelling might be noticeable upon palpation [12]. In athletes suspected of a stress fracture, it is imperative to inquire about recent increases in training load, type of footwear, and type of training surface. Furthermore, evaluation of limb length and axis, range of motion, muscular asymmetry, and gait should be performed [12]. When suspecting a stress fracture of the foot, investigation of the plantar arch should be undertaken [13].

Radiographic evaluation of stress fractures is not always reliable. During the first 2–3 weeks after onset of symptoms up to 87% of cases are not visible on radiographs [11, 14]. Computed Tomography (CT) imaging is a valuable alternative to detect stress fractures and may aid to distinguish stress injuries from stress fractures [11]. CT imaging has demonstrated 100% sensitivity and 90% specificity for tibial stress fractures [15].

Magnetic Resonance (MR) imaging is considered the golden standard in identifying stress fractures. MR imaging can accurately delineate the exact anatomic location and the extent of the stress injury, by detecting bone edema and changes in cortical density [16, 17].

28.5 General Treatment Concepts

Stress fractures can be subdivided into low-risk and high-risk stress fractures, based on their healing potential. Treatment should be tailored to the healing tendency of the stress fracture and the athletes' intrinsic and extrinsic risk factors (Table 28.1).

Low-risk stress fractures generally have a high healing propensity when treated conservatively. This includes modification of training regiments to reduce the load on the affected limb, adaptation of footwear or training surface, and evaluation of athletes' hormonal and nutritional status.

High-risk fractures often warrant surgical treatment due to poor healing propensity. In elite athletes, surgical treatment of high-risk stress fractures may be considered as a first-line treatment in order to improve return to sports. However, a recent systematic review showed that there was only low-quality evidence comparing surgery with conservative treatment for the treatment of high-risk stress fractures of the lower limb [18].

28.6 Tibial Stress Fractures

Tibial stress fractures are one of the most common stress fractures in athletes, with elite Track and Field athletes being particularly susceptible to this type of injury [3, 4]. In the current literature, tibial stress fractures have been stated to account for 19–63% of all stress injuries observed in athletes [19]. Tibial stress fractures can be categorized into two different entities; (1) low-risk posteromedial tibial cortex and (2) high-risk anterior tibial cortex stress fractures [20].

28.6.1 Etiopathogenesis

In the majority of the cases (~80%), stress injuries of the tibia affect the posteromedial cortex [21, 22]. This occurs as a result of repetitive impact forces and pulling of the calf muscles, experienced in long-distance runners. Anterior tibial cortex stress fractures occur only in 5–15% of all tibial stress fractures and are primarily associated with repetitive jumping [21, 22]. Anterior tibial stress fractures have a poor healing tendency as they occur on the tension side of the tibia [23, 24].

28.6.2 Clinical Assessment and Radiological Evaluation

Clinical symptoms include exercise-induced pain, swelling, and point tenderness. However, classic symptoms may be lacking. Conventional radiography is the primary imaging modality, despite limited sensitivity due to a delay in radiographic findings [25].

In case of negative radiographs, despite high clinical suspicion, Magnetic Resonance Imaging (MR imaging) should be considered. With a 82% sensitivity and 100% specificity, MR imaging is considered the golden standard and it can be used to classify tibial stress fractures according the modified Fredericson classification [26] (Table 28.2).

28.6.3 Management

It is imperative to differentiate between posterior and anterior tibial stress fractures. The management of the individual athlete should be tailored to the healing propensity of the fracture [23, 24].

Table 28.2 Modified Fredericson classification for tibial stress fractures

Grade 0	No abnormality
Grade 1	Periosteal edema with no associated bone marrow signal abnormalities
Grade 2	Periosteal edema and bone marrow edema visible only on T2-weighted images
Grade 3	Periosteal edema and bone marrow edema visible on both T1- and T2-weighted images
Grade 4A	Multiple focal areas of intracortical signal abnormality and bone marrow edema visible on both T1-weighted and T2-weighted images

Posteromedial stress fractures can often be managed successfully with conservative treatment. Conservative treatment consists of rehabilitation, load management, and continued weight-bearing as tolerated. Gradual return to sports can be commenced after the patient has been able to bear weight pain free for 2 weeks corresponding to return to sports after a mean period of 3 months [27].

Anterior tibial cortex stress fractures have demonstrated poor outcomes when treated conservatively, with a large subset of fractures (53%) resulting in nonunion [18, 20]. When successful, conservative treatment allows athletes to return to sports after a mean 6 months with a return to sports rate of 55% [20]. In case of persisting symptoms or nonunion, surgical intervention is indicated (compression plating, drilling, intramedullary nailing, excision of the lesion). Chaudhry et al. recently concluded that the different surgical interventions resulted in resolution of symptoms in 88% [28]. Return to sports was possible in 95% of the patients and return to preinjury level of sports in 73%. Nonunited stress fractures treated with subsequent surgery returned to sports at 28 weeks postoperatively, as concluded by Orava et al. [18, 20]. Despite reasonable outcomes after surgical treatment, operative treatment is associated with high complication rates (25%) and need for subsequent surgery (15%) [28].

Therefore, high quality evidence prospectively comparing primary surgical treatment with nonoperative treatment in the management of posteromedial and anterior cortex stress fractures is warranted [18, 29, 30].

28.7 Fifth Metatarsal Stress Fractures

Fifth metatarsal stress fractures usually occur in the proximal metaphysis and diaphysis of the fifth metatarsal and have a higher prevalence in athletes [31]. It is imperative to differentiate stress fractures from the proximal avulsion fractures of the fifth metatarsal, as stress fractures have a lower healing propensity [32, 33].

28.7.1 Etiopathogenesis

Stress fractures may occur as a result of forces that act upon the fifth metatarsal. Several tendons and ligaments insert on the base of the fifth metatarsal [34, 35]. The plantar fascia inserts on the plantaro-lateral aspect of the tuberosity while the peroneus brevis inserts on the dorso-lateral side. The peroneus tertius inserts on the dorsal metaphysis. The aformentioned structures mainly determine the tensile forces, while the capsulo-ligamentous structures around the cuboid, the fourth and fifth metatarsals determine rigidity. The diaphysis of fifth metatarsal is relatively mobile and therefore, stress fractures are more prone to form in the metaphyseal-diaphyseal junction of the bone.

28.7.2 Clinical Assessment and Radiological Evaluation

The clinical presentation of a fifth metatarsal stress fracture is characterized by pain on the lateral aspect of the forefoot (aggravated by weight-bearing), tenderness on palpation, and localized swelling. Often patients are unable to walk on tiptoes [1]. Resisted foot eversion can be used as a provocative clinical test to differentiate proximal avulsion fractures. Various classifications based on anatomical and radiological features have been described, including: (1) the Lawrence and Botte's Classification [36] (Fig. 28.2) and (2) Torg's Classification [37] (Table 28.3).

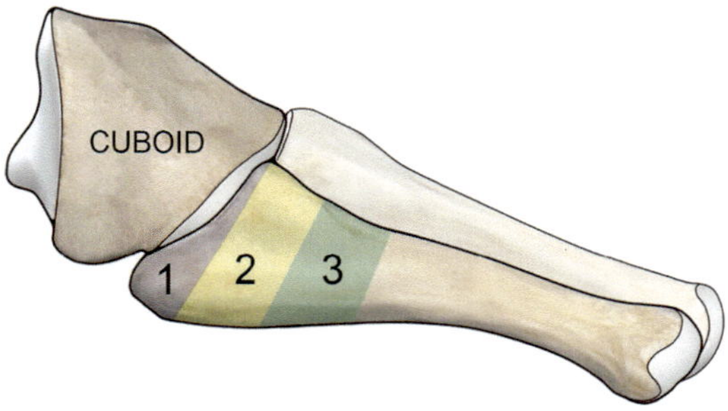

Fig. 28.2 Lawrence and Botte's Classification of Proximal Fifth Metatarsal Fractures (Zones 1, 2, and 3). This classification is the mostly used and it distinguishes the three Zones of proximal fifth metatarsal fractures based on the mechanism of injury, location, treatment options, and prognosis: Zone 1 (tuberosity area), Zone 2 (tuberosity—metaphyseal area), and Zone 3 (metaphyseal-diaphyseal area)

Table 28.3 Torg Classification of Proximal Fifth Metatarsal Stress Fractures

Type	Description
Type I	Acute
Type II	Delayed union with periosteal and intramedullary bone formation
Type III	Nonunion

This classification is based on the radiographic characteristics of the fracture and provides an additional treatment plan: Type 1 (acute fracture), Type 2 (delayed union and radiographs show periosteal new bone formation, resorption, and sclerosis at the fracture line), Type 3 (nonunion with complete obliteration of the medullary canal by the sclerotic bone). Type 1 can be treated conservatively by nonweight-bearing for approximately 10 weeks, whereas type 2–3 fractures are advised to be treated surgically in combination with an autologous bone augmentation (Type 3) to assist in fracture healing

28.7.3 Management

Fractures of the metaphysis and diaphysis of the fifth metatarsal are predisposed to delayed healing as a result of the vascular watershed zone between the insertion of peroneus brevis and the diaphyseal blood supply [33]. Fractures interrupt the vascular channels in this area, leading to a poor healing tendency.

Fifth metatarsal stress fractures can be treated both conservatively and surgically. Josefsson et al. [38] reported a 95% union rate and good functional results with nonoperative treatment. In athletes, the treatment is mostly surgical because of the prolonged recovery time and higher risk of nonunion with conservative treatment [39, 40].

Various surgical techniques have been described, but the most commonly applied technique is percutaneous screw fixation with a 5.5 mm intramedullary screw. Other techniques such as tension band wiring have been used with similar outcomes [39, 41]. Postoperative rehabilitation [39] consists of immobilization with a short leg cast or plaster splint for 1–2 weeks, followed by a walking boot for 2 weeks. After 6–8 weeks postoperatively, full weight-bearing is allowed and normal activities can be resumed. In general, full sport activities can be resumed 10 weeks postoperatively. The use of functional bracing or orthotics upon return to sports may reduce the rate of reinjury [42].

28.8 Navicular Stress Fractures

Stress fractures of the navicular bone account for up to 35% of all foot and ankle stress fractures [43]. Navicular fractures are typically seen in athletes engaged in explosive push-off activities such as track and field athletes (e.g., sprinting athletes), professional tennis players, and jumping athletes.

28.8.1 Etiopathogenesis

Biomechanically, the navicular bone is considered a keystone in the medial column, connecting the midfoot to the hindfoot. With sudden movements, such as sprinting, striking or cutting, the navicular bone undergoes maximal shear and compressive forces, thereby affecting the central third of the navicular bone. Its function can be influenced by anatomical variations in foot morphology such as: short first and long second metatarsal bones, pes cavovarus, limited ankle dorsiflexion, and metatarsus adductus. In addition, contraction of the tibialis posterior tendon increases the medial stress over the navicular bone. Due to a relatively poor vascular supply in the central third of the navicular bone, these stress fractures are considered to be at high-risk of nonunion [44].

28.8.2 Clinical Assessment and Diagnosis

Patients usually complain of exercise-induced pain over the dorsal aspect of the midfoot and/or the medial arch, with an insidious onset. Evident ecchymosis or swelling is rare. Provocative tests include a hop test on the affected foot and standing on tiptoes. The palpable point of tenderness (*"N-spot"*) is located between the tibialis anterior and extensor hallucis longus tendons, corresponding to the area of the central third of the navicular bone. Due to the atypical presentation, the diagnosis is often delayed 4–7 months from the onset of symptoms [45]. MR imaging is a sensitive method for diagnosis, although CT imaging is currently considered the golden standard [46].

28.8.3 Clinical Assessment and Radiological Evaluation

Navicular stress fractures are considered high-risk stress fractures [47]. However, conservative treatment in a nonweight-bearing plaster cast or boot for 5 weeks followed by 4–6 weeks of rehabilitation is recommended in the general population [45, 48, 49]. Currently, there is no consensus on the best therapeutic strategy in athletes. According to Saxena's classification [50], conservative treatment should be considered in type 1 stress fractures (involvement of the dorsal cortex only) and surgery in both type 2 (propagation of the fracture into the navicular body) and type 3 (bicortical disruption) stress fractures. Various case series [14, 48, 51] have demonstrated a 100% healing rate after 6 weeks of nonweight-bearing cast immobilization. However, similar studies contradict these results and reported persistent pain, delayed-union or nonunion with conservative management [45]. In elite and high-level athletes, primary surgical treatment can be considered in order to promote a rapid return to play [52]. Saxena et al. [53] reported that patients treated conservatively with nonweight-bearing cast had a 86% healing rate with a mean 5.6 months to return to activity, while patients treated surgically had a 83% healing rate with a mean 3.8 months to return to activity.

28.9 Medial Malleolus Stress Fractures

Medial malleolus stress fractures are relatively uncommon. They account for 0.6–4.1% of all stress fractures [31, 43, 53]. They typically occur in high-level runners and jumpers. They occur as a result of repetitive impingement of the talus on the medial aspect of the distal tibia during forced dorsiflexion of the ankle.

Shelbourne et al. [54] established three criteria useful in evaluation of medial malleolus stress fracture: (1) localized tenderness medial to the anterior tibialis tendon, (2) pain during activities, and (3) evidence of a vertical fracture line on diagnostic images. On X-rays, cortical or medullary radiolucency, regional osteopenia or callus formation (in advanced cases) can be noted. In case of acute onset of pain, with negative plain radiographs, MRI (more sensitive) or CT scan can be used to demonstrate an intramedullary fracture line [55, 56].

The treatment of medial malleolus stress fractures is controversial and various authors suggest contradicting methods of management [57]. In our experience, management should depend on various aspects, such as the presence of a fracture line, displacement, athletic level, and season status. Conservative management involves 4–8 weeks of functional rest with a gradual return to activities. Plaster cast and boot immobilization have been described as well as protected weight-bearing [58]. Mean return to sports of 7.6 weeks have been reported, although complete resolution of symptoms may take up to 4–5 months.

In case of a clear fracture line or displacement, especially in elite and "in season" athletes, Open Reduction and Internal Fixation (ORIF) is recommended [54, 55, 59, 60]. Operative management may allow faster and safer mobilization, considering that conservative treatment may result in delayed or nonunion in up to 10% of the cases. Time to return to sport after ORIF has been reported in two studies and ranged from 24 days to 6 months [59, 60].

28.10 Second Metatarsal

Metatarsal stress fractures represent 8.8% of all stress fractures of the lower limb and are often referred to as "march fractures" due to their high incidence in military recruits [1]. The second metatarsal is most frequently affected [1]. Stress fractures most commonly occur in the distal part of the second metatarsal, as a result of high bending forces in the meta-diaphyseal region. Although no direct link with a specific forefoot morphology has been reported, a shorter and hypermobile first metatarsal or a longer second metatarsal is hypothesized to increase the risk of a stress fracture [1, 61]. In general, distal fractures of the second metatarsal have a good prognosis with a relatively fast recovery when treated conservatively. In these patients, rest and partial to full weight-bearing in a Controlled Ankle Motion (CAM) boot is recommended.

28.11 Other Stress Fractures of the Foot

28.11.1 Calcaneal

Calcaneal stress fractures are rare and most studies report on the occurrence of these fractures in army recruits rather than in athletes. Symptoms include localized tenderness at the heel with increased activity which subsides with rest or immobilization. The diagnosis is often delayed as symptoms are often misinterpreted as plantar fasciitis, achilles tendinopathy, neuropathy of the inferior calcaneal nerve or calcaneal apophysitis. At radiographic evaluation, a thin radiolucent or sclerotic line may become apparent, 2–3 weeks after the onset of symptoms. MR imaging can be a useful tool to identify early bone morrow edema or fracture lines. Calcaneal stress fractures can be managed with nonoperative treatment and activity modification in most cases.

28.11.2 Talus

Talar stress fractures are a relatively rare entity but may present in athletes as a result of repetitive cycles of axial loading. Talar stress fractures are often associated with concomitant stress injuries of the foot. A study by Sormaala et al. revealed stress fractures of the talar head to be associated with navicular stress injuries in 60% of the cases and talar body stress fractures with calcaneal stress injury in 78% of the cases [62]. MR imaging is required as conventional radiographs are often unable to visualize talar stress fractures. When deciding on the treatment of talar stress fractures, the possibility of secondary displacement should be considered. Conservative management by 6 weeks of nonweight-bearing cast or boot immobilization is often advocated for undisplaced talar fractutres. In case of secondary displacement, surgical fixation is indicated to reduce the risk of avascular necrosis and to improve return to play [63].

References

1. Mayer SW, Joyner PW, Almekinders LC, et al. Stress fractures of the foot and ankle in athletes. Sports Health. 2014;6(6):481–91.
2. Miller TL, Jamieson M, Everson S, et al. Expected time to return to athletic participation after stress fracture in division I collegiate athletes. Sports Health. 2018;10(4):340–4.
3. Rizzone KH, Ackerman KE, Roos KG, et al. The epidemiology of stress fractures in collegiate student-athletes, 2004-2005 through 2013-2014 academic years. J Athlet Train. 2017;52(10):966–75.
4. Guermazi A, Hayashi D, Jarraya M. Sports injuries at the Rio de Janeiro 2016 Summer Olympics: use of diagnostic imaging services. Radiology. 2018;287(3):922–32.
5. Changstrom BG, Brou L, Khodaee M, et al. Epidemiology of stress fracture injuries among US high school athletes, 2005-2006 through 2012-2013. Am J Sports Med. 2014;43(1):26–33.
6. Bennell K, Matheson G, Meeuwisse W, et al. Risk factors for stress fractures. Sport Med. 1999;28(2):91–122.

7. Carter DR, Hayes WC. Compact bone fatigue damage – I: residual strength and stiffness. J Biomech. 1977;10:325–37.
8. Schaffler MB, Radin EL, Burr DB. Long-term behavior of compact bone at low strain magnitude and rate. Bone. 1990;11:321–6.
9. Mori S, Burr DB. Increased intracortical remodeling following fatigue damage. Bone. 1993;14:103–9.
10. Harrast MA, Colonno D. Stress fractures in runners. Clin Sports Med. 2010;29(3):399–416.
11. Kaeding CC, Yu JR, Wright R, Amendola A, Spindler KP. Management and return to play of stress fractures. Clin J Sport Med. 2005;15(6):442–7.
12. Kaiser PB, Guss D, DiGiovanni CW. Stress fractures of the foot and ankle in athletes. Foot Ankle Orthop. 2018;3(3):1–11.
13. Boden BP, Osbahr DC, Jimenez C. Low-risk stress fractures. Am J Sports Med. 2001;29:100–11.
14. Murray SR, Reeder M, Ward T, et al. Navicular stress fractures in identical twin runners: high-risk fractures require structured treatment. Phys Sportsmed. 2005;33(1):28–33.
15. Gaeta M, Minutoli F, Scribano E. CT and MR imaging findings in athletes with early tibial stress injuries: comparison with bone scintigraphy findings and emphasis on cortical abnormalities. Radiology. 2005;235(2):553–61.
16. Arendt EA, Griffiths HJ. The use of MR imaging in the assessment and clinical management of stress reactions of bone in high-performance athletes. Clin Sports Med. 1997;16(2):291–306.
17. Lee JK, Yao L. Stress fractures: MR imaging. Radiology. 1988;169(1):217–20.
18. Mallee WH, Weel H, van Dijk CN, et al. Surgical versus conservative treatment for high-risk stress fractures of the lower leg (anterior tibial cortex, navicular and fifth metatarsal base): a systematic review. Br J Sports Med. 2015;49(6):370–6.
19. Bennell KL, Brukner PD. Epidemiology and site specificity of stress fractures. Clin Sports Med. 1997;16:179–96.
20. Orava S, Karpakka J, Hulkko A, et al. Diagnosis and treatment of stress fractures located at the mid-tibial shaft in athletes. Int J Sports Med. 1991;12:419–22.
21. Liimatainen E, Sarimo J, Hulkko A, et al. Anterior mid-tibial stress fractures. Results of surgical treatment. Scand J Surg. 2009;98:244–9.
22. Orava S, Hulkko A. Stress fracture of the mid-tibial shaft. Acta Orthop Scand. 1984;55:35–7. https://doi.org/10.3109/17453678408992308.
23. Batt ME. Delayed union stress fractures of the anterior tibia: conservative management. Br J Sports Med. 2001;35:74–7.
24. Beals RK, Cook RD. Stress fractures of the anterior tibial diaphysis. Orthopedics. 1991;14:869–75.
25. Kijowski R, Choi J, Mukharjee R, de Smet A. Significance of radiographic abnormalities in patients with tibial stress injuries: correlation with magnetic resonance imaging. Skelet Radiol. 2007;36:633–40.
26. Kijowski R, Choi J, Shinki K, et al. Validation of MRI classification system for Tibial stress injuries. Am J Roentgenol. 2012;198:878–84.
27. Robertson GAJ, Wood AM. Return to sports after stress fractures of the tibial diaphysis: a systematic review. Br Med Bull. 2015;114:95–111.
28. Chaudhry ZS, Raikin SM, Harwood MI, et al. Outcomes of surgical treatment for anterior Tibial stress fractures in athletes: a systematic review. Am J Sports Med. 2019;47:232–40.
29. Borens O, Sen MK, Huang RC, et al. Anterior tension band plating for anterior Tibial stress fractures in high-performance female athletes. J Orthop Trauma. 2006;20:425–30.
30. Cruz AS, de Hollanda JPB, Junior AD, Neto JSH. Anterior tibial stress fractures treated with anterior tension band plating in high-performance athletes. Knee Surg Sport Traumatol Arthrosc. 2013;21:1447–50.
31. Tsukada S, Ikeda H, Seki Y, et al. Intramedullary screw fixation with bone autografting to treat proximal fifth metatarsal metaphyseal-diaphyseal fracture in athletes: a case series. Sports Med Arthrosc Rehabil Ther Technol. 2012;4(1):25.
32. Jones R. Fractures of the base of the fifth metatarsal bone by indirect violence. Ann Surg. 1902;35(6):697–700.2.
33. Smith JW, Arnoczky SP, Hersh A. The intraosseous blood supply of the fifth metatarsal: implications for proximal fracture healing. Foot Ankle. 1992;13(3):143–52.
34. DeVries JG, Taefi E, Bussewitz BW, et al. The fifth metatarsal base: anatomic evaluation regarding fracture mechanism and treatment algorithms. J Foot Ankle Surg. 2015;54(1):94–8.
35. Cheung CN, Lui TH. Proximal fifth metatarsal fractures: anatomy, classification, treatment and complications. Arch Trauma Res. 2016;5(4):e33298.
36. Lawrence SJ, Botte MJ. Jones' fractures and related fractures of the proximal fifth metatarsal. Foot Ankle. 1993;14(6):358–65.
37. Torg JS. Fractures of the base of the fifth metatarsal distal to the tuberosity. Orthopedics. 1990;13(7):731–7.
38. Josefsson PO, Karlsson M, Redlund-Johnell I, et al. Closed treatment of Jones fracture. Good results in 40 cases after 11-26 years. Acta Orthop Scand. 1994;65(5):545–7.
39. Delee J, Evans J, Julian J. Stress fracture of the fifth metatarsal. Am J Sports Med. 1983;11:349–53.
40. Clapper M, O'Brien T, Lyons P. Fractures of the fifth metatarsal: analysis of a fracture registry. Clin Orthop Relat Res. 1995;315:238–41.
41. Lee KT, Park YU, Young KW, et al. Surgical results of 5th metatarsal stress fracture using modified tension band wiring. Knee Surg Sports Traumatol Arthrosc. 2011;19:853–7.
42. Larson CM, Almekinders LC, Taft TN, et al. Intramedullary screw fixation of Jones fractures. Analysis of failure. Am J Sports Med. 2002;30:55–60.

43. Brukner P, Bradshaw C, Khan KM, et al. Stress fractures: a review of 180 cases. Clin J Sports Med. 1996;6:85–9.
44. McKeon KE, McCormick JJ, Johnson JE, Klein SE. Intraosseous and extraosseous arterial anatomy of the adult navicular. Foot Ankle Int. 2012;33(10):857–61.
45. Khan K, Fuller P, Brukner P, et al. Outcome of conservative and surgical management of navicular stress fracture in athletes: 86 cases proven with computerized tomography. Am J Sport Med. 1992;20:657–66.
46. Scott GB, Mahoney CM, Forster BB, et al. Tarsal navicular stress injury: long-term outcome and clinicoradiological correlation using both computed tomography and magnetic resonance imaging. Am J Sports Med. 2005;33(12):1875–81.
47. Kaeding CC, Yu JR, Wright R, et al. Management and return to play of stress fractures. Clin J Sport Med. 2005;15(6):442–7.
48. Fowler JR, Gaughan JP, Boden BP, et al. The non-surgical and surgical treatment of tarsal navicular stress fractures. Sports Med. 2011;41(8):613–9.
49. Torg JS, Moyer J, Gaughan JP, et al. Management of tarsal navicular stress fractures: conservative versus surgical treatment: a meta-analysis. Am J Sport Med. 2010;38(5):1048–53.
50. Saxena A, Fullem B, Hannaford D. Results of treatment of 22 navicular stress fractures and a new proposed radiographic classification system. J Foot Ankle Surg. 2000;39(2):96–103.
51. Towne L, Blazina M, Cozen L. Fatigue fracture of the tarsal navicular. A retrospective review of twenty-one cases. J Bone Joint Surg Am. 1970;52:376–8.
52. Jacob KM, Paterson RS. Navicular stress fractures treated with minimally invasive fixation. Indian J Orthop. 2013;47(6):1048–53.
53. Iwamoto J, Takeda T. Stress fractures in athletes: review of 196 cases. J Orthop Sci. 2003;8(3):273–8.
54. Shelbourne KD, Fisher DA, Rettig AC, McCarroll JR. Stress fractures of the medial malleolus. Am J Sports Med. 1988;16:60–3.
55. Sherbondy PS, Sebastianelli WJ. Stress fractures of the medial melleolus and distal fibula. Clin Sports Med. 2006;25:129–37.
56. Steinbronn DJ, Bennett GL, Kay DB. The use of magnetic resonance imaging in the diagnosis of stress fractures of the foot and ankle: four case reports. Foot Ankle Int. 1994;15(2):378–9.
57. Miller TL, Kaeding CC. Stress fractures in athletes: diagnosis and management. New York: Springer; 2014.
58. Okada K, Senma S, Abe E, et al. Stress fractures of the medial malleolus: a case report. Foot Ankle Int. 1995;16(1):49–52.
59. Shabat S, Sampson KB, Mann G, et al. Stress fractures of the medial malleolus: review of the literature and report of a 15-year-old elite gymnast. Foot Ankle Int. 2002;23(7):647–50.
60. Orava S, Karpakka J, Taimela S, et al. Stress fracture of the medial malleolus. J Bone Joint Surg Am. 1995;77(3):362–5.
61. Gross TS, Bunch RP. A mechanical model of metatarsal stress fracture during distance running. Am J Sports Med. 1989;17:669–74.
62. Sormaala MJ, Niva MH, Kiuru MJ, et al. Bone stress injuries of the talus in military recruits. Bone. 2006;39:199–204.
63. D'Hooghe P, Wiegerinck JI, Tol JL, et al. 22-year-old professional soccer player with atraumatic ankle pain. Br J Sport Med. 2015;49(24):1589–90.

Part VII

Special Considerations

Injury Prevention in Track and Field

29

Pascal Edouard

29.1 Introduction

The practice of track and field leads to a risk of injuries [1]. During a track and field season, about two-third of athletes occur an injury [2–4]. During an international championships, about 10% of athletes occur an injury [5–9]. The consequences of injury will depend on the injury location, type, and severity according to the track and field disciplines, but injury has always a negative impact on practice, because it can decrease training participation, decrease performance, and lead to pain [10]. Even if the injury is a minor anatomical lesion or leads to minor resounding on practice, there will be at least an impact on the musculoskeletal and psychological aspects, and can also negatively impact other domains of the life (e.g., social, professional, family, school, financial) at the short- or long-term [1]. Therefore, the prevention of injuries in track and field represents an important area for athletes and all stakeholders, such as coaches, health professionals, family, sports scientists, managers, sponsors, as well as international and national govern bodies [1, 11–13].

29.2 Prevention: A Multisteps Challenge!

In order to reach this injury prevention challenge, Van Mechelen et al. [14] described a four-steps methodological sequence of evidence-based injury prevention (Fig. 29.1): (1) determine the extent of the problem in terms of the incidence, severity, and characteristics of the sports injuries; (2) determine the risk factors (intrinsic and extrinsic) and injury mechanisms that play a role in the occurrence of sport injuries; (3) develop preventive measures that are likely to reduce the future risk and/or severity of injuries, based in particular on the knowledge acquired during the second step; and (4) evaluate the effectiveness of prevention measures especially developed in the third step.

In 2006, Finch [15] proposed a new sports injury research framework: the Translating Research into Injury Prevention Practice framework (TRIPP). This model was based on the fact that only research that can, and will, be adopted by sports participants, their coaches and sporting

P. Edouard (✉)
Inter-university Laboratory of Human Movement Science (LIBM EA 7424), University of Lyon, University Jean Monnet, Saint Etienne, France

Department of Clinical and Exercise Physiology, Sports Medicine Unit, Faculty of Medicine, University Hospital of Saint-Etienne, Saint-Etienne, France

European Athletics Medical & Anti Doping Commission, European Athletics Association (EAA), Lausanne, Switzerland
e-mail: Pascal.Edouard@univ-st-etienne.fr

G. L. Canata et al. (eds.), *Management of Track and Field Injuries*,
https://doi.org/10.1007/978-3-030-60216-1_29

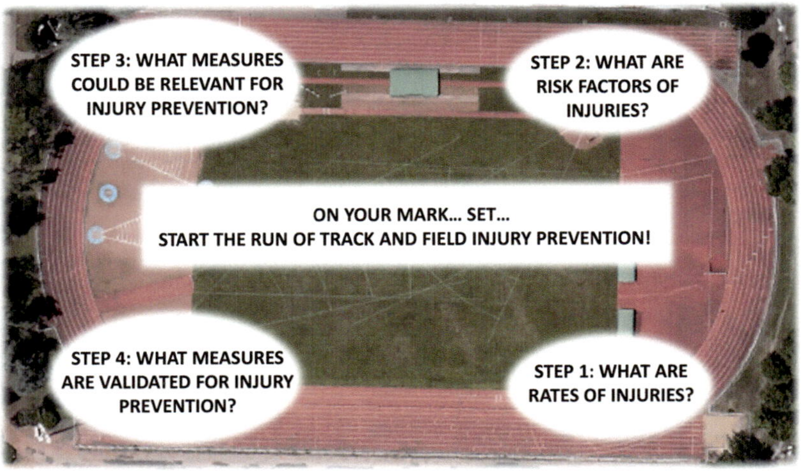

Fig. 29.1 The four-steps injury prevention sequence inspired from van Mechelen et al. [14]

bodies will reduce the occurrence of injuries. This means that studies on injury prevention should include information on key implementation factors (e.g., athletes' recruitment, reasons for use/nonuse the implementation). Based on the four-step sequence from Van Mechelen et al. [14], the TRIPP added two steps: (5) describe intervention context to inform implementation strategies; and (6) evaluate effectiveness of preventive measures in implementation context [15]. This proposed framework highlights that the use and thus the efficacy of an injury prevention measure in real life needs that the injury prevention measure should be developed by thinking and taking into account the acceptability, feasibility, and implementability in real life. The context of experimental research could be different than the context of the real life. There is thus a need to take into account the real life context and barriers from real life to develop injury prevention measure than will be use in practice.

In agreement with this proposed framework [15], Bolling et al. [16] recently revisited the first step of the "sequence of prevention" of sports injuries from Van Mechelen et al. [14]. Given the complex nature of the sports injuries, they suggested that the first step of the sequence should be improved by better understanding this complex nature by a more global approach. They proposed an alternative approach to explore and understand the context of the sports injuries at multiple levels, i.e., individual, sociocultural, and environmental [16]. Indeed, a better understanding of the context of the injury problem will guide more context-sensitive studies [16], and thus can improve implementation and use of the injury prevention measures.

Given the complex nature of sports injuries, the sports injury prevention measures should be appropriated to this complex nature and to the context of the sports injury in order to be efficient [17]. A step-by-step approach allows simplifying this complex challenge. This step-by-step approach aimed to understand and describe all components of the sports injury in order to build, develop, or create measures, strategies, and/or programs that can reduce the occurrence of injuries.

For track and field injury prevention, the magnitude of the injury problem was described in the chapter "The Burden and Epidemiology of Injury in Track and Field" of the present book, and there is now need to better understand the context of the track and field injuries as recommended by Bolling et al. [16]. For the second step, studies on track and field injuries reported that some factors seem to be associated with higher injury rates: a first episode of injury [4, 18–21], male sex [2–4, 6], increased age [2, 3, 7], participation in certain disciplines [5–9], training load [4], or maladaptive coping practice of self-blame [22]. However, work in this area should continue through specific studies on populations of athletes, taking into account the differences between disciplines and the large variety of potential risk factors (intrinsic, extrinsic, physical, psychological, social…) [12, 13]. This information can help to propose some ideas for injury prevention in track and field described in the next paragraph, as well as the current knowledge on the steps three and four.

29.3 What Can we Do to Reduce the Risk of Injuries in Track and Field?

Unlike other sports [23–25], currently and to the best of my knowledge, there is no scientific published evidence proven by randomized controlled trials or other high-quality studies on the efficacy of injury prevention measure, program or strategy in track and field. This thus represents an important challenge and perspective for track and field injury prevention.

It is however to note that a 40-week prospective cohort study (level of evidence 2), was conducted by Edouard et al. [26], including 63 inter-regional and national-level athletes. Athletes were asked to regularly perform an athletics injury prevention program (AIPP) including eight exercises addressing core stability, hamstring, leg and pelvic muscles strengthening and stretching, and balance exercises. These exercises have been chosen to target the most common athletics injuries [1–5, 7, 8, 12, 13, 27]: hamstring muscle injuries, Achilles and patellar tendinopathies, low back pain, ankle sprains, while being time-efficient and feasible. The program was based on the literature on the epidemiology of athletics injuries, injury risks factors, and current evidence-based injury prevention programs. Exercises used successfully for primary and/or secondary prevention were selected: eccentric strengthening to prevent hamstring injuries [28, 29], Achilles tendinopathies [30], and patellar tendinopathies [31]; strengthening and neuromuscular control to prevent ankle sprains [32]; and core stability to guard against low back pain [33]. The AIPP included eight exercises with levels of progression: core stability (plank and side plank), postural control (one-leg balance), pelvic strengthening (lunges and hip abductor strengthening), hamstring exercises (stretching and isometric, concentric and eccentric strengthening), and lower leg exercises (stretching and eccentric strengthening). At 12 weeks of follow-up, performing the AIPP was associated with a significant lower risk of participation restriction injury complaint, with hazard ratio of 0.29 (95% CI: 0.12–0.73). After 40 weeks of follow-up, there was no significant association. These results are encouraging and are in favor of the use in practice of this program. However, they should be taken with caution before promoting its use, given some limitations of the study (e.g., it is not a randomized controlled trial leading to selection bias, there was a small sample size, the choice in performing the program or not can also influence the outcome) [26].

Therefore, a controlled randomized trial called PREVATHLE has been conducted during a 40-week period in a population of track and field athletes aged from 16 to 40 years. It was reviewed and approved by the Committee for the Protection of Persons (CPP Ouest II—Angers, number: 2017-A01980-53), and was registered at ClinicalTrials.gov (ClinicalTrials.gov Identifier: NCT03307434). It was aimed at including 880 athletes randomly divided into two groups: one control group continuing its usual training and one intervention performing the AIPP at least two times a week in addition to its usual training. We expect that the results of this PREVATHLE controlled randomized trial will help to define whether the AIPP is relevant to help reducing the occurrence of injuries in track and field.

According to these results, this athletics injury prevention program can be considered as a first step in the development of an exercise-based injury prevention program. One way of improvement can be to individualize the program to the sex and the disciplines of athletes. Indeed, since injury characteristics varied according to sex and disciplines [8], it seems relevant to adapt the selection of exercises of the injury prevention program in order to target the main injuries incurring for a discipline and by sex. For example, the main injuries in female long-distance runners will be different than in male sprinters [8]. Consequently, it is logical to think that exercises included in an exercises-based program, which can help to reduce the occurrence of these injuries, will be different. Thus, the next step when reflecting at an injury prevention program will be to adapt it to the discipline and sex. After that, another next step will be to individualize it to the individual characteristics of each athlete. This can be reached by individual screening of athlete's deficiencies [34], in order to develop exercises-based injury prevention program appropriate to discipline, sex, and individual characteristics.

In addition, the preventative approach should not only consider exercises aiming at improving strength, flexibility, neuromuscular control. The preventative approach should be global, multimodal, and multifactorial. Since there is no scientific published evidence proven by randomized controlled trials or other high-quality studies on the efficacy of injury prevention measure, program or strategy in track and field, injury prevention measures could be proposed based on evidence-based approach combining evidences from other sports and expert experience in track and field. In this way, Edouard et al. [12, 13, 35] proposed, based on a nonexhaustive review and brainstorming between the coauthors, some measures that may help for injury prevention:

1. Physical conditioning of athletes for improvement of sensorimotor control by, for instance stretching, muscular strengthening particularly eccentric, proprioceptive, balance, increased resistance to fatigue.
2. Technical movement and biomechanics improvements to avoid technopathies and/or technical mistakes that may result in injury.
3. Sports equipment and rules (e.g., modification of rules to improve safety, changes in competition schedules according to weather conditions, the circadian cycle).
4. Lifestyle (e.g., improved recovery, sleep, and/or nutrition).
5. Psychological approach (e.g., mental preparation, mental imagery, psychological follow).
6. Coordinated and consistent medical care of athletes (e.g., medical staff, early and correct care of injury, athletes' health monitoring).
7. Systematic and sustained approach by all stakeholders: the top management of national and international athletics federations should support injury prevention and safety promotion initiatives.

Finally, as for the general injury and illness prevention at major athletics championships, the 10 tips "PREVATHLES" proposed by Edouard et al. [36] could be relevant to help to reduce the occurrence of injuries in track and field:

1. When there is a travel, it is important to anticipate and prepare it (e.g., medical checking, vaccine, time-zone, jet lag, culture, food habits).
2. As stated above, it is relevant to respect athlete characteristics and discipline specificity when developing injury prevention program or strategy (e.g., sex, endurance/explosive).
3. Education of athletes and their entourages is important to make them actively participate in athlete's health protection and athlete's injury prevention; being vigilant of painful symptoms and subclinical illness markers.
4. Prevent illness can limit new injuries, so avoiding infection risk by, for instance washing hands, safe food and drink, avoid contact with sick people, could be of help.
5. Train appropriately and optimally (not too much and not too less), including for instance physical conditioning, technical training, load management, psychological preparation.
6. Taking into account the health status (e.g., history of previous injuries, well-being in the month before championships) seems relevant to individualize injury prevention strategies.
7. Improving lifestyle is relevant to reduce the risk of injuries, e.g., good sleep, regular hydration and nutrition with safe water/food, regular fruits and vegetables, improve recovery strategies.
8. It seems relevant to take into consideration the environmental conditions (e.g., heat, cold, air cleaning, changes or climatic conditions).

9. Finally, it is important to have a safety practice and lifestyle (e.g., equipment, rules, own-practice in athletics and extra-sport activities).

29.4 Conclusions

Given the risk of injuries lead by the track and field practice, the prevention of injuries in track and field represents an important area for athletes and all stakeholders, such as coaches, health professionals, family, sports scientists, managers, sponsors, as well as international and national govern bodies. Using a step-by-step approach that aims to understand and describe all components of the sports injuries seems relevant to develop measures, strategies, and/or programs that can reduce the occurrence of injuries. Unlike other sports, currently and to our knowledge, there is no scientific published evidence proven by randomized controlled trials or other high-quality studies on the efficacy of injury prevention measure, program or strategy in track and field. Injury prevention approach should thus target the main injuries, taking into account the specific injury characteristics by disciplines and sex, and if possible, of each individual athlete's characteristic. In addition, the preventative approach should be global, multimodal, and multifactorial, including but not limited to, improvements of physical conditioning, technical movement and lifestyle, psychological approach, adaptation of sports equipment and rules, coordinated and consistent medical care of athletes, and systematic and sustained approach by all stakeholders to support and promote injury prevention and safety practice.

References

1. Edouard P, Morel N, Serra J-M, Pruvost J, Oullion R, Depiesse F. [Prévention des lésions de l'appareil locomoteur liées à la pratique de l'athlétisme sur piste. Revue des données épidémiologiques]. Sci Sports. 2011;26:307–15.
2. D'Souza D. Track and field athletics injuries - a one-year survey. Br J Sports Med. 1994;28:197–202.
3. Bennell KL, Crossley K. Musculoskeletal injuries in track and field: incidence, distribution and risk factors. Aust J Sci Med Sport. 1996;28:69–75.
4. Jacobsson J, Timpka T, Kowalski J, Nilsson S, Ekberg J, Dahlström Ö, et al. Injury patterns in Swedish elite athletics: annual incidence, injury types and risk factors. Br J Sports Med. 2013;47:941–52.
5. Feddermann-Demont N, Junge A, Edouard P, Branco P, Alonso J-M. Injuries in 13 international Athletics championships between 2007–2012. Br J Sports Med. 2014;48:513–22.
6. Edouard P, Feddermann-Demont N, Alonso JM, Branco P, Junge A. Sex differences in injury during top-level international athletics championships: surveillance data from 14 championships between 2007 and 2014. Br J Sports Med. 2015;49:472–7.
7. Edouard P, Branco P, Alonso J-M. Muscle injury is the principal injury type and hamstring muscle injury is the first injury diagnosis during top-level international athletics championships between 2007 and 2015. Br J Sports Med. 2016;50:619–30.
8. Edouard P, Navarro L, Branco P, Gremeaux V, Timpka T, Junge A. Injury frequency and characteristics (location, type, cause and severity) differed significantly among athletics ('track and field') disciplines during 14 international championships (2007–2018): implications for medical service planning. Br J Sports Med. 2020;54:159–67.
9. Edouard P, Richardson A, Navarro L, Gremeaux V, Branco P, Junge A. Relation of team size and success with injuries and illnesses during eight international outdoor athletics championships. Front Sport Act Living. 2019;1:8.
10. Bolling C, Delfino Barboza S, van Mechelen W, Pasman HR. How elite athletes, coaches, and physiotherapists perceive a sports injury. Transl Sport Med. 2019;2:17–23.
11. Edouard P, Branco P, Alonso J-M. Challenges in Athletics injury and illness prevention: implementing prospective studies by standardised surveillance. Br J Sports Med. 2014;48:481–2.
12. Edouard P, Alonso JM, Jacobsson J, Depiesse F, Branco P, Timpka T. Injury prevention in athletics: the race has started and we are on track! New Stud Athl. 2015;30:69–78.
13. Edouard P, Alonso JM, Jacobsson J, Depiesse F, Branco P, Timpka T. On your marks, get set, go! A flying start to prevent injuries. Aspetar Sport Med J. 2019;8:210–3.
14. van Mechelen W, Hlobil H, Kemper HCG. Incidence, severity, aetiology and prevention of sports injuries. A review of concepts. Sports Med. 1992;14:82–99.
15. Finch C. A new framework for research leading to sports injury prevention. J Sci Med Sport. 2006;9:3–9. https://doi.org/10.1016/j.jsams.2006.02.009.
16. Bolling C, van Mechelen W, Pasman HR, Verhagen E. Context matters: revisiting the first step of the 'sequence of prevention' of sports injuries. Sport Med. 2018;48:2227–34.

17. Edouard P, Ford KR. Great challenges toward sports injury prevention and rehabilitation. Front Sport Act Living. 2020;2:80.
18. Timpka T, Jacobsson J, Bargoria V, Périard JD, Racinais S, Ronsen O, et al. Preparticipation predictors for championship injury and illness: cohort study at the Beijing 2015 International Association of Athletics Federations World Championships. Br J Sports Med. 2017;51:272–7.
19. Edouard P, Jacobsson J, Timpka T, Alonso JM, Kowalski J, Nilsson S, et al. Extending in-competition Athletics injury and illness surveillance with preparticipation risk factor screening: a pilot study. Phys Ther Sport. 2015;16:98–106.
20. Alonso J-MJ-M, Jacobsson J, Timpka T, Ronsen O, Kajenienne A, Dahlström Ö, et al. Preparticipation injury complaint is a risk factor for injury: a prospective study of the Moscow 2013 IAAF Championships. Br J Sports Med. 2015;49:1118–24.
21. Rebella GS, Edwards JO, Greene JJ, Husen MT, Brousseau DC. A prospective study of injury patterns in high school pole vaulters. Am J Sports Med. 2008;36:913–20.
22. Timpka T, Jacobsson J, Dahlström Ö, Kowalski J, Bargoria V, Ekberg J, et al. The psychological factor "self-blame" predicts overuse injury among top-level Swedish track and field athletes: a 12-month cohort study. Br J Sport Med. 2015;49:1472–7.
23. Lauersen JB, Bertelsen DM, Andersen LB. The effectiveness of exercise interventions to prevent sports injuries: a systematic review and meta-analysis of randomised controlled trials. Br J Sports Med. 2014;48:871–7.
24. Lauersen JB, Andersen TE, Andersen LB. Strength training as superior, dose-dependent and safe prevention of acute and overuse sports injuries: a systematic review, qualitative analysis and meta-analysis. Br J Sports Med. 2018;52:1557–63.
25. Al Attar WSA, Alshehri MA. A meta-analysis of meta-analyses of the effectiveness of FIFA injury prevention programs in soccer. Scand J Med Sci Sport. 2019;29:1846–55.
26. Edouard P, Cugy E, Dolin R, Morel N, Serra J-M, Depiesse F, et al. The athletics injury prevention programme can help to reduce the occurrence at short term of participation restriction injury complaints in athletics: a prospective cohort study. Sports. 2020;8:84.
27. Carragher P, Rankin A, Edouard P. A one-season prospective study of illnesses, acute, and overuse injuries in elite youth and junior track and field athletes. Front Sport Act Living. 2019;1:1–12.
28. Arnason A, Andersen TE, Holme I, Engebretsen L, Bahr R. Prevention of hamstring strains in elite soccer: an intervention study. Scand J Med Sci Sport. 2008;18:40–8.
29. Petersen J, Thorborg K, Nielsen MB, Budtz-Jørgensen E, Hölmich P. Preventive effect of eccentric training on acute hamstring injuries in men's soccer: a cluster-randomised controlled trial. Am J Sports Med. 2011;39:2296–303.
30. Beyer R, Kongsgaard M, Hougs Kjaer B, Ohlenschlaeger T, Kjaer M, Magnusson SP. Heavy slow resistance versus eccentric training as treatment for achilles tendinopathy: a randomized controlled trial. Am J Sport Med. 2015;43:1704–11.
31. Kongsgaard M, Kovanen V, Aagaard P, Doessing S, Hansen P, Laursen AH, et al. Corticosteroid injections, eccentric decline squat training and heavy slow resistance training in patellar tendinopathy. Scand J Med Sci Sport. 2009;19:790–802.
32. Kerkhoffs GM, van den Bekerom M, Elders LA, van Beek PA, Hullegie WA, Bloemers GM, et al. Diagnosis, treatment and prevention of ankle sprains: an evidence-based clinical guideline. Br J Sport Med. 2012;46:854–60.
33. Coulombe BJ, Games KE, Neil ER, Eberman LE. Core stability exercise versus general exercise for chronic low back pain. J Athl Train. 2017;52:71–2.
34. Verhagen E, Van Dyk N, Clark N, Shrier I. Do not throw the baby out with the bathwater; Screening can identify meaningful risk factors for sports injuries. Br J Sports Med. 2018;52:1223–4.
35. Edouard P, Alonso JM, Branco P. New insights into preventing injuries and illnesses among elite athletics athletes. Br J Sports Med. 2018;52:4–5.
36. Edouard P, Richardson A, Murray A, Duncan J, Glover D, Kiss-Polauf M, et al. Ten tips to hurdle the injuries and illnesses during major athletics championships: practical recommendations and resources. Front Sport Act Living. 2019;1:12.

30 Management of Track and Field Injuries: Insights into Energy Availability in Athletes

Tom G. H. Wiggers, John IJzerman, and Petra Groenenboom

30.1 Introduction

Elite athletes and trainers are constantly looking for the optimal amount of training. Aiming to optimize performance, there is a delicate balance between executing a high training load and simultaneously not exceeding the athlete's physical capabilities. Continuity in training is key in long-term athlete development, so it is of utmost importance to avoid long interruptions of training. To secure continuity of training and optimal training adaptation, adequate energy intake by the athlete is key. Athletes performing high training loads not matching this energy expenditure with sufficient energy intake are prone to have low energy availability. Low energy availability is a systemic problem which affects many aspects of physiological function and consequently the athlete's health and sports performance.

Effects of insufficient energy availability in elite athletes are first recognized in female athletes. This phenomenon is called the Female Athlete Triad (FAT) and consists of disordered eating, amenorrhea, and osteoporosis [1]. We discuss the interconnection of these three triad components later. The FAT has been extensively studied in the 80s and 90s of the last century. Especially in athletes with a high energy flux (middle-long distance runners) and/or low body weight advantage (e.g., high jumpers), health issues and underperformance are frequently caused by low energy availability. This complex phenomenon, which incorporates the FAT, has recently been described in scientific literature as the model of Relative Energy Deficiency in Sport (RED-S) [2–4].

Trainers, physiotherapists, physicians, and others working with (elite) athletes must be familiar with the concepts of FAT and RED-S, because it is important to emphasize that together we have the responsibility to protect the health of the athlete. Moreover, prevention, early

T. G. H. Wiggers (✉)
Department of Orthopaedic Surgery, Amsterdam Movement Sciences, Amsterdam UMC, University of Amsterdam, Amsterdam, The Netherlands

Academic Center for Evidence-Based Sports Medicine (ACES), Amsterdam, The Netherlands

Amsterdam Collaboration for Health and Safety in Sports (ACHSS), International Olympic Committee (IOC) Research Center Amsterdam UMC, Amsterdam, The Netherlands
e-mail: t.g.wiggers@amsterdamumc.nl

J. IJzerman
Medical Committee Royal Dutch Athletics Federation, Arnhem, The Netherlands

P. Groenenboom
Department of Sports Medicine, Gelderse Vallei Hospital, Ede, The Netherlands

Royal Dutch Athletics Federation, Arnhem, The Netherlands
e-mail: groenenboomp@zgv.nl

G. L. Canata et al. (eds.), *Management of Track and Field Injuries*,
https://doi.org/10.1007/978-3-030-60216-1_30

recognition, and treatment of this problem have a direct effect on sports performance because good health is a prerequisite for elite performance. The aim of the present chapter is to provide an overview of the underlying principles of energy availability and discuss the clinical approach in prevention and treatment of this condition.

30.2 Low Energy Availability in Sports

The International Olympic Committee (IOC) expert working group defined the syndrome of RED-S as follows [2]:

> The syndrome of RED-S refers to impaired physiological function including, but not limited to, metabolic rate, menstrual function, bone health, immunity, protein synthesis, cardiovascular health caused by relative energy deficiency.

The key problem of FAT and RED-S is low energy availability. This is characterized by a mismatch between the athlete's energy intake (diet) and the energy expended in exercise. Low energy availability for a prolonged time results in down-regulation of physiological systems that are essential for growth, development, and health in order to fulfill the physical demands of the training [1–3]. It is important to realize that low energy availability can be present with normal energy balance because of the energy shift between biological systems [5]. Energy availability is calculated by the following formula [2, 3]:

$$\text{energy availability} = \text{energy intake} - \text{energy cost of exercise relative to fat free mass}$$

30.2.1 Energy Intake

Energy intake (in kcal/day) is best estimated by food records, preferably for 4–7 days [6, 7]. The expertise of a (sports) dietician is useful in making the most accurate estimation of energy intake.

30.2.2 Exercise Energy Expenditure

Exercise energy expenditure (kcal/day) is best estimated via training diaries, including exact mode, duration, and intensity of the training; preferably including data as heart rate or running pace [6, 7]. Exercise energy expenditure is expressed relative to fat-free mass (FFM) representing the most metabolically active tissues [3].

30.2.3 Interpretation

Energy availability is categorized as low, moderate or optimal [3, 6].

- Low energy availability: <30 kcal/kg FFM/day.
- Moderate energy availability: 30–45 kcal/kg FFM/day.
- Optimal energy availability: >45 kcal/kg FFM/day.

It should be stated that these values suggest a specific threshold of energy availability below which problems arise; however, this is not always the case. Rather, the concept can be considered as a continuum with a linear increase of negative health consequences as the energy availability decreases [8, 9].

In discussing the components of low energy availability, we adopt the structure of the updated 2018 IOC consensus statement on RED-S distinguishing health aspects and performance consequences [3, 10]. We start by reviewing the classical FAT which is still regarded as the fundament of the concept of low energy availability. After that, we discuss the development from FAT as a triad to a continuum and to the more comprehensive concept of RED-S. Finally, we outline strategies for prevention, early recognition, and management of low energy availability in athletes.

30.3 Health Aspects of Low Energy Availability–Female Athlete Triad

The relation between disordered eating, low body fat, menstrual irregularities, and bone health in elite and collegiate female athletes had already been recognized before the term Female Athlete Triad was introduced in 1992 [8, 11, 12]. We start reviewing the three components of the classical FAT, namely disordered eating, amenorrhea, and osteoporosis.

30.3.1 Disordered Eating

Disordered eating is a key component of the FAT. The prevalence of eating disorders like anorexia nervosa and bulimia nervosa is higher in athletes than in nonathletes, especially in sports emphasizing low body weight and esthetic sports [1]. In a Norwegian study, prevalence of subclinical or clinical eating disorders was 13.5% in elite athletes and 4.6% in the general population ($p < 0.001$) [13]. More prevalent than clinical eating disorders are athletes with disordered eating patterns, characterized by deliberate attempts to lose weight, the elimination of specific foods from their diet or obsessive attention to their diet [1]. Disordered eating as a component of the FAT was substituted by low energy availability with or without an eating disorder in updated definitions [14, 15]. This shows that low energy availability can occur in an unintentional manner when increased training load is not matched with increased energy intake. Mismatch between intake and expenditure is essential, because insufficient energy intake in relation to the training load results in low body fat mass, thereby starting the cascade leading to the clinical syndrome of the FAT.

30.3.2 Amenorrhea

Amenorrhea is the second component of the FAT. Prevalence of menstrual irregularities is higher in athletes than in nonathletes [1]. Prevalence rates vary depending on the definition of menstrual irregularities and type of sport. However, prevalence rates have been reported to occur between 7% and 37% [9]. The highest prevalence has been consistently found in dancers and runners [1]. Menstrual irregularities can be classified as oligomenorrhea, primary amenorrhea, and secondary amenorrhea [14, 16]. The menstrual cycle is directly influenced by release of gonadotropin-releasing hormone (GnRH) by the hypothalamus which consequently stimulates the pituitary gland to release luteinizing hormone (LH) and follicle stimulating hormone (FSH) [1, 3]. Low energy availability leads to lower pulsatility of GnRH and LH and directly affects the ovaries as their main target organ [7, 8]. This results in lower levels of circulating estrogen and progesterone and consequently menstrual irregularities or postponement of menarche occurs [1, 3]. Therefore, low energy availability is an important etiological factor in menstrual irregularities. Female athletes losing their periods or not presenting to have the menarche at the age of 16 can be considered alarming signals.

30.3.3 Osteoporosis

The third component of the FAT is osteoporosis. Osteoporosis is defined by a bone mineral density (BMD) equal to or more than 2.5 standard deviations below the BMD of young adults. Bone is an active tissue, constantly being turned over by activity of bone forming osteoblasts and bone resorbing osteoclasts ("bone remodeling") [17]. The net result of this process is dependent on the amount of physical activity (axial load), dietary intake, and hormonal levels [18, 19]. Estrogens are one of these hormones and low estrogen levels caused by amenorrhea have direct consequences for bone health. Estrogens have an inhibitory effect on number and activity of osteoclasts which pushes the bone remodeling in favor of bone formation [1, 8, 17]. Low level of estrogens pushes this balance the other way and increases the risk of stress fractures and osteoporosis [1, 12]. Athletes normally have a higher BMD than nonathletes thereby resulting in significant lower fracture risk at older age [19]. However, late menarche and secondary amenorrhea resulting from low energy intake and/or excessive exercise are associated with low BMD [17, 18]. In a study of female collegiate distance runners, athletes with a history of amenorrhea were found to have a statistically significant higher lifetime stress fracture risk than female distance athletes with history of regular periods (32% vs. 6%) [16]. In general, risk of stress fractures is 2–4 times higher for amenorrheic athletes than regularly menstruating athletes [12, 16, 19]. Moreover, BMD of athletes with amenorrhea or oligomenorrhea was similar for nonathletes, suggesting that these athletes do not take advantage of the bone forming effect of physical exercise [2, 16]. This is critical, as adolescence and early adulthood are essential periods for building peak bone mass [20]. A lower peak bone mass at age 25 is associated with higher risk (about 50% higher relative risk) of osteoporosis and accompanying fractures at older age [20].

30.4 Further Development of the FAT

Introduced in 1992 as a classical triad, the FAT was updated to a continuum in 2007 [14]. This continuum starts in an optimal situation in all three components and gradually develops through a subclinical phase to a situation in which one or more components of the FAT are present (Fig. 30.1) [14, 15]. Key is to detect athletes who deteriorate from the optimal situation before they develop the full clinical syndrome of the FAT. Despite the development of the FAT to a continuum, about 20 years after the introduction, this clinical syndrome was regarded as to narrow [2, 3]. One major factor was that similar problems were recognized in men. Men are, by definition, not included in the FAT [2, 3]. This is why in 2014 a more comprehensive concept called the syndrome of Relative Energy Deficiency in Sport (RED-S) was introduced [2]. However, this introduction was criticized because there was insufficient evidence that all the items of the RED-S model are in a direct relationship with energy deficiency [8, 15]. Another point of criticism was that the new concept would minimize the importance of the FAT while women experience the most severe medical consequences. The key component of both concepts however, is low energy availability. The FAT consists of three aspects of the spectrum, while the RED-S model focuses on all physiological systems that can be hampered by a shortage of energy.

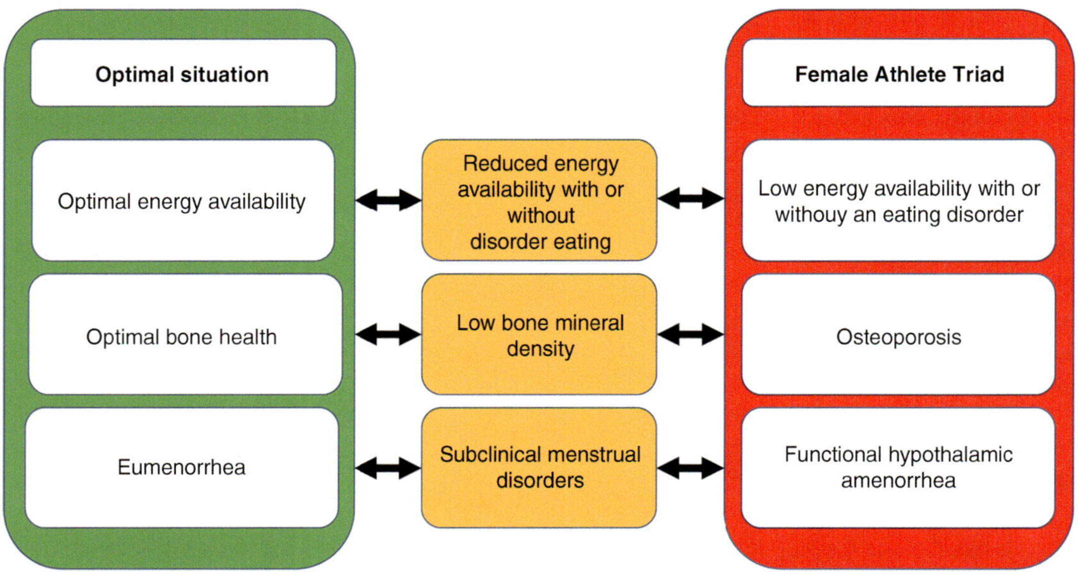

Fig. 30.1 Female Athlete Triad as a continuum. Adapted from Nattiv et al. [14]

30.5 Health Consequences of Low Energy Availability–Specifically in Men

The three key components of the FAT have been touched upon in the beginning of this chapter. Due to the fact that these are only applicable to women, we will now shortly discuss the effect of low energy availability on reproductive function, bone health, and eating disorders in men.

Male athletes with low energy availability have lower testosterone concentration and consequently lower libido [5, 6]. Same as in women, the responsible mechanism is probably the lower LH pulsatility, which is found in trained marathon runners (running 125–200 km/week) [21].

Low energy availability has unfavorable consequences on bone in male as well and is known to be a major risk factor in the development of stress fractures [3, 5]. Testosterone stimulates the activity of bone forming osteoblasts and this is why hypogonadal men experience rapid bone loss [8, 17]. In a study with elite male distance athletes, those with low testosterone were found to have significantly more career stress fractures in comparison to athletes with normal testosterone [6].

Eating disorders are less prevalent in male athletes compared to women. In a cohort of elite adolescent Norwegian athletes, prevalence of eating disorders was 14.0% and 3.2% ($p < 0.001$) for female and male athletes, respectively [22].

30.6 Health Consequences of Low Energy Availability–specific Components of the Syndrome of RED-S

Health consequences of low energy availability for almost all physiological systems are described in the syndrome of RED-S [2, 3]. We highlight some key points on the hematological system, cardiovascular system, and the immunological system.

Hematological system. Low nutritional intake makes the athlete prone for deficiencies in macronutrients and micronutrients, such as iron. Iron is essential for erythrocytes and in muscle contraction. Iron deficiency with or without anemia is frequent in athletes [5, 23]. Suboptimal hemoglobin concentrations can be improved by supplementation but certainly also by increasing overall energy intake.

Cardiovascular system. Secondary to a hypoestrogenic state, unfavorable changes in lipid profile (higher total cholesterol and higher LDL cholesterol) and endothelial function are reported [3, 24]. However, the exact clinical implications of these potentially negative cardiovascular effects are unknown [24].

Immunological system. Athletes in prolonged periods of low energy availability are at increased risk of illness and infections [25]. Additionally, athletes during periods of heavy training already have an increased susceptibility for infections, especially upper respiratory tract infections [5, 26].

30.7 Performance Consequences of Low Energy Availability

Inadequate energy availability for a prolonged time has detrimental effects on sport performance and occurs in athletes in all kind of sports [27, 28]. One of the important effects of low energy availability is a loss in lean mass due to reduced protein balance under influence of a lower anabolic environment. Reduced total muscle mass leads to reduced muscle force, stability, and neuromuscular control thereby increasing the risk of exercise-related injuries [29]. Muscle power is related to better running economy by improvement of coordination, timing, and trunk kinematics [30]. Therefore, loss of muscle mass and strength due to low energy availability may also hamper running performance. Another mechanism that reduces endurance capacity of athletes in low energy availability is a decrease in intramuscular and hepatic glycogen storage capacity [25]. This is mediated among others by low estrogen levels and the anorectic effect of chronic inflammation by cytokines which act on the hypothalamus and decrease hunger [31]. Low energy availability works on brain function as well, resulting in impaired judgement and decreased concentration [25].

30.8 Prevention and Early Recognition of Low Energy Availability

Knowing the components of energy availability in athletes, we continue with discussing practical aspects of this concept. In the prevention and recognition of low energy availability, the sports medicine physician has a major role. In the beginning, low energy availability presents with small decreases in athletes capacities. One can think of minor injuries, infections, sleep disturbances, irritability, and unexplained underperformance. This can occur when there are still normal weight, no change in menstruation or changes in eating pattern. Nonvoluntary energy deficit can also occur because there is no strong biological drive to match energy intake to activity-induced energy expenditure [7]. Whereas food deprivation increases hunger, the same energy deficit produced by exercise training does not [32]. Inadvertent energy deficit occurs in particular with a low fat, high carbohydrate diet, which is common in endurance athletes [7]. Complicating factor is, however the fact that lower body weight can have performance enhancing effects in short term. On the other side, increasing energy intake will not always lead to a decrease in performance, in contrast to what athletes often think.

One of the key elements in prevention is creating awareness, communication, and organization. Athletes at risk have to be detected and included in a well-organized health surveillance program (periodic sports medical examinations; discussed below). The collection of longitudinal data on personalized health and performance tests is of vital importance in monitoring athletes [33]. Ideally, data collection starts when the athletes enter the performance program. Benefits of monitoring the athlete are understanding of training response and explaining changes in performance. Based on that analysis, modifications in training and competition program can be addressed [34].

Prevention of eating disorders in athletes in weight-dependent sports can be associated with a number of complicating factors. First, athletes know the performance-enhancing effect of a lower body weight to a certain level; however, the athlete cannot judge the optimal balance. Secondly, young athletes regard world-class athletes at major competitions as role models without realizing how these athletes may look like off-season and when they were junior athletes. Their present appearance is the result of years of training and of accurate periodization of their body weight and fat percentage throughout the year [5]. An intervention program aiming for prevention of disordered eating and eating disorders in elite high school athletes found promising results [35]. Primary focus of this program was enhancing self-esteem of the athletes by teaching mental training techniques on motivation and goal setting [35].

30.8.1 Monitoring Tools

The periodic sports medical examination includes several components and can be individualized based on athlete characteristics, sporting event, and medical background.

The examination starts with an evaluation of the previous months and discussing the actual situation of the athlete. The athlete should be encouraged to come forward with his/her own questions. Questionnaires as the POMS (Profile of Moods States) can help to reveal important issues to discuss. It is important to talk about the athlete's beliefs about body weight, weight loss, diet, and performance. Changes in the athlete's ideas about his/her diet and/or preferable weight for competition can be a first signal in developing a disordered eating pattern. In screening for eating disorders and disordered eating patterns, there are several questionnaires available (e.g., Brief Eating Disorder in Athletes Questionnaire (BEDA-Q) and Eating Disorder Screen for Primary Care (ESP)) [3, 25]. Discussing menstrual cycle in female athletes and libido in male athletes is essential. It should be mentioned that menstrual status can be masked by use of oral contraceptives. Therefore, other markers must be used to evaluate energy availability [6].

In physical examination, it is vital to define the athlete's somatotype, body weight, and body composition (body fat and fat-free mass).

Periodic blood tests can provide evidence for low energy availability and underlying problems. This includes red and white blood cell count, thrombocytes, kidney function (creatinine, urea), thyroid function (TSH, triiodothyronine (T3) and thyroxine (T4)), liver enzymes, iron status (ferritin, transferrin), vitamins (folic acid, B12, and D), cholesterol, insulin growth factor-1 (IGF-1), and hormones (cortisol, testosterone, estrogen, progesterone). Preferably, a couple of parameters which are essential in monitoring the particular athlete are defined instead of always ordering the same (complete) package of blood tests. Measurement of ketone bodies in urine as a marker for carbohydrate availability may additionally be considered.

After this, an (as accurately as possible) estimation of energy availability has to be made by determination of energy intake and exercise energy expenditure. Several questionnaires can be helpful in the determination of energy availability, e.g., LEAF-Q (Low Energy Availability in Females Questionnaire).

Cardiopulmonary exercise testing as a monitoring tool can be used in cases of (possible) overtraining syndrome.

30.9 Management of Low Energy Availability in Sports Medical Practice

Early signs of low energy availability can be detected when accurately monitoring the athlete. In the management of this condition, key is to make changes in both components of the energy availability balance. This entails reducing exercise energy expenditure by reducing training load and on the other side increasing energy intake. The extent of these interventions must be determined individually, dependent on the severity of the situation. Especially athletes who are not monitored periodically can present with severe symptoms and must be managed aggressively.

The sports medicine physician can be considered the case manager in management of low energy availability. He/she can define the extent of the current problem(s) and has to decide which other health care providers are necessary to include in the management, such as a (sports) dietician, a psychologist and/or a psychiatrist. Relatively "simple" cases of disordered eating can be managed by the sports medicine physician, sports dietician, and sports psychologist. Athletes with clinical eating disorders or resistant patterns of disorder eating have to be referred to specialized psychiatric centers. In suspicion of serious medical conditions or organic problems, the athlete can be referred to a pediatrician, internal medicine specialist or gynecologist. Furthermore, it is of utmost importance to explain to the athlete what exactly the problem is and make her/him aware of the seriousness and possible consequences of this condition. This can be a difficult process and can take some time to get full understanding and cooperation of the athlete. In this process, the role of the coach should not be overlooked. Take time to educate the coach about energy availability and find out his/her beliefs about weight, health, and sports performance. Several psychosocial factors were found to be helpful in female collegiate athletes recovering from an eating disorder: support from others (friends and professionals), the desire to be healthy to participate in sport, and change in values/beliefs about their body, diet, and sport [36].

The sports medicine physician makes the definite treatment plan, using a shared decision-making model. This treatment plan must contain clear guidelines about training and diet, monitoring tools, and evaluation moments. Usually, monitoring tools are a selection of the tools discussed before and depend on the athlete and available resources. The RED-S clinical assessment tool can be useful as a guide in return to play decisions [37].

30.10 Final Remarks

This present chapter gives an overview of low energy availability in sports and their concepts of the Female Athlete Triad and the syndrome of Relative Energy Deficiency in Sport. Low energy availability is a key concept in the athlete's health and has a wide range of health and performance consequences. It is of paramount importance that everyone involved in the team around the athlete is familiar with this concept and that the involved team is aware of the key principles in prevention, recognition, and management of the condition.

References

1. IOC Medical Commission Working Group Women in Sport. Position Stand on The Female Athlete Triad. 2005. https://www.olympic.org/news/ioc-consensus-statement-on-the-female-athlete-triad. Accessed 20 Jan 2020.
2. Mountjoy M, Sundgot-Borgen J, Burke L, Carter S, Constantini N, Lebrun C, et al. The IOC consensus statement: beyond the female athlete triad—relative energy deficiency in sport (RED-S). Br J Sports Med. 2014;48:491–7.
3. Mountjoy M, Sundgot-Borgen JK, Burke LM, Ackerman KE, Blauwet C, Constantini N, et al. IOC consensus statement on relative energy deficiency in sport (RED-S): 2018 update. Br J Sports Med. 2018;52:687–97.
4. Burke LM, Close GL, Lundy G, Mooses M, Morton JP, Tenforde AS. Relative energy deficiency in sport in male athletes: a commentary on its presentation among selected groups of male athletes. Int J Sport Nutr Exerc Metab. 2018;28(4):364–74.
5. Burke LM, Castell LM, Casa DJ, Close GL, Costa RJS, Desbrow B, et al. International Association of Athletics Federations Consensus Statement 2019: nutrition for athletics. Int J Sport Nutr Exerc Metab. 2019;29(2):73–84.
6. Heikura IA, Uusitalo ALT, Stellingwerff T, Bergland D, Mero AA, Burke LM. Low energy availability is difficult to assess but outcomes have large impact on bone injury rates in elite distance athletes. Int J Sport Nutr Exerc Metab. 2018;28:403–11.
7. Loucks AB. Low energy availability in the marathon and other endurance sports. Sports Med. 2007;37(4–5):348–52.
8. De Souza MJ, Williams NI, Nattiv A, Joy E, Misra M, Loucks AB, et al. Misunderstanding the female athlete triad: refuting the IOC consensus statement on relative energy deficiency in sport (RED-S). Br J Sports Med. 2014;48(20):1461–5.
9. Lieberman JL, De Souza MJ, Wagstaff DA, Williams NI. Menstrual disruption with exercise is not linked to an energy availability threshold. Med Sci Sports Exerc. 2018;50(3):551–61.
10. Keay N, Rankin A. Infographic. Relative energy deficiency in sport: an infographic guide. Br J Sports Med. 2019;53:1307–9.
11. Myburgh KH, Hutchins J, Fataar AB, Hough SF, Noakes TD. Low bone density is an etiologic factor for stress fractures in athletes. Ann Intern Med. 1990;113(10):754–9.
12. Barrow GW, Saha S. Menstrual irregularity and stress fractures in collegiate female distance runners. Am J Sports Med. 1988;16:209–16.
13. Sundgot-Borgen J, Torstveit MK. Prevalence of eating disorders in elite athletes is higher than in the general population. Clin J Sport Med. 2004;14(1):25–32.
14. Nattiv A, Loucks AB, Manore MM, Sanborn CF, Sundgot-Borgen J, Warren MP. American College of Sports Medicine position stand. The female athlete triad. Med Sci Sports Exerc. 2007;39(10):1867–82.
15. Williams NI, Koltun KJ, Strock NCA, De Souza MJ. Female Athlete triad and relative energy deficiency in sport: a focus on scientific rigor. Exerc Sport Sci Rev. 2019;47(4):197–205.
16. Ackerman KE, Cano Sokoloff N, De Nardo MG, Clarke HM, Lee H, Misra M. Fractures in relation to menstrual status and bone parameters in young athletes. Med Sci Sports Exerc. 2015;47(8):1577–86.
17. Compston JE. Sex steroids and bone. Physiol Rev. 2001;81(1):419–47.
18. Ackerman KE, Nazem T, Chapko D, Russell M, Mendes N, Taylor AP, et al. Bone microarchitecture is impaired in adolescent amenorrheic athletes compared with eumenorrheic athletes and nonathletic controls. J Clin Endocrinol Metab. 2011;96:3123–33.
19. Nichols DL, Sanborn CF, Essery EV. Bone density and young athletic women. An update. Sports Med. 2007;37(11):1001–14.
20. Tveit M, Rosengren BE, Nilsson JA, Karlsson MK. Exercise in youth: high bone mass, large bone size, and low fracture risk in old age. Scand J Med Sci Sports. 2015;25(4):453–61.
21. MacConnie SE, Barkan A, Lampman RM, Schork MA, Beitins IZ. Decreased hypothalamic gonadotropin-releasing hormone secretion in male marathon runners. N Engl J Med. 1986;315:411–7.
22. Martinsen M, Sundgot-Borgen J. Higher prevalence of eating disorders among adolescent elite athletes than controls. Med Sci Sports Exerc. 2013;45(6):1188–97.
23. Sim M, Garvican-Lewis LA, Cox GR, Govus A, McKay AKA, Stellingwerff T, et al. Iron considerations for the athlete: a narrative review. Eur J Appl Physiol. 2019;119(7):1463–78.
24. Rickenlund A, Eriksson MJ, Schenck-Gustafsson K, Hirschberg AL. Amenorrhea in female athletes is associated with endothelial dysfunction and unfavorable lipid profile. J Clin Endocrinol Metab. 2005;90(3):1354–9.

25. Ackerman KE, Holtzman B, Cooper KM, Flynn EF, Bruinvels G, Tenforde AS, et al. Low energy availability surrogates correlate with health and performance consequences of relative energy deficiency in sport. Br J Sports Med. 2019;53:628–33.
26. Lancaster GI, Halson SL, Khan Q, Drysdale P, Jeukendrup AE, Drayson MT, et al. Effect of acute exhaustive exercise and a 6-day period of intensified training on immune function in cyclists. J Physiol. 2003;548:96.
27. Sygo J, Coates AM, Sesbreno E, Mountjoy ML, Burr JF. Prevalence of indicators of low energy availability in elite female sprinters. Int J Sport Nutr Exerc Metab. 2018;28(5):490–6.
28. Van Heest JL, Rodgers CD, Mahoney CE, De Souza MJ. Ovarian suppression impairs sport performance in junior elite female swimmers. Med Sci Sports Exerc. 2014;46(1):156–66.
29. Tornberg ÅB, Melin A, Koivula FM, Johansson A, Skouby S, Faber J, et al. Reduced neuromuscular performance in amenorrheic elite endurance athletes. Med Sci Sports Exerc. 2017;49:2478–85.
30. Lima LCR, Blagrove R. Infographic. Strength training–induced adaptations associated with improved running economy: potential mechanisms and training recommendations. Br J Sports Med. 2020;54:302–3.
31. Smith LL. Cytokine hypothesis of overtraining: a physiological adaptation to excessive stress? Med Sci Sports Exerc. 2000;32(2):317–31.
32. Stubbs RJ, Hughes DA, Johnstone AM, Whybrow S, Horgan GW, King N, et al. Rate and extent of compensatory changes in energy intake and expenditure in response to altered exercise and diet composition in humans. Am J Physiol Regul Integr Comp Physiol. 2004;286:R350–8.
33. Bourdon PC, Cardinale M, Murray A, Gastin P, Kellmann M, Varley MC, et al. Monitoring athlete training loads: consensus statement. Int J Sports Physiol Perform. 2017;12(2):S2161–70.
34. Soligard T, Schwellnus M, Alonso JM, Bahr R, Clarsen B, Dijkstra HP, et al. How much is too much? (part 1) International Olympic Committee consensus on load in sport and risk of injury. Br J Sports Med. 2016;50:1030–41.
35. Martinsen M, Bahr R, Børresen R, Holme I, Pensgaard AM, Sundgot-Borgen J, et al. Preventing eating disorders among young elite athletes: a randomized controlled trial. Med Sci Sports Exerc. 2014;46(3):435–47.
36. Arthur-Cameselle JN, Quatromoni PA. Eating disorders in collegiate female athletes: factors that assist recovery. Eat Disord. 2014;22:50–61.
37. Mountjoy M, Sundgot-Borgen J, Burke L, Carter S, Constantini N, Lebrun C, et al. The IOC relative energy deficiency in sport clinical assessment tool (RED-S CAT). Br J Sports Med. 2015;49:1354.

31 The Aging Athlete: Influence of Age on Injury Risk and Rehabilitation

Thaisa Lazari Gomes, Larissa Oliveira Viana, Daniel Miranda Ferreira, Mauro Mitsuo Inada, Gerson Muraro Laurito, and Sergio Rocha Piedade

31.1 Introduction

In the last 50 years, life expectancy of the world population has risen by 20 years, and the number of elderly has increased dramatically. As a consequence, many countries have changed the base of their social pyramid. With this new global scenario, the literature has moved its focus to a comprehensive approach to the aging process due to the new social demands related to older people [1].

Along time, the human body suffers from a gradual process of deterioration, manifested by biomechanical and physiological changes that impact the biological and metabolic systems negatively. Regarding the cardiovascular system, the aging process causes an increase in heart size as the myocardium becomes thicker and more rigid with bigger cardiac chambers. In addition, there is less vasodilation in response to beta-adrenergic stimuli, contributing to an increase in afterload and a decrease in the cardiac response to physical exertion [2, 3].

T. L. Gomes · L. O. Viana · D. M. Ferreira
Exercise and Sports Medicine, School of Medical Sciences, University of Campinas—UNICAMP, Campinas, SP, Brazil

M. M. Inada · G. M. Laurito · S. R. Piedade (✉)
Exercise and Sports Medicine, Department of Orthopedics, Rheumatology, and Traumatology, School of Medical Sciences, University of Campinas—UNICAMP, Campinas, SP, Brazil
e-mail: piedade@unicamp.br

Moreover, at rest, there may be a slight decrease in heart rate, which demanding additional efforts and resulting in lower heart ability to increase heart rate and cardiac output, and in a reduction in the induced ejection fraction by exercise demand [2].

Furthermore, the literature has pointed out that the aging process promotes a progressive loss of tendon and muscle elasticity, and a decrease in the size and number of muscle fibers that will be manifested by lower muscle strength, reflecting the reduction in lean mass, muscle power, and strength. Additionally, due to joint stiffness, biomechanical changes may result in gait imbalance that affects walking and running, favoring the occurrence of injuries [4–6].

The literature has explored the biological process of aging in athletes' performance and how sports practice and regular physical activity impact their quality of life [7–9].

Every single athlete faces the same process in their career: They go through recruitment and selection processes, physical adaptation to their sports modality, long periods of training, competitions, impairment of social and family relationships, and socialization in the sports environment to reach the highest level they can. All of this enhances the interface between musculoskeletal units and neuromotor control, playing a vital role in the achievement of optimal outcomes in their sports career. Athletes' performance is guided by their level of physical conditioning, specific

G. L. Canata et al. (eds.), *Management of Track and Field Injuries*,
https://doi.org/10.1007/978-3-030-60216-1_31

training program to sports modality, adequate nutrition and hydration, and psychological balance. The sports career comprises several phases from the beginning to the high performance until the ending of the competitive career. In this chapter, we will focus on **physiological aging of the musculoskeletal system, the role of physical exercise and sports activity on aging, patterns of performance decline in master athletes, psychological and social aspects, injuries in track and field athletes, previous injuries, novice athletes, and rehabilitation of sports injuries.**

31.2 Physiological Aging of the Musculoskeletal System

The rise of life expectancy is not a guarantee of comparable quality of life due to the inexorable process of aging. Along time, aging follows a progressive and natural biological decline that causes biomechanical changes in the musculoskeletal system: bone, muscle, and tendon.

Healthy bones are essential for general health, functioning as a reservoir of minerals, and vital for specific physiological functions such as hematopoiesis and regulation of endocrine organs. Mechanical stress has a positive effect on strengthening of bone structure, but there are biological and physiological limitations. The body mass density (BMD) changes throughout life, and its peak occurs from 20 to 30 years old and decreases about 1% per year. Around the age of 80, a person will have lost about 40% of the original BMD. When the original BMD losses are higher than 25%, a scenario for spontaneous fractures from minimal or no trauma may occur [9].

Besides bone loss, there are clinical conditions that may potentialize the harmful effect of the aging processes such as genetic predisposition, use of steroids, lack of dietary calcium, vitamin D deficiency, systemic diseases, diseases that cause malabsorption, kidney disease, administration of heparin or oral anticoagulants, hyperparathyroidism, hyperthyroidism, diabetes mellitus, excessive use of alcoholic beverages, and prolonged immobilization, especially when body mass index (BMI) is less than 20 kg/m^2 [10]. Also, in cases of athletes using anabolic steroids to improve their sports performance, a "deliberate" weight control, requiring dietary restrictions and conditioning and resistance exercises, affects the metabolism and bone quality and, consequently, its ability to withstand the load. And of course, this harmful effect may be potentialized by the aging process.

Moreover, **osteoarthritis** is a common problem related to aged athletes. In the early stages, the aged athletes with osteoarthritis in the hip or knee with little symptomatic may be benefited by an exercise program that includes stretching, strengthening, flexibility, and stability exercises or aquatic exercises, with reduced load in the affected region [11].

In recent years, the participation of elderly athletes trained in endurance races, such as 5 km, 10 km, half-marathons, and marathons, has increased significantly, with a decrease in running times, suggesting that runners probably have not yet reached their performance limits of the races [12–15].

Intense and long-lasting resistance exercises, such as the half-marathon and the marathon in the elderly, result in high cardiovascular tension and also musculoskeletal overloads, with both beneficial and harmful clinical repercussions, which have been little studied in this age group [16, 17].

The elderly marathoners are at the opposite end of the spectrum of health and functional evaluation compared to the frail and sedentary elderly. These older athletes are endowed with substantial physical capacity, long-term health, high motivation and psychosocial perspective, fighting dogma, and negative stereotypes of being elderly and aging [12–15].

Another harmful effect of the physiological aging is sarcopenia, which pathophysiology comprises metabolic, endocrine, and nutritional factors that, together with cellular aging, lead to muscle mass loss. The reduction in anabolic hormonal secretion (growth hormone, testosterone, and insulin-like growth factor) and low degradation of pro-inflammatory cytokines potentialize the catabolic action and, consequently, muscle loss [18, 19].

Moreover, the protein absorption and synthesis decrease in muscle cells, causing a progressive fat deposition and lower muscle mass volume per body mass, increasing the harmful effects of obesity in the population called "sarcopenic obesity" [10, 19].

31.3 The Role of Physical Exercise and Sports Activity on Aging

The benefits of being physically active and practicing sports along lifespan are well-established in the literature and strengthen the concept that these activities should be integrated into the arsenal of medical treatment [20, 21].

Regular physical exercise and sports practice are health promoters as they boost overall psychological health and well-being, and are clinically related to a decreased risk of clinical depression and anxiety, cardiovascular and metabolic risk, and muscle aging delay among active athletes [22, 23].

Chodzko-Zajko et al. [24] have found significant associations of exercise and sports practice with a reduced coronary risk profile, less cardiovascular and metabolic stress during exercise, relative preservation of muscle mass in the limbs and bone mineral density, less total and abdominal body fat, an improved capacity to transport and use oxygen (consequently less muscle fatigue), and slower development of established disability in old age (Fig. 31.1).

The literature has confirmed a tendency of less pronounced physiological aging changes in physically active individuals; however, the needed volume, intensity, and frequency of physical and sports activity to interfere in these previous markers remain unclear.

Although physical exercises do not seem to have any influence on the size of type I and type II skeletal fiber, they may carry out adaptations. These can improve the contractile function, type I fiber power, and preserve the power of the fast-twitch type II fibers, increasing the muscle contractile speed.

Therefore, the plasticity of skeletal fibers, at the myocellular level, resulting from continuous physical stimulation throughout life, seems to be able to partially compensate for the biological muscle aging in the group of athletes [25].

The type of physical exercise practiced during life seems to influence the preservation of mass capacity in advanced age. Elderly athletes, who trained strength during life, demonstrated a higher muscle mass and 30–50% more strength than sedentary elderly. They also showed more muscle mass and more preserved bone mineral density when compared to elderly practitioners of aerobic activities such as running and swimming [26].

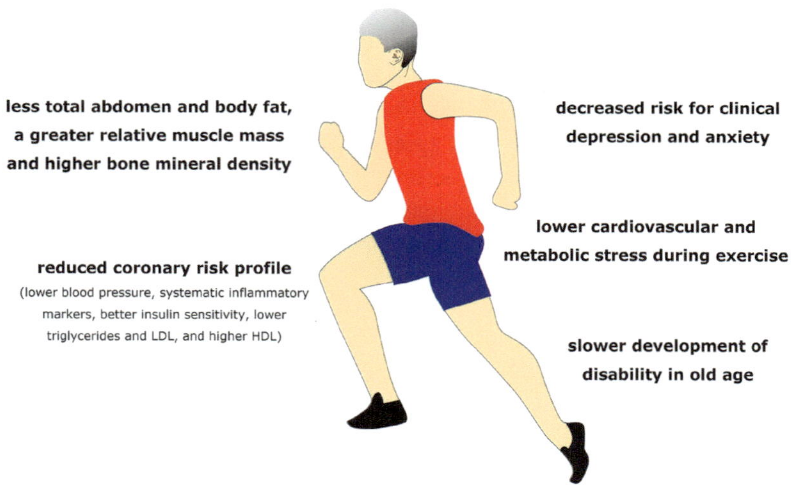

Fig. 31.1 Advantages of continuous exercise in the aging process (comparing masters athletes with their sedentary peers). Data based on literary review by Chodzko-Zajko et al. [24]. Credit: Ana Karina Piedade

However, when compared to younger athletes, veteran athletes have a reduction in exercise tolerance, an increased risk of heat and cold illnesses, and a change in the perception of thirst [27]. A decrease in maximum cardiac output is also expected with age, and as a result, there is a reduction in maximum oxygen consumption (O2 max) in the order of 0.4–0.5 ml/kg/min/year (1% per year in adults). Lower heart rate associated with changes in oxygen consumption in the elderly can result in a less favorable demand for oxygen to the myocardium. Thus, being more alert to the warm-up period is vital to better prepare the athlete for the demands that the exercise will require [28].

31.4 Patterns of Performance Decline in Master Athletes

Studies state that an individual can maintain his maximum resistance performance until approximately 30–35 years of age [9, 29, 30], when physiological transformations inherent to aging become predominant (better explained in the previous topics of this chapter).

Goodpaster et al. [31] investigated strength and muscle mass in 3075 healthy, nonathletes elderly for 3 years. Despite noticing a decline in strength and muscle mass over the years, the study observed that the loss of muscle mass was more significant than the loss of strength, suggesting some components in addition to the amount of muscle mass, such as the quality of these fibers and other extrinsic factors [31].

There are several cited determiners to justify a reduction in performance. One of the biggest influencers in this drop is the reduction in maximum oxygen consumption with aging, together with the reduction in maximum aerobic capacity, resulting in the decreased endurance capacity [32, 33]. Master athletes in regular training might have these falls possibly lessened [34, 35].

31.4.1 Age and Modality

Reviewing the literature, Siparsky et al. [18] found an average decline in strength estimated at 10–15% per decade until the age of 70, and accelerating to 25–40% after age 70 [18]. Ganse et al. [36] studied athletes practicing track and field in Germany, during the years 2001–2014, finding an association in the decrease in the athletes' overall performance with advancing age, but varying according to the modality practiced within the track and field.

Admitting that each activity differs concerning power, speed, endurance, coordination, and others, the decline patterns can vary between runners and field athletes. A greater decline in performance is observed in medium- and long-distance runners when compared to jumping and throwing athletes [36]. Similarly, other studies compared performance in strength modalities (throwing, jumping) and aerobic modalities (swimmers, marathoners), showing a greater decrease in the performance of aerobic modalities over the years [30, 37, 38]. A possible explanation proposed for the performance difference in these groups is the greater preservation of muscle mass and bone mineral density noticed in athletes who practiced strength activity for most of their life [26, 28]. However, these data are not consistent in the literature, and, in some studied populations, the performance did not differ according to the modality [36]. Future studies on the current topic are, therefore, required.

31.4.2 Peak Performance Age

The age of the peak performance can be estimated by modalities (as taken from Ganse et al. [36] and illustrated in Table 31.1). However, they are approximate measures, especially considering that the peak performance in athletes is influenced by many biopsychosocial variables, as pointed out in this topic [29, 34].

31.4.2.1 Runners

Once again, we highlight some particularities observed in a wide group of runners. Knechtle et al. 2009 [39] collected data that positively associate personal experience and older age as predictors for better performance, specifically for marathon and ultra-marathon runners, a modality

Table 31.1 Peak age of performance according to gender in track and field modalities

Peak of age performance		
Gender	Female	Male
100 m	22.86	18.78
200 m	22.83	19.53
400 m	27.15	23.91
8000 m	23.63	22.39
1500 m	26.06	22.43
5000 m	23.41	23.75
Shot put	18.61	15.75
Discus throw	19.70	13.06
Javelin throw	19.60	18.85
Long jump	17.21	18.57
High jump	19.92	19.54
Pole vault	18.23	23.09
Average	21.60	19.97

Source: Adapted from Ganse et al. [36]

that demands longer preparation. Therefore, it is estimated that the age of the best performance is between 39 and 41 years (older than the age presented in the other modalities).

Furthermore, it is possible to observe that older runners have a different pace than younger runners, though they present similar running times. Older athletes maintain a more constant pace, with no major oscillations during the race. Additionally, athletes in older age groups have a relatively more uniform pace compared to athletes in younger age groups [40, 41].

The predominance of injuries associated with physical training can also contribute to variations in the performance of these athletes [29, 42]. The interruption of aerobic training leads to a rapid loss of cardiovascular fitness, similarly for all ages [43].

31.4.3 Psychological and Social Aspects

The alterations in the training rhythm can be a contributing factor to the performance decrease with aging, even in master athletes. Numerous factors can affect these modifications: (1) time available for training; (2) support from clubs, family members, and sports organizations; and (3) changes in motivation for physical training concerning the intensity, duration, and weekly frequency [35, 44, 45].

31.5 Injuries in Track and Field Athletes

The epidemiological and traumatological understanding is crucial for the prevention of sports injuries [46]. The variety of activities involved in the track and field and the large number of participants partially hinder an adequate and unanimous analysis in the literature [47].

The risk of injury and the most affected site of the body seem to vary with each discipline in track and field. Additionally, several contributing factors were identified as the responsible ones for facilitating injuries in competitors beyond age, as gender, personal history, modality, and others [9, 42, 48].

31.5.1 Age and Gender

In various stages of the physical aging process, the forces transmitted to the athlete's body differ in terms of their intensity and their influence in the body. There is conflicting evidence about whether older age is a risk factor for injuries in track and field athletes [46].

A systematic review analyzed six high-quality studies, and in four of them, older age was reported as a significant risk factor for the occurrence of running injuries. However, this relationship in the other two studies did not prove to be statistically significant [48].

A significant association between genders was observed for hamstring tension in an athletic championship, with a greater predominance of it in men rather than in women. However, further work needs to be done on the interference of gender and age in the prevalence of sports injuries [42].

31.5.2 Championship

Almost 10–14% of all track and field athletes incurred an injury during international competitions (mainly in the finals), and half of these were expected to be temporarily unfitting for sport [49, 50]. The risk of injury is about four times greater during competition when compared to the training period [47].

A team of researchers analyzed the incidence and characteristics of injuries in athletes participating in the 13th World Championship of the International Association of Athletics Federations in 2011, in Daegu, Korea. A total of 1851 athletes were followed during Daegu 2011, and 13.4% of injuries were reported, while 48% of them resulted in lost time in the sport. The most frequent types of injury were found in the lower limb (~74%), and overuse was the predominant cause (59%). Posterior thigh injuries (hamstring) were the main diagnosis, involving 23.3% of all injuries. The most frequent types of injury were strains (30.9%), sprains (21.7%), muscle cramps (17.3%), and skin laceration (9.2%)—hamstring strain was the main. These results were similar to those reported in the Berlin (2009) and Osaka (2007) Athletics World championships. Most injuries occurred in athletes over 30 years of age. Differences in the lesion location by age were observed, although they were similar in relation to the type and severity of the lesion [49].

Athletes practicing more than one sport and medium- and long-distance runners had a higher incidence of injuries [49].

31.5.3 Runners

When studied separately from the other track and field modalities, the group of long-distance runners usually presents a different pattern of injuries. Two reviews of the literature on injuries in long-distance runners concluded that the most common site of lower extremity injuries was the knee, lower leg, the foot, and the upper leg. Knee injuries were the main ones reported [51, 52].

Indeed, running produces long periods of repetitive stress on the musculoskeletal system, leading to an overuse of this system, predisposing to injuries. On the other hand, field events depend on the generation of maximum strength in a short-time period, producing intense muscle contractions which also admit injury risks [46].

Limited evidence in a study of marathon athletes accused that older age was positively associated with front thigh injuries, but protective against calf injuries. Female athletes were more associated with hip injuries and male athletes with a risk of getting hamstring or calf injuries [53].

Training for more than 64 km/week was a significant risk factor for runners, most evident in the male group [53]. However, increased training distance per week was a protective factor, but it was significant only for knee injuries [53, 54].

31.5.3.1 Previous Injuries

Another significant risk factor for newly reported injuries is previous injuries [45]. Benca et al. identified that 67.2% of the injured patients had already presented a similar injury in the past, notably in iliotibial band syndrome [55].

We must consider that almost three-quarters of recurrent injuries might lead to withdrawal from training the sport [49]. The Vienna Study [55] with runners in injuries recuperation demonstrates many contributing factors: (1) Scoliosis and a higher body mass index (BMI) were the main risks for lower back injuries; (2) presence of planus foot deformity related to more knee injuries; (3) previous injury history was a contributing factor for knee injuries and iliotibial band syndrome; and (4) knee malalignment associated with more knee injuries, patellofemoral pain syndrome, and patellar tendinopathy. Age did not have a significantly positive association with the increased risk of injury in this study.

31.5.3.2 Stress Fractures

Track and field athletes also have the highest incidence of stress fractures when compared to athletes from other sports (basketball, football, and others) [56]. The average time to recover after a stress fracture was 12.8 weeks [57]. The sites of stress fractures vary according to modality and age: Stress fractures of the navicular, tibia, and metatarsal are more common in track and field athletes; however, in distance runners, it is the fibula and tibia [58] (Fig. 31.2).

Significant associations have been documented from the interaction between age and site of injury. Femoral and tarsal stress fractures were more common in older athletes, while tibial and fibular stress fractures in younger athletes [57].

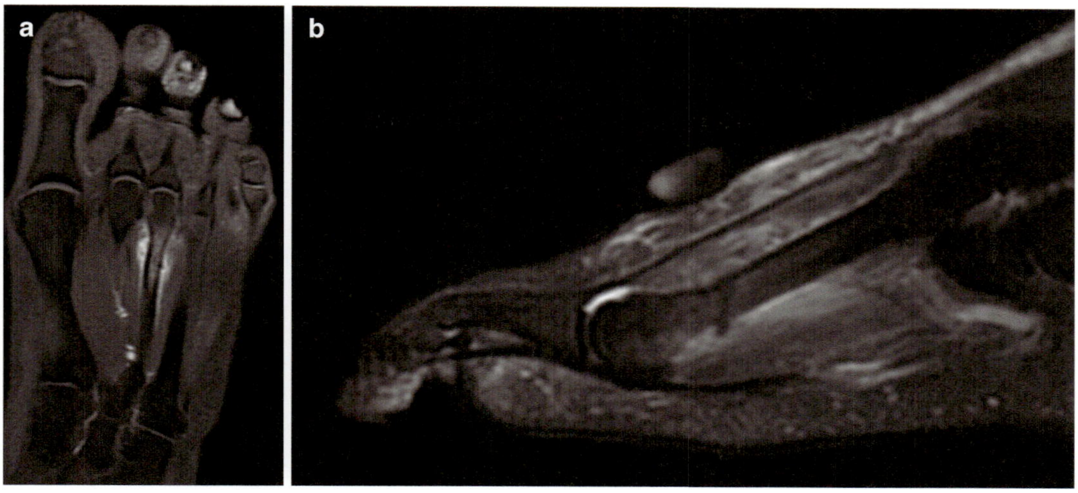

Fig. 31.2 MRI views showing a stress diaphysis fracture of the third metatarsal of the right foot in (**a**) sagittal and (**b**) coronal DP FAT SAT images, in a 55-year-old runner

In adolescents, fractures, sprains, and strains are more common, whereas inflammatory problems such as tendonitis or bursitis gradually increase in importance after the age of 30 [46].

31.5.3.3 Novice Athletes

There are a rising number of middle-aged runners among the participants of running events. For instance, the number of athletes older than 50 years that participated in the New York City marathon (involving recreational and professional athletes) increased 119% from 1983 to 1999 [59].

There was conflicting evidence for an association between inexperience in running, and more injuries, as well as the sites of injury in the body, differ in these two groups [51, 60].

For 4 years, a Dutch race and its participants were analyzed (2010–2013). Over the years, the average age has increased in novices and experienced runners (from 30.0 to 31.3 years in beginners and from 34.1 to 36.3 in experienced runners). Moreover, the absolute number of new injuries in all runners almost doubled from 350,000 in 2010 to 640,000 in 2013 [60].

In the group of novice athletes, women are the majority, and we must consider that female sex was statistically related to a higher risk of injuries in this group [60]. The knee is the most commonly injured site, both in experienced runners as novice ones [55, 60] A significant difference for injuries at the Achilles tendon and hip injuries was shown, with more Achilles tendon injuries prevailing in experienced runners and more hip injuries in novice runners (group or runners). However, the study reveals the low prevalence of these injuries normally in the scenario as a limiting factor (but again, an underlying number of injuries was too small for reliable analysis) [60].

In a group of military personnel, injuries to calcaneus and metatarsals had a higher incidence in novice recruits, and they were related to the sudden increase in running and marching without adequate preparation [57]. As Kemler et al. observed, novice runners train less than experienced runners over the year (median of 14.6 h and 20 weeks for the former ones compared to a median of 25 h and 36 weeks for the experienced runners) [60].

A training distance of <40 km a week was a strong protective factor of future calf injuries in recreational male marathon runners [52], while a training distance of <60 km a week appears to be a protective factor in professional runner athletes [54]. Regular interval training proved to be a strong protective factor for knee injuries for all groups, novices, and experienced athletes [52–54].

31.6 Rehabilitation of Sports Injuries

With aging, the biological competence of muscle tissue repair and regeneration worsen. The mechanism of satellite stem cell activation, migration to the site of injury, proliferation, fusion with the damaged fiber to regenerate the sarcomeric structure, and synthesis of myofibrillar and non-myofibrillar proteins are less competent [9, 61, 62]. Moreover, there is an age-related decline in the density of satellite cells surrounding type II muscle fibers and an increase in the density of satellite cells surrounding type I muscle fibers [9, 31, 63, 64]. Variations in the muscle fiber composition and regenerative ability may result in reduced strength and make older people more susceptive to contraction-induced injuries, even elite older athletes [61].

Besides that, aging causes loss in the elasticity of tendons and muscles, a decrease in the number of muscle fibers, and decreasing muscle strength, which justifies the reduction in lean mass, muscle power, and strength. It is followed by joint stiffness, causing biomechanical changes to walking and running, and contributing to the increase in musculoskeletal injuries in elderly runners.

Although the protocols of rehabilitation in sports injuries in aged athletes seem to be similar to the ones for young athletes, the biological process of aging plays an important role in the strategies of treatment and decision-making of when to return to play.

31.7 Take-Home Message

- Physiological aging produces a decrease in muscle mass and a loss of muscle strength estimated at the rate of 1% per year after the third decade of life, notably after 50 years, preferentially through the loss of type II fast fibers
- The plasticity of skeletal fibers as a result of continuous physical stimulation throughout life seems to be able to partially compensate for the biological muscle aging in the group of athletes
- Variation in the composition and regeneration capacity of muscle fibers makes older peoples more susceptible to contraction-induced injuries, even elite older athletes
- Elderly athletes strength-trained during life demonstrated more muscle mass and bone mineral density when compared to elderly practitioners of aerobic activities (runners and swimmers). The complexity of track and field in its various modalities and many additional factors still partially known are related to the performance of an athlete throughout life. The inherent decline in the aging process does not seem to behave evenly among track and field athletes. Not only strength and metabolism were identified as determinants in performance, but also technique, biopsychosocial factors, personal history, practiced sports, and others.

References

1. ACSM. American college of sports medicine position stand. Exercise and physical activity for older adults. Med Sci Sports Exerc. 1998;30:992–1008.
2. Obas V, Vasan RS. The aging heart. Clin Sci (Lond). 2018;132(13):1367–82.
3. Wang JC, Bennett M. Aging and atherosclerosis: mechanisms, functional consequences, and potential therapeutics for cellular senescence. Circ Res. 2012;111(2):245–59.
4. Fukuchi RK, Stefanyshyn DJ, Stirling L, Duarte M, Ferber R. Flexibility, muscle strength and running biomechanical adaptations in older runners. Clin Biomech (Bristol, Avon). 2014;29(3):304–10.
5. McGregor RA, Cameron-Smith D, Poppitt SD. It is not just muscle mass: a review of muscle quality, composition and metabolism during ageing as determinants of muscle function and mobility in later life. Longev Healthspan. 2014;3(1):9.
6. Schöne D, Freiberger E, Sieber CC. Influence of skeletal muscles on the risk of falling in old age. Internist (Berl). 2017;58(4):359–70.
7. Dziechciaż M, Filip R. Biological psychological and social determinants of old age: bio-psycho-social aspects of human aging. Ann Agric Environ Med. 2014;21(4):835–8.
8. Gava P, Ravara B. Master World Records show minor gender differences of performance decline with aging. Eur J Transl Myol. 2019;29(3):8327.
9. Faulkner JA, Davis CS, Mendias CL, Brooks SV. The aging of elite male athletes: age related changes in performance and skeletal muscle structure and function. Clin J Sport Med. 2008;18(6):501–7.

10. Piedade SR, Carvalho LM, Mendes LA, Possedente M, Ferreira DM. Physical activity at adulthood and old age. In: Rocha Piedade S, Imhoff AB, Clatworthy M, Cohen M, Espregueira-Mendes J, editors. The sports medicine physician. Cham: Springer International Publishing; 2019. p. 59–69.
11. Coughlan T, Dockery F. Osteoporosis and fracture risk in older people. Clin Med (Lond). 2014;14(2):187–91.
12. Amoako AO, Pujalte GGA. Osteoarthritis in young, active, and athletic individuals. Clin Med Insights Arthritis Musculoskelet Disord. 2014;7:27–32.
13. Tanaka H. Aging of competitive athletes. Gerontology. 2017;63(5):488–94.
14. Akkari A, Machin D, Tanaka H. Greater progression of athletic performance in older Masters athletes. Age Ageing. 2015;44(4):683–6.
15. Ahmadyar B, Rosemann T, Rüst CA, Knechtle B. Improved race times in marathoners older than 75 years in the last 25 years in the world's largest marathons. Chin J Physiol. 2016;59(3):139–47.
16. Marck A, Antero J, Berthelot G, Johnson S, Sedeaud A, Leroy A, Marc A, Spedding M, Di Meglio JM, Toussaint JF. Age-related upper limits in physical performances. J Gerontol A Biol Sci Med Sci. 2019;74(5):591–9.
17. Lee PG, Jackson EA, Richardson CR. Exercise prescriptions in older adults. Am Fam Physician. 2017;95(7):425–32.
18. Siparsky PN, Kirkendall DT, Garrett WE Jr. Muscle changes in aging: understanding sarcopenia. Sports Health. 2014;6(1):36–40.
19. Trouwborst I, Verreijen A, Memelink R, Massanet P, Boirie Y, Weijs P, Tieland M. Exercise and nutrition strategies to counteract sarcopenic obesity. Nutrients. 2018;10(5):605.
20. Distefano G, Goodpaster BH. Effects of exercise and aging on skeletal muscle. Cold Spring Harb Perspect Med. 2018;8(3):a029785.
21. Harridge SDR, Lazarus NR. Physical activity, aging, and physiological function. Physiology. 2017;32:152–61.
22. Cartee GD, Hepple RT, Bamman MM, Zierath JR. Exercise promotes healthy aging of skeletal muscle. Cell Metab. 2016;23(6):1034–47.
23. Garatachea N, Pareja-Galeano H, Sanchis-Gomar F, Santos-Lozano A, Fiuza-Luces C, Morán M, Emanuele E, Joyner MJ, Lucia A. Exercise attenuates the major hallmarks of aging. Rejuvenation Res. 2015;18(1):57–89.
24. Chodzko-Zajko WJ, Proctor DN, Fiatarone SMA, Minson CT, Nigg CR, Salem G, Skinner JS. Exercise and physical activity for older adults. Med Sci Sports Exerc. 2009;41(7):1510–30.
25. Gries KJ, Minchev K, Raue U, et al. Single-muscle fiber contractile properties in lifelong aerobic exercising women. J Appl Physiol (1985). 2019;127(6):1710–9.
26. Suominen H. Muscle training for bone strength. Aging Clin Exp Res. 2006;18:85–93.
27. Kenney WL, Munce TA. Invited review: aging and human temperature regulation. J Appl Physiol. 2003;95:2598–603.
28. Nóbrega ACL, Freitas EV, Oliveira MAB, Leitão MB, Lazzoli JK, Nahas RM. Posicionamento Oficial da Sociedade Brasileira de Medicina do Esporte e da Sociedade Brasileira de Geriatria e Gerontologia: Atividade física e saúde no idoso. Rev Bras Med Esporte. 1999;5:207–11.
29. Tanaka H, Seals DR. Age and gender interactions in physiological functional capacity: insight from swimming performance. J Appl Physiol (1985). 1997;82(3):846–51.
30. Baker AB, Tang YQ, Turner MJ. Percentage decline in masters superathlete track and field performance with aging. Exp Aging Res. 2003;29:47–65.
31. Goodpaster BH, Park SW, Harris TB, et al. The loss of skeletal muscle strength, mass, and quality in older adults: the health, aging and body composition study. J Gerontol A Biol Sci Med Sci. 2006;61:1059–64.
32. Larsson L, Grimby G, Karlsson J. Muscle strength and speed of movement in relation to age and muscle morphology. J Appl Physiol. 1979;46:451–6.
33. Allen WK, Seals DR, Hurley BF, Ehsani AA, Hagberg JM. Lactate threshold and distance-running performance in young and older endurance athletes. J Appl Physiol (1985). 1985;58(4):1281–4.
34. Tanaka H, Seals DR. Endurance exercise performance in Masters athletes: age-associated changes and underlying physiological mechanisms. J Physiol. 2008;586(1):55–63.
35. Tanaka H, Seals DR. Invited review: dynamic exercise performance in Masters athletes: insight into the effects of primary human aging on physiological functional capacity. J Appl Physiol (1985). 2003;95(5):2152–62.
36. Ganse B, Ganse U, Dahl J. Degens Linear decrease in athletic performance during human lifetime. Front Physiol. 2018;9:1100.
37. Baker AB, Tang YQ. Aging performance for masters records in athletics, swimming, rowing, cycling, triathlon, and weightlifting. Exp Aging Res. 2010;36(4):453–77.
38. Grimby G, Nilsson NJ, Saltin B. Cardiac output during submaximal and maximal exercise in active middle-aged athletes. J Appl Physiol. 1966;21:1150–6.
39. Knechtle B, Wirth A, Knechtle P, Zimmermann K, Kohler G. Personal best marathon performance is associated with performance in a 24-h run and not anthropometry or training volume. Br J Sports Med. 2009;43(11):836–9.
40. Nikolaidis PT, Knechtle B. Effect of age and performance on pacing of marathon runners. Open Access J Sports Med. 2017;8:171–80.
41. de Leeuw AW, Meerhoff LA, Knobbe A. Effects of pacing properties on performance in long-distance running. Big Data. 2018;6(4):248–61.
42. Lexell J, Taylor CC, Sjostrom M. What is the cause of the ageing atrophy? Total number, size and proportion

of different fiber types studied in whole vastus lateralis muscle from 15- to 83-year-old men. J Neurol Sci. 1988;84:275–94.
43. Coyle EF, Martin WHr, Sinacore DR, Joyner MJ, Hagberg JM, Holloszy JO. Time course of loss of adaptations after stopping prolonged intense endurance training. J Appl Physiol. 1984;57:1857–64.
44. Medic N, Starkes JL, Young BW. Examining relative age effects on performance achievement and participation rates in Masters athletes. J Sports Sci. 2007;25(12):1377–84.
45. Eskurza I, Donato AJ, Moreau KL, Seals DR, Tanaka H. Changes in maximal aerobic capacity with age in endurance-trained women: 7-year follow-up. J Appl Physiol. 2002;92:2303–8.
46. Jakobsen BW, Nielsen AB, Yde J, Kroner K, Moller-Madsen B, Jensen J. J3 Epidemiology and traumatology of injuries in track athletes. Scand J Med Sci Sports. 1992;3:57–61.
47. Zemper ED. Track and field injuries. Med Sport Sci. 2005;48:138–51.
48. Campbell MJ, McComas AJ, Petito F. Physiological changes in ageing muscles. J Neurol Neurosurg Psychiatry. 1973;36:174–82.
49. Alonso JM, Edouard P, Fischetto G, Adams B, Depiesse F, Mountjoy M. Determination of future prevention strategies in elite track and field: analysis of Daegu 2011 IAAF Championships injuries and illnesses surveillance. Br J Sports Med. 2012;46(7):505–14.
50. Bahr R, Krosshaug T. Understanding injury mechanisms: a key component of preventing injuries in sport. Br J Sports Med. 2005;39:324–9.
51. van Gent RN, Siem D, van Middelkoop M, et al. Incidence and determinants of lower extremity running injuries in long distance runners: a systematic review. Br J Sports Med. 2007;41:469–80.
52. van Mechelen W. Running injuries. A review of the epidemiological literature. Sports Med. 1992;14(5):320–35.
53. Satterthwaite P, Norton R, Larmer P, Robinson E. Risk factors for injuries and other health problems sustained in a marathon. Br J Sports Med. 1999;33(1):22–6. https://doi.org/10.1136/bjsm.33.1.22.
54. Walter SD, Hart LE, McIntosh JM, et al. The Ontario cohort study of running-related injuries. Arch Intern Med. 1989;149:2561–4.
55. Benca E, Listabarth S, Flock FK, Pablik E, Fischer C, Walzer SM, Dorotka R, Windhager R, Ziai P. Analysis of running-related injuries: the Vienna study. J Clin Med. 2020;9:438.
56. Fredericson M, Jennings F, Beaulieu C, Matheson GO. Stress fractures in athletes. Top Magn Reson Imaging. 2006;17(5):309–25.
57. Matheson GO, Clement DB, McKenzie DC, Taunton JE, Lloyd-Smith DR, MacIntyre JG. Stress fractures in athletes. A study of 320 cases. Am J Sports Med. 1987;15(1):46–58.
58. Brukner P, Bradshaw C, Khan KM, White S, Crossley K. Stress fractures: a review of 180 cases. Clin J Sport Med. 1996;6(2):85–9.
59. Jokl P, Sethi PM, Cooper AJ. Master's performance in the New York City Marathon 1983–1999. Br J Sports Med. 2004;38:408–12.
60. Kemler E, Blokland D, Backx F, Huisstede B. Differences in injury risk and characteristics of injuries between novice and experienced runners over a 4-year period. Phys Sportsmed. 2018;46(4):485–91.
61. Faulkner JA. Terminology for contractions of muscles during shortening, while isometric, and during lengthening. J Appl Physiol. 2003;95:455–9.
62. Conboy IM, Rando TA. Aging, stem cells and tissue regeneration: lessons from muscle. Cell Cycle. 2005;4:407–10.
63. Petrella JK, Kim JS, Cross JM, et al. Efficacy of myonuclear addition may explain differential myofiber growth among resistance-trained young and older men and women. Am J Physiol Endocrinol Metab. 2006;291:E937–46.
64. Verdijk LB, Koopman R, Schaart G, et al. Satellite cell content is specifically reduced in type II skeletal muscle fibers in the elderly. Am J Physiol Endocrinol Metab. 2007;292:E151–7.

Growth and Development

32

Adam D. G. Baxter-Jones

Although most children are involved in sport on a casual or recreational basis, a growing number do devote many hours to intensive physical training and this reflects in part the younger age at which athletes today take part in international competition. Children and adolescents taking part in high-level competition are likely to have undergone several years of intensive training. During the period of rapid growth, adolescents have been reported to be particularly vulnerable to injuries and as such intensive training at a young age may cause long-term harmful effects. Given the possible interaction between intensive training and growth during adolescence, some adolescent athletes may be particularly vulnerable to repetitive microtraumatic injury [1]. This highlights the importance of monitoring both an athletes' chronological and maturational age.

Interest in the effect that intensive training at an early age has on a child's growth and development has a long history [2]. This interest highlights the "catch them young" philosophy [3], and the widespread belief that achieving international success at the senior level requires starting intensive training prior to puberty [4]. Of course, the negative side to this philosophy is the issue of burnout during the pubertal years, where young athletes may retire prematurely from sport because of physical (e.g., injury) and psychological issues [5, 6].

It should be emphasized that regular physical training is only one of many factors that can affect growth in the growing child and that it is difficult to define the precise influences that training programs have on growth and by that inflection injury. Problems arise when attributing growth differences to physical training despite the fact that young athletes are likely to have been selected as much for physique as for skill [7].

Germane to the sport selection issues, Stephan Hall [8] published a book entitled "*Size Matters*" in which he argued that although the childhood hierarchy primarily involved age (i.e., who is older) when it came to playing games in the schoolyard, it was size rather than age that mattered. Except for gender, and possibly skin color, size is probably the first thing others notice about each other [8]. Size matters from the time of birth, when birthweight is used to predict adult health problems. It is also apparent that size matters in sports throughout childhood, as physical size often translates into physical superiority and athletic dominance. The alignment of competition by maturity rather than chronological age warrants further investigation.

Figure 32.1 illustrates the problem that many coaches and sports professionals face when working with child athletes of the same age but

A. D. G. Baxter-Jones (✉)
College of Kinesiology, University of Saskatchewan, Saskatoon, SK, Canada
e-mail: baxter.jones@usask.ca

G. L. Canata et al. (eds.), *Management of Track and Field Injuries*,
https://doi.org/10.1007/978-3-030-60216-1_32

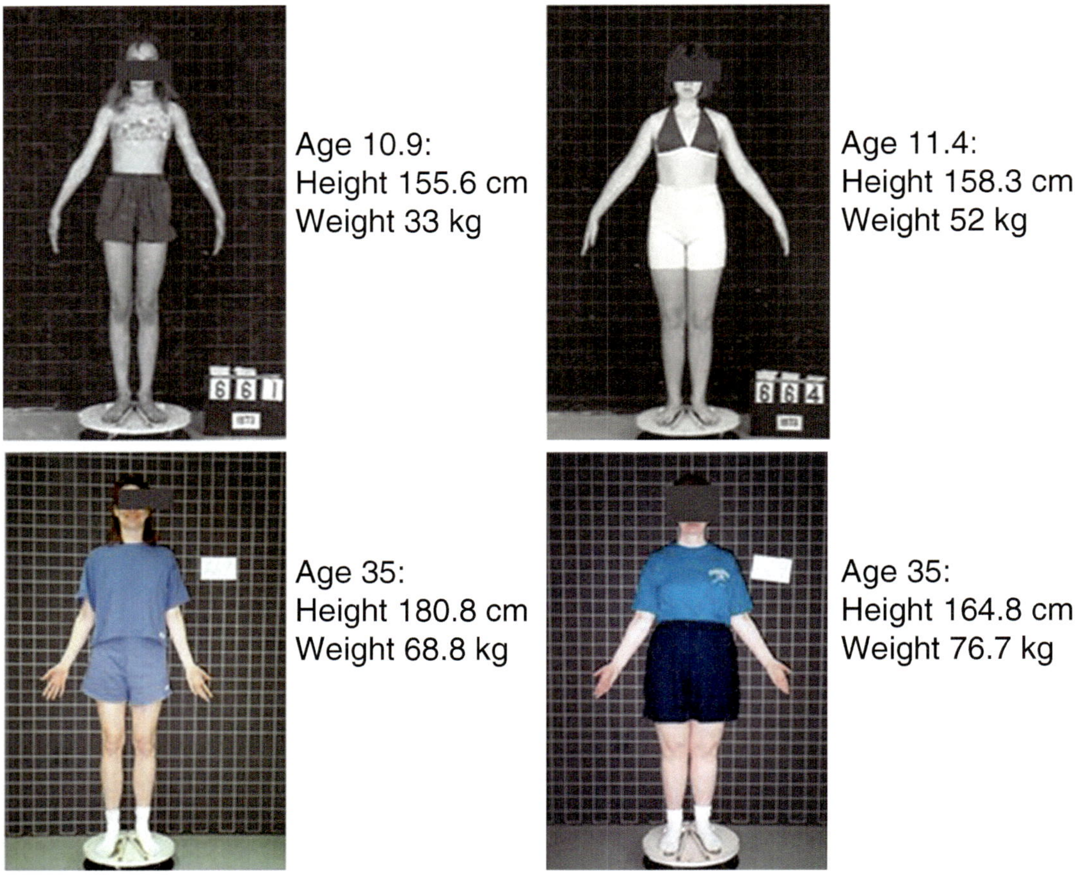

Fig. 32.1 Stature of two girls at age of 11 and 35 years. Data taken from 2 individuals who participated in the Saskatchewan Growth and Development Study [9]

with different maturity statuses. It is a photograph of two girls aged 11 and 35 years. At a chronological age of 11 years, they are, from left to right, 155.6 and 158.3 cm in height, respectively. The girl on the left is 6 months younger. The average height for 12-year-old girls when they are at the peak of their adolescent growth spurt is 156.8 cm. The distribution of height within such chronological age bands is not normally distributed. So, rather than expressing an average height for an age, the frequency of distribution of height is specified in terms of empirical centiles. A centile is a point on the distribution that splits populations into specified fractions; thus, both girls are approximately on the 50th centile for height for their age. It is also important to note that the girl on the left, who is 2.7 cm taller, is in fact much closer to her final adult height than the girl on the right, and she has 6.5 cm of growth remaining and is thus more mature than the girl on the left who has 25.2 cm of growth remaining. This photograph highlights the dilemma for many coaches and youth sports practitioners who work with children during periods of rapid growth where they use chronological age to band training and competition. It illustrates the great variations in growth and development at this age. If, as suggested, the observed physiques of youth athletes need to mirror the physiques of the successful adult athlete, then selection of such traits at a relatively young age is likely preferable and the girl of the right would be chosen. This suggests that greater size can trump or neutralize greater athletic skill.

32.1 Normal Patterns of Growth

To understand why some children are tall for their age and others are small for their age, an understanding of how children grow is required. Growth refers to measurable changes in size, physique, and body composition, whereas biological development, used interchangeably with biological maturation, refers to progress toward the mature state. Figure 32.2a shows the growth of a boy measured from birth to 18 years of age, with his height plotted at successive ages. If you think of growth in the form of a train journey and each age representing a train station, then you can imagine that growth takes the form of motion and the speed across the distance traveled is different between ages, indicated by the differences in slopes of lines between ages. Since the shape of the curve is nonlinear, this shows that the speed, or velocity, between ages is different. The data for Fig. 32.2 are taken from the oldest known record of the curve of human growth, which was published in a supplement to volume 14 of the "Histoire Naturelle, Generale et Particuliere" in 1778 [11]. It is the record of the growth of the son of Philibert Gueneau De Montbeillard, a natural scientist during the period of the Enlightenment. De Montbeillard measured the height of his son about every 6 months from his birth, in 1759, until he was 18 years of age, in 1777, using the French units of the time which were subsequently translated into centimeters by the American anatomist R.E. Scammon [10]. The first graph is known as a height distance or height-for-age curve (Fig. 32.2a). In terms of our train journey, it is apparent that we do not travel at the same speed between stations and so do not gain the same amount of height each calendar year. Although these data are over 250 years old, it is important to note that children today still show the same pattern of growth. This height distance curve shows 4 distinct phases: rapid growth (decreasing from 22 to 6 cm/year) in infancy (up to 4 years of age), steady growth (4–6 cm/year) in childhood (between 4 and 12 years), rapid growth (increasing from 6–12 cm/year) in adolescence (12–16 years), and slow growth (decreasing from 6 to 2 cm/year) as adulthood approaches (16–18 years). There are also two other spurts not shown in Fig. 32.2b: the prenatal spurt and juvenile growth spurt. Although these two spurts vary in magnitude, they occur at roughly the same age, both within and between the sexes, and Fig. 32.2a also highlights the dramatic increase in size during growth, from approximately 60 cm at birth to 180 cm in adulthood. By 2 years of age a boy (18 months for a girl) is roughly half their adult height, highlighting the fact that the majority of

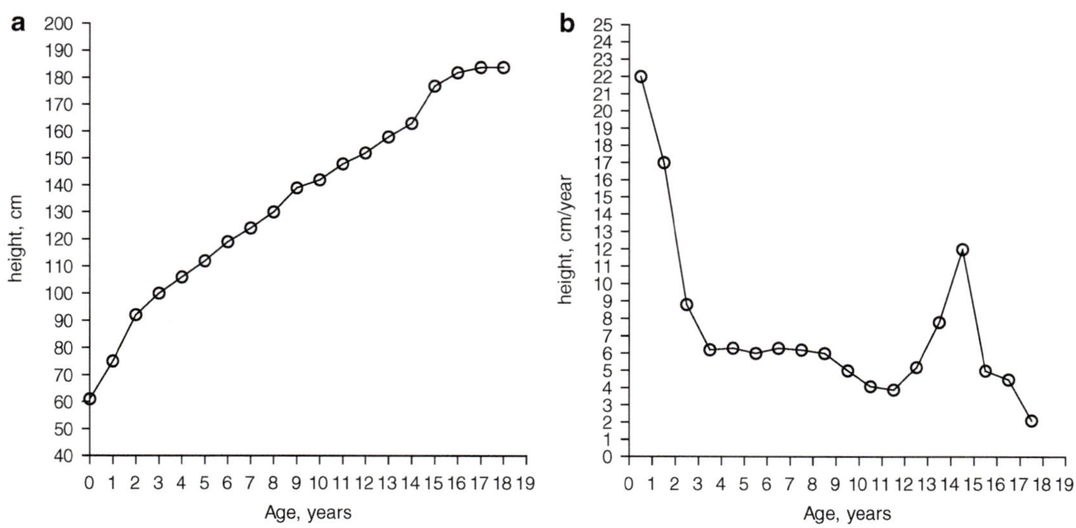

Fig. 32.2 Growth of De Montbeillar's son 1759–1777 (**a**) Distance of height by age, (**b**) Velocity of height between ages (redrawn from [10])

growth occurs during infancy and childhood, with an additional marked acceleration during adolescence [11].

The acceleration during adolescence is known as the adolescent growth spurt. It is a constant phenomenon and occurs in all children, though it varies in intensity and duration from one child to the next (illustrated in Fig. 32.1). The actual patterns of growth change between age time points are more clearly seen by visualizing the height–distance curve as a rate of change in size. Figure 32.2b shows the height velocity graph for De Montbeillard's son and emphasizes that during growth children show a succession of varying velocities. The graph shows that following birth, there is a decrease in velocity until 4 years of age, followed by a period of steady growth and then after 12 years of age an obvious spurt in growth between 12 and 14 years (adolescent growth spurt). The adolescent growth spurt varies in both magnitude and timing within and between sexes. Boys enter their adolescent growth spurt almost 2 years later than girls, at approximately 14 years of age, and have a slightly greater magnitude of height gain at peak (11 cm/year compared to 9 cm/year for boys and girls, respectively). At the same time, other skeletal changes are occurring that result in wider shoulders in boys and wider hips in females. Boys also demonstrate a rapid increase in muscle mass compared to girls, who accumulate greater amounts of fat mass.

Growth is affected by both genetic and environmental conditions and the interactions between the two. Lifestyle characteristics are transmitted from parents to their children through education and economic status and can have effects on the child's phenotype. A genetic effect is associated with a gene or set of genes encoded in the DNA of the chromosomes in the nucleus of the cells. Parent–child studies of stature have shown that parent–child correlations at birth are low but increase progressively with age, reaching 0.50 after the adolescent growth spurt. Since the expected correlation between parents and offspring would be 0.5 if the heritability of the trait was 1.0 (the heritability of a trait is a measure of the degree of genetic control of a phenotype), then it can be concluded that the population variation in height is highly determined by genetic factors. Using parental data, it is therefore possible to predict a target height for a child. Child height can be calculated as the sum of the father's height in cm (−13 cm if a girl) plus mother's height in cm (+13 cm if a boy) divided by two, with an error of 9 cm [12].

Variations in the intensity and duration of statural growth between children are illustrated in Fig. 32.3. The graph shows the statural growth of my two sons, brothers born 2 years apart, from the same biological parents, plotted on reference centile charts (personal data). The smoothed centile lines depict the normal range of heights for boys from 2 to 18 years of age. The normal range is bound by outer centile limits of the 3rd and 90th centiles; normal heights are thought of as heights that fall between these limits. Most healthy children exhibit patterns of growth that fall steadily and continuously parallel to the centile line from 2 years of age. However, as the adolescent growth spurt takes place, they depart this parallel pattern and a crossing of centile lines is observed. In early developers, the height-for-age curve rises through the centiles and levels off early. In contrast, late developers initially appear to fall away from their peer's centiles but then accelerate into adolescence crossing centile lines their peers have already crossed. The target height for the brothers in Fig. 32.3 is predicted to be 179 cm, predicted from parental heights where the father's height is 183 cm and the mothers 162 cm; thus, the boys have target heights of 179 ± 9 cm (i.e., target height = 183 + (162 + 13)/2). At 2 years of age, child 1 was on the 50th centile for age compared to child 2 who was on the 75th centile. By 20 years of age, and the cessation of growth, both boys ended up being the same height (179 cm); however, the journeys they took to get there were different. At 11 years of age, child 1 height was on the 50th centile compared to child 2 who was on the 90th centile, with child 1 reaching the 90th centile at 15 years of age. However, both boys' heights drop back to the 50th centile by 20 years of age. What this illustrates is that although the brothers end up the same height (genetically determined), the timing and tempo of growth are different

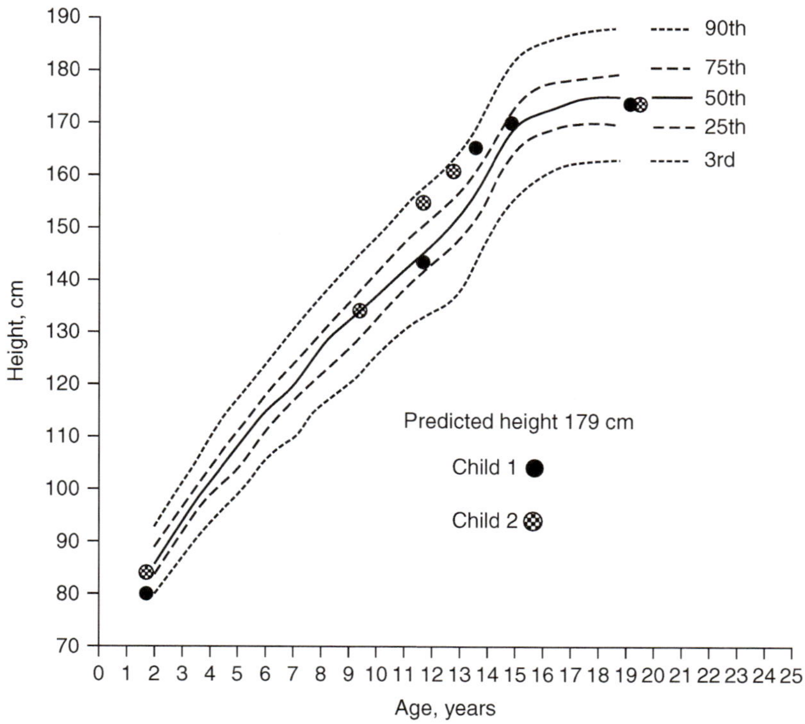

Fig. 32.3 Growth in stature from 2 to 20 years of age of two brothers

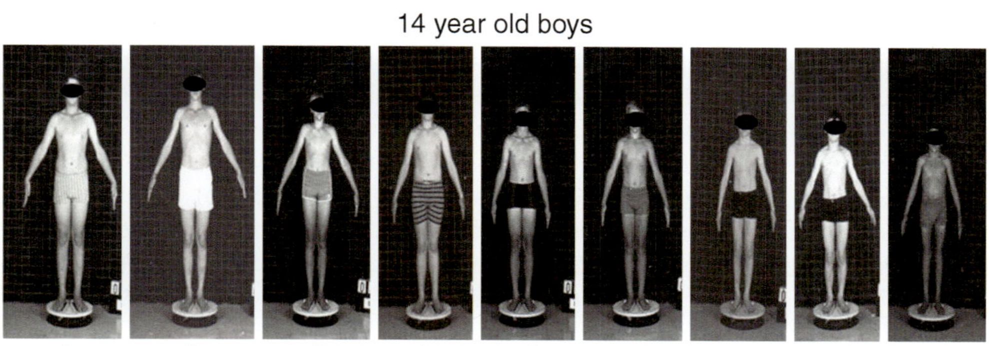

	A	B	C	D	E	F	G	H	I
Age	13.7	14.0	14.3	13.7	13.6	13.6	14.0	14.2	13.6
Ht	180.3	175.0	167.8	162.2	158.0	155.7	154.7	145.1	140.6
APHV	13.5	12.8	14.7	13.5	14.9	15.2	14.3	15.0	15.7
Yrs from PHV	+0.2	+1.2	-0.4	+0.2	-1.3	-1.6	-0.3	-0.8	-2.1
Adult Ht	180.5	178.0	187.5	174.7	185.0	177.1	176.4	168.7	166.7
% Adult ht	99.8	98.0	89.5	92.8	85.4	87.9	87.7	86.0	84.3

Fig. 32.4 Nine 14-year-old boys aligned by height. Data taken from 3 individuals who participated in the Saskatchewan Growth and Development Study [9]

between them. Child 2 attained his adolescent growth spurt 2 years earlier than child 1, illustrated by crossing from the 75th to 90th centile between 9 and 11 years in contrast to child 1 who crossed between 11 and 13 years. Thus, child 2 displays advanced maturation compared to his brother. Similar patterns are also observed between sister pairings.

Differences in timing and tempo of growth are also illustrated in the boys depicted in Fig. 32.4 These boys were participants of the Saskatchewan Growth and Development Study (SGDS), which was initiated in 1963 and ran continuously to 1973 [9]. The boys were all born in 1956, and by 1970 were 14 years of age, with heights falling within normal ranges for their age—from the 90th centile for the tallest boy (boy A) to the third centile for the smallest boy (boy I). Although all the boys had their ages rounded up to 14 years within this 12-month chronological age band, they were in fact not the exact same age when testing was performed. The youngest boys (boys E and I) had 7 months less time to grow than the oldest boy (boy C) when testing took place.

32.2 Chronological Versus Maturational Age

As previously emphasized, there is wide variation among children both within and between genders as to the exact timing and tempo of biological maturation. When considering how to assess biological maturation, or biological age as it is often termed, it is important to understand that 1 year of chronological time does not equal 1 year of maturational time. So, rather than considering comparisons between chronological age and biological age, comparisons should be thought of as between years from birth and years from maturity. While every individual passes through the same stages of maturity, they do so at differing rates, resulting in children of the same chronological age differing in their degree of maturity. This is reflected in Fig. 32.4, where boy A's maturity appears to be far more advanced than that of boy I.

To adequately control for maturity, an indicator of maturity needs to be assessed. The maturity indicator chosen should be any definable and sequential change in any part of the body that is characteristic of the progression of the body from immaturity to maturity [11]. The most commonly used methods to assess maturity are skeletal maturity, sexual maturity, biochemical and hormonal maturity, somatic or morphological maturity, and dental maturity [13]. The technique of choice depends on the study design [14]. Descriptions of each method, with their associated limitations, are described in detail in other publications [15]. Correlations between the timing of maturity indicators are generally moderate to high, suggesting that there is a general maturity factor underlying the tempo of growth and maturation during adolescence in both boys and girls. However, there is sufficient variation to suggest that no single system (i.e., sexual, skeletal, or somatic) provides a complete description of the tempo of maturation during adolescence. Furthermore, although sexual maturation and skeletal development are associated, an individual in one stage of a secondary sexual characteristic cannot be assumed to be in the same stage of skeletal development [13]. The apparent discord among the aforementioned indicators reflects individual variation in the timing and tempo of sexual and somatic maturity, and the methodological concerns in the assessment of maturity.

One method that has become increasingly popular in recent years is the measurement of the adolescent growth spurt or peak height velocity (PHV), a measure of somatic maturity. To obtain age at PHV, whole year height velocity (cm/year) increments are plotted and mathematical curve fitting procedures are used to identify the age when the maximum velocity in statural growth occurs (see Fig. 32.2b). The timing of this event in relation to chronological age shows great variance. The average age for girls is 12 years (range 9.5–14.5) and for boys 14 years (range 10.5–17.5) [16]. Once age at PHV has been determined, individuals can be aligned by biological age (years from age at PHV) rather than chronological age (years from birth); in other words, a measure of maturity offset is centered on age at PHV. For example, at age of PHV an individual has a biological age equal to 0.0 years from PHV. At 11.8 years, an individual who reached PHV at 13.8 years will have a biological age of −2.0 years from PHV. Age at PHV (APHV) and years from PHV (or maturity offset) are shown for the 9 boys in Fig. 32.4. The tallest boy (boy A) has already reached and passed his adolescent growth spurt; his APHV of 13.5 years is

0.2 months earlier than his age at measurement of 13.7 years. In contrast, the smallest boy (boy I) who is similar in age to the tallest boy at 13.6 years is still 2.1 years from obtaining his peak adolescent growth spurt.

Alternatively, individuals can be characterized as early, average, or late maturers depending on the age at which PHV is attained. Early maturers are those whose age at PHV is earlier than 1 year of the average age, while late maturers have an age at PHV later than 1 year of the average age, and the remainder are classified as average maturers. In Fig. 32.4, if the average APHV is taken to be 14 years then boy B would be identified as an early maturer, boys A, C, D, E, and G would be labeled as average maturers and boys F, H, and I would be classified as late maturers.

To obtain the years from age at PHV, serial data are required, and therefore, this indicator of maturity has previously been limited to longitudinal studies. However, there are now a number of gender-specific multiple regression equations, based on segmental growth patterns, which predict the maturity offset age parameter [17–19]. The prediction equations require measures such as stature, trunk length, and leg length, as well as body mass and chronological age. Using growth indicators, age from PHV can be predicted within ±1 year in 95% of cases [18] or the maturity offset can be used as a categorical (pre- or post-PHV) measure of maturity. These predicted maturity offset ages are quick, noninvasive to administer and can be used in cross-sectional studies. The added advantage to these techniques is that they can predict a maturity benchmark that exists in both boys and girls. Therefore, they allow for between-sex comparisons. The accuracy of such non-intrusive prediction equations has been questioned, and results showed that prediction methods can influence the APHV ascertained, and thus, caution is stressed when using these methods [20].

The height attained at any given chronological age can also be compared to reference norms to assess maturity. An individual is assigned a morphological age based on height for age classifications. The major disadvantage of this method is that it does not take into account the variability of height related to heritability and the amount of growth remaining (Fig. 32.1).

Another method of utilizing somatic growth is to express measured height in terms of percentage of final adult height [21]. This is illustrated in both Figs. 32.1 and 32.4. In Fig. 32.1, although the girls appear similar in height at 11 years of age, the girl on the left has reached 86% of their adult height compared to 96% achieved by the girl on the right. In Fig. 32.4, although in absolute terms boy B appears to be small for his age, when presented as a percentage of final adult height there is no difference between boys B and C at 7 and 14 years of age. This is because at 40 years of age, boys A and B are the same height and boy C is 15 cm taller. Because roughly 92% of adult stature is reached at PHV [22], individuals can be classified into pre- or post-PHV maturity groups. Thus, with the average age of PHV in boys being 14 years, boy A in Fig. 32.4 would be classified as an early maturer (percentage adult stature >92%) and boys B and C as average maturers (percentage adult stature <92%). This classification is not apparent just from height measures alone because it is impossible at a single measurement occasion to know the amount of growth that has occurred. Using this approach, the nine boys in Fig. 32.4 would be classified as boys C, E, F, G, H, and I being pre-PHV and boys A, B, and D as being post-PHV. The disadvantage of this technique is that an adult value is required, and a maturity status can only be applied retrospectively.

Expressing current height as a percentage of adult height can, however, be used in cross-sectional studies if adult height is predicted. Many equations have been developed to predict adult height [12, 21, 23–26]. The most commonly used methods are those of Bayley and Pinneau [23], Roche et al. [25], and Tanner [12]. However, these methods all require an assessment of skeletal age and are thus not practical outside of a clinical setting. Recently, predictive equations have been developed that do not require a measure of skeletal age [21, 24, 26], and have the potential for use in pediatric studies.

32.3 Summary

Although it has often been assumed that regular physical activity or exercise is important to support normal growth and development, most healthy children will grow and mature whether or not they are physically active. Currently available data do not support the assertion that intensive physical activity and/or training for sport will affect a child's statural growth. However, regular activity or training is important for the regulation of body mass—increasing muscle size and bone density and reducing fat accrual, all of which can impact injury risk. Diet, nutrition, and socioeconomic resources are considered the prime environmental influences on growth. However, you could add to this list seasonality, altitude, pollutants, pharmaceuticals, and noise [27]. For example, studies of birthweights of children born close to airports and who were exposed to noise stress have been found to consistently have birthweights that are depressed [28]. This suggests that the endocrine system is being compromised and growth altered. So, although it is probably not necessary to continue to investigate the effect of training on the young athlete's body physique, there is still the unanswered question as to whether maturity is attenuated by sports involvement. Erlandson, Sherar, Mirwald, Maffulli, and Baxter-Jones [29] found that although final adult height was not compromised in gymnasts, swimmers, or tennis and soccer players, gymnasts' maturation was attenuated. Lindholm, Hagenfeldt, and Hagman [30] also working with gymnasts suggested that gymnasts were malnourished and that this influenced their growth. Other work by Caine, Bass, and Daly [31] observed that growth spurts in gymnasts occurred after an incidence of injury. These studies highlight the fact that while stature may not be compromised in youth athletes, the speed of their growth and maturation could be influenced by various other factors. Another area that is understudied is the effect of psychological stress on growth, and in particular its effects on the endocrine system of the young athlete. Finally, with the introduction of maturational age alignment (bio-banding) to youth sports [32], the long-term effects of such classifications on injury prevalence warrant investigation. Successful banding of young athletes will likely involve a delicate interplay of matching levels of physical, psychological, and social maturation.

32.4 Recommendations

When matching children and youth for sports competitions, it is important that consideration is made for inherited characteristics and growth in terms of both timing (chronological age) and tempo (biological age). Those working with young athletes need to be aware of why a child is of a particular stature. There are now a number of quick and easy methods that can be used to predict both a child's final adult height and current maturity status. To ensure that all children are given an equal chance to perform, those working with children need to look at, in addition to a child's chronological age (timing of growth), the heights of the child's parents, the child's month of birth, and the child's biological maturity (tempo of growth). To avoid unnecessary injury and potential drop-out from sport, those working with children need to be continuously monitoring a child over time rather than making selection and other decisions related to one-off assessments.

References

1. Maffulli N, Baxter-Jones ADG, Grieve A. Long-term sport involvement and sport injury rate in elite young athletes. Arch Dis Child. 2005;90:525–7.
2. Malina RM, Baxter-Jones ADG, Armstrong N, Beunen GP, Caine D, Daly RM, Lewis RD, Rogol AD, Russell K. Role of intensive training in the growth and maturation of artistic gymnasts. Sports Med. 2013;43:783–802.
3. Rowley S. The effects of intensive training on young athletes: a review. London: The Sports Council; 1986.
4. Maffulli N, Helms P. Controversies about intensive training in young athletes. Arch Dis Child. 1988;63:1405–7.
5. Feigley DA. Psychological burnout in high-level athletes. Phys Sportsmed. 1984;12:109–19.

6. Smith AL, Pacewicz CE, Raedeke TD. Athlete burnout in competitive sport. In: Horn TS, Smith AL, editors. Advances in sport and exercise psychology. 4th ed. Champaign, IL: Human Kinetics; 2019. p. 409–24.
7. Beunen G. Biological age in pediatric exercise research. In: Bar-Or O, editor. Advances in pediatric sports sciences. Champaign, IL: Human Kinetics; 1989. p. 1–39.
8. Hall SS. Size matters: how height affects the health, happiness, and success of boys and the men they become. New York, NY: Houghton Mifflin Company; 2006.
9. Bailey DA, Shephard RJ, Mirwald RL, McBride GA. A current view of Canadian cardiorespiratory fitness. Can Med Assoc J. 1974;111:25–30.
10. Scammon RE. The first seriatim study of human growth. Am J Phys Anthropol. 1927;10:329–36.
11. Cameron N. Human growth curve, canalization, and catch-up growth. In: Cameron N, editor. Human growth and development. San Diego, CA: Academic Press; 2002. p. 1–20.
12. Tanner JM. Foetus into man: physical growth from conception to maturity. London: Castlemead Publications; 1989.
13. Baxter-Jones ADG, Eisenmann JC, Sherar LB. Controlling for maturation in pediatric exercise science. Pediatr Exerc Sci. 2005;17:18–30.
14. Kemper HC, Verschuur R. Maximal aerobic power in 13- and 14-year-old teenagers in relation to biologic age. Int J Sports Med. 1981;2:97–100.
15. Baxter-Jones ADG. Growth and development of young athletes: should competition levels be age related? Sports Med. 1995;20:59–64.
16. Tanner JM. A history of the study of human growth. Cambridge: Cambridge University Press; 1981.
17. Fransen J, Bush S, Woodcock S, Novak A, Deprez D, Baxter-Jones ADG, Vaeyens R, Lenoir M. Improving the prediction of a maturity from anthropometric variables using a maturity ratio. Pediatr Exerc Sci. 2018;30:296–307.
18. Mirwald RL, Baxter-Jones ADG, Bailey DA, Beunen GP. An assessment of maturity from anthropometric measurements. Med Sci Sports Exerc. 2002;34:689–94.
19. Moore SA, Brasher PMA, Macdonald H, Nettlefold L, Baxter-Jones ADG, Cameron N, McKay HA. Enhancing a maturity prediction model that uses anthropometric variables. Med Sci Sports Exerc. 2015;47:1755–64.
20. Sherar LB, Cumming SP. Human biology of physical activity in the growing child. Ann Hum Biol. 2020;47:313–5.
21. Khamis HJ, Roche AF. Predicting adult stature without using skeletal age: the Khamis-Roche method. Pediatrics. 1994;94:504–7.
22. Tanner JM. The physique of the Olympic athlete. London: George Allan and Unwin Ltd.; 1964.
23. Bayley N, Pinneau SR. Tables for predicting adult height from skeletal age: revised from and for use with the Greulich-Pyle hand standards. J Pediatr. 1952;40:423–41.
24. Beunen GP, Malina RM, Lefevre J, Claessens AL, Renson R, Simons J. Prediction of adult stature and noninvasive assessment of biological maturation. Med Sci Sports Exerc. 1997;29:225–30.
25. Roche AF, Wainer H, Thissen D. The RWT method for the prediction of adult stature. Pediatrics. 1975;56:1026–33.
26. Sherar LB, Mirwald RL, Baxter-Jones ADG, Thomas M. Prediction of adult height using maturity based cumulative height velocity curves. J Pediat. 2005;14:508–14.
27. Schell LM, Knutsen KL. Environmental effects on growth. In: Cameron N, editor. Human growth and development. San Diego, CA: Academic Press; 2002. p. 165–96.
28. Schell LM. Environmental noise and human prenatal growth. Am J Phys Anthropol. 1981;56:63–70.
29. Erlandson M, Sherar LB, Mirwald RL, Maffulli N, Baxter-Jones ADG. Growth and maturation of adolescent female gymnasts, swimmers and tennis players. Med Sci Sports Exerc. 2008;40:34–42.
30. Lindholm C, Hagenfeldt K, Hagman U. A nutrition study in juvenile elite gymnasts. Acta Paediatr. 1995;84:273–7.
31. Caine D, Bass SL, Daly R. Does elite competition inhibit growth and delay maturation in some gymnasts? Quite possibly. Pediatr Exerc Sci. 2003;15:360–72.
32. Rogol AD, Cumming SP, Malina RM. Biobanding: a new paradigm for youth sports and training. Pediatrics. 2018;142:e201180423.

Optimizing Training and Performance

33

Claudio Gaudino, Renato Canova, Marco Duca, Nicola Silvaggi, and Paolo Gaudino

33.1 General Training Concepts

"**Sports training** is a complex pedagogical-educational process based on the organization of repeated physical exercise. Volume and intensity must progressively increase stimulating the physiological processes of supercompensation of the organism and favour the increase in the athlete's physical, mental, technical and tactical abilities, in order to enhance and consolidate his performance in the competition" [1].

This definition simply summarizes the aim of sports training that is to allow the athlete to achieve the best result throughout his career and to reiterate it on scheduled occasions. In practice, it includes all the principles that regulate sports training and determine its final result, emphasizing the essence of this process: adaptation. Adaptation is the consequence of the supercompensation process, and it consists of the growth of all conditional, coordinative, psychic and mental qualities, which in fact allow the achievement of the best result [1].

In addition, the following clarification that characterizes sports activities in which the coordinating factors are very important, and among them, also various athletic disciplines are important: "Training is a complex pedagogical-educational process based on the organization of repeated physical exercise in quantities, intensities, forms and degrees of difficulty such as to favour and consolidate the assimilation of skills (general and specific), which are progressively more complex and effective" [1]. Coordinating factors must interact with the various expressions of strength in order to reach the best execution of complex technical action.

This training consideration can be applied to athletes of the highest level and to who do not reach the highest level, but who nevertheless intend to improve their results according to the possibilities, time and energy to devote to the chosen sport activity. Genetic factors and individual qualities are the other cornerstones that determine the training result [2].

Volume and intensity (articulated and measurable in different ways depending on the disci-

C. Gaudino (✉)
Speed and Hurdles Track and Field Coach and Football Italian National Team Fitness Coach (2006), Milan, Italy

R. Canova
IAAF (Worldathletics) Lecturer and Coach of the World Record Holder 3000 m Steeple Saaeed Seif Shaheen, Quai Antoine, Monaco

M. Duca
Universita' degli Studi di Milano, Milan, Italy
e-mail: marco.duca@unimi.it

N. Silvaggi
Universita' degli Studi dell'Aquila, L'Aquila, Italy

P. Gaudino
Manchester United first team fitness and rehabilitation coach (UK), Manchester, UK

G. L. Canata et al. (eds.), *Management of Track and Field Injuries*,
https://doi.org/10.1007/978-3-030-60216-1_33

pline) combined with the coordinating factors determine the external workload. This is the stimulus from which the body's response derives. This response represents the internal load: it is individual, complex (since it involves different apparatuses and systems of the organism) and can change according to the moment [1].

The challenge for every coach is to define a short-, medium- and long-term programme, as suitable and specific as possible for each individual athlete. In practice, it is a matter of organizing the training following a method, using certain exercises, combining them with each other, have the athlete involved and aware of it and get the best possible response in order to achieve the best result. An example of training exercises categorization is presented in the throws training paragraph.

First of all, the performance model of the specific track and field discipline needs to be analysed according to different aspects [1]:

- Technical
- Biomechanical
- Physiological (metabolic).

The example of a relatively simple athletic speciality like the 100 m race can be explanatory (Figs. 33.1 and 33.2):

Technical aspects are more difficult to represent in a graph, but some indications can also be given in this regard:

- Start from the blocks pushing simultaneously with both feet;
- Be in a clear pushing phase until the end of the acceleration phase;
- Keep your feet taut when run and look for maximum relaxation of the cutaneous and shoulder muscles especially in the high-speed phase.

Even more important than the performance model of a discipline (from which the choice of the exercises to be used and therefore the training programming derives) is the individual performance model. This is based on the individual characteristics of the athlete, which takes into account the level of its qualities at that moment in time and all the variations that may occur, including the morphological ones.

A peculiarity of the training is its complexity. The relationship between the proposed training and the result obtained can be explained by the "supercompensation" concept. This reaction is complex because it represents a set of responses provided by various physiological systems stimulated by the training stimulus: for this reason,

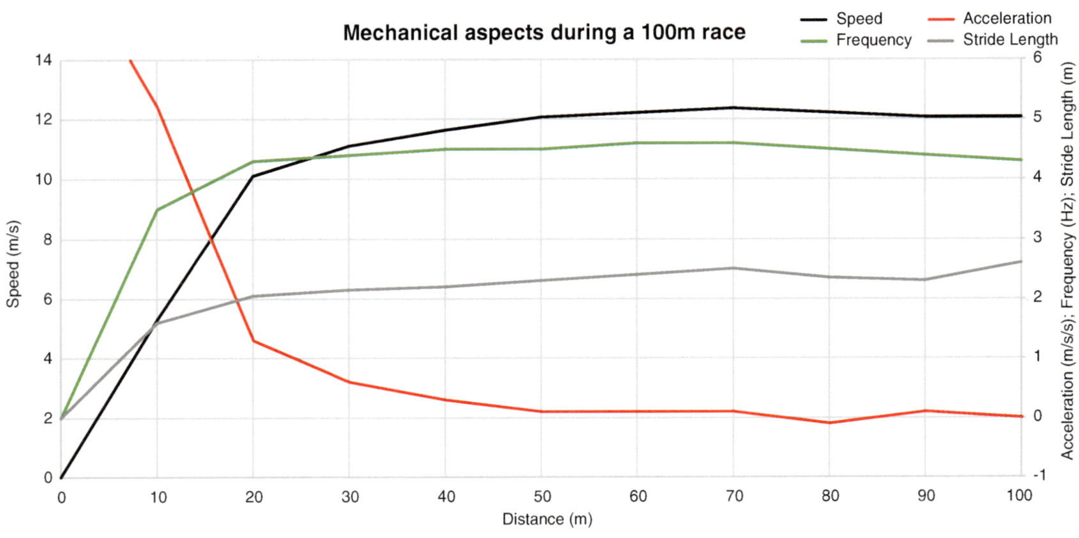

Fig. 33.1 Mechanical aspects during a 100 m race: speed in m/s; acceleration in m/s/s; stride frequency in Hz; and stride length in m

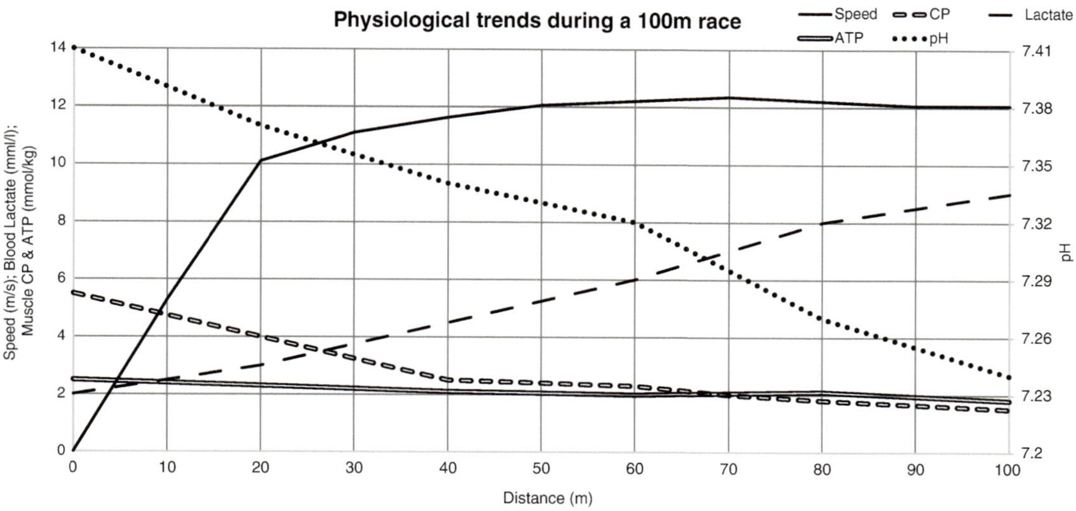

Fig. 33.2 Metabolic and biochemical trends during a 100 m race. Muscle CP and ATP in mmol/kg; blood lactate in mmol/l and pH. Speed in m/s is represented as well as a reference [3]

we generically talk about a "sum of responses". Therefore, a training stimulus produces not only a direct physiological adaptation, but also an indirect adaptation on other conditional and coordinative factors, which must be taken into account.

With regard to the training indirect adaptation (transfer), it manifests itself to a very significant extent especially in youth athletes, on both conditional and coordinative abilities. The lower the age (from 8 or 9 years old), the greater is the effect. It happens in fact that when a conditional capacity is stimulated, there is a positive impact on others as well. The same happens with regard to coordination skills: in this case, the objective is to take advantage of the "sensitive" phases, to constitute a good "motor expertise". By this term, we mean the set of motor experiences (suitable for the age) conducted in a global way and not necessarily aimed at the specificity of a discipline. From this interference of multiple motor experiences, gradually supported by an increase in conditional capacities, an expansion of the "motor expertise" derives, which will allow the athlete to acquire very complex skills. What has not been done in certain moments of great reception capability by the organism ("sensitive" phases) will no longer be fully recoverable later on. This underlines the importance of acquiring the widest range of motor skills possible that will be essential for subsequent technical specialization. The optimization of training must also take into account these intermediate steps [1].

Another aspect to consider is the heterochrony of the body's responses to the training stimulus. This aspect affects the recovery times of the various systems, and it must be taken into account when planning the training [4].

All these needs and other equally important factors characterizing the training (load increase in different times and modalities, alternation and variability of the load, evaluation of individual responses and athlete perceptions) must be taken into account with an adequate training plan in the short, medium and long term, and it must be the most specific and suitable for each individual athlete [4]. The main goal is to achieve their best physical condition at the time of the most relevant competitive events (tapering) during the season. **Periodization** consists in dividing the season into various training and competition periods, in order to achieve the aforementioned objective. Normally, the competitive season consists of an annual or semi-annual periodization (double periodization) and each macrocycle (annual or half-yearly) is characterized by a preparatory period, a competitive period and a transition period. Double periodization has become common in athletics, and it allows to reduce time between one competitive phase and the next one. Sometimes in a double periodization, the first period of competitions has a subordinate function to the second, where the most important competitive events are concentrated. Classic subdivision into microcycles (1 week), mesocycles (3/4 microcycles) and macrocycles (more mesocycles, up to an entire season) favours the alternation of load and recovery with all the benefits that derive from it. An example of throws training periodization is reported in the throws training paragraph.

Between all the conditional qualities, strength plays a role of primary importance in all athletic disciplines. According to Vittori, the prerogative of the muscle is to contract and its strength depends on the functional fibers. The same methodologist and athletic coach accurately defined this quality as follows: "Strength is a physical quality which is the foundation of human motility, responsible for bodies or objects movement and their speed" [5].

In his methodology, Vittori defined the different strength expressions with appropriate terminology, which does not always coincide with the most widespread (and less accurate) terms that have now become fashionable (Fig. 33.3).

The differentiation between active and reactive strength implies that the first one (active) occurs as an effect of the muscle shortening phase only (concentric phase only: e.g. an action carried out starting from a standstill position), while the second one (reactive) occurs as an effect of the stretching shortening cycle (with the eccentric phase followed by the concentric one, therefore with reference to the elastic component). Two examples of high jumps exercises can simply clarify the difference:

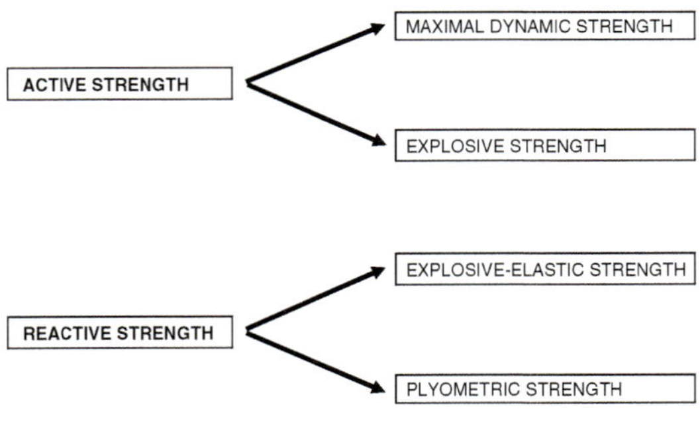

Fig. 33.3 Differentiation of strength expressions according to Vittori [5]

1. Squat jump (active strength): starting from a half-squat stationary position and jump as high as possible by solely extend the legs.
2. Countermovement jump (reactive strength): starting from an upright standing position, make a preliminary downward movement by flexing knees and hips, and then immediately extend knees and hips to jump vertically up off the ground.

Active strength includes both maximal dynamic strength and explosive strength:

- *Maximal dynamic strength* is what is needed to move the highest possible load. It is defined as dynamic in order to differentiate it from the isometric strength;
- *Explosive strength* can be expressed at the maximal speed allowed by the resistance (which can be represented by the body weight, an overload or any other tool) starting from a static situation so that the muscle contraction is purely concentric.

Reactive strength includes both explosive elastic strength and plyometric strength:

- *Explosive elastic strength* is expressed by the stretching shortening cycle that consists of an eccentric muscle contraction quickly followed by a concentric muscle contraction. In this case, the elastic mechanism is mainly due to the SEC (series elastic component).
- *Plyometric strength* is a particular expression of explosive elastic strength with a reduced stretching phase in terms of both articular range excursion and time. In this way, the effect of the myotatic reflex is more marked and more profitable, which further increases the extent of the elastic response. In addition, the quickness and the reduced amplitude of the eccentric phase also improve the stiffness effect.

An example of the combination of the aforementioned expressions of strength can be found in the analysis of a 100 m race (Fig. 33.4):

In summary, maximum dynamic strength and explosive strength ("explosive strength" in Fig. 33.4) are those most used in the starting phase, taking into account that the athlete starts from a stationary position. Successively, the explosive elastic strength comes into play during the acceleration phase when the ankle, knee and hip angles are initially marked and gradually become smaller at the end of the acceleration itself. Finally, during the maximal speed phase, articular excursions are smaller, and the plyometric strength becomes the most important (Fig. 33.4). Obviously, none of these expressions of strength completely replace the other ones at any point. They combine between themselves in a mix where, depending on the moment, one prevails over the other [5].

A fundamental part of training is also all the prevention activities, which, although not neglected in the past, have now taken on a more precise configuration, substantially affecting the workload [6]. Core stability, in essence, is the joint and balanced reinforcement of the deep and superficial abdominal and back-lumbar muscles that guarantee the stability and mobility of the vertebral column. The vertebral column represents a force transmission axis and because of that it must be protected and put in a position to function at its best.

A general and sectoral research for concentric and eccentric strength balance between agonist and antagonist muscles not only represents a guarantee of injury prevention but also leads to a higher level of effectiveness. The actual sport practice leads to the strengthening of the agonist muscles that perform the movement, while the antagonists are normally less stressed: Therefore, rebalancing becomes necessary. Nevertheless, the proprioceptive regulation that is stimulated through unstable equilibrium must be taken into account. The kinaesthetic sense that automatically allows to evaluate the position of the body segments and their movement is stimulated by different types of receptors stimulated precisely by instability.

The control of training has always been a priority in track and field. Obviously, over the course of the last few years, significant improvements have been made thanks to the most modern technologies (lasers, cameras, GPS, accelerometers, etc.). However, all these tools do not replace the

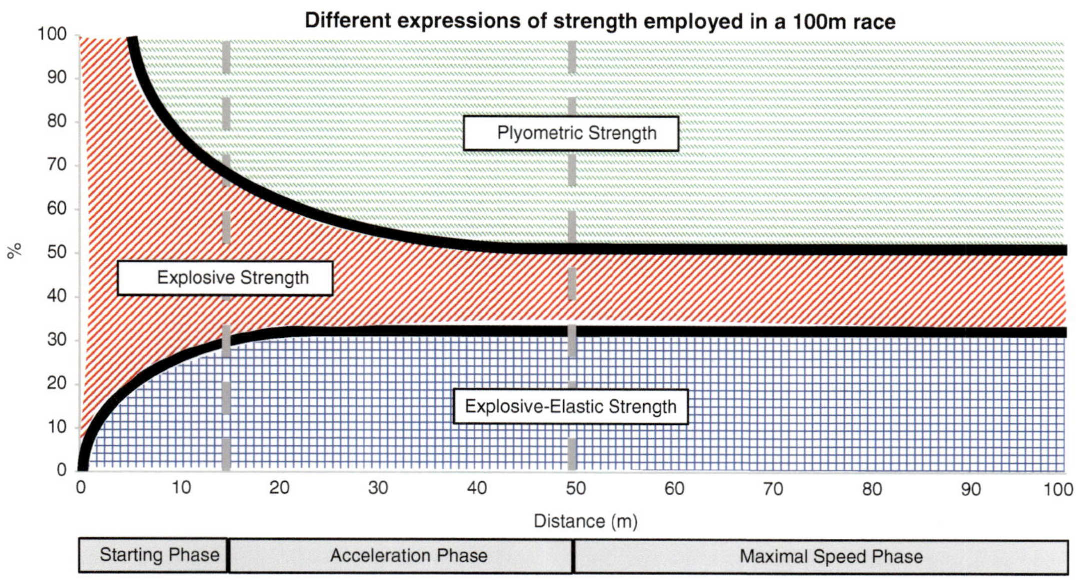

Fig. 33.4 This graph represents the influence (as percentage) of the different expressions of strength during a 100 m race [5]

attention, the observation and that attitude called "speculative" of the track and field coach. The ability of the coach consists of data evaluation, observing training sessions details, comparing the athlete over time (longitudinal analysis) and making deductions in order to modify the training sessions when necessary.

Directly linked to training is the nutrition. Perhaps in the past its importance has not been recognized as much as it is now. The individual characteristics, the nature of the discipline practised and in particular the type of training carried out day by day with the related energy requirements contribute to structuring the nutrition strategy. It must meet the needs of restoration and accumulation of glycogen reserves, the intake of water and electrolytes in their best combination and ensure protein intake not only as a function of building muscle cells, but also for the synthesis of hormones and enzymes [2, 4].

Finally, with the recent increase in length of many athlete's career, in some athletic disciplines there is a relative reduction in the use of very specific exercises in favour of the use of more general exercises aimed at guaranteeing the physical condition.

33.2 Speed and Hurdles Training

Track and field speed (100 m, 200 m and 400 m) and hurdles (100/110 m Hs and 400 m Hs) training follow some fundamental guidelines:

- The development of strength as a function of speed.
- The technique and the rhythm combined with rapidity in order to reach the maximal speed.
- The distribution of the effort.
- The specific endurance.

The development of **strength** follows a fairly linear direction that starts from working with more or less heavy load through classic exercises (such as squat, half-squat and half-squat jump) with all their variations. Afterwards, it moves on to the special exercises for strength (link between strength and speed) performed with light loads (e.g. sled sprints), with additional resistance to body weight (e.g. uphill sprints) or performed as bounds that allow a progressive approach to the technical gesture. With reference to the subdivision previously made with regard to the different expressions of strength, it can be stated that explosive strength, explosive elastic strength and plyometric strength are all involved. Therefore, these are solicited through the use of the aforementioned exercises to a different extent based on the time of the season, the characteristics of the race (race distances) and the individual qualities of the athlete.

Unlike speed, which can be considered a capacity derived from strength, **rapidity** is normally identified as a coordinative conditional quality. It is stimulated through specific exercises carried out in conditions particularly favourable to its development. It is associated with running technique and rhythm to help increasing speed. Therefore, it can be deduced that technical and rhythmic exercises of speed and rapidity are essential. They require neuromuscular freshness and complete recovery to be performed with the right intensity and quality.

Among the **speed** and hurdle races, only the 60 m indoors can be performed without really the necessity to dose the effort that must be maximal from the beginning to the end of the race due to its short duration. On the contrary, during all the other speed races (100–400 m), the distribution of effort is important in order to achieve the best result. This means that the 100 m, for example, will not be run at maximal speed, otherwise it will not be possible to achieve the optimal result. The maximal speed reached during the competition will be equal to the 98–99% of the personal maximal speed. This will allow the athlete to maintain it almost until the end of the race. It is obvious that by extending the distance from 100 to 400 m, the percentage of maximum speed reached will tend to decrease and it will be adjusted according to the consistency of the intervention of the various energy-producing mechanisms requested (anaerobic alactacid and anaerobic lactacid above all).

In order to improve the efficiency of these mechanisms and in particular their power, the athlete specialized in speed and hurdles disciplines must perform an adequate training based on short and long distances (from 60 m to 400–500 m). These distances must be run at certain speeds (not maximal) with incomplete recoveries (increasing the mechanisms capacity) and at higher speeds with almost complete recoveries (in order to increase the mechanisms power). The current trend is to favour high-intensity training sessions in order to stimulate and improve power rather than the mechanism capacity.

33.3 Long-Distance Running Training

Endurance running training changed crucially over the course of the last century. At the beginning of 1900, the only known procedure was to run long distance following the athlete feelings. Training methodological fundamentals did not exist. At the beginning of 1930, an epochal turning point happened in Freiburg: track and field coach Woldemar Gershler together with doctor Herbert Reindell studied a new training method on more than 3000 University students. Their study showed how alternating short distances run at high speed (in particular 200 and 400 m, heart rate 180–190 bpm) with slow recovery run (heart rate 120 bpm) was the most effective training method to improve the cardiac activity. This method is known as "Freiburg Interval Training". The most emblematic product of that method was the German Rudolf Harbig who established in 1939 the 400 m race European record (46″) in Frankfurt and the 800 m race World record (1′46″6) in Milan during an epic race with Mario Lanzi.

During the same period, Swedish track and field coach Gosta Holmer studied a variation of that method, which had an important impact in longer distance runs. Gosta trained the best Swedish athletes (Gunder Hägg and Arne Andersson) introducing during their continuous run long periods of running at competition speed with recovery periods running at 85% of competition speed. This method called "Fartlek" (literally "Run Game") allowed Gunder Hägg to be the first man in the World to run 5000 m race below 14′ (13′58″2 in 1942). German doctor Ernst Van Aaken was the first person to understand that beyond the cardiac work, there were peripheral circulatory limits that had to be overcome, in order to increase oxygen transport capacity. Van Aaken set long periods of training on continuous running at low intensity, in order to increase the number of capillaries (therefore the aim was called "capillarization"). His idea was followed in New Zealand by Arthur Lydiard and in Australia by Percy Cerutty. The two Oceanic coaches produced the best athletes of that time, leaving an indelible imprint in training methodology. Australian Herbert Elliot won the 1500 m race in Rome Olympic Games and established the World record (3′35″6) when he was just 22 years old, and this was the last race of a short but dazzling career. Peter Snell, New Zealander, won the 800 m race in both Rome and Tokyo in 1964. On the second occasion, he doubled the gold medal with the victory in the 1500 m race and was able to improve the two World records in the 800m (1′44 ″3) and in the mile (3′51″ 3) races. Peter Snell was not the only Lydiard top athlete: in fact, for many years the trio composed of John Walker (first man in the World to run the mile more than 100 times under 4′), Dick Quax and Rod Dixon (which eventually also managed to win the New York marathon) remained at the highest levels in the track and field disciplines from 1500 m to 5000 m races. However, the Lydiard method, called "Marathon Training", produced striking results in the disciplines up to 5000 m race, while, despite the name, it proved absolutely unsuccessful on the marathon.

The period from 1970 to 1985 saw an exasperation of the volume, which allowed the athletes to bring themselves slightly below 27′30″ on 10000 m race and 13′10″ on 5000 m race, when the limits of 800 m race (1′41″73) and 1500 m race (3′29″77) were already at the same level as the best current athletes. The search for superior quality initially led to a contraction of the top results, to the point that, in 2003, the

Table 33.1 Progression of the World record in endurance disciplines since 1970

	1970	1980	1990	2000	2010	2020
800 m	1′44″3	1′42″33	1′41″73	1′41″11	1′41″01	1′40″91
1500 m	3′33″1	3′31″36	3′29″46	3′26″00	3′26″00	3′26″00
5000 m	13′16″6	13′08″4	12′58″39	12′39″36	12′37″35	12′37″35
10,000 m	27′39″69	27′22″47	27′08″23	26′22″75	26′17″53	26′17″53
3000 m SC	8′21″98	8′05″40	8′05″35	7′55″72	7′53″63	7′53″63
HM	1:03′53″	1:02′16″	1:00′10″	59′17″	58′23″	58′01″
Marathon	2:08′34″	2:08′34″	2:06′50″	2:05′42″	2:03′59″	2:01′39″

Table 33.2 Male World record improvements during the last 30 years (since the professionalization of African athletes has taken place)

800 m	(1′41″73 → 1′40″91) = 0″82	(0.80%)
1500 m	(3′29″46 → 3′26″00) = 3″46	(1.65%)
5000 m	(12′58″39 → 12′37″35) = 21″04	(2.70%)
10,000 m	(27′08″23 → 26′17″53) = 50″70	(3.11%)
3000 m SC	(8′05″35 → 7′53″63) = 11″72	(2.41%)
HM	(1:00′10″ → 58′01″) = 2′09″	(3.57%)
Marathon	(2:06′50″ → 2:01′39″) = 5′11″	(4.09%)

best British marathon runner was Paula Radcliffe, with no man able to run under 2 h 15′ (Table 33.1).

Table 33.2 clearly shows how modern **training methodologies** for short endurance distances (800 m and 1500 m races) have not produced substantial improvements, while current long-distance training methodologies have led to very significant progress, particularly in the last 10 years. What has essentially changed in the current advanced methodology?

1. Modulation in the intensity of training in the various sessions: training with specific high intensity is more frequent and the recovery between them is longer.
2. Balance in the total distance run: decrease in the total volume (180–220 km per week instead of 280–320 km usually run in the 1980s) and simultaneous percentage increase in km run at specific race speed (30–35% per week, equal to 60–70 km, compared to 20% in the past, equal to 55–60 km).
3. Clarification of the role of low-intensity running, as a simple support for running at specific race speed.
4. Maintenance of what has already been achieved with training, even during the fundamental period (never lose what the athlete already has, in terms of aerobic power).
5. Promote the intensity (therefore starting from the concept of speed, obviously relative to the race distance), rather than the volume as it happened in the past. In other words, nowadays athletes run "fast" over distances of 5–10 km and then try to run longer distances at a similar speed, looking at the "extension" of the intensity, while, on the contrary, in the past it was required first to reach a great general resistance, running 40–50 km at moderate pace, to then try to "speed up" the athlete. From a methodological and mental point of view, it is easier to extend the speed than to speed up the distance.
6. Use of speed variations, both short and long, which allow to improve the permeability of cell membranes in order to favour the clearance of lactate produced in shorter times. Since lactate can be considered a limiting factor in performance, if the level of saturation in the muscle fibres is too high, but at the same time a percentage of it is capable of producing energy, it is obvious that, if the athlete carries out a training capable of speed up the clearance action then the athlete can run faster, according to the equations:
   (a) Faster lactate clearance = Less lactate accumulation in muscle fibres
   (b) Less lactate accumulation in muscle fibres = Possibility of producing more lactate by running faster
   (c) Higher lactate production = Higher percentage of energy available.

This means that nowadays there is the possibility of running the entire marathon faster, increasing the resistance coefficient. Up to 10 years ago, the best athletes could run the marathon at 94–95% of the half-marathon speed. Currently, the resistance coefficient has risen to 96–97%, also thanks to the new energy gels that allow a quick energy recharge.

Some examples of **specific training** currently adopted with the best World athletes are reported here:

1. 5 × 5 km at the race pace, alternated with 1 km run at 90% of the race pace. For example, if an athlete runs the marathon at 3′/km = 2:06′36″, 5 × 15′ with 1 km recovery at 3′15″/3′20″, for a total of 30 km in 1: 31′15″.
2. 20 km on the track: 2 × 3000 m at 105% of the marathon rhythm (MR), in the previous case in 8′33″, + 3 × 2000 m at 107% MR (in that case, 5′36″) + 5 × 1000 m at 108% MR (in that case, 2′45″) + 6 × 500 m at 112% MR (in that case, 1′19″).
3. 24 km alternating speed every km (2′55″/3′05″).
4. Continuous run at even pace for 40 km at 97% MR (to be performed 4–5 weeks before the competition).
5. "Special block", which consists of prolonged training of specific quality, both in the morning and in the afternoon. Example, 10 km at 90% MR in 33′ + 15 km MR in 45′ in the morning, 10 km at 90% MR in 33′ + 6 × 2000 m on the track at 103% MR in 5′48″ with 2′ recovery jogging in the afternoon, for a total of 47 km of specific training +8 km of warming up on the same day.

33.4 Jumps Training

In track and field, **jumping events** are characterized by the presence of a run-up, a take-off (three in the case of the triple jump), a flight phase and a landing phase [7]. During the run-up, the athlete builds up horizontal velocity. Later, part of that horizontal velocity is converted into vertical velocity during the take-off. In all the events but pole vault, the jumper's stance leg is planted in front of the athlete and applies a force to the ground that generates a reaction force in the opposite direction (GRF). This GRF acting on the athlete's body is generated in a very short time (150–200 ms) and, although partially reducing the horizontal velocity, thrusts the athlete centre of mass (CM) upward. It has to be noted that, during the take-off, the athlete stance lag is unable to convert horizontal velocity into vertical velocity without a loss of energy [7], but this can be minimized by planting the take-off leg faster and straighter [8]. The resultant velocity and projection angle of the CM dictate the jumping performance achievable by the athlete. Alternately, considering pole vault, the pole acts as the stance leg of the jumper and converts horizontal velocity in vertical velocity and during the take-off there is a net energy gain, thanks to the muscular actions performed by the upper body of the athlete on the pole [9].

Another factor to be considered is the horizontal and vertical distance travelled by the athlete's CM during the take-off, which can be controlled by the athletes by purposely swinging their arms forward and/or upward and [7, 10]. In the horizontal jumps, measuring starts from the end of the take-off board; therefore, the athlete must be precise in their run and plant their foot as close as possible to the end of it. In all jumping events, the athletes' ability to control the position of body segments, while in the air is also a contributing factor. In vertical jumps, it allows for clearing bar set higher than the athletes' CM and in horizontal jumps it allows for a further reach when landing in the sandpit.

The most important characteristics for an athlete to succeed in the jumping events are speed, showing always the greatest predicting power towards performance, and strength [11, 12]. Therefore, **speed** development should be prioritized over strength development [13] and can be pursued by means of sprint training. The emphasis should be placed on top speed and step length awareness and control (e.g. 30- to 60-m sprints or 10 m fly-ins with 3 to 6 min of recovery). Pole vaulter should perform sprint training carrying the pole, as it alters sprint kinematics and reduces sprint velocity.

Table 33.3 Example of strength training programme for a horizontal jumper

Phase	Hypertrophy	Strength	Power
Duration (weeks)	0–4	4–8	2 – 4
Sessions/week (n)	3	3	2
Exercises/session (n)	5–6	4–5	3 – 4
Sets x repetitions (n)	5 × 10/3 × 10	5 × 5/3 × 5/3 × 3	3 × 3/3 × 2/2 × 2
Intensity (%1RM)	60–70%	70–85%	40–60%/80–95%

1RM One repetition maximum

Strength and the ability to generate large GRF in a brief time can be developed effectively by resistance training (2–3 sessions per week) and plyometrics (1–2 sessions per week) [14]. The implementation of a block periodization paradigm (consisting of the sequential development of hypertrophy, strength and power) is to be preferred, as it leads to improved maximal and explosive strength adaptations over other periodization paradigms (Table 33.3) [15, 16]. Resistance training should prioritize multi-joint movements involving lower limb triple extension (e.g. squats, pulls), and exercise selection should allow for a variation in range of motion, muscle action and specificity throughout the training plan (e.g. squat, ½ squat, ¼ squat and countermovement jump). Regarding pole vaults, additional emphasis should be put on shoulder girdle strength (e.g. horizontal bar gymnastic derivatives exercises), especially so for women.

Alongside strength and speed development, **jumping skill** can effectively be trained with varying emphasis through the training phases. A way to improve jumping skill consists of the use of dynamic drills, which replicate the take-off or action with a reduced run-up (three strides). The lower speed allows the athlete to elicit a greater control over his body segments, without a substantial alteration in the kinetic of the movement [17]. An effective training method to obtain straighter and stiffer plant leg consists in the use of raised flat and inclined boards at take-off [18]. When jumping off the flat boards, the athlete enhanced the pivot of their body over the stance leg and reduced flexion at the knee. This can be effectively transferred to the standard take-off condition.

To allow for optimal performance, the coach should select and integrate the proper means for speed, strength and skill development based on the biological, psychological and technique level of the athlete being trained.

33.5 Throws Training

Training is represented by the different physical exercises that directly or indirectly influence the improvement of sports performance. Many authors have divided sports training exercises into categories that characterize the development of the qualities related to the specific sports disciplines [19]. Training exercises can be divided into three main groups:

- **Exercises for general (conditional) preparation.**
- Exercises that do not represent any element of the technical model and which differ in terms of execution time, position and movement with respect to the competition.
- **Exercises for special preparation.**
- Exercises that represent the technical model but modify the spatiotemporal characteristics of the technique and reduce or increase the speed of it compared to the competition.
- **Competition or specific exercises.**
- Exercises that correspond to the technical actions carried out in conditions close to the competition ones.

In throws disciplines, **exercises for general preparation** are not very correlated or even have no correlation at all with the competition action. For this reason, sometimes the use of some of these exercises can lead to the development of physical qualities that are not very solicited in the competition, limiting the possibility of improving specific qualities. In order to have maximum effectiveness, sports training must respect an important principle: it must be highly specific. It must have a high correlation in its exercises (stimuli) with the competition exercise. This means that each exercise must have at least one technical component that makes it correlated with the competition action. By following this principle, competition or specific exercises are those with the highest correlation as they consist of performing exercises identical to the competition ones, or exercises that are extremely close to it, with respect to the rules and condition of the competition itself.

General exercises do not correspond to the competition actions; however, they promote the development of the organism's functional capacities. Their goal is to increase the training effect on certain physiological systems and on certain functions of the organism [20]. It is evident that in order to increase the effectiveness of these exercises and to increase the correlation with special exercises, general exercises must respect an important principle: they must have correlation with the physical characteristics of the discipline.

In sports characterized by neuromuscular factors such as throwing, general preparation contents have three very important parameters: the expressions of maximal strength, explosive strength and explosive elastic strength. These three parameters are very important for the athlete's functional status and must be constantly monitored.

Among the exercises for the development of maximal strength, there are:

- ½ squats, deep lounges, squats, deadlifts, snatch, upright barbell row, inclined bench and horizontal bench.

Among the exercises for the development of explosive and explosive elastic strength, there are:

- ½ squat jump performed from standing still position (explosive), continuous, with countermovement, with countermovement jump (explosive elastic) and continuous jumps. With regard to the development of explosive strength, we can also consider all forms of bounds since they have no correlation with the throws technical action. On the contrary, with regard to jumps training, these exercises would have been considered as special exercises.

Exercises for special preparation have a high correlation with the technical model since they contain elements of the competition itself but ensure the possibility of expressing higher or lower strength commitments compared to the one expressed in the competition making its speed to decrease or increase.

In throws, for example, special exercises are throws with tools of a different weight from the standard (competition one) or throws with overload such as weighted belts or weighted vests. In addition, are also considered special exercises in throwing those exercises with overload that reproduce only a part of the whole technical action such as only the hips movement or only the transaction in shot put.

Specific exercises are those exercises of global and segmental technique without overloads and performed with standard equipment. Throws made with tools that are slightly lighter and heavier than the standard weight also fall into this category as well as those with reduced actions like the standing throws.

Most of the track and field disciplines are classified as power activities since during those performances there is a high development in explosive strength such as in throws, jumps, sprints and hurdles races. All these disciplines have in common a single objective: to improve the **speed of execution**. That means to run faster, to increase the exit speed of the tool in the throws or the take-off speed in jumps. The difference between the various disciplines is the modality of developing speed in cyclic or acyclic movements, but the concept is that speed is the only parameter able to improve the performance. Therefore, a modification of the athlete's functional status must lead to increase in this parameter. To be able to do that all the training contents (general, special and specific, mentioned above) must lead to an increase in speed. This factor is the only one that can, over the years and for many disciplines (in particular for throws), continuously vary and influence the performance.

The most important part of throws training plan is the special physical preparation. Increasing maximal strength for example carrying out bench press exercises or squat exercise does not mean that there is an improvement in the throwing performance. There is no correlation between those exercises and the throw. In order to make the most of all the adaptations obtained with the exercises of maximal and explosive strength, it is necessary, without anticipating or delaying the development of speed, to selectively intensify the work regime through the special preparation.

The objectives of the special physical preparation are to improve intra- and intermuscular coordination and thereby to create better conditions for technical improvement. Special strength exercises must have the following characteristics:

- high correlation between the strength exercise (special) movement and the competition movement (complete movement).
- high correlation between the strength exercise (special) movement and one or more elements of the technical action (segmental movements).

An example of throws **training periodization** leading to a competition is shown below (Fig. 33.5). The objective is to bring the athlete to his best competitive condition in 17 weeks. These are divided into a first period (first 8 weeks, in red) mainly focused on the development of maximal strength and explosive strength by using general exercises. In the following period (from the 6th week to the 14th week, in yellow), the percentage of special work prevails over the general one and the specific work increases. In the competitive period (last 3 weeks, in light blue), specific work prevails over special work and only a small percentage of general work remains.

In Fig. 33.5, five mesocycles are schematized, the first two are 4 weeks each (in red) while the other three are formed by 3 weeks each (in yellow and light blue). Each column represents a week that makes up the cycle and the height of the column shows the training load of the entire week. The first week of each cycle is the one where the maximal volume of work is expected. The volume of work in the first week is dictated by the intensity used in the respective period and the level of development of the subject's physical abilities. In the second and third weeks, for the 4-week mesocycle and only the second for the 3-week one, the volume of work is reduced by 20%, while the number of exercises and the methods used in the respective cycle remain the same. The exercises must remain the same for 3 or 4 weeks (a mesocycle) since that allows the athlete to obtain the best effects and effective physiological adaptations lasting over time. The 20% reduction in training load must be implemented to respect the ratio between external load and internal load. At the beginning of the second week of work of the cycle, the organism of the athlete is at a lower performance level if compared to the starting level, due to the stresses suffered in the first week. As a consequence, to have an internal response equal to the first week a slightly lower training volume is sufficient. The

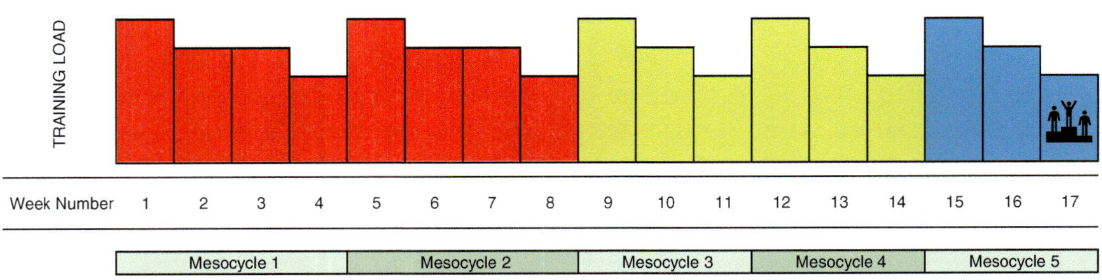

Fig. 33.5 Example of 17 weeks of throws training periodization leading to a competition

fourth week for the first two cycles and the third week for the others refer to the unloading week where work is reduced by up to 60% compared to the first. This is to allow the body to recover and have the effect of supercompensation.

Going into more details, general exercises carried out during the first 8 weeks (in red in Fig. 33.5) include three sessions of strength of which 70% is maximal strength and 30% is explosive strength with prevalent pyramidal programmes and fixed repetitions. The most used exercises are horizontal bench, inclined bench, snatch, upright barbell row, squats and half-squats. During this phase, special exercises are carried out three times per week and they include exercises with heavy load that mainly reproduce segmental technical movements, for example exercises with barbells, weighted belts, weighted vests and very heavy throwing. Specific exercises in this phase are very limited, and only a few throws are performed.

In the following 6 weeks (in yellow in Fig. 33.5), special physical preparation prevails and specific work increases. In this phase, general exercises are reduced to two sessions per week of which 50% maximal strength and 50% explosive strength. The exercises remain the same as in the previous period. There are four sessions per week focused on special physical preparation in which the speed of execution during the exercises increases considerably. Complete throws are performed. In shot put, for example, in this phase, the weight of the shot can range from 9 to 6 kg for men and from 6 to 3 kg for women. At the same time, specific training increases. The number of throws increases including the use of competition tools and great attention is paid to the throwing technique.

Finally, the last 3 weeks (in light blue in Fig. 33.5) represent the competitive period. In this phase, general exercises are still used for two times per week but with a percentage of 30% maximum strength and 70% explosive strength. Special exercises are performed three times per week. Complete throws are carried out with heavy and light tools. In shot put, throws are carried out with heavy and light tools ranging from 8.30 kg to 6.26 kg for males and from 5 to 3 kg for women. Official competition weights in shot put are 7.25 kg for men and 4.00 kg for women. Specific exercises prevail over the others especially in order to refine the technical movement.

References

1. Bellotti P, Donati A. L'organizzazione dell'allenamento sportivo. Roma: Societa' Stampa Sportiva; 1992.
2. Weineck J. Biologia dello sport. Torgiano: Ed. Calzetti e Mariucci; 2013.
3. Arcelli E. Acido lattico e prestazione. Palermo: Cooperativa Dante Editrice; 1995.
4. Weineck J. L'allenamento ottimale. Torgiano: Ed. Calzetti e Mariucci; 2009.
5. Vittori C. Le gare di velocita'. Roma: FIDAL Centro studi e ricerche; 1995.
6. Broussal A, Bolliet O. La preparation physique moderne. Counter Movement Collection; 2010.
7. Zatsiorsky V. Biomechanics in sport: performance enhancement and injury prevention, vol. IX. Oxford: Blackwell Science Ltd.; 2000. https://doi.org/10.1097/00005768-200105000-00033.
8. Greig MP, Yeadon MR. The influence of touchdown parameters on the performance of a high jumper. J Appl Biomech. 2000;16(4):367–78. https://doi.org/10.1123/jab.16.4.367.
9. Linthorne NP, Weetman AHG. Effects of run-up velocity on performance, kinematics, and energy exchanges in the pole vault. J Sports Sci Med. 2012;11(2):245–54.
10. Hay JG. The biomechanics of the triple jump: A review. J Sports Sci. 1992;10(4):343–78. https://doi.org/10.1080/02640419208729933.
11. Dapena J, McDonald C, Cappaert J. A regression analysis of high jumping technique. Int J Sport Biomech. 1990;6(3):246–61. https://doi.org/10.1123/ijsb.6.3.246.
12. Morin JB, Jeannin T, Chevallier B, Belli A. Spring-mass model characteristics during sprint running: correlation with performance and fatigue-induced changes. Int J Sports Med. 2006;27(2):158–65. https://doi.org/10.1055/s-2005-837569.
13. Schiffer J. The Horizontal Jumps. New Stud Athl. 2011;26(3/4):7–24.
14. de Villarreal ESS, González-Badillo JJ, Izquierdo M. Low and moderate plyometric training frequency produces greater jumping and sprinting gains compared with high frequency. J Strength Cond Res. 2008;22(3):715–25. https://doi.org/10.1519/JSC.0b013e318163eade.
15. DeWeese BH, Hornsby G, Stone M, Stone MH. The training process: planning for strength-power training in track and field. Part 1: Theoretical aspects. J Sport Health Sci. 2015;4(4):308–17. https://doi.org/10.1016/j.jshs.2015.07.003.

16. DeWeese BH, Hornsby G, Stone M, Stone MH. The training process: planning for strength-power training in track and field. Part 2: Practical and applied aspects. J Sport Health Sci. 2015;4(4):318–24. https://doi.org/10.1016/j.jshs.2015.07.002.
17. Wilson C, Simpson S, Hamill J. Movement coordination patterns in triple jump training drills. J Sports Sci. 2009;27(3):277–82. https://doi.org/10.1080/02640410802482433.
18. Koyama H, Muraki Y, Ae M. Athletics: Effects of an inclined board as a training tool on the take-off motion of the long jump. Sports Biomech. 2005;4(2):113–29. https://doi.org/10.1080/14763140508522858.
19. Werchoshanskij Y. La moderna programmazione dell'allenamento sportive. Roma: Societa' Stampa Sportiva; 2001.
20. Werchoshanskij Y. La preparazione fisica speciale. Roma: Societa' Stampa Sportiva; 2001.

Management of Track and Field Injuries: Rehabilitation and Return to Sport After Injury in Track and Field Athletes

34

Tom G. H. Wiggers, Peter Eemers, Luc J. Schout, and Gino M. M. J. Kerkhoffs

34.1 Introduction

"When can I return to sport?" is the golden question from athletes and coaches to healthcare professionals working in sports. This chapter focuses on the process from injury to return to sport (RTS) for the track and field athlete, aiming at giving an evidence-based outline of the principles of returning back to sport after injury.

34.2 Definition of Return to Sport

Different terms and definitions are used to define the moment of the athlete's ability to "sport" again: return to sport, return to play, return to participation, and return to performance. The definitions of the 2016 *British Journal of Sports Medicine* consensus statement are generally accepted (Fig. 34.1) [1]:

"Return to participation" is reached when the athlete is participating in sport, however at a lower level than his or her ultimate goal. This means the athlete is in the final part of

Fig. 34.1 Return to sport continuum. Adapted and adjusted from Ardern et al. [1]

T. G. H. Wiggers (✉)
Department of Orthopaedic Surgery, Amsterdam Movement Sciences, Amsterdam UMC, University of Amsterdam, Amsterdam, The Netherlands

Academic Center for Evidence Based Sports medicine (ACES), Amsterdam, The Netherlands

Amsterdam Collaboration for Health and Safety in Sports (ACHSS), International Olympic Committee (IOC) Research Center Amsterdam UMC, Amsterdam, The Netherlands
e-mail: t.g.wiggers@amsterdamumc.nl

P. Eemers
MMFysio, Nijmegen, The Netherlands
e-mail: peter@mmfysio.nl

L. J. Schout
Royal Dutch Athletics Federation, Arnhem, The Netherlands

Het Fysiolab, Nijmegen, The Netherlands
e-mail: luc.schout@atletiekunie.nl

G. M. M. J. Kerkhoffs
Department of Orthopaedic Surgery, Amsterdam Movement Sciences, Amsterdam UMC, University of Amsterdam, Amsterdam, The Netherlands

Academic Center for Evidence Based Sports medicine (ACES), Amsterdam, The Netherlands

Amsterdam Collaboration for Health and Safety in Sports (ACHSS), International Olympic Committee (IOC) Research Center Amsterdam UMC, Amsterdam, The Netherlands
e-mail: g.m.kerkhoffs@amsterdamumc.nl

G. L. Canata et al. (eds.), *Management of Track and Field Injuries*,
https://doi.org/10.1007/978-3-030-60216-1_34

rehabilitation and is doing sport-specific training but not physically, conditionally, and/or psychologically ready for return to competition. "Return to sport" means that the athlete has returned to his or her sport in competition. "Return to performance" is an extension of the return to play continuum in which the athlete has returned to his or her sport at the desired performance level. This can be at or above his or her pre-injury level of sports [1]. For track and field, we can consider achieving a personal best (PB) performance as successful return to performance.

34.3 Injury Characteristics

Prior to discussing the aspects of rehabilitation, we will first touch upon injury characteristics as training load, biomechanics, and other contributing factors for the present injury. These aspects are essential for defining the complete scope of the injury and establishing successful return to performance.

34.3.1 Injury Analysis

When an athlete gets injured, the first goal is to establish an accurate structural diagnosis of the injury and to identify factors having contributed to the current injury [2]. As many factors are responsible for an injury, it is essential to identify the mutual relationship between different factors. Several multifactorial etiology injury models are developed to show the interconnection of the contributing factors [3, 4]. The comprehensive model for injury causation described by Bahr et al. distinguishes different types of risk factors: internal nonmodifiable risk factors (e.g., age), internal modifiable risk factors (e.g., strength, coordination, neuromuscular control), and external risk factors (e.g., playing schedule, opponent behavior) [4]. Moreover, it is of clinical importance to identify which inciting event has given rise to occurrence of the injury [4].

34.3.2 Sports Science and Medicine Team

In the early phase of an injury, the team around the athlete has to determine who will be regarded as the "case manager" of the injured athlete. The case manager makes a rough timeline for the rehabilitation and decides which medical team members (or external specialists) are needed for consultation in each phase. This is dependent on specific injury characteristics and whether the injury will be treated conservatively or operatively. It is the responsibility of the case manager to ensure adequate communication and collaboration between the different team members. The case manager continuously informs the athlete and his or her coach and educates them about the current situation, rehabilitation plan, and risks of proposed management strategies [5]. This results in the greatest involvement of the athlete in the rehabilitation process and fosters his/her autonomy [6]. Another important aspect to mention is the pressure from various sources (coach, club, parents, manager, and/or press) that can be experienced by the athlete. The case manager should discuss this with the athlete because this can have a major impact on the athlete and can be an obstructive factor in recovery.

34.3.3 Training Load

Analysis of training diaries provides insight into the training load of the athlete. Poor management of the training program and training periodization is a major risk factor for injury [7]. This includes the planning and sequence of different types of training sessions within a week. Accurate study of training diaries is essential in the analysis of training planning and periodization. In addition to these training-related factors, the competition schedule has become busier over the years and is a factor in the occurrence of injuries [7]. High training load and especially spikes in training load are strongly associated with injuries [8]. High training load induces fatigue, which consequently diminishes coordination, neuromuscular function, and decision-making ability, thereby making the athlete vulnerable to sustaining injury [2, 7]. On the other hand, high training load is required to develop full potential of the athlete.

The acute/chronic ratio (A/C ratio) is a recently developed model to quantify training load and can be helpful in planning of training and competitions [9]. The A/C ratio is calculated by dividing the acute training load (the average training load of the last week) by the chronic training load (the average training load in the last 3–6 weeks) [9]. First studies in cricket, rugby, and Australian rules football found an optimal ("sweet spot") A/C ratio of 0.8–1.3 for the lowest injury risk [9, 10]. Highest injury risk was found with an A/C ratio >1.5 ("danger zone") [9]. These results show that athletes with high chronic training load seemed more resistant to injury in periods of acute high load, compared to athletes with low chronic load [11]. For injury prevention, moderate changes in A/C ratio within the "sweet spot" range seem advisable [9].

Monitoring of load should always be done individually [7, 12]. This means that one must try to quantify the specific training load factor being the most relevant for the athlete in question. In track and field, one could regard the number of jumps for a long jumper and the number of throws for a javelin thrower as training load factors. To quantify training load more specifically, several training load factors should be used in one athlete. For example, in long-distance running, one can quantify training load by the total number of kilometers that the athlete has run, the amount of high-speed running and the number of kilometers run at the track [13]. Of note is the most relevant factor in training load can differ over time in the same athlete depending on training period and type and localization of the current or previous injuries. This is especially the case in heptathletes and decathletes, for whom it is extremely difficult to coordinate the training program for all the different events. In general, monitoring training load is best executed through a combination of internal and external training load factors [7, 12, 14].

34.3.4 Biomechanics

Analysis of the biomechanics of the athlete in his/her particular event can give insight into causative factors of the injury. This can be performed by analyzing optimal biomechanics of this particular athlete in injury-free top shape and comparing this to the athlete's biomechanics prior to the injury. In some events, specific injuries are more prevalent due to specific demands of that event. In a study of British track and field athletes, sprinters had a significantly higher incidence of plantaris tendon injury (tendinopathy or (partial) rupture) compared to endurance athletes [15]. Moreover, bend running sprinters (200 m and 400 m) had significantly more right-sided than left-sided plantaris injuries. It was hypothesized that there are higher load and higher rotational forces on the right leg in running counterclockwise, especially on the plantaris muscle as plantar flexor in high-speed running [15, 16]. This aspect can be used in rehabilitation to introduce (high speed) bend running in the final phase of rehabilitation. This example shows that integrating biomechanical, medical, and sport-specific knowledge is essential to really understand and explain the current injury and consequently reduce the risk of reinjury.

34.4 Athlete Characteristics

Specific athlete strengths and weaknesses should be taken into account in making the rehabilitation plan. Considering athlete characteristics can highlight factors that need more attention and/or demand involvement of a specific specialist. In doing this, an injury gives the athlete the opportunity to work on (hidden) weaknesses and factors not directly related to the current injury. Working on these weaknesses can have a performance benefit in the long term. Injury history can give insight into vulnerable body parts, and insufficient recovery of previous injuries can be a recurrent contributing factor in new injuries.

Psychological state is an important factor to consider thoughts, feelings, and athlete's behaviors influence sports injury rehabilitation outcomes [17]. It is advised to ask about the athlete's ideas on the injury, recovery process, and fear of reinjury. Moreover, reflecting on the injury period as an opportunity for growth and development can have positive rehabilitation outcomes [17].

Lastly, it should be stated that creating clarity about the end goal of rehabilitation by defining what successful RTS entails for this particular athlete is also a vital aspect of the athlete's characteristics.

34.5 Aspects of Rehabilitation

Defining the injury and athlete characteristics gives an overview of the present injury. It is advised to sketch a rough timeline to return to sport and highlight points for attention. Now, we will review aspects of rehabilitation starting with general principles followed by strength and sport-specific exercises and nutritional aspects during rehabilitation.

34.5.1 General Principles

In rehabilitation, a shift is going on from a pure time-based approach to a criteria-based approach [18]. In criteria-based rehabilitation, different phases from the injury to final return to sport are run through step-by-step, without setting a specific time period for each phase. The biggest advantage of criteria-based rehabilitation is that one phase is fully completed before moving on to the next. This prevents insufficient rehabilitation when time-based phases are too fast and prevents unnecessary time loss when recovery goes (too) fast. However, a factor that always should be taken into account is biological healing of the injured tissue [19].

The basic principle of rehabilitation is stepwise progression of injury-specific exercises in strength and muscle fiber recruitment toward sport-specific exercises (Fig. 34.2) [18]. Besides that, the rehabilitation plan has to include injury nonspecific exercises and general conditioning. Aerobic exercise training can be performed on

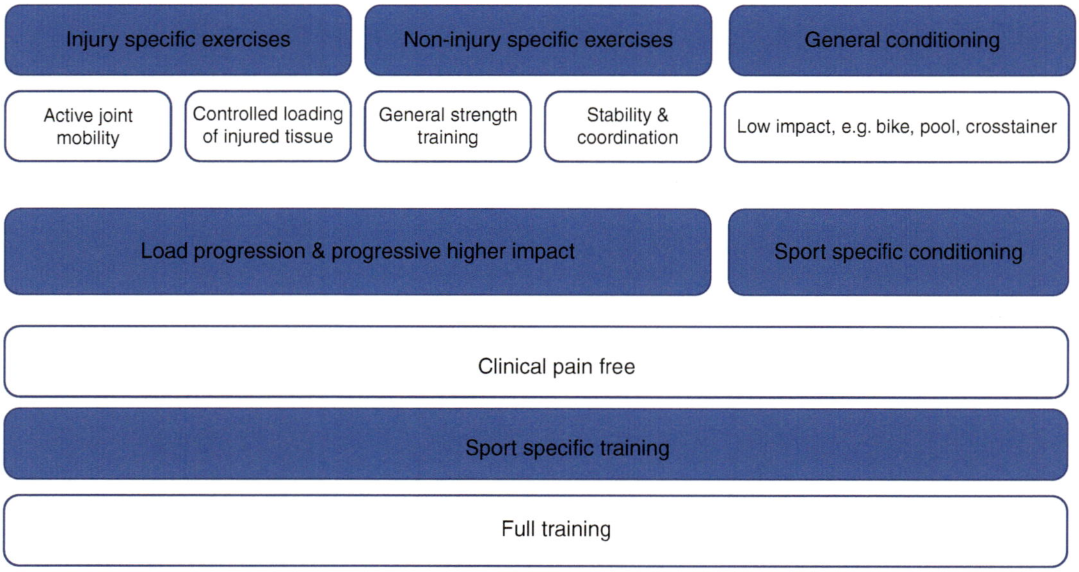

Fig. 34.2 Structure of a rehabilitation program. Adapted and adjusted from Serner et al. [18]

the bike, on the antigravity treadmill, or in the pool, dependent on the specific athlete and injury characteristics and, naturally, available facilities.

34.6 Strength and Sport-specific Exercises

Optimal loading is key during rehabilitation. Gradual buildup of tissue load is essential, and in each phase of the rehabilitation, it is a continuous search to determine what exactly is regarded as optimal loading. In monitoring load, the previously discussed A/C ratio can be used. In acute injury, there is a selective inhibition of the injured muscle as a protective mechanism. In the first phase of rehabilitation, this selective inhibition must be reduced in order to avoid chronic activation deficits in that specific muscle group [2]. It is proposed that high load isometric exercises are most effective because this activates the biggest amount of motor units [2]. Therefore, in the beginning of rehabilitation, focus must be put on careful loading and recruitment of the injured muscle(s) with basic, non-sport-specific, closed chain exercises. In later phases, progression can be made to open chain exercises and more dynamic exercises with gradual progress in velocity of movement. Proprioceptive and neuromuscular training both reduce the incidence and recurrence of several injuries, such as acute ankle sprains and acute knee injuries [20, 21].

Rehabilitation gradually progresses to sport-specific training during which it is essential to restore normal movement patterns. Altered hip and pelvis kinematics post-hamstring injuries is found and should be a main focus in early phases of sport-specific rehabilitation [2]. In making a sport-specific rehabilitation plan, the specific injury and athlete's characteristics have to be taken into account. Contributing factors to the current injury and factors known as high impact (e.g., spikes vs. normal running shoes) should be introduced carefully and gradually. Technical aspects should receive careful attention during rehabilitation, especially when the athlete's technique can be regarded as a contributing factor in the present injury (e.g., introducing changes in running technique or technique of foot placement in jumping events).

Specifically in training buildup for track and field athletes, one can think of sprinters and middle–long- distance runners who gradually progress from walking to easy jogging and doing progressive running drills. Jumpers can perform jumping exercises on blocks and can slowly progress to jumps with higher force (in height and/or distance). Throwing athletes start throwing with lighter or heavier material (javelins/discus). Heavier material during rehabilitation is useful when a slower movement and better movement control is preferred and executed. When optimal biomechanics are established, gradual training progression can be made in volume, intensity, and sport-specific aspects (e.g., using hurdles and starting blocks).

34.6.1 Nutritional Aspects

Adequate nutritional intake is essential in reaching optimal training effect. Due to the fact that the training program of the athlete is changed during rehabilitation, it is the case that also the nutritional requirements are subject to change. Where a carbohydrate-rich diet is normally eaten during high-intensity training weeks, there may be a greater need for protein in rehabilitation phases focusing on strength [22, 23]. It is known that dietary protein supplementation enhances muscle protein synthesis following resistance exercise [22]. Total energy intake should therefore be matched to the training load during rehabilitation in order to avoid gain of body weight. The sports medicine physician and dietician can determine an optimal diet (with or without nutritional supplements) based on injury characteristics, rehabilitation phase, and blood tests. Again, especially in elite sports, this highlights the importance of shared decision making, communication, and teamwork.

34.7 Aspects in Final RTS Decision

Return to play is a continuum; however, the final RTS decision should be made at one specific moment. RTS clearance is a multi-faceted clinical decision and is ideally made in the performance team in a shared decision-making process [5]. We discuss several aspects of the final RTS decision, which in essence is finding the right balance between returning an athlete too early and suffering a reinjury or delaying RTS and missing competitions unnecessarily.

34.7.1 Clinical Testing

Injury-specific and sport-specific clinical tests can be useful in RTS decisions. Basic clinical tests such as palpation, stretch, and manual strength testing should be pain-free and similar to the contralateral side. For example, in acute hamstring injuries, pain on posterior thigh palpation and isometric knee flexion force deficit at 15° within 7 days after return to play are associated with a higher hamstring reinjury rate [24]. Ideally, test results of this particular athlete are available before the injury occurred and are performed several times during rehabilitation. For example, reduced hip adductor strength is found to be a significant risk factor for groin injuries in male soccer players [25]. This makes that the presence of reduced adductor strength and/or asymmetry in comparison with the contralateral leg can push the decision to postponement of RTS.

Sport-specific clinical tests are exercises and activities that mimic the athlete's sporting event as closely as possible. These tests aim at making the transition from rehabilitation to RTS small and at giving the athlete confidence in sport-specific function and skill. This particular part of decision whether to make the final decision of RTS can be assessed by performance-based tests of sport-specific movement patterns, muscle strength, and reactive agility [19].

The physical reaction to sport-specific training should be monitored carefully and is essential for evaluation of the readiness for RTS. Parameters that can be used for this purpose are pain, morning stiffness, joint effusion, and joint mobility.

34.7.2 Psychological Readiness

Assessment of psychological state of the athlete is essential in the decision of RTS [17]. It is found that psychological factors at return to sport focused on performance-related and reinjury-related anxiety and fear [17]. Psychological readiness, sport-related confidence (self-efficacy), and social support are found to be important factors in successful return to sport and should therefore be taken into account [17].

34.7.3 Imaging

In making a final decision about RTS, imaging is not recommended. In a study of acute hamstring injuries, 89% of athletes with clinically recovered acute hamstring injuries still had increased signal intensity on MRI [26]. In bone stress injuries, bone marrow edema can be present on MRI months after successful RTS and is widely present in asymptomatic runners [27].

34.7.4 Decision to Return to Sport

Taking results of clinical tests, reaction to sport-specific training, and psychological readiness into account, decisions about return to sport still can be difficult. The StARRT (Strategic Assessment of Risk and Risk Tolerance) framework is developed for structural assessment of the decision to return to sport [1, 28]. This framework consists of three components: assessment of health risk, assessment of activity risk, and assessment of risk tolerance. In the assessment of health risk, one has to determine tissue health (patient characteristics, symptoms, and special tests) and tissue stresses (sport characteristics, competitive level, ability to protect, and psychological readiness). In the assessment of risk tolerance, risk tolerance modifiers have to be considered (timing (season), pressure from athlete and/or external pressure, and possible conflict of interest (financial aspects)). RTS should be allowed when the risk assessment (health and activity risk) is below the acceptable risk tolerance threshold [28]. Assessment of risk tolerance is based on risk tolerance modifiers, and this is why athletes presenting with the same risk assessment can have different moments of RTS depending on the situation (e.g., national competition or Olympics). Ideally, RTS is a multidisciplinary decision to create the greatest support for the final decision. Factors healthcare professionals primarily have to take into account are physiological and psychological readiness, risk of reinjury, and possible long-term health risks [5].

34.7.5 Secondary Prevention

Having had a specific injury is one of the major risk factors for reinjury. Secondary preventive interventions reduce the reinjury risk. Athletes have to be aware of their increased injury risk and should continuously work on injury prevention. In sports injury prevention programs, strength training is essential because this reduces injury risk in a dose–response relation and improves performance [29]. Protection of the previously injured tissue is found effective in some injuries, for example, using an ankle brace or tape in the secondary prevention of ankle sprains [30]. Monitoring load is in particular important in the first phase after RTS in order to secure gradual progression in load. This is essential in reaching the end goal of injury rehabilitation for elite athletes: performing at a higher level than ever before.

References

1. Ardern CL, Glasgow P, Schneiders A, Witvrouw E, Clarsen B, Cools A, et al. 2016 consensus statement on return to sport from the first world congress in sports physical therapy, Bern. Br J Sports Med. 2016;50:853–64.
2. Macdonald B, McAleer S, Kelly S, Chakraverty R, Johnston M, Pollock N. Hamstring rehabilitation in elite track and field athletes: applying the British athletics muscle injury classification in clinical practice. Br J Sports Med. 2019;53:1464–73.
3. Meeuwisse WH, Tyreman H, Hagel B, Emery C. A dynamic model of etiology in sport injury: the recursive nature of risk and causation. Clin J Sport Med. 2007;17(3):215–9.
4. Bahr R, Krosshaug T. Understanding injury mechanisms: a key component of preventing injuries in sport. Br J Sports Med. 2005;39:324–9.
5. Dijkstra HP, Pollock N, Chakraverty R, Ardern CL. Return to play in elite sport: a shared decision-making process. Br J Sports Med. 2017;51(5):419–20.
6. King J, Roberts C, Hard S, Ardern CL. Want to improve return to sport outcomes following injury? Empower, engage, provide feedback and be transparent: 4 habits! Br J Sports Med. 2019;53:526–8.
7. Soligard T, Schwellnus M, Alonso JM, Bahr R, Clarsen B, Dijkstra HP, et al. How much is too much? (part 1) International Olympic Committee consensus on load in sport and risk of injury. Br J Sports Med. 2016;50:1030–41.
8. Hulin BT, Gabbett TJ, Blanch P, Chapman P, Bailey D, Orchard JW. Spikes in acute workload are associated with increased injury risk in elite cricket fast bowlers. Br J Sports Med. 2014;48:708–12.
9. Gabbett T. The training—injury prevention paradox: should athletes be training smarter and harder? Br J Sports Med. 2016;50:273–80.
10. Blanch P, Gabbett TJ. Has the athlete trained enough to return to play safely? The acute:chronic workload ratio permits clinicians to quantify a player's risk of subsequent injury. Br J Sports Med. 2016;50:471–5.
11. Hulin BT, Gabbett TJ, Lawson DW, Caputi P, Sampson JA. The acute:chronic workload ratio predicts injury: high chronic workload may decrease injury risk in elite rugby league players. Br J Sports Med. 2016;50:231–6.
12. Bourdon PC, Cardinale M, Murray A, Gastin P, Kellmann M, Varley MC, et al. Monitoring athlete training loads: consensus statement. Int J Sports Physiol Perform. 2017;12(2):S2161–70.
13. Duhig S, Shield AJ, Opar D, Gabbett TJ, Ferguson C, Williams M, et al. Effect of high-speed running on hamstring strain injury risk. Br J Sports Med. 2016;50:1536–40.
14. Windt J, Gabbett TJ. How do training and competition workloads relate to injury? The workload—injury aetiology model. Br J Sports Med. 2017;51:428–35.
15. Pollock N, Dijkstra P, Calder J, Chakraverty R. Plantaris injuries in elite UK track and field athletes over a 4-year period: a retrospective cohort study. Knee Surg Sports Traumatol Arthrosc. 2016;24:2287–92.
16. Chang YH, Kram R. Limitation to maximum running speed on flat curves. J Exp Biol. 2007;210:971–82.
17. Forsdyke D, Smith A, Jones M, Gledhill A. Psychosocial factors associated with outcomes of sports injury rehabilitation in competitive athletes: a mixed studies systematic review. Br J Sports Med. 2016;50:537–44.

18. Serner A, Weir A, Tol JL, Thorborg K, Lanzinger S, Otten R, et al. Return to sport after criteria-based rehabilitation of acute adductor injuries in male athletes. A prospective cohort study. Orthop J Sports Med. 2020; https://doi.org/10.1177/2325967119897247.
19. Filbay SR, Grindem H. Evidence-based recommendations for the management of anterior cruciate ligament (ACL) rupture. Best Pract Res Clin Rheumatol. 2019;33(1):33–47.
20. Zech A, Hübscher M, Vogt L, Banzer W, Hänsel F, Pfeifer K. Neuromuscular training for rehabilitation of sports injuries: a systematic review. Med Sci Sports Exerc. 2009;41(10):1831–41.
21. Hübscher M, Zech A, Pfeifer K, Hänsel F, Vogt L, Banzer W. Neuromuscular training for sports injury prevention: a systematic review. Med Sci Sports Exerc. 2010;42(3):413–21.
22. Morton RW, Murphy KT, McKellar SR, Schoenfeld BJ, Henselmans M, Helms E, et al. A systematic review, meta-analysis and meta-regression of the effect of protein supplementation on resistance training-induced gains in muscle mass and strength in healthy adults. Br J Sports Med. 2018;52(6):376–84.
23. Burke LM, Castell LM, Casa DJ, Close GL, Costa RJS, Desbrow B, et al. International Association of Athletics Federations Consensus Statement 2019: nutrition for athletics. Int J Sport Nutr Exerc Metab. 2019;29(2):73–84.
24. De Vos RJ, Reurink G, Goudswaard GJ, Moen MH, Weir A, Tol JL. Clinical findings just after return to play predict hamstring re-injury, but baseline MRI findings do not. Br J Sports Med. 2014;48:1377–84.
25. Engebretsen AH, Myklebust G, Holme I, Engebretsen L, Bahr R. Intrinsic risk factors for groin injuries among male soccer players: a prospective cohort study. Am J Sports Med. 2010;38(10):2051–7.
26. Reurink G, Goudswaard GJ, Tol JL, Almusa E, Moen MH, Weir A, et al. MRI observations at return to play of clinically recovered hamstring injuries. Br J Sports Med. 2014;48(18):1370–6.
27. Bergman AG, Fredericson M, Ho C, Matheson GO. Asymptomatic tibial stress reactions: MRI detection and clinical follow-up in distance runners. AJR Am J Roentgenol. 2004;183(3):635–8.
28. Shrier I. Strategic assessment of risk and risk tolerance (StARRT) framework for return-to-play decision-making. Br J Sports Med. 2015;49:1311–5.
29. Lauersen JB, Andersen TE, Andersen LB. Strength training as superior, dose-dependent and safe prevention of acute and overuse sports injuries: a systematic review, qualitative analysis and meta-analysis. Br J Sports Med. 2018;52:1557–63.
30. Vuurberg G, Hoorntje A, Wink LM, Van der Doelen BFW, Van den Bekerom MP, Dekker R, et al. Diagnosis, treatment and prevention of ankle sprains: update of an evidence-based clinical guideline. Br J Sports Med. 2018;52:956.

Correction to: Management of Track and Field Injuries

Gian Luigi Canata, Pieter D'Hooghe, Kenneth J. Hunt, Gino M. M. J. Kerkhoffs, and Umile Giuseppe Longo

Correction to: G. L. Canata et al. (eds.), *Management of Track and Field Injuries*, https://doi.org/10.1007/978-3-030-60216-1

This book was inadvertently published with a typo in the book title: the term "injures" has been corrected to "Injuries".

The updated online version of this book can be found at
https://doi.org/10.1007/978-3-030-60216-1

G. L. Canata et al. (eds.), *Management of Track and Field Injuries*,
https://doi.org/10.1007/978-3-030-60216-1_35

Printed by Printforce, the Netherlands